Geophysical Monograph Series

Including

IUGG Volumes
Maurice Ewing Volumes
Mineral Physics Volumes

Geophysical Monograph 110

The Controlled Flood in Grand Canyon

Robert H. Webb
John C. Schmidt
G. Richard Marzolf
Richard A. Valdez
Editors

American Geophysical Union
Washington, DC

Published under the aegis of the AGU Books Board

Library of Congress Cataloging-in-Publication Data

The controlled flood in Grand Canyon / Robert H. Webb...[et al.],
   editors.
      p. cm. -- (Geophysical monograph ; 110)
      Includes bibliographical references.
      ISBN 0-87590-093-3
         1. Floods--Arizona--Grand Canyon National Park. 2. Steamflow-
-Arizona--Grand Canyon National Park. 3. Sedimentation and
deposition--Arizona--Grand Canyon National Park. 4. Aquatic
biology--Arizona--Grand Canyon National Park. 5. Floods--Colorado
River (Colo.-Mex.) 6. Streamflow--Colorado River (Colo.-Mex.)
7. Sedimentation and deposition--Colorado River (Colo-Mex.)
I. Webb, Robert H. II. Series.
GB1399.4A6C65   1999
551.48'9'0979132--dc21   99-14123
                                                       CIP

ISBN 0-87590-093-3
ISSN 0065-8448

Copyright 1999 by the American Geophysical Union
2000 Florida Avenue, N.W.
Washington, DC 20009

Cover
Moonset over Vermillion Cliffs at Lees Ferry (background, photograph by Carl
J. Bowser); Mile ¯9 in Glen Canyon (top, photograph by Carl J. Bowser); Glen
Canyon Dam with jet tubes open during peak of the flood (upper middle, U.S.
Bureau of Reclamation); campers on a typical Grand Canyon sandbar with
tamarisk trees (lower middle, U.S. Bureau of Reclamation); humpack chub [*Gila
cypha*] (bottom, photograph by Richard A. Valdez).

Printed in the United States of America.

CONTENTS

CONTENTS

Riverine and Aquatic Biology

Management Implications and Societal Costs

FOREWORD

The Glen Canyon flood began on March 26, 1996, as I stood on a catwalk right in front of the dam, turning a valve that opened the jet tubes, allowing water to surge through the dam and down into the river at 45,000 cubic feet per second. I watched in wonder as the river surged, cascading a fountain of mist hundreds of feet in the air and raising the water level the entire length of the Canyon. Our purpose in flooding the Grand Canyon was to restore the beaches that had not been replenished and habitat that had been damaged by the artificial river flows, which fluctuated erratically from day to day and week to week in response to power demands from cities as distant as Phoenix and Salt Lake City.

The problem was that Glen Canyon Dam, one of the two great keystone structures on the Colorado River, was built in the early 1960s. In that era—prior to federal environmental laws like the Endangered Species Act and the National Environmental Policy Act—no one, not even the National Park Service, paused to consider, much less analyze, how those hydropower-driven fluctuations might impact the downstream habitat of one of the world's great national parks. That lack of consideration was not exclusive to the Dam. Until very recently, in fact, we have lived on and used the land and its forests and rivers as if the landscape were merely an assemblage of unrelated parts, each to be used, removed, or substituted without regard to the others.

Today we know better. With the insights of modern ecological science, we have come to understand that neither Grand Canyon nor any other protected place is an island unto itself, that every part of the landscape is tributary to the whole; and that the operation of Glen Canyon Dam has consequences for the entire watershed. Grasping this connection is one thing; reestablishing it as a basis for policy is something else entirely. It involves taking the many frayed ecological strands that were torn loose by the dam and weaving them back together into a cohesive fabric. To that end, the restoration of the Grand Canyon would never even have begun were it not for the careful and steady application of science and the many extensive comments of stakeholders.

When the Interior Department first considered a different way of operating the dam (including nature-mimicking floods) back in 1982, the idea failed to take root. True, many people in the West understood that the floods that occurred nearly every spring before the dam was constructed had been beneficial to habitat while damaging to human occupation of

flood plains. Re-creating floods in Grand Canyon after the dam was built would, they thought, jeopardize their own interests. When serious proposals to flood Grand Canyon were made in the early 1990s, opposition was stiff:

- Hydroelectric power users in six states opposed any plan in which water would have to be passed around the generators, and thus reduce their power revenue.
- The water users in the four states of the upper Colorado River basin threatened to sue on grounds that the proposed water releases would violate the storage provisions of the Colorado River Compact.
- Trout fishermen and the Arizona Game and Fish Department complained that an artificial flood would wipe out the trophy trout fishery below the dam; rafting outfitters worried they'd lose business from public fears and schedule changes.
- All eight Indian tribes that border the Grand Canyon voiced fears that rising waters would destroy petroglyphs, burial sites, and other sacred archaeological remains.
- Even Interior's own Fish and Wildlife Service fretted that floods might damage the habitat of several endangered species, including the southwestern willow flycatcher, the Kanab ambersnail, and the humpback chub.

Meanwhile, in response to this mounting skepticism and outright opposition, we had nothing to offer but charts and theoretical models. There was simply no precedent on the Colorado River—or as far as we know anywhere in the history of civilization—for what Interior was proposing to do. But while opposition to the proposed floods gridlocked the process, it also revealed a silver lining. Gridlock gave Interior time to gather more and more useful information and, eventually, to assemble an interdisciplinary team of scientists from the Glen Canyon Environmental Studies Program. Biologists, hydrologists, archaeologists, geologists, and ecologists began to integrate their data toward a common base. Additionally, in the process of all this data gathering and modeling, our scientific team began to do something pivotal to reverse the direction of the project. They began to share.

Once the scientists opened up their notebooks and models, the hydropower users took a second look at their own economic models. Soon they discovered that their own initial estimates of power revenue loss were at least two times too high. And as the discussions widened, the utilities also realized that many power consumers were also sportsmen and environmentalists who would favor restoring the river. Then came the fishermen. Using a video display, scientists were able to create a virtual flood, allowing angling and outfitter groups to watch the water progress down the Canyon, submerging their favorite sandbars and shoreline camping spots. They were assured that the flood was hardly a cataclysmic event, but that a water surge might actually stir up nutrients in the system, and boost *Cladaphora glomerata*, the algae that has developed in the clear cold water below the dam and forms the base of the aquatic food chain.

The Fish and Wildlife Service—still concerned about danger to endangered southwestern willow flycatcher nests and humpback chub—used flow models to determine that the flood effects would be minimal. And when at the last minute they discovered an entirely new population of the endangered Kanab Ambersnail, they used the models again to show that habitat damage wouldn't adversely affect them. For additional mitigation, they marked each snail and moved them, one by one, farther up the bank to safety. In fact, scientific research helped the Fish and Wildlife Service move forward with a non-jeopardy opinion that the flood would not cause lasting damage to endangered species in Grand Canyon.

The Indian tribes, having lived on the land from time immemorial, were understandably skeptical of any more manipulations of a river already compromised by modern technology. One week before the scheduled flood, the Hualapai tribe threatened to seek a court injunction. But after looking in detail at the hydrologic work of the U. S. Geologi-

cal Survey, tribal leaders concluded that archaeological sites would actually receive more protection because additional sediment deposited were expected to protect the sites from erosion.

Finally, the camera crews and helicopters and national correspondents began to arrive, all full of healthy skepticism about how a deluge of biblical proportions could benefit anyone or anything. They asked, "Hey, if the environment wins from this flood, then who are the losers?" Turning the valve to release the waters, I could smile and answer, "Good luck in finding one!" A week after the flooding, aerial photos of the river showed dozens of sparkling new sandbars protruding above the water, and lining the riverbanks. Some were piled twelve feet high. Habitat was improved; fishing, rafting, power and water use quickly resumed without skipping a beat. Indian cultural resources remained undisturbed. In short, scientists found the results exceeded their highest expectations.

Our experience with the artificial flood from Glenn Canyon Dam illustrates, in a most spectacular and complex landscape, how we, as a people, have begun to change the way we make decisions. The flood shows how stakeholders, working with scientists, can make a difference to protecting our environment. And more than anything, it shows how objective science can benefit and serve management of our national parks.

Bruce Babbitt
U.S. Secretary of the Interior

PREFACE

The natural flow of almost every river in the United States has been modified to meet various socioeconomic goals—navigation, irrigation, power generation and flood control. The success of the dams and reservoirs built to achieve these goals has been accompanied by changes in the status of riverine resources downstream, a cause of growing environmental and ecological concern. For example, before Glen Canyon Dam was completed, the Colorado River transported large quantities of sediment in floods as large as 8500 m^3/s. After the dam was closed in 1963, dam releases typically were less than the powerplant capacity of 890 m^3/s and exhibited large daily flow fluctuations. The river carried little sediment. The daily fluctuations in flow eroded sand bars, and the smaller, controlled flow did not redeposit them. The clear, cold water resulted in increased aquatic productivity such that rainbow trout and other nonnative fishes thrived while most native species were lost or endangered.

Evolving scientific knowledge led to the realization that high flow from the Glen Canyon Dam was essential to maintaining desirable river resources in Grand Canyon. This book describes the research results from the 1996 controlled flood, the largest scientific experiment ever conducted on a regulated river. When the controlled flood began, more than 100 scientists waited downstream to begin data collection concerning all aspects of river resources—hydraulics, sediment transport, geomorphology, aquatic ecology, riparian vegetation, impacts to archaeological resources, and safety of boating in rapids. This event was well-publicized but has not been comprehensively summarized.

Our purpose is to report technical results from the flood in the context of adaptive management of the Colorado River and Glen Canyon Dam, with relevance for regulated rivers worldwide. The first two chapters survey the ecological and hydrological research in Grand Canyon leading up to the flood. Subsequent chapters detail the effects of the flood on Glen Canyon and Grand Canyon. The concluding chapters synthesize effects of the flood and discuss the economic and scientific implications of controlled floods as management tools in regulated rivers.

Geomorphologists, stream ecologists, natural resource managers, resource economists, and land management policy makers should find this volume important and useful because it documents in detail the planning, execution, and analysis of this massive field manipulation. Students and researchers will read it for the new knowledge it reports. Additionally, all readers should appreciate the candor with which it acknowledges uncertainty and the difficulty that researchers experienced as they attempted to maintain objectivity in the face of the contentious water resource issues at the core of the enterprise.

In the early 1980s, the Bureau of Reclamation, the federal agency responsible for the construction and operation of the dam, was required to rewind the generators in the dam's powerplant. This led to the possibility of wider variations in daily releases. The public's response to the prospect of further change to the river in the national park resulted in the implementation of the Glen Canyon Environmental Studies (GCES) in 1982. It was to have been a three-year study to document these effects. The scope and objectives of the program quickly expanded to encompass a broad range of resources and issues related to the operation of the Glen Canyon Dam powerplant to meet peak electrical demand. The first phase of this program resulted in numerous publications and was completed in 1987. In 1989, the studies were expanded to a GCES II program, which in 1992 evolved into a monitoring program to support data needs of an environmental impact analysis.

The resource management decisions required by this analysis were most successful and defensible when based on sound information learned from relevant scientific inquiry. The research centered on hydraulic phenomena of flow as controlled by the dam, and included the consequent transport and redeposit of sediment that were considered the basic independent variables in the river corridor system. These studies were accompanied by attempts to couple new physical understanding with the status of dependent habitat variables associated with changes in populations of endangered species. The science led to an appreciation of system interactions and novel management options. GCES then concluded with the controlled flood in 1996, the subject of this volume.

In this sense, the 1996 controlled flood was the culmination of a 14-year research effort on the effects of operations on Glen Canyon Dam on the riverine resources of lower Glen Canyon and Grand Canyon. These studies represent one of the most comprehensive and in-depth investigations of the effects of reservoir and dam operations on the downstream physical and biological environment undertaken for any river.

As long-time researchers of the Colorado River, we welcomed the opportunity to assemble 25 independently written and peer-reviewed chapters that represent a culmination of much of the current state of knowledge on the riverine ecosystem in Grand Canyon. We partitioned the chapters of this book into categories of sediment transport and geomorphology, riverine and aquatic biology, and management implications and societal costs to insure comprehensive treatment of all resources evaluated during the controlled flood.

We learned a great deal from assimilating the research findings of the various disciplines, and we hope that we have been able to clearly convey these findings to the readers of this book so that others may go forward and apply the findings of this landmark experiment to other river systems.

We thank the authors of the various chapters for their untiring dedication and willingness to share their findings. We thank numerous peer reviewers who helped us refine the content of this book. We particularly thank David L. Wegner, the program director for GCES, for his support of our efforts and his long-term commitment to scientific research in the cause of better river management. Donna Opocensky and Peter Griffiths of the U.S. Geological Survey provided invaluable help in the editorial and production phases of this project.

Robert H. Webb
U.S. Geological Survey
Tucson, Arizona

John C. Schmidt
Utah State University
Logan, Utah

G. Richard Marzolf
U.S. Geological Survey
Reston, Virginia

Richard A. Valdez
SWCA, Inc.
Logan, Utah

Downstream Effects of Glen Canyon Dam on the Colorado River in Grand Canyon: A Review

R.H. Webb[1], D.L. Wegner[2], E.D. Andrews[3], R.A. Valdez[4], and D.T. Patten[5]

Glen Canyon Dam, completed in 1963, has altered geomorphic and ecological processes and resources of the Colorado River in Grand Canyon. Before the dam was completed, the river transported large quantities of sediment during spring floods as large as 8500 m^3/s. After 1963, dam releases typically were less than 900 m^3/s with large diurnal fluctuations and little sediment. The 2-yr peak discharge decreased by a factor of 2.5, resulting in aggraded rapids and a large increase in riparian vegetation. The clearwater releases from the dam eroded sand deposited on the bed and banks. Although pre-dam water temperatures varied seasonally, dam releases typically are about 8°C year round. Because of the clear, cold water and reduced flooding, post-dam aquatic productivity is considerably higher in the tailwater. Rainbow trout and other non-native fishes are now common, 3 native species have been extirpated, and the remaining species, including the endangered humpback chub, cannot successfully reproduce in the river.

1. INTRODUCTION

Construction of Glen Canyon Dam on the Colorado River has affected a number of aquatic and terrestrial resources downstream in lower Glen Canyon and in Grand Canyon (Figure 1). The Bureau of Reclamation manages the dam and its powerplant, which produces 3% of the summer power demand in the region [*Harpman*, this volume]. Flood control and diurnally fluctuating releases of clear, cold water are blamed for narrowing of rapids, widespread beach erosion, invasion of nonnative riparian

[1]U.S. Geological Survey, Tucson, Arizona
[2]Ecosystem Management International, Flagstaff, Arizona
[3]U. S. Geological Survey, Boulder, Colorado
[4]SWCA, Inc., Logan, Utah
[5]Department of Plant Biology, Arizona State University, Tempe, Arizona

The Controlled Flood in Grand Canyon
Geophysical Monograph 110
Copyright 1999 by the American Geophysical Union

vegetation, and losses of native fishes. The river passes through Grand Canyon National Park and Glen Canyon National Recreation Area and is on the boundary of the Navajo and Hualapai Reservations in Arizona. These management entities, as well as environmental and recreational groups, have a vested interest in managing the Colorado River to protect its resources.

Responding in part to pressure from conservationists, and to meet legal requirements for rewinding the generators in the dam's powerplant, the Bureau of Reclamation initiated the Glen Canyon Environmental Studies (GCES) Program in 1982 [*Wegner*, 1991; *Schmidt et al.*, this volume]. The studies conducted under the GCES Program are one of the most comprehensive and in-depth investigations of the effects of reservoir operations on the downstream physical and biological environment ever undertaken for a river. The purpose of this chapter is to review the salient results of numerous researchers who had worked in the riverine environment of the Colorado River in Grand Canyon prior to the 1996 controlled flood.

Initially, the GCES program was to be a 3-year effort. The scope and objective of the program quickly expanded to encompass the broad range of resources affected by the

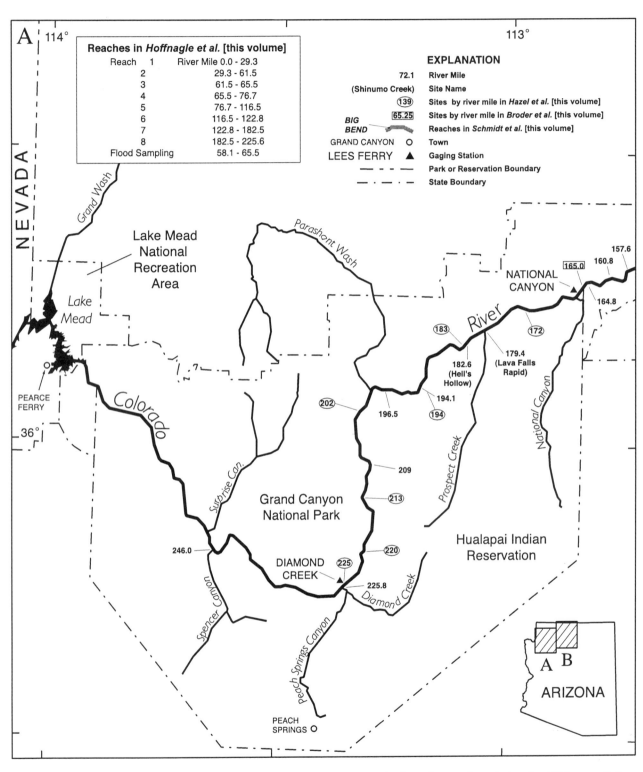

Figure 1. Map of the Colorado River in Grand Canyon showing primary study sites of researchers during the 1996 controlled flood on the Colorado River. A. Western Grand Canyon. B. Eastern Grand Canyon.

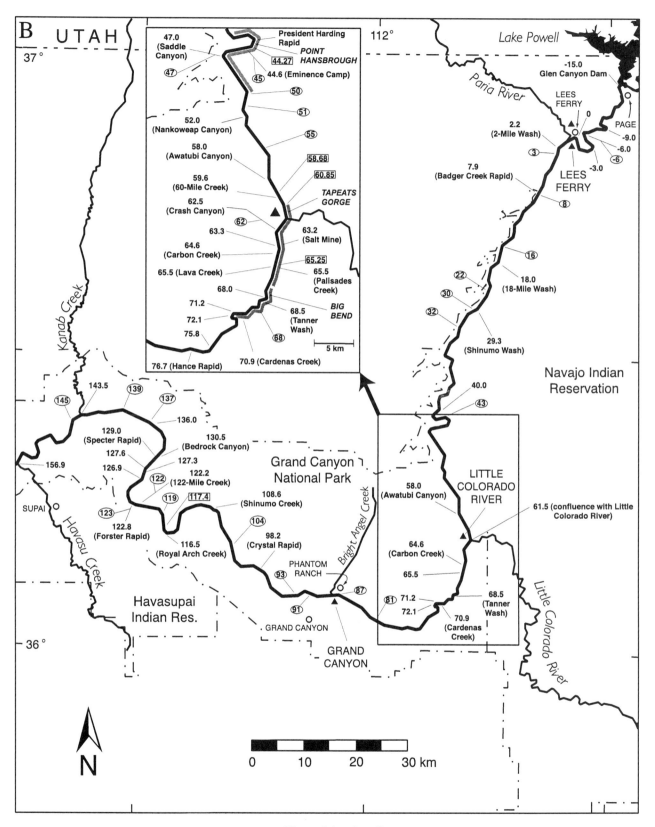

Figure 1 (continued)

operation of Glen Canyon Dam. GCES Phase I, as it was known, resulted in numerous publications and summary volumes [*Bureau of Reclamation*, 1988a, 1988b] and was completed in 1987 [*Patten*, 1991]. The studies of GCES I were expanded to GCES Phase II in 1989 after the Department of Interior initiated the environmental impact statement process for operation of Glen Canyon Dam. Major features of GCES II were "research flows" (1990-1991) from Glen Canyon Dam to evaluate various alternatives being considered for the EIS [*Patten*, 1991], and in-depth studies to describe cause-effect relations of dam operations. Expansions of the program included studies of recreation and power-production economics, impacts to the power market, and the potential impacts to the Native American cultural resources along the river corridor. In 1992, the advent of restrictions on dam operations, known as the Interim Operating Criteria [*U.S. Department of the Interior*, 1995], changed some of the mission of GCES to monitoring. GCES II concluded with the 1996 controlled flood in Grand Canyon, and was reviewed in *National Research Council* [1996]. The 1996 controlled flood in Grand Canyon resulted from 14 yrs of research under the GCES Program.

2. THE RIVERINE RESOURCES OF GRAND CANYON

Many natural features of the river corridor are important because of their value to society in this, one of the most visited national parks in the United States. Riverine resources create public interest and generate public support for environmentally sensitive management of Glen Canyon Dam. The attention that scientific inquiry has focussed on how operations of Glen Canyon Dam affect these resources represents an essential connection between science and natural resource management [*Marzolf et al.*, this volume].

2.1. Recreation and Rapids

When Congress authorized construction of Glen Canyon Dam on April 11, 1956, fewer than 500 people had navigated the Colorado River through Grand Canyon [*Lavender*, 1985]. The popularity of whitewater recreation increased dramatically during the two decades following development of the commercial river-running industry. In 1971, the National Park Service imposed a limit of 22,000 river runners per year through Grand Canyon. These recreationists are attracted by the world-class whitewater in the Colorado River. A plan-and-profile map of the Colorado River, surveyed in 1923 [*Birdseye*, 1924], documents the water-surface fall through major riffles and rapids and continues to be used to describe rapids [*Stevens*, 1990].

Leopold [1969] found that rapids account for only 10% of the distance, but most of the drop, through Grand Canyon.

Most rapids in Grand Canyon are created by debris fans, which control the hydraulics of rapids and locations of sand bars [*Dolan et al.*, 1978; *Howard and Dolan*, 1981]. Debris flows from tributary canyons create and maintain these debris fans [*Webb et al.*, 1988, 1989]. Debris fans are central to the fan-eddy complex that creates environments for sand-bar deposition upstream and downstream [*Schmidt and Rubin*, 1995]. Pre-dam floods reworked most of the smaller particles deposited by debris flows, leaving large boulders to form rapids [*Graf*, 1979; *Howard and Dolan*, 1981]. Debris flows after 1963 have aggraded many debris fans and have altered flow through major rapids [*Howard and Dolan*, 1981; *Melis et al.*, 1994; *Webb et al.*, 1997].

2.2. Sand Bars

Sand bars, which river runners use as campsites, are an important resource in Grand Canyon [*U.S. Department of Interior*, 1995]. Historical photographs and aerial photography provide the only definitive evidence of sand-bar size and location in Grand Canyon before closure of Glen Canyon Dam. From analysis of historical photography, *Schmidt et al.* [1995] and *Webb* [1996] reported that sand bars were in the same locations in the 1890s and 1990s, indicating that the eddies in which sand is deposited are persistent. *Schmidt et al.* [1995] analyzed sand height around persistent rocks at Badger Creek Rapid and found considerable variation in the size of pre-dam sand bars. Pre-dam flood deposits, which contained mostly sand-sized particles [*McKee*, 1938], had higher silt and clay contents than those deposited after the dam was built [*Howard and Dolan*, 1981; *Schmidt and Graf*, 1990], which might confer internal strength to the deposit as well as a higher nutrient content for growth of riparian vegetation [*Stevens*, 1989]. Regular scouring and inundation limited riparian vegetation on most pre-dam sand bars [*Turner and Karpiscak*, 1980; *Webb*, 1996].

2.3. Riparian Vegetation

The riparian vegetation of the pre-dam Colorado River is known from several floral surveys [*Clover and Jotter*, 1944; *Martin*, 1971] and repeat photography [*Turner and Karpiscak*, 1980; *Stephens and Shoemaker*, 1987; *Webb*, 1996]. Marshes were not present along the unregulated Colorado River [*Stevens et al.*, 1995; *Webb*, 1996] except where perennial springs discharged into or near the channel. *Clover and Jotter* [1944] described the "margin of moist sand" as a scoured zone devoid of perennial vegetation. Most riparian vegetation grew around the "pre-dam floodline" [*Turner and Karpiscak*, 1980] or the "old high-water line" [*Carothers and Brown*, 1991; *Johnson*, 1991]. The species comprising what is now called the old high-water zone change through the river corridor, but the most common species is *Prosopis glandulosa* (mesquite) [*Turner and Karpiscak*, 1980].

Certain nonnative plant species, particularly *Tamarix* sp. (saltcedar or tamarisk), were widely distributed in the Colorado River basin but were not common in Grand Canyon before construction of Glen Canyon Dam. *Tamarix* has become naturalized in all the major river systems in the Southwest, reproducing prolifically and establishing dense stands after its introduction in the late 1800s [*Horton*, 1964; *Robinson*, 1965; *Harris*, 1966]. It expanded rapidly throughout the western United States in the 1920s and 1930s [*Christensen*, 1962]. *Graf* [1978] proposed that *Tamarix* spread through Grand Canyon upstream from the Grand Wash Cliffs between 1900 and 1910. Repeat photography does not support this early arrival date but instead suggests invasion from tributaries such as the Paria River [*Webb*, 1996]. *Clover and Jotter* [1944] noted *Tamarix* at several sites in Grand Canyon in 1938; in 1936, it was present along the river between Nankoweap Creek and Tanner Rapid (miles 52 to 69) and near Phantom Ranch [*Patraw*, 1936; *Dodge*, 1936].

Little is known about wildlife usage of habitat in the old high-water line before Glen Canyon Dam. Southwestern willow flycatcher (*Empidonax trailli extimus*), an endangered species, nests in dense stands of riparian vegetation, which were not present along the unregulated river. Specimens of this species were collected near Lees Ferry and at the mouth of the Little Colorado River before Glen Canyon Dam [*Brown*, 1988].

2.4. Aquatic Resources

Because of the seasonally low light penetration caused by the high sediment load, low primary production, pool-rapid hydraulics, and long geological isolation from other drainage basins, the native fishes in the Colorado River basin have a 74% level of species endemism, which is the highest in North America [*Miller*, 1959]. Eight native fishes occurred in Grand Canyon; of these, the humpback chub (*Gila cypha*), speckled dace (*Rhinichthys osculus*), flannelmouth sucker (*Catostomus latipinnis*), bluehead sucker (*C. discobolus*), and razorback sucker (*Xyrauchen texanus*) remain [*Minckley*, 1991]. Both the humpback chub and razorback sucker are federally listed endangered species. The native species likely spawned in both the Colorado River and in major tributaries, such as the Little Colorado

and Paria rivers [*Valdez and Ryel*, 1995]. Warm waters in pre-dam backwaters are thought to be where some native fishes spent part of their first year of life. Important habitats formed in the mouths of perennial tributaries during spring floods, when river water impounded warmer tributary waters into sizeable pools that provided seasonal refugium.

Populations of native fishes declined before completion of Glen Canyon Dam [*Miller*, 1961]. Fragmentation of the river by dams, introduction of nonnative fishes, pollution, and water extraction decreased native fish densities. A total of 24 nonnative fish species have made their way into the Colorado River in Grand Canyon [*Valdez and Ryel*, 1995]; many of these were introduced in the Colorado River system in the 19th century [*Minckley*, 1991]. *Valdez and Ryel* [1995] provide a list of introduction dates for nonnative species including rainbow trout (*Oncorhynchus mykiss*), carp (*Cyprinus carpio*), channel catfish (*Ictalurus punctatus*), red shiners (*Cyprinella lutrensis*), and fathead minnows (*Pimephales promelas*). Anglers highly value the trout fishery between Lees Ferry and Glen Canyon Dam (Figure 1). Many of these species compete directly with native species for food or are piscivores of native fish eggs, larvae, and young of the year and displace native fish from their habitat [*Minckley*, 1991].

3. HYDROLOGY OF THE COLORADO RIVER IN GRAND CANYON

3.1. The Colorado River drainage

The Colorado River and its major tributaries, the Green and San Juan rivers, begin in the mountains of Colorado, Utah, and Wyoming in snowfields at elevations of over 4300 m. Over 70% of the total flow of the Colorado River originates in these states, mostly during spring runoff. Upstream from Grand Canyon, the river drains approximately 627,000 km^2 and flows for over 2200 km to its delta in Mexico. Extensive diversions prevent water from reaching the Sea of Cortez in most years. Development of water resources in the Colorado River basin began in the late 1800s and was essentially completed in the 1970s. Large scale land and river development began with passage of the Reclamation Act of 1902 and signing of the Colorado River Compact in 1922 [*Stevens*, 1988]. Distances along the Colorado River in Grand Canyon are traditionally measured in river miles [*Stevens*, 1990], with river mile 0, in both the upstream and downstream directions, at Lees Ferry (Figure 1).

3.2. Gaging Records of the Colorado River and its Tributaries

Gaging stations have recorded discharges for the Colorado River at Lees Ferry since May 1921, and the Colorado River near Grand Canyon since October 1922 (Figure 2a). The Colorado River near Grand Canyon gaging station is located just upstream from the confluence with Bright Angel Creek (river mile 87.5; Figure 1). Discharges at Lees Ferry were affected to varying degrees by the early construction phases of Glen Canyon Dam between 1956 and 1962; flow at both gaging stations has been completely regulated by Glen Canyon Dam since March 13, 1963 [*Garrett and Gellenbeck, 1989*]. Streamflow gaging on the Paria River at Lees Ferry began in October 1923; the Paria River is unregulated. The legally defined point dividing the upper and lower basins is 0.4 km downstream from the mouth of the Paria River, and discharges of both the Colorado and Paria rivers are required for compliance with the Colorado River Compact. The primary gaging station on the Little Colorado River, near Cameron, Arizona, has operated continuously since June 1947; the Little Colorado River is partially regulated by several small, upstream reservoirs.

In addition to gaging stations at Lees Ferry and Grand Canyon, GCES funded 4 gaging stations between Glen Canyon Dam and Diamond Creek (Figure 1). The Colorado River below Glen Canyon Dam station (river mile -15) operated between October 1989 and March 1993. Gaging stations at the Colorado River just upstream from the mouth of the Little Colorado River (river mile 61.1); the Colorado River above National Canyon (river mile 166.5); and the Colorado River above Diamond Creek (river mile 225.0), operated at various times between 1983 and 1996 [*Garrett et al.*, 1993; *Rote et al.*, 1997].

Before completion of Glen Canyon Dam, seasonal peak discharges occurred between May and July, fed by snowmelt in the headwaters. For the Colorado River near Grand Canyon, the 2-yr and 10-yr floods in the pre-dam period were 2160 and 3950 m^3/s, respectively (Figure 2). The unregulated mean annual peak discharge was 2420 m^3/s; the maximum peak discharge in the gaging record is 6230 m^3/s in 1921 and a flood in 1884 was estimated to be about 8500 m^3/s [*Garrett and Gellenbeck*, 1989]. Four water years —1931, 1934, 1954, and 1955 — had peak discharges less than the 1996 controlled flood peak of 1345 m^3/s. The smallest annual peak discharge was only 720 m^3/s in 1934. After the snowmelt flood subsided, flow in the Colorado River typically fell to less than 200 m^3/s except during brief but occasionally substantial summer tributary flashfloods. The largest tributary floods increased the flow of the

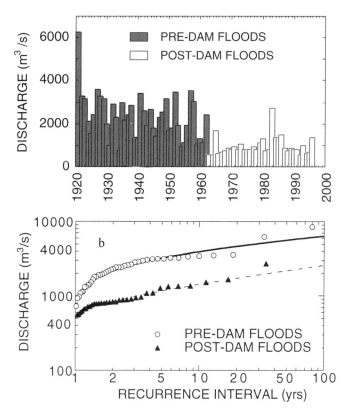

Figure 2. A. Annual peak flood series for the Colorado River near Grand Canyon, Arizona. B. Flood frequency for pre- and post-dam periods for the Colorado River near Grand Canyon, Arizona.

Colorado River to the magnitude of the annual snowmelt peaks, notably in September 1923.

The unregulated flow of the Colorado River was highly variable. The highest and lowest mean annual flow volume passing Lee Ferry was 761 m^3/s in 1924 and 171 m^3/s in 1934, respectively [*Anderson and White*, 1979]. The average unregulated flow at Lees Ferry from 1922 to 1962 was 476 m^3/s; most of the high flow years were during the wet period between 1896 and 1930 where the average was approximately 666 m^3/s [*Dawdy*, 1991]. The difference in annual flow volume between Lees Ferry and Grand Canyon gages — 16 m^3/s — is primarily the contribution of the Paria and Little Colorado rivers [*Turner and Karpiscak*, 1980]. From tree-ring evidence in the headwaters, the long-term average natural flow is 539 m^3/s [*Stockton and Jacoby*, 1976].

3.3. Sediment Data Collection

The sediment in the Colorado River mostly comes from tributaries that drain the semiarid sections of the basin.

These sections are composed primarily by Mesozoic and Cenozoic sandstones, mudstones, and shales from the Wingate and Navajo Sandstone; the Entrada, Morrison, Chinle, and Moenkopi Formations; and the Tropic and Mancos Shale [*Howard*, 1947; *Irons et al.*, 1965; *Howard and Dolan*, 1981; *Andrews*, 1986]. Snowmelt runoff had sediment concentrations of less than 10,000 parts per million (ppm). During summer flashfloods, sediment concentrations were higher than 20,000 ppm, and much of the sediment was silt and clay.

Daily sampling of the suspended-sediment concentration began at Lees Ferry in 1928 and at Grand Canyon in 1925. At Lees Ferry, daily sediment samples were collected between 1922 and 1933, 1942 and 1944, 1947 and 1965, and at various times after 1983 [*Garrett et al.*, 1993; *Rote et al.*, 1997]; the average sediment load between 1947 and 1957 was $60 \cdot 10^6$ metric tons/yr. At Grand Canyon, the average sediment load between 1941 and 1957 was $78 \cdot 10^6$ metric tons/yr.

From 1925 to 1941, the sediment load passing the Grand Canyon gaging station averaged $177 \cdot 10^6$ metric tons/yr [*Andrews*, 1990, 1991], or 2.3 times higher than the average from 1941 to 1957. As discussed by *Graf* [1987], *Andrews* [1991] and *Gellis et al.* [1991], sediment inflow may have decreased because regional arroyo cutting greatly slowed in the 1940s. *Hereford and Webb* [1992] and *Graf et al.* [1991] explained arroyo cutting and decreasing flood frequency, and their effects on sediment yield and flood-plain formation, by changes in regional climatic variability, such as changes in storm frequency and rainfall intensity.

Suspended sediment was sampled daily from 1947 to 1976 at the Paria River at Lees Ferry and 1947 to 1972 at the Little Colorado River near Cameron; samples also have been collected at various times in the 1980s and 1990s [*Garrett et al.*, 1993; *Rote et al.*, 1997]. On average, $2.74 \cdot 10^6$ metric tons/yr of sediment enters the Colorado River from the Paria River, mostly in August [*Andrews*, 1991]. The Little Colorado River delivers $8.4 \cdot 10^6$ metric tons/yr of sediment, and Little Colorado River tributaries downstream from Cameron, primarily Moenkopi Wash, contribute an additional $2.7 \cdot 10^6$ metric tons/yr.

4. EFFECTS OF GLEN CANYON DAM

4.1. Dam Construction

The Colorado River Compact led the way to authorization and construction of Hoover Dam, completed in 1935 [*Stevens*, 1988]. Hoover Dam was the first dam to significantly impact Grand Canyon. Lake Mead reservoir, which completely filled in 1939, impounds water into the lower 64 km of Grand Canyon. In addition, the reservoir serves as a refuge for nonnative fish species, which migrate from its relatively warm waters upstream into the Colorado River for spawning. One species, striped bass (*Morone saxatilis*), migrates upstream and may be a significant predator of young native fish [*Valdez and Leibfried*, 1999].

Congress authorized the Colorado River Storage Project and Glen Canyon Dam in 1956, and the diversion tunnels at the dam were sealed in March 1963 [*Martin*, 1989]. The dam is the major flow regulation structure controlling delivery of water from the upper-basin states of Wyoming, Colorado, Utah, and New Mexico to the lower-basin states of Arizona, Nevada, and California. Long-term average outflow of Glen Canyon Dam is limited to $10 \cdot 10^9$ m^3 of water [*U.S. Department of the Interior*, 1995]. Because of the annual outflow constraints, and the desire to maximize hydropower production, releases from Glen Canyon Dam rarely exceed the maximum powerplant capacity of 940 m^3/s when the reservoir is full [*U.S. Department of the Interior*, 1995].

Lake Powell reservoir filled to its capacity of $30 \cdot 10^9$ m^3 for the first time on June 17, 1980, 17 yrs after storage began. The reservoir extends 300 km upstream at its maximum pool elevation, inundating the lower part of Cataract Canyon in east-central Utah and much of the San Juan Canyon in southeastern Utah [*Potter and Drake*, 1989]. The reservoir is in excess of 150 m deep at the base of the dam at full-pool elevation. The reservoir stratifies seasonally with a well-defined epilimnion and hypoliminion from late spring to fall. The hypolimnion consists of a large mass of cold, dense water, which can have high specific conductance (1100-1500 siemens) owing to large amounts of salts in the inflow water. The surface of the reservoir has never frozen. The combination of great depth, small fetch, and higher salinity at depth prevents complete mixing near the dam during the isothermal periods of most winters. Water is drawn from a depth of 70 m below the full-pool elevation of the reservoir for the 8 generators in the powerplant of Glen Canyon Dam.

4.2. River Discharge

Glen Canyon Dam has profoundly changed the hydrology of the Colorado River in Grand Canyon. Flow regulation has greatly reduced interannual flow variability, although load-following hydroelectric power has increased the typical range of flow during a day [*Turner and Karpiscak*, 1980; *Howard and Dolan*, 1981; *Dawdy*, 1991]. Pre- and post-dam hydrographs for the Colorado River at Grand Canyon are significantly different (Figure 3a). Before Glen Canyon Dam, the annual flood peak typically

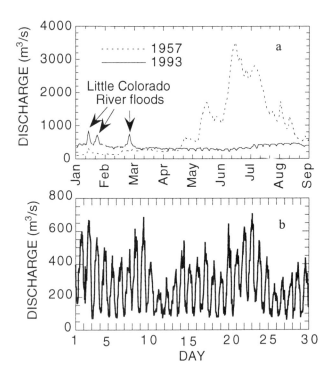

Figure 3. Hydrographs for the Colorado River near Grand Canyon, Arizona. A. Hydrograph of daily discharges for calendar years 1957 and 1993. B. Instantaneous discharge, measured every 15 minutes, in September 1982.

occurred between May and July with fluctuations and secondary peaks occurring in response to variability in snowmelt and summer flashfloods on major tributaries, particularly the Green and San Juan rivers. After construction of Glen Canyon Dam, discharges in Grand Canyon are more seasonally uniform, occasionally raised above powerplant releases by floods on either the Paria or Little Colorado rivers (Figure 3a). The peak discharge in the post-dam river typically occurs either in December-January or July-August when tributary floods coincide with high power production.

Flow regulation by Glen Canyon Dam has substantially reduced the annual range of river discharge at Lees Ferry. After 1963, the mean annual peak discharge of the Colorado River at Lees Ferry is 920 m³/s. In 26 of 32 years of flow regulation, the annual peak discharge was less than the powerplant capacity of approximately 930 m³/s. Legally mandated release of water from Lake Powell to Lake Mead caused a high release in 1965; similarly, a temporary legal limit on the surface elevation of Lake Powell caused the 1980 high release [*Martin*, 1989]. Flow substantially greater than powerplant capacity occurred throughout water years 1983 to 1986, as a result of unusually large runoff into

a full reservoir [*U.S. Department of the Interior*, 1995]. The 2-yr and 10-yr floods for the Colorado River near Grand Canyon are 851 and 1440 m³/s, respectively, which are reductions of about 38 and 40% from the historic frequency (Figure 2b). In the post-dam period, flow releases in excess of powerplant capacity, such as in the 1996 release, are considered "floods."

Water stored in Lake Powell during the spring runoff is released throughout the remainder of the year. The volume of water released in a given month varies only by a factor of 2 throughout the year and reflects demands for electrical power and water in the Southwest and California. Typically, highest monthly releases occur in December and January when electricity is needed for heating and during July and August when it is needed for air conditioning and irrigation. Releases in September 1982 (Figure 3b), typical of the 1970s and early 1980s, reflect the daily and weekly variation in electrical power demand. Peak demand occurs in the morning and early evening. Flow through the power-plant is decreased to a minimum in the late evening as power usage diminishes [*U.S. Department of the Interior*, 1995]. Significantly less electrical power is needed on weekends, and flow releases from Glen Canyon Dam accordingly show a 7-day periodicity (Figure 3b). Only relatively large, infrequent tributary floods produced similar daily changes in discharge in the unregulated river.

Reservoir operations have significantly altered the duration of daily mean stream flows at Lees Ferry (Figure 4). The magnitude of relatively large, infrequent flows (those equalled or exceeded less than 10% of the time) have been reduced by 50%, and release of stored snowmelt runoff during the remainder of the year has increased the magnitude of relatively common flows (those equalled or exceeded between 30 and 99% of the time). The discharge equalled or exceeded 50% of the time since 1965, compared to the period between 1922 and 1957, increased 60%.

Interim flow operations [*U.S. Department of the Interior*, 1995] were implemented in 1992 after the GCES research flow studies were completed. These flows, with maximum peaks of 566 m³/s and minimums of 142 m³/s, were designed to optimize sediment storage in the eddies and main channel from tributary inflows. After completion of the EIS and implementation of the Record of Decision, maximum powerplant releases were increased to 708 m³/s.

4.3. Aggradation of Debris Fans

Because Glen Canyon Dam is operated as a *de facto* flood control structure for water conservation, flow compe-tence has been greatly reduced in Grand Canyon. *Graf* [1980] first called attention to the problem of aggradation of

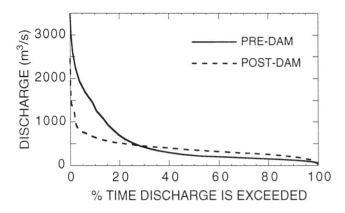

Figure 4. Duration of daily discharge for the Colorado River near Lees Ferry for the pre- and post-dam periods.

rapids downstream from dams on the Green River in Utah. For the post-dam period in Grand Canyon, the first significant debris flows occurred during an extraordinary storm in December 1966 [*Cooley et al.*, 1977]. Crystal Rapid changed from a minor rapid to one of the largest rapids in Grand Canyon [*Cooley et al.*, 1977; *Kieffer*, 1985; *Webb et al.*, 1989]. The same storm caused debris flows in Nankoweap Creek, Lava Canyon, Bright Angel Creek, and Prospect Canyon at Lava Falls Rapid [*Webb et al.*, 1989, 1997]. The Bright Angel Creek debris flow changed the stage-discharge relation by more than 1 m at the Grand Canyon gaging station several hundred meters upstream; the 1983 high releases washed out the constriction [*Burkham*, 1986]. By analyzing historical aerial photography, *Howard and Dolan* [1979, 1981] reported aggradation of 25% of the debris fans in Marble Canyon since the 1963 closure of Glen Canyon Dam.

Aggradation of debris fans was systematically documented after 1984 [*Melis et al.*, 1994]. An average of two debris flows occurred each year between 1984 and 1997 in Grand Canyon, creating 2 new rapids and enlarging debris fans at 9 existing riffles and rapids [*Webb et al.*, this volume]. *Griffiths et al.* [1996] developed a statistical model of debris flows in eastern and western Grand Canyon. They concluded that the frequency of debris flows is related to the location of shale in the drainage basin, as well as morphometric variables such as basin area. The frequency of debris flows is greatest in reaches where the Colorado River trends to the southwest; the highest frequency is in Marble Canyon.

Pre-dam floods removed all but the largest particles from debris fan and reshaped their configuration [*Graf*, 1979; *Howard and Dolan*, 1981; *Kieffer*, 1985; *Webb et al.*, 1997]. Using historical photographs, *Webb* [1996] documented

only a few changes to major rapids before closure of Glen Canyon Dam in 1963. Certain rapids, particularly Lava Falls, changed considerably owing to repeated debris flows. Subsequent reworking by the pre-dam Colorado River was insufficient to widen the river to its former condition [*Webb et al.*, 1997]. *Graf* [1980], *Howard and Dolan* [1981], *Kieffer* [1985], and *Webb et al.* [1997] presented conceptual models of Colorado River reworking. *Kieffer* [1985, 1990] defined the constriction ratio of rapid width to upstream width, and *Melis* [1997] found that 444 debris fans constrict the river between 1 and 75%, with a median of 44.5%.

The amount of debris-fan reworking that could be achieved by low dam releases has been an area of scientific controversy. *Dolan et al.* [1974] stated that significant reworking occurred above 1410 m^3/s. *Kieffer* [1985], studying reworking of the Crystal Creek debris fan during the 1983 flood, concluded that a discharge of more than 11,300 m^3/s would be required to reduce the constriction to 50%. *Webb et al.* [1997] documented the nearly complete removal of debris fans at Lava Falls Rapid by pre- and post-dam floods as low as 1000 m^3/s. *Melis et al.* [1994] documented reworking of other aggraded debris fans by powerplant releases less than 850 m^3/s. The latter two studies indicate that floods of similar magnitude to the 1996 controlled flood would significantly rework aggraded debris fans in Grand Canyon.

4.4. Sediment Mass Balance and Bed Scour

Glen Canyon Dam releases are essentially clear because nearly all of the formerly prodigious sediment load entering Grand Canyon is deposited in Lake Powell. Sediment coarser than 0.5 mm comprised less than 1% of the pre-dam sediment load [*Smith et al.*, 1960]; the regulated river transports a higher percentage of coarser sand, depending on tributary influxes. Post-dam suspended-sediment transport is approximately 5% of the pre-dam value downstream from Lees Ferry and the Paria River and approximately 25% of the pre-dam value at Grand Canyon. Tributaries downstream from the dam supply limited but significant quantities of fine sediment to the Colorado River and the annual sediment load increases downstream. Suspended-sediment transport from the Paria and Little Colorado rivers is approximately 75% of the post-dam suspended sediment entering the river between Lees Ferry and Phantom Ranch; these tributaries represent 94% of the contributing drainage area in this reach [*Andrews*, 1991]. About 219 small ungaged tributaries [*Melis et al.*, 1994] supply the remaining 25% of the post-dam sediment input.

Bed scour begins downstream from major dams after closure [*Williams and Wolman*, 1984], eventually armoring the bed with coarser sediment. Following closure of the bypass tunnels at Glen Canyon Dam, fine sediment was scoured at progressively further distances away from the dam. By 1963, the scour zone extended 11 km downstream [*Pemberton*, 1976]. The first significant post-dam change in bed elevation at Lees Ferry, located 25 km downstream, occurred during the 1965 flood (Figure 5). Comparing the lowest point in cross section data collected during discharge measurements, *Burkham* [1986] documented 8.3 m of scour, followed by a rise of 3.7 m, resulting in a persistent, permanent scour of 4.6 m (Figure 5). By 1970, the bed of the Colorado River from the dam to the Paria River was armored in shallower reaches by coarse gravel and cobbles, and fine sediment was scoured from pools. Erosion slowed after 1975, but nearly $8.5 \cdot 10^6$ m^3 of sediment — mostly sand — had been eroded in the reach between the dam and Lees Ferry [*Pemberton*, 1976].

In 1984, *Wilson* [1988] collected data on the type of bed material, its location on the bed, and the depth of the river, and concluded that the percentage of the Colorado River bed composed of bedrock or boulders varied between 30 and 81%. The percentage of bedrock or boulders was highest in narrow reaches and lowest in wider reaches [see *Bureau of Reclamation*, 1988a, Table A-26]. At a discharge of 700 m^3/s, *Wilson* [1988] reported an "average thalweg depth" that ranged from 1.5 to 32.7 m. In 1965, *Leopold* [1969] measured a maximum river depth of 33.9 m near river mile 114.3 at a discharge of 1374 m^3/s.

Most researchers have concluded that sand accumulates on the bed between Lees Ferry and Grand Canyon at most dam releases because of sediment added by the Paria and Little Colorado rivers. *Howard and Dolan* [1981] calculated a net increase in sediment storage from 1965 to 1977 for the Lees Ferry to Grand Canyon reach (Figure 1). Using a modeling approach that did not account for the possibility of large dam releases, *Laursen et al.* [1976] concluded that emergent sand bars would only persist for 200 yrs after completion of Glen Canyon Dam.

Orvis and Randle [1988] and *Randle and Pemberton* [1988] developed a one-dimensional sediment transport model to calculate sediment transport through Grand Canyon based on *Wilson's* [1988] 209 cross sections and estimates of bed particle size. *Randle and Pemberton* [1988] concluded that high-flow years, such as 1983-1985, resulted in significant loss of sand, but most scenarios of dam operations caused aggradation of sand between the Paria and Little Colorado rivers. *Bennett* [1993] modelled release scenarios proposed for the EIS [*U.S. Department of the Interior*, 1995] and noted trade-offs between sand stored on the bed versus on the margins as emergent sand bars; higher fluctuating flows deposited sand in usable locations

at the expense of channel storage. Using a sediment mass-balance approach, *Smillie et al.* [1993] presented dam-release scenarios where sand could accumulate in the reach between the Paria and Little Colorado rivers; accumulation occurred if peak discharges were less than 566 m^3/s. These calculations suggested that sufficient sand was stored in the bed of the Colorado River to allow a regulated flood that would redistribute sand from the channel to its banks.

Additional sediment-transport modeling began to shed light on the underlying processes associated with sediment transport. Increasingly more complex operations of the dam required more sophisticated models, and data requirements increased. For example, attempts to predict unsteady flow hydrographs [*Lazenby*, 1988] and travel times [*Dawdy*, 1991] of water through Grand Canyon had considerable uncertainty. The travel time of water at various discharges is an important means of verifying the channel-geometry data used in most sediment-transport models. *Graf* [1995] used the dye-injection technique to determine travel times through Grand Canyon at several steady and unsteady dam releases. High-accuracy monitoring of cross sections [*Graf et al.*, 1995a, 1997] and extensive bathymetric mapping [*Graf et al.*, 1995b] supported development and verification of sediment-transport models.

Eventually, two types of models were developed to predict sediment transport. *Wiele and Smith* [1996] presented an improved one-dimensional sediment-transport model of Grand Canyon and found that less than 50% coverage of the bed with sand was required to maintain equilibrium sand transport through Grand Canyon. Video imagery and side-scan surveys since 1990 indicate that large areas of the bed are composed of gravel and cobbles [*Rubin et al.*, 1994]. To understand reach-specific changes, *Wiele at al.* [1996] developed a two-dimensional, vertically averaged flow model to predict deposition and cross-section changes just downstream from the mouth of the Little Colorado River after its 1993 floods (Figure 3a). They concluded that eddies immediately downstream from the Little Colorado River filled within 3 days during a flood whose suspended load had a concentration similar to floods before the dam; their work is the first quantification of the rates of sand accumulation in eddies during high flows.

4.5. Sand-Bar Erosion

By the early 1970s, campsites along the river had been lost because of sand-bar erosion and the encroachment of thick stands of riparian vegetation [*Dolan et al.*, 1974, 1977; *Carothers and Brown*, 1991]. Without replenishment of sand by floods, erosion was inevitable. Many factors contribute to sand-bar erosion, including 1) low suspended-

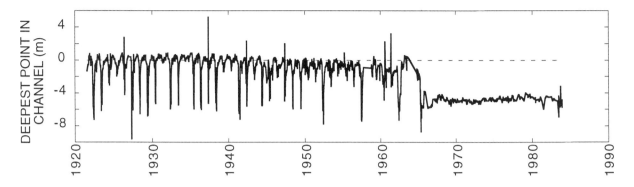

Figure 5. Deviation of low point in the channel cross section for the Colorado River at Lees Ferry, Arizona [*Burkham*, 1986].

sand concentrations in diurnally varying flow, 2) wave action [*Bauer and Schmidt*, 1993], 3) seepage erosion during downramping of diurnal fluctuations [*Budhu and Gobin*, 1994], 4) heavy trampling and downslope displacement by river runners [*Valentine and Dolan*, 1979], 5) wind deflation [*Dolan et al.*, 1974], 6) tributary floods and debris flows [*Melis et al.*, 1994], and 7) periodic slump failures [*Cluer*, 1995].

National Park Service concerns prompted the first systematic study of sand bars in 1974 and 1975 [*Howard*, 1975; *Howard and Dolan*, 1976, 1979, 1981]. Their work documented significant sand-bar erosion and encroachment of vegetation onto 20 formerly barren bars. Erosion, which generally reduced bar area and elevation, was greatest upstream from the Little Colorado River, although the trend was pervasive throughout Grand Canyon. *Borden and Weeden* [1973] identified and classified 443 potential camping beaches and began long-term monitoring.

The 1983 flood (Figure 2a) temporarily rebuilt sand bars through much of Grand Canyon. In October 1983, *Brian and Thomas* [1984] examined 227 of the 443 sand bars monitored by *Borden and Weeden* [1973]. Erosion after 1973 but before the 1983 releases had claimed 26 campsites, mostly in narrow reaches. Of the comparable camping beaches, 28% had decreased, 30% had increased, and 42% were unchanged in size. The 1983 flood completely removed 24 beaches, created 50 new ones, and aggraded many others, mostly in western Grand Canyon. Most of the new sand bars were unstable and rapidly eroded during high releases between 1984 and 1986. Generally, sand bars in narrow reaches were more severely eroded while those in wide reaches aggraded.

Kearsley et al. [1994] continued the evaluation of campsites begun by *Borden and Weeden* [1973] by directly measuring beach size in the field and in aerial photographs. *Kearsley et al.* [1994] concluded that 52% of campsites

were smaller, 46% of campsites were the same size, and only 2% were larger in 1990 than in 1965. They concluded that the "benefit" of sand aggradation from the 1983 flood, noted by *Brian and Thomas* [1984], had not been long lasting. Campsites changed by vegetation encroachment or removal and by deposition or erosion of sand; some of the changes occurred because of tributary floods or debris flows [*Melis et al.*, 1994].

A large amount of research concentrated on measurement of changes in sand bars. Many researchers focused on how Glen Canyon Dam could be operated to minimize sand-bar erosion, ignoring the conclusions of *Dolan et al.* [1974, 1977] that the problem stemmed from lack of replenishment of beaches. Because most studies under GCES I were conducted in the high-release period of 1983-1985, they advised avoidance of intentional and uncontrolled floods from Glen Canyon Dam to minimize sand-bar erosion [*Bureau of Reclamation*, 1988, p. A-21].

Some researchers revisited the original data set established by *Howard* [1975]; others monitored new sites and sand-bar types, creating a mosaic of different sand bars monitored at different times using different methods. For example, *Beus et al.* [1985] remeasured *Howard's* [1975] profile lines; *Ferrari* [1988] reported the resurvey of 20 of these lines with 4 additions made for ecological reasons; *Schmidt and Graf* [1990] remeasured profile lines and began detailed topographic and bathymetric surveys of beaches and eddies; and *Cluer* [1995] used high-frequency repeat photography and photogrammetry to measure sand-bar change at 6 sites. Technological advances improved the analyses, particularly the addition of bathymetric measurements [*Schmidt and Graf*, 1990]. Many of the monitored sand bars differed in morphology, configuration, and location, leading to a complex array of responses to operation of Glen Canyon Dam and conflicting conclusions in reports. Effects of antecedent conditions, particularly during the research flows of 1990-1991, were not adequately considered. In their study of sand-bar response to fluctuating flow, *Beus and Avery* [1993] concluded that "no single test flow affected all sand bars in the same manner."

A hydrodynamic classification of sand bars added an important tool to understanding the dynamics of sand bars. *Schmidt and Rubin* [1995] defined the fan-eddy complex (Figure 6), after preliminary work by *Howard and Dolan* [1981] and following *Schmidt* [1990] and *Schmidt and Graf* [1990]. Later modification resulted in the definition of three types of sand bars: (1) separation bars, which typically occur on the downstream side of debris fans on the upstream side of recirculation zones; (2) reattachment bars, which typically form beneath the primary eddy and extend to the stagnation point on the downstream side of the recirculation zone; and (3) channel-margin deposits, which typically form along channel banks and minor flow obstructions (Figure 6).

The classification of sand bars led to immediate gains in understanding the origins and stability of deposits with respect to operations of Glen Canyon Dam. *Schmidt and Graf* [1990] and *Schmidt* [1990] measured changes in eddy size and tracked changes in the flow reattachment point with respect to sand-bar deposition. *Melis* [1997] distinguished debris fans in terms of the discharge range that submerges them and thereby minimizes or eliminates the eddies downstream [*Schmidt*, 1990]. *Melis* [1997] also evaluated the range of river flood stages under which sand bars would be deposited. *Rubin et al.* [1990] made hydrodynamic interpretations of bedforms that indicated the nature of eddy recirculation that deposits reattachment bars. *Schmidt* [1993] correlated the type, extent, and stage of erosion with general dam-operating regimes; he noted the apparent trade-off between high-stage and low-stage sand bars, and aggradation of separation bars and degradation of reattachment bars during high dam releases. *Schmidt* [1993] stressed the importance of overall spatial and temporal history in understanding short-term changes in sand bars. *Schmidt et al.* [1995] and *Webb* [1996] found that reattachment bars were more unstable than separation or channel-margin bars, but that 60% of all sand bars had eroded between 1890 and the early 1990s. Using repeat photography, these researchers documented the systematic decrease in net sand-bar erosion with increasing distance from Glen Canyon Dam.

Research on changes in sand bars generally established that: (1) all sand bars do not respond in similar ways to operations of Glen Canyon Dam; (2) reattachment bars respond differently than separation and channel-margin bars; (3) steady flows generally erode sand bars and redistribute sand to lower elevations in the eddy; (4) high fluctuating flows aggrade some sand bars while eroding others; and (5) sand-bar response may vary with distance downstream from Glen Canyon Dam. In the course of research, the management paradigm shifted from emphasis on lack of replenishment, to minimizing erosion by minimizing or altering flow fluctuations, to rebuilding sand bars while minimizing erosion in the intervening periods.

Lack of annual flooding, caused by operation of Glen Canyon Dam, allowed colonization by native and nonnative riparian species on sand-mantled banks. Lack of flooding led to concern that trees in the old high-water zone were becoming senescent and not reproducing because of the drought imposed by the dam. Subsequent research indicates that the old high-water zone will survive despite the lack of annual inundation because most of the species are facultative riparian plants that can utilize precipitation, although their growth rates decrease [*Anderson and Ruffner*, 1988]. Seedling establishment of native species is expected to shift downslope from the old high-water zone, which is becoming increasingly fragmented in their new, primarily xeric, environment [*Carothers and Brown*, 1991].

The "new high-water zone" [*Johnson*, 1991], which formed between the 850-2830 m^3/s river stages, is a critical resource that supports extensive vegetation and 5-10 times the number of breeding birds than occurred along the pre-dam river [*Brown et al.*, 1987]. For example, the black-chinned hummingbird (*Archilochus alexandri*), which is common in the southwestern United States, nests almost exclusively in *Tamarix* trees in the river corridor [*Brown*, 1992]. *Turner and Karpiscak* [1980] documented the decreased frequency of river-bank inundation, pre- to post-dam, up to 1977 and used repeat photography to document the increase in riparian vegetation.

In the late 1960s, *Martin* [1971], following the *Clover and Jotter* [1944] survey, found *Tamarix* "abundantly distributed" along the river corridor and was alarmed at its "explosive spread" through Grand Canyon. *Tamarix* along with the native coyote willow (*Salix exigua*), Goodding willow (*Salix gooddingii*), cottonwood (*Populus fremontii*), and several native species of shrubs aggressively colonized the newly available substrate [*Turner and Karpiscak*, 1980]. The large increases, which occurred in both the old and new high-water zones, were significant. The 1983 flood, as well as increases in the beaver population, have destroyed many young *Populus*. *Brown and Trossett* [1989] estimated that 500 ha of new riparian habitat was created in 20 yrs of dam operations. In an analysis of aerial photography taken between 1965 and 1985, *Pucherelli* [1988] found *Prosopis* cover increased in the old high-water zone but at a rate 5 times slower than its increase in the new

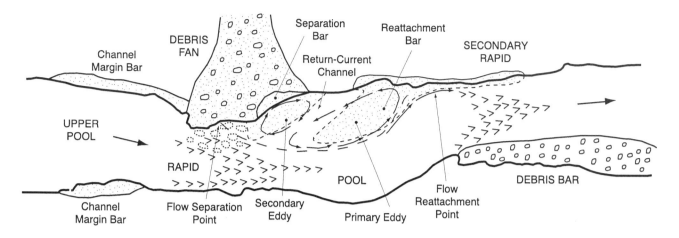

Figure 6. Schematic diagram of a typical fan-eddy complex in Grand Canyon [*Schmidt*, 1990].

high-water zone. The rate of the invasion slowed as the new high-water zone filled with plants.

As vegetation becomes denser and senescent, problems emerge as *Tamarix* makes its local environment more saline and less productive and the number of insects it supports declines [*Stevens*, 1989]. *Stevens* [1989] presented a competition model in which self-shading effects may increase the probability of eventual replacement of *Tamarix* with *Salix exigua*. Researchers concluded that the continued recruitment of riparian species requires periodic disturbance (i.e., flooding) to open patches [*Stevens and Ayers*, 1993] and suggested to managers desiring more riparian vegetation that the optimal dam-release scenario included periodic flooding with lower fluctuating and less damaging powerplant releases. The new high-water zone vegetation is highly valued by some managers as wildlife habitat, particularly for the endangered southwestern willow flycatcher [*U.S. Department of the Interior*, 1995] and other neotropical birds [*Brown et al.*, 1987]. The population of this species in Grand Canyon, while relatively small in total numbers, is believed to be the largest population remaining in Arizona [*Unitt*, 1987]; the birds heavily use dense stands of *Tamarix*. Prolonged flooding and inundation damage or kill *Tamarix* as well as native species. The flood of 1983 destroyed large numbers of *Tamarix* [*Stevens and Waring*, 1988]; continued high water in the mid-1980s minimized *Tamarix* reestablishment. Therefore, flooding may help minimize an undesirable, nonnative species while damaging a habitat desirable for endangered species.

Post-dam marshes developed in wide reaches of the Colorado River (Figure 1), particularly between Nankoweap Rapid (river mile 52) and the Little Colorado River; in the Furnace Flats reach of river miles 65-72; and

in western Grand Canyon, particularly near Parashant Canyon (river mile 198). Marsh vegetation cannot withstand scouring or lengthy inundation; the 1983 flood, which peaked at 2720 m³/s, killed half of the vegetation below the 1420 m³/s stage by drowning, burial, or erosion [*Stevens and Waring*, 1988]. *Stevens and Ayers* [1993] documented 65 sites containing 4.8 ha of marsh vegetation; the 1983 flood decreased the number of sites to 17. Despite the damage, seedlings became established within several years of the cessation of the high releases from the dam [*Stevens et al.*, 1995].

Riverine riparian vegetation in Grand Canyon poses a management dilemma [*Schmidt et al.*, 1998]. On the one hand, lack of flooding has created a ribbon of lush riparian vegetation along the Colorado River that is habitat for wildlife, but continued lack of flooding may result in a low-productivity, senescent environment. Restoration of periodic flooding could damage what is considered extremely valuable wildlife habitat, particularly for bird species, and damage marsh habitat, the most biologically productive part of the new high-water zone. One of the primary goals of the 1996 controlled flood was creating some disturbance to the new high-water zone without risking major habitat destruction for endangered species.

4.7. Temperature, Clarity, and Aquatic Productivity

Before completion of Glen Canyon Dam, the Colorado River was thought to be largely heterotrophic with little primary production in the sediment-laden water [*Carothers and Brown*, 1991]. The green algae *Cladophora glomerata* was present but not abundant in the river; numerous algal species were found in tributaries in Glen Canyon, but few studies were conducted in the Colorado River itself [*Blinn and Cole*, 1991]. Aquatic insect populations may have been low in the mainstem [*Blinn and Cole*, 1991], and certain terrestrial insects were likely to have been a part of the diets of native fishes [*Carothers and Brown*, 1991]. Debris fans and associated debris bars in the pre-dam river provided stable substrate for abundant aquatic insects, while shifting sand beds supported little benthic productivity.

The temperature of the Colorado River through Grand Canyon varied seasonally from 0 to 29°C before construction of Glen Canyon Dam [*Rote et al.*, 1997]. The surface of the river froze in the low-velocity reaches near Lees Ferry every 10-20 yrs [*Webb*, 1996]. The annual range in water temperature steadily declined after 1963 as Lake Powell filled. The temperature of water released through the dam has averaged about 8°C, with a range of 7-15°C between 1977 and 1995.

The combination of cold water and low sediment concentrations have dramatically altered the aquatic ecosystem of the Colorado River, converting a historic heterotrophic system to an autotrophic system dependent on autochthonous production. These conditions have resulted in high photosynthetic productivity downstream from the dam, making ideal conditions for a blue-ribbon trout fishery in the tailwaters [*Carothers and Brown*, 1991]. Productivity decreases downstream as sediment input from tributaries increases and reduces light available for photosynthesis. The clear, cold water favored development of an abundant, but moderately diverse trophic structure supported by *Cladophora* and a large assemblage of ephiphytic diatoms. The more upright forms of are more available to grazing macroinvertebrates and fish; higher temperatures favor more sessile diatom forms that presumably are less available as food to the nonnative amphipod, *Gammarus lacustris*, and other consumers. Higher productivity takes place under low to moderate river fluctuations and when the wetted channel perimeter is greatest; *Blinn et al.* [1995] concluded that twice the amount of energy was available at flows of 807 m^3/s than at 142 m^3/s. Wetting and drying, caused by wide-ranging fluctuating flows, caused significant loss of *Cladophora* biomass with attendant losses of diatoms and *Gammarus* [*Angradi and Kubly*, 1993].

Although few collections provide details on algal and macroinvertebrate species composition of the pre-dam river, changes in species composition appear to be dramatic, with numerous species disappearing and several new or uncommon species becoming more abundant [*Blinn and Cole*, 1991]. Using Cataract Canyon, upstream from Lake Powell in Utah, as a surrogate for pre-dam conditions in Grand Canyon, *Haden* [1997] found the primary consumers were collector/filterer insects that rely on allochthonous

detritus as an energy source. *Haden* [1997] found considerable differences in species composition of benthic invertebrates related to changes in trophic structure and temperature. The primary consumers in Cataract Canyon (and presumably the historic river) were dominated by collector/filterer insects which required a range of temperatures to complete their life cycles and relied on allochthonous detritus as an energy source. The benthic community below Glen Canyon Dam was dominated by *Gammarus*, and nearctic dipterans that are closely associated with *Cladophora* [*Stevens et al.*, 1997]. Although macroinvertebrate composition differed, total macroinvertebrate biomass was similar in Cataract Canyon and below Glen Canyon Dam.

Little is known about the food supply available to native fishes in the heterotrophic, pre-dam river, although most of it likely was supplied in drift. *Vanicek* [1967] found that aquatic and terrestrial insects, plant debris, filamentous algae, and fish were the primary foods of roundtail chub, humpback chub, and bonytail in the Green River prior to completion of Flaming Gorge Dam. *Tyus and Minckley* [1988] reported gorging by humpback chub on migrating Mormon crickets (locusts) in the Green and Yampa rivers, suggesting opportunistic feeding and the importance of terrestrial food sources. In the post-dam river in Grand Canyon, many fish species rely heavily on the abundant macroinvertebrate biomass found in the dam tailwaters that decreases in abundance downstream. In the tailwaters, rainbow trout ate primarily *Gammarus*, filamentous algae, midges, and blackflies [*Carothers and Minckley*, 1981; *Maddux et al.*, 1987; *Leibfried*, 1988]. *Valdez and Ryel* [1997] reported that humpback chub near the Little Colorado River (125 km downstream from the dam) ate primarily *Gammarus* (45%) and blackflies (40%), with some use of midges and terrestrial invertebrates. Further downstream (220 km downstream from the dam), humpback chub ate primarily blackflies (49%) and terrestrial invertebrates (30%), with less use of *Gammarus* and midges. Hence, despite a vastly different trophic structure, native as well as nonnative fishes have to rely heavily on *in situ* primary and secondary production, since upstream sources of organic material have been interrupted by impoundment of the river.

Inherent in all of the research on aquatic productivity is the assumption that high aquatic productivity is a desirable management goal for the post-dam river [*Carothers and Brown*, 1991]. Research on the aquatic community has focused on minimizing productivity losses by minimizing daily fluctuations and implementing large dam releases. Implementation of interim flows and the preferred alternative, low modified fluctuating flows, have benefitted, maintained or increased aquatic productivity and are, therefore, thought to be desirable. Benefits of controlled flooding to biological productivity manifest themselves over a longer time period than, for example, sand-bar deposition, and those benefits can only be determined through long-term monitoring. Because of the high societal value placed on the blue-ribbon trout fishery upstream from Lees Ferry, maximizing productivity of *Cladophora*, and hence *Gammarus*, is considered to be valuable. One of the primary reasons that minimum discharges from Glen Canyon Dam were raised from 85 to 227 m^3/s in summer months was to minimized stranding of trout and loss of incubating eggs and newly hatched fry [*U.S. Department of the Interior*, 1995].

4.8. Adaptation of Native Fish Populations to the Post-Dam River

The native fishes of the Colorado River were in decline as early as the 1950s, primarily as a result of water diversion, degraded water quality, and the invasion of competing and predaceous nonnative fishes [*Miller*, 1961]. By the time Glen Canyon Dam was completed in 1963, fish assemblages were dominated by nonnative carp, channel catfish, and red shiners in the mainstem and rainbow trout and brown trout in the tributaries [*Valdez and Ryel*, 1995]. Of the 8 native mainstem fish species, 3 were extirpated at about the time of, or shortly after, dam completion; roundtail chub and bonytail were last reported in Grand Canyon in the 1950s, and Colorado squawfish were last reported in the mid-1970s. Razorback suckers are now extremely rare, humpback chub are endangered, and flannelmouth suckers, bluehead suckers, and speckled dace are confined to local populations, primarily near seasonally-warmed tributaries. Razorback suckers and Colorado squawfish were probably transient species through Grand Canyon.

By the early 1970s, water temperatures in Grand Canyon had dropped below the range of 16-22°C needed for successful mainstem reproduction by the warmwater native fishes. Successful mainstem reproduction has not been documented for these species, although attempted spawning has been reported in warm springs for humpback chub [*Valdez and Masslich*, 1999]. Humpback chub spawn primarily in the Little Colorado River and flannelmouth suckers, bluehead suckers, and speckled dace spawn in tributaries, including the Little Colorado River [*Robinson et al.*, 1996], Paria River [*Weiss*, 1993], and Bright Angel and Kanab creeks [*Otis*, 1994]. Flannelmouth suckers are also reported spawning in the dam tailwaters, possibly as a result of more stable flows from implementation of interim flows

in 1992. The cold water, while appropriate for spawning of nonnative salmonids, suppressed mainstem reproduction by most nonnative warm-water species, as well. Like the native warm-water fishes, these species also ascend seasonally-warmed tributaries for spawning.

Native fishes have faced substantial competition and predation from nonnative fishes since before completion of Glen Canyon Dam. In the post-dam river, nonnative fish predation may account for substantial mortality of a year class [*Valdez and Ryel*, 1997; *Douglas and Marsh*, 1996]. Carp are voracious predators of fish eggs and larvae, and can quickly locate and scavenge newly deposited eggs, particularly in small tributaries. Channel catfish consume large numbers of young native fishes when they are most active, at night and during high turbidity. Rainbow and brown trout consume large numbers of juvenile native fishes descending from natal streams, and fathead minnows and red shiners consume newly-hatched larvae in shallow-nursery habitats, such as backwaters.

The cold clear riverine conditions have influenced fish distribution and habitat use. *Maddux et al.* [1987] and *Valdez and Ryel* [1995] demonstrated that mainstem fishes in Grand Canyon are distributed primarily near seasonally warmed tributary mouths. Habitat use, particularly for humpback chub, reflects a combination of canyon geomorphology, local cover, and food supply. Preferred channel reaches have abundant debris fans and are typically in proximity to warm springs or perennial tributaries [*Valdez and Ryel*, 1995]. Young-of-year and juveniles (generally <200 mm long) prefer backwaters (i.e., eddy return channels), river banks with vegetation overhanging from the new high-water zone, or shorelines of talus or debris fans [*Converse et al.,* 1998]. Adults prefer offshore habitats, especially large recirculating eddies associated with debris fans. The quality of these shoreline habitats is affected by flow stage changes, and high fluctuating flows can destabilize these habitats, forcing fish to relocate at great risk of predation, excessive energy expenditure, and starvation. High water clarity significantly reduces movement by humpback chub, possibly affecting feeding and suggesting use of turbidity as cover [*Valdez and Ryel*, 1995]. This underscores another negative effect of lowered suspended sediment in Grand Canyon.

Backwaters in Grand Canyon are used extensively as nurseries by young of both native and nonnative fishes, and as rearing and holding areas by small forms such as native speckled dace and nonnative fathead minnows and red shiners [*Arizona Game and Fish Department*, 1997]. Backwaters are not extensively used as nurseries by young humpback chub, flannelmouth suckers, and bluehead suckers in less regulated reaches of the upper basin [*Valdez and Wick*, 1983], primarily because the fish are hatched shortly after the peak of runoff, before sand bars that form these backwaters are exposed by receding flows. The strong interdependence of Colorado squawfish and backwaters in the upper basin results from fish hatching in late summer when backwaters are available. Backwaters in Grand Canyon, however, are important because their temperatures are slightly higher than water moving downstream. Because the mainstem is consistently cold and backwaters are available under most interim flows, backwaters serve as thermal refugia for fish by entraining water and allowing warming from solar radiation. Although the warm, sheltered water attracts fish and results in high primary and macroin-vertebrate production, high water clarity in the presence of predators often precludes daytime use by many fishes. Hence, except when turbidity is high, many fish use backwaters transiently. During periods of high sediment input and low fluctuating flows, however, backwaters can become filled with sediment. Periodic high controlled releases can increase water velocities through eddy return channels and flush these sediments, restoring capacity and reinitiating productivity in backwaters.

Humpback chub evolved in an environment of excep-tionally large floods with high sediment concentrations. Controlled high releases from Glen Canyon Dam are not expected to significantly affect their survival, behavior, or status. Floods may improve the habitat of native fishes by increasing turbidity, restoring backwater space and produc-tivity, redistributing nutrients, and importing a large mass of terrestrial organisms and debris used as food by these fishes. In the post-dam river, floods will temporarily scour large amounts of moribund *Cladophora* and possibly reduce available food supplies for a short time to both native fishes and nonnative trout in the dam tailwaters. *Cladophora* has proven to be resilient and the associated macroinvertebrate assemblages recover quickly, often exceeding pre-flood densities, following floods.

5. CONCLUSIONS FROM THE GLEN CANYON ENVI-RONMENTAL STUDIES PROGRAM

The following conclusions were drawn from scientific studies completed under the GCES program:
1. Despite blockage of most sediment that formerly entered Grand Canyon by Glen Canyon Dam, most researchers concluded that sediment input from tributaries downstream from the dam was being stored on the channel bed or in eddies by most dam releases because of the reduced sediment-transport capacity.
2. *De facto* flood control resulting from dam operations prevents deposition of sand on high-stage bars and

reworking of debris fans. The absence of floods, therefore, is a disturbance to pre-dam geomorphic processes.
3. The results of sand-bar monitoring during the research flows, designed to determine the effects of various fluctu-ating releases, were inconclusive, in part because antecedent conditions were not properly analyzed.
4. Encroachment of riparian vegetation on sand bars used as campsites was almost as important as erosion in reducing usable campsite size.
5. Some debris fans have aggraded because of continuing tributary debris flows and reduced flood frequency in the Colorado River. Debris flows erode or bury existing sand bars, and aggradation affects navigability of rapids and minimizes deposition of replacement sand bars.
6. Riparian vegetation, much of it nonnative but valued as wildlife habitat, was encroaching and filling backwaters and marsh areas that support dam-related biota.
7. During many flows, sand moved from emergent sand bars to the eddies, thereby filling backwater aquatic habitats.
8. Clear, cold water changed the trophic structure of the aquatic ecosystem from a heterotrophic to an autotrophic system dominated by *Cladophora*, which supports *Gammarus*, now a major food source for native and nonnative fishes.
9. Native fish populations, already seriously threatened or endangered by introductions of nonnative species and blockages of migration routes, can no longer spawn in most of the mainstem, placing increasing dependence on seasonally warmed tributaries.
10. Maximizing the rainbow trout fishery and the endan-gered populations of humpback chub may be mutually exclusive because of their strongly contrasting life-history strategies and reproduction requirements.

Acknowledgments. The authors thank John C. Schmidt, Sheila David, William Lewis, and G. Richard Marzolf for their critical and insightful reviews of this chapter.

REFERENCES

Anderson, L.S., and G.A. Ruffner, Effects of the post-Glen Canyon Dam flow regime on the old high-water zone plant community along the Colorado River in Grand Canyon, in *Glen Canyon Environmental Studies: Executive Summaries of Technical Reports*, pp. 271-286, U.S. Dept. Int., Bur. Reclam., Upper Colo. River Reg., Salt Lake City, UT, 1988.
Anderson, T.W., and N.D. White, *Statistical summaries of Arizona streamflow data*, U.S. Geol. Surv. Water-Res. Invest. Rept. 79-5, 1979.

Andrews, E.D., Downstream effects of Flaming Gorge Reservoir on the Green River, Colorado and Utah, *Geol. Soc. Amer. Bull.*, 97, 1012-1023, 1986.

Andrews, E.D., Effects of stream flow and sediment on channel stability of the Colorado River: A perspective from Lees Ferry, Arizona, in *The Geology of North America: v. 0-1; Surface Water Hydrology*, ed. M.G. Wolman and H.C. Riggs, pp. 304-310, Geol. Soc. Amer., 1990.

Andrews, E.D., Sediment transport in the Colorado River basin, in *Colorado River ecology and dam management*, ed G.R. Marzolf, pp. 54-74, Natl. Acad. Press, Washington, DC, 1991.

Angradi, J.D., and D.M. Kubly, Effects of atmospheric exposure on chlorophyll *a*, biomass, and productivity of the epilithon of a tailwater river, *Regl. Rivers, Res. Manage.*, 8, 345-358, 1993.

Arizona Game and Fish Department, *Glen Canyon Environmental Studies 1996 Annual Report*, Glen Canyon Environ. Studies, Bur. Reclam., Flagstaff, AZ, Ariz. Game Fish Dept., Phoenix, AZ, 1997.

Bauer, B.O., and J.C. Schmidt, Waves and sandbar erosion in the Grand Canyon: Applying coastal theory to a fluvial system, *Ann. Ass. Amer. Geogr.*, 83, 475-497, 1993.

Bennett, J.P., *Sediment transport simulations for two reaches of the Colorado River, Grand Canyon, Arizona*, U.S. Geol. Surv. Water-Res. Invest. Rept. 93-4034, 1993.

Beus, S.S., S.W. Carothers, and C.C. Avery, Topographic changes in fluvial terrace deposits used as campsite beaches along the Colorado River in Grand Canyon, *J. Ariz. - Nev. Acad. Sci.*, 20, 111-120, 1985.

Beus, S.S., and C.C. Avery, The influence of variable discharge regimes on Colorado River sand bars below Glen Canyon Dam, Chapter 6 in *The influence of variable discharge regimes on Colorado River sand bars below Glen Canyon Dam, Final Report*, ed. S.S. Beus and C.C. Avery, N. Ariz. Univ., Natl. Park Serv. Coop. Agree. No. CA 86006-8-0002, Flagstaff, AZ, 1993.

Birdseye, C.H., *Plan and profile of the Colorado River from Lees Ferry, Arizona, to Black Canyon, Arizona-Nevada, and the Virgin River, Nevada*, U.S. Geol. Surv. Map Publ., 21 sheets (A-U), scale 1:36,680, 1924.

Blinn, D.W., and G.A. Cole, Algal and invertebrate biota in the Colorado River: Comparison of pre- and post-dam conditions, in *Colorado River ecology and dam management*, ed. G.R. Marzolf, pp. 102-123, Natl. Acad. Press, Washington, DC, 1991.

Blinn, D.W., J.P. Shannon, L.E. Stevens, and J.P. Carter, Consequences of fluctuating discharge for lotic communities, *J. N. Amer. Benth. Soc.*, 14, 233-248, 1995.

Borden, F.Y., and W.A. Weeden, *Design and methodology for carrying capacity estimation for camping beaches along the Colorado River in the Grand Canyon region*, Natl. Park Serv., Proj. Rept., Contr. No. CX0001-3-0061, 1973FY, Denver, CO, 1973.

Brian, N.J., and J.R. Thomas, *1983 Colorado River beach campsite inventory*, Natl. Park Serv. Rept., Grand Canyon, AZ, 1984.

Brown, B.T., Breeding ecology of a willow flycatcher population along the Colorado River in Grand Canyon, Arizona, *West. Birds*, 19, 25-33, 1988.

Brown, B.T., Nesting chronology, density, and habitat use of Black-Chinned Hummingbirds along the Colorado River, Arizona, *J. Field Ornith.*, 63, 393-400, 1992.

Brown, B.T., S.W. Carothers, and R.R. Johnson, *Grand Canyon birds*, Univ. Ariz. Press, Tucson, AZ, 1987.

Brown, B.T., and M.W. Trossett, Nesting-habitat relationships of riparian birds along the Colorado River in Grand Canyon, Arizona, *Southwest. Natur.*, 34, 260-270, 1989.

Budhu, M., and R. Gobin, Instability of sand bars in Grand Canyon, *J. Hydraul. Engr.*, 120, 919-933, 1994.

Bureau of Reclamation, *Glen Canyon Environmental Studies, Final Report*, U.S. Dept. Interior, Bur. Reclam., Salt Lake City, UT, 1988a.

Bureau of Reclamation, *Glen Canyon Environmental Studies, Executive Summaries of Technical Reports*, U.S. Dept. Interior, Bur. Reclam., Salt Lake City, UT, 1988b.

Burkham, D.E., *Trends in selected hydraulic variables for the Colorado River at Lees Ferry and near Grand Canyon for the period 1922-1984*, Springfield, Virg., U.S. Dept. Comm., National Tech. Info. Serv., Rept. PB88-216098, GCES/07/87, 1986.

Carothers, S.W., and W.L. Minckley, *A survey of the fishes, aquatic invertebrates and aquatic plants of the Colorado River and selected tributaries from Lees Ferry to Separation Rapid*, Final rept, Bur. Reclam. Contract 7-07030-C0026, Mus. N. Ariz., Flagstaff, AZ, 1981.

Carothers, S.W., and B.T. Brown, *The Colorado River through Grand Canyon: Natural history and human change*, Univ. Ariz. Press, Tucson, AZ, 1991.

Christensen, E.M., The rate of naturalization of Tamarisk in Utah, *Amer. Midl. Natur.*, 68, 51-57, 1962.

Clover, E.U., and L. Jotter, Floristic studies in the canyon of the Colorado and tributaries, *Amer. Midl. Natur.*, 32, 591-642, 1944.

Cluer, B.L., Cyclic fluvial processes and bias in environmental monitoring, *J. Geol.*, 103, 411-421, 1995.

Converse, Y.K., C.P. Hawkins, and R.A. Valdez, Habitat relationships of subadult humpback chub in the Colorado River through Grand Canyon: Spatial variability and implications of flow regulation, *Regul. Rivers*, 14, 267-284, 1998.

Cooley, M.E., B.N. Aldridge, and R.C. Euler, *Effects of the catastrophic flood of December, 1966, North Rim area, eastern Grand Canyon, Arizona*, U.S. Geol. Surv. Prof. Paper 980, 43 pp., 1977.

Dawdy, D.R., Hydrology of Glen Canyon and the Grand Canyon, in *Colorado River ecology and dam management*, ed. G.R. Marzolf, pp. 40-53, Natl. Acad. Press, Washington, DC, 1991.

Dodge, N.N., *Trees of Grand Canyon National Park*, Grand Can. Natur. Hist. Assoc., Bull. No. 3, 1936.

Dolan, R., A. Howard, and A. Gallenson, Man's impact on the Colorado River in the Grand Canyon, *Amer. Sci.*, 62, 392-401, 1974.

Dolan, R., B. Hayden, and A. Howard, Environmental management of the Colorado River within the Grand Canyon, *Environ. Manage.*, 1, 391-400, 1977.

Dolan, R., A. Howard, and D. Trimble, Structural control of the rapids and pools of the Colorado River in the Grand Canyon, *Science*, 202, 629-631, 1978.

Douglas, M.E., and P.C. Marsh, Population estimates/population movements of *Gila cypha*, an endangered cyprinid fish in the Grand Canyon region of Arizona, *Copeia*, 15-28, 1996.

Ferrari, R., Sandy beach area survey along the Colorado River in the Grand Canyon National Park, in *Glen Canyon Environmental Studies: Executive summaries of technical reports*, pp. 55-65, U.S. Dept. Int., Bur. Reclam., Upper Colo. River Reg., Salt Lake City, UT, 1988.

Garrett, J.M., and D.J. Gellenbeck, *Basin characteristics and streamflow statistics in Arizona as of 1989*, U.S. Geol. Surv. Water-Res. Invest. Rept. 91-4041, 1989.

Garrett, W.B., E.K. Van De Vanter, and J.B. Graf, *Streamflow and sediment-transport data, Colorado River and three tributaries in Grand Canyon, 1983 and 1985-86*, U.S. Geol. Surv. Open-File Rept. 93-174, 1993.

Gellis, A., R. Hereford, S.A. Schumm, and B.R. Hayes, Channel evolution and hydrologic variations in the Colorado River basin: Factors influencing sediment and salt loads, *J. Hydrol.*, 124, 317-344, 1991.

Graf, J.B., Measured and predicted velocity and longitudinal dispersion at steady and unsteady flow, Colorado River, Glen Canyon Dam to Lake Mead, *Water Res. Bull.* 31, 265-281, 1995.

Graf, J.B., R.H. Webb, and R. Hereford, Relation of sediment load and floodplain formation to climatic variability, Paria River drainage basin, Utah and Arizona, *Geol. Soc. Amer. Bull.*, 103, 1405-1415, 1991.

Graf, J.B., J.E. Marlow, G.G. Fisk, and S.M.D. Jansen, *Sand-storage changes in the Colorado River downstream from the Paria and Little Colorado Rivers, June 1992 to February 1994*, U.S. Geol. Surv. Open-File Rept. 95-446, 61 pp., 1995a.

Graf, J.B., S.M.D. Jansen, G.G. Fisk, and J.E. Marlow, *Topography and bathymetry of the Colorado River, Grand Canyon National Park, Little Colorado River to Tanner Rapids*, U.S. Geol. Surv. Open File Rept. 95-726, 7 sheets, 1995b.

Graf, J.B., J.E. Marlow, P.D. Rigas, and S.M.D. Jansen, *Sand-storage in the Colorado River downstream from the Paria and Little Colorado Rivers, April 1994 to August 1995*, U.S. Geol. Surv. Open-File Rept. 97-206, 41 pp., 1997.

Graf, W.L., Fluvial adjustments to the spread of Tamarisk in the Colorado Plateau region, *Geol. Soc. Amer. Bull.* 89, 1491-1501, 1978.

Graf, W.L., Rapids in canyon rivers, *J. Geol.*, 87, 533-551, 1979.

Graf, W.L., The effect of dam closure on downstream rapids, *Water Res. Res.*, 16, 129-136, 1980.

Graf, W.L., Late Holocene sediment storage in canyons of the Colorado Plateau, *Geol. Soc. Amer. Bull.*, 99, 261-271, 1987.

Griffiths, P.G., R.H. Webb, and T.S. Melis, *Initiation and frequency of debris flows in Grand Canyon, Arizona,* U.S. Geol. Surv. Open-File Rept. 96-491, 35 pp., 1996.

Haden, A., *Benthic ecology of the Colorado River system through the Colorado Plateau region*, N. Ariz. Univ., unpubl. M.S. thesis, 141 pp., Flagstaff, AZ, 1997.

Harris, D.R., Recent plant invasions in the arid and semi-arid Southwest of the United States, *Ann. Assoc. Amer. Geogr.*, 65, 408-422, 1966.

Hereford, R., and R.H. Webb, Historic variation in warm-season rainfall on the Colorado Plateau, U.S.A., *Clim. Change*, 22, 239-256, 1992.

Horton, J.S., *Notes on the introduction of deciduous Tamarisk*, U.S. Dept. Agri., For. Serv. Res. Note RM-16, 1964.

Howard, A.D., *Establishment of benchmark study sties along the Colorado River in Grand Canyon National Park for monitoring of beach erosion caused by natural forces and human impacts*, Univ. Virg., Grand Can. Study, Tech. Rept. 1, 182 p., Charlottesburg, VA, 1975.

Howard, A.D., and R. Dolan, *Alteration of terrace and beaches of the Colorado River in the Grand Canyon caused by Glen Canyon Dam and by camping activities during river float trips: Summary, management implications, and recommendations for future research and monitoring*, Natl. Park Serv., Grand Can. Natl. Park, Colo. River Res. Ser. No. 36, Tech. Rept. No. 2, 1976.

Howard, A.D., and R. Dolan, Changes in the fluvial deposits of the Colorado River in the Grand Canyon caused by Glen Canyon Dam, in *Proceedings of the First Conference on Scientific Research in the National Parks, Volume II*, ed. R.M. Linn, pp. 845-851, Washington, DC, 1979.

Howard, A., and R. Dolan, Geomorphology of the Colorado River in Grand Canyon, *J. Geol.*, 89, 269-297, 1981.

Howard, C.S., *Suspended sediment in the Colorado River, 1925-41*, U.S. Geol. Surv. Water-Supply Paper 998, 1947.

Irons, W.V., C.H. Hembree, and G.L. Oakland, *Water resources of the upper Colorado River basin*, U.S. Geol. Surv. Prof. Paper 441, 370 p., 1965.

Johnson, R.R., Historic changes in vegetation along the Colorado River in the Grand Canyon, in *Colorado River ecology and dam management*, ed. G.R. Marzolf, pp. 178-206, Natl. Acad. Press, Washington, DC, 1991.

Kearsley, L.H., J.C. Schmidt, and K.D. Warren, Effects of Glen Canyon Dam on Colorado River sand deposits used as campsites in Grand Canyon National Park, USA: *Regul. Rivers, Res. Manage.*, 9, 137-149, 1994.

Kieffer, S.W., The 1983 hydraulic jump in Crystal Rapid: Implications for river-running and geomorphic evolution in the Grand Canyon, *J. Geol.*, 93, 385-406, 1985.

Kieffer, S.W., Hydraulics and geomorphology of the Colorado River in the Grand Canyon, in *Grand Canyon Geology*, ed. S.S. Beus and M. Morales, pp. 333-383, Oxford Univ. Press, New York, 1990.

Laursen, E.M., S. Ince, and J. Pollack, On sediment transport through Grand Canyon, *Third Federal Interagency Sedimentation Conference Proceedings*, 4-76 to 4-87, 1976.

Lavender, D., *River runners of the Grand Canyon*, Univ. Ariz. Press, Tucson, AZ, 1985.

Lazenby, J.F., Unsteady flow modeling of the releases from Glen

Canyon Dam at selected locations in Grand Canyon, in *Glen Canyon Environmental Studies: Executive summaries of technical reports*, pp. 91-99, U.S. Dept. Int., Bur. Reclam., Upper Colo. River Reg., Salt Lake City, UT, 1988.

Leibfried, W.C., *The utilization of Cladophora glomerata and epiphytic diatoms as a food resource by rainbow trout in the Colorado River below Glen Canyon Dam, Arizona*, M.S. thesis, 41 pp., N. Ariz. Univ., Flagstaff, AZ, 1988.

Leopold, L.B., The rapids and the pools — Grand Canyon, in *The Colorado River region and John Wesley Powell*, U.S. Geol. Surv. Prof. Paper 669, pp. 131-145, 1969.

Maddux, H.R., D.M. Kubly, J.C. deVox, W.R. Persons, R. Staedicke, and R.L. Wright, *Effects of varied flow regimes on aquatic resources of Glen and Grand Canyons*, Ariz. Game Fish Dept., final report to the Bur. Reclam., Phoenix, AZ, 1987.

Martin, P.S., *Trees and shrubs of the Grand Canyon, Lees Ferry to Diamond Creek*, Univ. Ariz., Desert Lab., unpubl. manu., Tucson, AZ, 1971.

Martin, R., *A story that stands like a dam*, Henry Holt Company, New York, 1989.

McKee, E.D., Original structures in Colorado River flood deposits of Grand Canyon, *J. Sed. Pet.,* 8, 77-83, 1938.

Melis, T.S., *Geomorphology of debris flows and alluvial fans in Grand Canyon National Park and their influences on the Colorado River below Glen Canyon Dam, Arizona*, Ph.D dissertation, Univ. Ariz., Tucson, AZ, 1997.

Melis, T.S., R.H. Webb, P.G. Griffiths, and T.J. Wise, *Magnitude and frequency data for historic debris flows in Grand Canyon National Park and vicinity, Arizona*, U.S. Geol. Surv. Water Res. Invest. Rept. 94-4214, 285 pp., 1994.

Miller, R.R., Origin and affinities of the freshwater fish fauna of western North America, in *Zoogeography*, ed. C.L. Hubbs, pp. 187-222, Amer. Assoc. Adv. Sci. Publ. 51, 1959.

Miller, R.R., Man and the changing fish fauna of the American Southwest, *Pap. Mich. Acad. Sci, Arts, Let.,* 46, 365-404, 1961.

Minckley, W.L., Native fishes of the Grand Canyon region: An obituary? in *Colorado River ecology and dam management*, ed. G.R. Marzolf, pp. 124-177, Natl. Acad. Press, Washington, DC, 1991.

National Research Council, *River resource management in the Grand Canyon*, Natl. Acad. Press, Washington, DC, 1996.

Orvis, C.J., and T.J. Randle, Sediment transport and river simulation model, in *Glen Canyon Environmental Studies: Executive summaries of technical reports*, pp. 101-114, U.S. Dept. Int., Bur. Reclam., Upper Colo. River Reg., Salt Lake City, UT, 1988.

Otis, E.O., *Distribution, abundance, and composition of fishes in Bright Angel and Kanab Creeks, Grand Canyon National Park, Arizona*, Tucson, Univ. Ariz., unpubl. M.S. thesis, 1994.

Patraw, P.M., *Check-list of plants of Grand Canyon National Park*, Grand Canyon, Grand Can. Natur. Hist. Assoc., Bull. No. 6, 1936.

Patten, D.T., Glen Canyon Environmental Studies research program: past, present, and future, in *Colorado River ecology and dam management*, ed. G.R. Marzolf, pp. 85-104, Natl. Acad. Press, Washington, DC, 1991.

Pemberton, E.L., Channel changes in the Colorado River below Glen Canyon Dam, p. 5-61 to 5-73 in *Proceedings of the Third Federal Interagency Sedimentation Conference*, Water Res. Council, Sediment. Com., Denver, CO, 1976.

Potter, L.D., and C.L. Drake, *Lake Powell, virgin flow to dynamo*, Univ. New Mexico Press, Albuquerque, NM, 1989.

Pucherelli, M.J., Evaluation of riparian vegetation trends in the Grand Canyon using multitemporal remote sensing techniques, in *Glen Canyon Environmental Studies: Executive summaries of technical reports*, pp. 217-228, U.S. Dept. Int., Bur. Reclam., Upper Colo. River Reg., Salt Lake City, UT, 1988.

Randle, T.J, and E.L. Pemberton, Results and analysis of STARS modeling efforts of the Colorado River in Grand Canyon, in *Glen Canyon Environmental Studies: Executive summaries of technical reports*, pp. 115-128, U.S. Dept. Int., Bur. Reclam., Upper Colo. River Reg., Salt Lake City, UT, 1988.

Robinson, T.W., *Introduction, spread, and areal extent of saltcedar (Tamarix) in the western states*, U.S. Geol. Surv. Prof. Paper 491-A, 1965.

Robinson, A.T., R.W. Clarkson, and R.E. Forrest, *Spatiotemporal distribution, habitat use, and drift of early life stage native fishes in the Little Colorado River, Grand Canyon, Arizona, 1991-1994*, Final Rept., Bur. Reclam., Glen Canyon Environ. Studies, Flagstaff, AZ, 1996.

Rote, J.J., M.E. Flynn, and D.J. Bills, *Hydrologic data, Colorado River and major tributaries, Glen Canyon Dam to Diamond Creek, Arizona, water years 1990-95*, U.S. Geol. Surv. Open-File Rept. 97-250, 1997.

Rubin, D.M., J.C. Schmidt, and J.N. Moore, Origin, structure, and evolution of a reattachment bar, Colorado River, Grand Canyon, Arizona, *J. Sed. Pet.,* 60, 982-991, 1990.

Rubin, D.M., R.A. Anima, and R. Sanders, *Measurements of sand thicknesses in Grand Canyon and a conceptual model for characterizing changes in sandbar volume through time and space*, U.S. Geol. Surv. Open-File Rept., 94-597, 16 pp., 1994.

Schmidt, J.C., Recirculating flow and sedimentation in the Colorado River in Grand Canyon, Arizona, *J. Geol.,* 98, 709-724, 1990.

Schmidt, J.C., Temporal and spatial changes in sediment storage in Grand Canyon, Chapter 8 in *The influence of variable discharge regimes on Colorado River sand bars below Glen Canyon Dam, Final Report,* ed. S.S. Beus and C.C. Avery, Coop. Park Serv. Unit, Rept. No. CA 86006-8-0002, N. Ariz. Univ., Flagstaff, AZ, 1993.

Schmidt, J.C., and J.B. Graf, *Aggradation and degradation of alluvial sand deposits, 1965-1986, Colorado River, Grand Canyon National Park, Arizona*, U.S. Geol. Surv. Prof. Paper 1493, 74 pp., 1990.

Schmidt, J.C., P.E. Grams, and R.H. Webb, Comparison of the magnitude of erosion along two large regulated rivers, *Water Res. Bull.*, 31 (4), 617-631, 1995.

Schmidt, J.C., and D.M. Rubin, Regulated streamflow, fine-grained deposits, and effective discharge in canyons with abundant debris fans, in *Natural and anthropogenic influences in fluvial geomorphology*, ed. J.E. Costa, A.J. Miller, K.W.

Potter, and P.R. Wilcock, pp. 177-195, Amer. Geophys. Union, Geophys. Mono. 89, Washington, DC, 1995.

Schmidt, J.C., R.H. Webb, R.A. Valdez, G.R. Marzolf, and L.E. Stevens, The roles of science and values in river restoration in the Grand Canyon, *Bioscience*, 48 (9), 735-747, 1998.

Smillie, G.M., W.L. Jackson, and D. Tucker, *Colorado River sand budget: Lees Ferry to Little Colorado River*, Natl. Park Serv., Tech. Rept. NPS/NRWRD/NRTR-92/12, 1993.

Smith, W.O., C.P. Vetter, and G.B. Cummings, *Comprehensive survey of sedimentation in Lake Mead, 1948-49*, U.S. Geol. Surv. Prof. Paper 295, 254 pp., 1960.

Stephens, H.G., and E.M. Shoemaker, *In the footsteps of John Wesley Powell*, Johnson Books, Boulder, CO, 1987.

Stevens, J.E., *Hoover Dam, an American adventure*, Univ. Okla. Press, Norman, OK, 1988.

Stevens, L.E., *Mechanisms of riparian plant community organization and succession in the Grand Canyon, Arizona*, N. Ariz. Univ., unpubl. Ph.D. diss., Flagstaff, AZ, 1989.

Stevens, L., *The Colorado River in Grand Canyon, A guide*, Red Lake Books, 115 pp., Flagstaff, AZ, 1990.

Stevens, L.E., and G.L. Waring, Effects of post-dam flooding on riparian substrate, vegetation, and invertebrate populations in the Colorado River corridor in Grand Canyon, in *Glen Canyon Environmental Studies: Executive summaries of technical reports*, pp. 229-243, U.S. Dept. Int., Bur. Reclam., Upper Colo. River Reg., Salt Lake City, UT, 1988.

Stevens, L.E., and T.J. Ayers, *The impacts of Glen Canyon Dam on riparian vegetation and soil stability in the Colorado River corridor, Grand Canyon, Arizona*, Natl. Park Serv., Final Rept. Coop. Agree. CA 80000-8-0002, Flagstaff, AZ, 1993.

Stevens, L.E., J.S. Schmidt, T.J. Ayers, and B.T. Brown, Flow regulation, geomorphology, and Colorado River marsh development in the Grand Canyon, Arizona, *Ecol. Applic.*, 5, 1035-1039, 1995.

Stevens, L.E., J.P. Shannon, and D.W. Blinn, Colorado River benthic ecology in Grand Canyon, Arizona, USA: Dam, tributary, and geomorphic influences, *Regul. Rivers*, 13, 129-149, 1997.

Stockton, C.W., and G.C. Jacoby, Jr., *Long-term surface-water supply and streamflow trends in the upper Colorado River Basin based on tree-ring analyses*, Univ. Calif. Los Angeles, Lake Powell Res. Proj. Bull. No. 18, Los Angeles, CA, 1976.

Turner, R.M., and M.M. Karpiscak, *Recent vegetation changes along the Colorado River between Glen Canyon Dam and Lake Mead, Arizona*, U.S. Geol. Surv. Prof. Paper 1132, 125 pp, 1980.

Tyus, H.M. and W.L. Minckley, Migrating Mormon crickets, Anabrus simplex (Orthoptera-Tettigoniidae), as food for stream fishes, *Great Basin Natur.*, 48, 25-30, 1988.

Unitt, P., *Empidonax trailli extimus*: an endangered species, *W. Birds*, 18, 137-162, 1987.

U.S. Department of the Interior, *Operation of Glen Canyon Dam - Final Environmental Impact Statement*, Colorado River Storage Project, Coconino County, AZ, 337 pp. + appendices, Salt Lake City, UT, 1995.

Valdez, R.A., and E.J. Wick, Natural versus manmade backwaters as native fish habitat, in Aquatic resources management of the Colorado River ecosystem 1983, ed. V.D. Adams and V.A. Lamarra, pp. 519-536, Ann Arbor Science, Ann Arbor, MI, 1983.

Valdez, R.A., and R.J. Ryel, *Life history and ecology of the humpback chub* (Gila cypha) *in the Colorado River, Grand Canyon, Arizona*, Logan, Utah, BIO/WEST Inc., Rept. No. TR-250-08, Rept. Bur. Reclam. Cont. No. 0-CS-40-09110, 1995.

Valdez, R.A., and R.J. Ryel, Life history and ecology of the humpback chub in the Colorado River in Grand Canyon, Arizona, in *Proceedings of the Third Biennial Conference of Research on the Colorado Plateau*, ed. C. Van Riper, III., and E.T. Deshler, pp. 3-31, Natl. Park Serv., Trans. Proc. Ser. NPS/NRNAU/NRTP 97/12, 1997.

Valdez, R.A., and W.J. Masslich, Evidence of humpback chub reproduction in a warm spring of the Colorado River in Grand Canyon, Arizona, *Southwest. Natur.*, in press, 1999.

Valdez, R.A., and W.C. Leibfried, Captures of striped bass in the Colorado River in Grand Canyon, Arizona, *Southwest. Natur.*, in press, 1999.

Valentine, S., and R. Dolan, Footstep-induced sediment displacement in the Grand Canyon, *Environ. Manage.*, 3, 531-533, 1979.

Vanicek, C.D., *Ecological studies of native Green River fishes below Flaming Gorge Dam, 1964-1966*, Ph.D. diss., Utah State Univ., Logan, UT, 1967.

Webb, R.H., *Grand Canyon, a century of change*, 290 pp., Univ. Ariz. Press, Tucson, AZ, 1996.

Webb, R.H., P.T. Pringle, S.L. Reneau, and G.R. Rink, Monument Creek debris flow, 1984, Implications for formation of rapids on the Colorado River in Grand Canyon National Park, *Geol.*, 16, 50-54, 1988.

Webb, R.H., P.T. Pringle, and G.R. Rink, *Debris flows from tributaries of the Colorado River, Grand Canyon National Park, Arizona*, U.S. Geol. Surv. Prof. Paper 1492, 39 pp., 1989.

Webb, R.H., T.S. Melis, P.G. Griffiths, J.G. Elliott, T.E. Cerling, R.J. Poreda, T.W. Wise, and J.E. Pizzuto, *Lava Falls Rapid in Grand Canyon: Effects of late Holocene debris flows on the Colorado River*, U.S. Geol. Surv. Prof. Paper 1591, 90 pp., 1997.

Wegner, D.L., A brief history of the Glen Canyon Environmental Studies, in *Colorado River ecology and dam management*, pp. 226-238, Natl. Acad. Press, Washington, DC, 1991.

Weiss, S.J., *Spawning, movement, and population structure of flannelmouth sucker in the Paria River*, unpubl. M.S. thesis, Univ. Ariz., Tucson, AZ, 1993.

Wiele, S.M., and J.D. Smith, A reach-averaged model of diurnal discharge wave propagation down the Colorado River through the Grand Canyon, *Water Res. Res.*, 32, 1375-1386, 1996.

Wiele, S. M., J.B. Graf, and J.D. Smith, Sand depositions in the Colorado River in the Grand Canyon from flooding of the Little Colorado River, *Water Res. Res.*, 32, 3579 - 3596, 1996.

Williams, G.P., and M.G. Wolman, *Downstream effects of dams*, U.S. Geol. Surv. Prof. Paper 1286, 83 pp., 1984.

Wilson, R.P., Sonar patterns of the Colorado riverbed in the Grand Canyon, in *Glen Canyon Environmental Studies: Executive Summaries of technical reports*, pp. 31-41, U.S. Dept. Int., Bur. Reclam., Upper Colo. River Reg., Salt Lake City, UT, 1988.

R.H. Webb, U.S. Geological Survey, 1675 W. Anklam Road, Tucson, AZ 85745; email: rhwebb@usgs.gov

D.L. Wegner, Ecosystem Management International, P.O. Box 23369, Flagstaff, AZ 86002

E.D. Andrews, U. S. Geological Survey, 3215 Marine St., MS-458, Boulder, CO 80303-1060

R.A. Valdez, SWCA, Inc., 172 W. 1275 S., Logan, UT 84321

D.T. Patten, Department of Plant Biology, Box 871601, Arizona State University, Tempe, AZ 85287-1601

Origins of the 1996 Controlled Flood in Grand Canyon

John C. Schmidt[1], Edmund D. Andrews[2], David L. Wegner[3], Duncan T. Patten[4],

G. Richard Marzolf[5], Thomas O. Moody[6]

The March 1996 controlled flood in Grand Canyon resulted from a decade-long evolution in scientific thinking about the appropriate role of floods in management of the Colorado River in Grand Canyon. The flood was implemented after 5 consecutive years in which proposals to conduct a similar event were rejected; final implementation of the 1996 flood necessitated revision of the definition of the appropriate basin-wide runoff conditions that would trigger such a flood. The flood partly resulted from a multi-year effort to reform the Colorado River Storage Project Act that had culminated in passage of the Grand Canyon Protection Act in 1992. The flood itself consisted of a 4-day period of steady discharge of 227 m^3/s, an 11-hr period of increasing discharge to a peak of 1274 m^3/s that lasted for 7 days, a 45-hr period of recession, and a 4-day period of steady discharge at 227 m^3/s. This event was partly a demonstration of the potential role of floods in regulated river management and also provided an opportunity for scientists to make measurements about physical and biological processes during flood conditions.

1. INTRODUCTION

At 0615 hrs Mountain Standard Time (MST) on March 26, 1996, Secretary of the Interior Bruce Babbitt opened one of the four valves of Glen Canyon Dam's river outlet works, thereby beginning to increase the discharge of the Colorado River beyond 891 m^3/s. This intentional increase of dam releases beyond powerplant capacity was pre-planned, was not in response to a hydrologic emergency, and had the purposes of improving the condition of downstream river resources and providing an opportunity for scientific measurements during a planned period of controlled high discharge. This action established a precedent that will probably be repeated at this dam and elsewhere.

The decision to undertake this action did not come about casually; it resulted from years of sustained effort by many individuals, including scientists, citizen activists, and government officials. It took 12 hrs for dam releases to reach the peak discharge of 1274 m^3/s, because the 4 valves were opened sequentially [Collier et al., 1997]. The opening of the other valves symbolized the cooperation that was necessary to implement this controlled flood. Bureau of Reclamation and Western Area Power Administration officials opened the second valve. The third valve was opened by 3 present and former officers of the Grand Canyon River Guides organization, a representative of the

[1] Utah State University, Logan, Utah
[2] U.S. Geological Survey, Boulder, Colorado
[3] Ecosystem Management International, Flagstaff, Arizona
[4] Department of Plant Biology, Arizona State University, Tempe, Arizona
[5] U.S. Geological Survey, Reston, Virginia
[6] Northern Arizona University, Flagstaff, Arizona

The Controlled Flood in Grand Canyon
Geophysical Monograph 110
Copyright 1999 by the American Geophysical Union

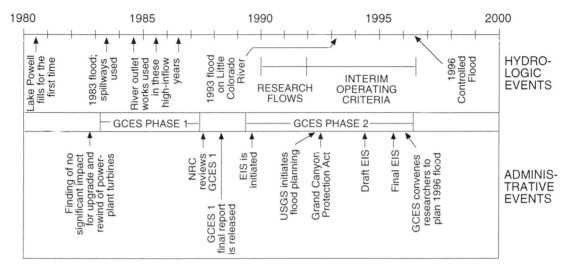

Figure 1. Timeline of important events that led to the 1996 controlled flood in Grand Canyon.

environmental organization American Rivers, and the Superintendent of Glen Canyon Dam. The fourth valve was opened by scientists associated with the Bureau of Reclamation's Glen Canyon Environmental Studies (GCES) program.

The many individuals who were responsible for the occurrence of the flood had not all originally supported this river management concept, but they eventually agreed that a controlled flood should be an essential component of dam operations. They collectively created the political pressure that resulted in the administrative approval and occurrence of the flood.

The purposes of this chapter are to describe the evolution of scientific thinking about floods in Grand Canyon, the events that led to the flood, the objectives of the flood, and the hydrologic characteristics of the flood. The theme of this chapter is that scientific understanding is an evolving venture, and that implementation of management actions derived from scientific information not only requires the development of a clear scientific recommendation but also the persistence of managers and constituents to ensure that those recommendations are implemented. Persistence, focus, and tenacity had to be maintained by scientists, managers, agency officials, and citizens alike.

2. THE EVOLUTION OF ADMINISTRATIVE OPINIONS ABOUT THE ROLE OF FLOODS IN STRUCTURING THE RIVER ECOSYSTEM

The role of floods in forming and maintaining the alluvial deposits of the Colorado River in Grand Canyon

has been discussed by geomorphologists for decades [*Howard and Dolan,* 1981]. The geomorphic role of floods in restructuring alluvial deposits in canyons where coarse bed and bank materials dominate is well known, because only large floods can move that material. In ecology, the role of floods as essential disturbances in restoring habitat and production is long-standing [*Poff et al.,* 1997]. Despite this widespread scientific understanding about the role of floods in forming and maintaining riverine ecosystems, the GCES Phase I research program concluded that floods were to be avoided in the operations of Glen Canyon Dam [*Bureau of Reclamation,* 1988]. The Phase I summary report stated, "Flood releases have [the] greatest potential for long-term and irreversible impacts" (p. 58) and "floods degrade main channel pools [and] cause long-term loss of camping beaches" (p. A-45).

The GCES Phase I investigations had been initiated by the Bureau of Reclamation in 1982 (Figure 1) in response to the agency's proposal to increase the capacity of the dam's powerplant from 891 to 933 m^3/s. The original objective of the GCES was to determine the relationship between the condition of downstream river resources and characteristics of powerplant operations. The studies ultimately examined the effects of higher magnitude dam releases, because the studies were conducted during a period of unprecedented high runoff. Annual runoff in the upper Colorado River basin between 1983 and 1986 was very high, and runoff in 1983 and 1984 was the second and third largest since 1896. High reservoir levels and high runoff throughout the Colorado River basin forced the Bureau of Reclamation to use the dam's river outlet works annually from 1983 to 1986 and to use the emergency spillways for a brief time in 1983 (Figure 2). In 1983, the peak discharge of the

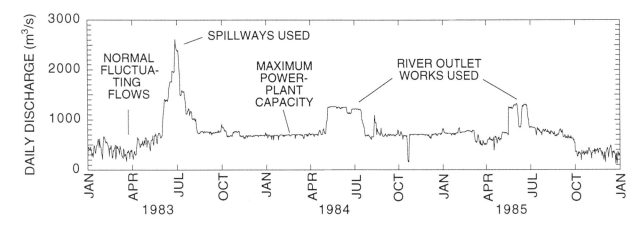

Figure 2. Hydrograph showing mean daily discharge for a period between 1982 and 1986 when runoff in the Colorado River basin was very high.

Colorado River approximately equalled the mean annual flood of the pre-dam river [*Webb et al.,* this volume]. Despite initial administrative pressure to only study the effects of discharges less than powerplant capacity, the environmental impacts caused by the high discharges of the mid-1980s forced the GCES to widen its scope.

Approximately 2 months after recession from the 1983 high releases, the number of sand bars large enough to be used as campsites was greater and their average size was larger than before these high flows [*Brian and Thomas,* 1984]. Subsequent measurements, however, showed that these flood-created bars were eroded by the sequence of flows that occurred between 1984 and 1986 [*Kearsley et al.,* 1994]. The net effect of the high releases between 1983 and 1986 was to reduce the area of sand bars in some of the narrow reaches of the river channel where campsites were already infrequent [*Schmidt and Graf,* 1990]. Sediment mass balance calculations suggested that a large volume of sand had been eroded from the Colorado River's bed and banks during this period [*Randle et al.,* 1993].

The potential beneficial effects of floods on the ecosystem had not gone unrecognized in the GCES Phase I report:

> Floods appear to benefit humpback chub [*Gila cypha*]. Younger age classes are well represented in the high-water years of 1984 and 1985, indicating good reproduction in those years... Floods do temporarily eliminate low-velocity, near-shore habitat for juvenile trout and common native fish, increasing mortality and energy expended on survival [by these species]. However, floods do not appear to have long-term [negative] effects on the aquatic system ... Infrequent

flooding may open areas for colonization by younger individuals of the same plant species or different species, thus increasing vegetation habitat diversity, and in turn increasing animal diversity ... floods can move sand from low elevations to high elevations where it is more useful for campsites. Redistribution of sand during floods cleans it of refuse and scours away any encroaching vegetation which may make camping more difficult [*Bureau of Reclamation,* 1988, p. 53, 59].

The primary negative effect of floods was thought to be on the introduced rainbow trout (*Oncorhynchus mykiss*) fishery in the 25 km between Glen Canyon Dam and Lees Ferry.

Although the unusual hydrologic conditions that caused the high flows between 1983 and 1986 were recognized, the management implications drawn from the Phase I research assumed that the environmental changes caused by these floods were representative of the potential effects of any high releases from the dam. The Phase I research team stated: "overall beach area will be lost during any flood, and much of the gain in sand at high elevations will be temporary" [*Bureau of Reclamation,* 1988, p. 59]. After publication of the Phase I summary report, the management guidelines for Glen Canyon Dam were revised to reduce the likelihood of high releases in excess of powerplant capacity. The possibility of floods of short duration for environmental management purposes, such as occurred in 1996, was never considered.

By the time the Final Environmental Impact Statement (EIS) for the operation of Glen Canyon Dam was issued in March 1995 [*U.S. Department of the Interior,* 1995], the opinion about floods had changed greatly. Floods were viewed as a restoration tool, although the EIS retained a strategy of avoiding long-duration floods by recommending (1) raising the height of the 4 spillway gates by 1.4 m and thereby increasing the storage capacity of Lake Powell reservoir and (2) changing the operating rules regarding the criteria for filling the reservoir. The EIS proposed that "high releases of short duration, designed to rebuild high elevation sand bars, deposit nutrients, restore backwater channels, and provide some of the dynamics of a natural system" occur "when sufficient quantities of sediment are available, but not following a year in which a large population of young humpback chub is produced" [*U.S. Department of the Interior,* 1995, p. 40-41].

For purposes of environmental and operations analysis, the EIS assumed that such a flood would occur in 1 of every 5 yrs. These floods would occur when the level of Lake Powell was relatively low so that there would be a low probability that a restoration flood would be followed by an unanticipated flood during a hydrologic emergency. Such a succession of floods might cause net erosion of sand bars, such as had happened between 1983 and 1986. Clearly, a major change in thinking about the role of floods as a restoration tool had occurred between the publication of the GCES Phase I summary report and the issuance of the final EIS. How had such a change in thinking occurred?

3. THE EVOLVING SCIENTIFIC UNDERSTANDING OF THE ROLE OF FLOODS IN THE GRAND CANYON ECOSYSTEM

Results from several research programs converged to ease the fear of floods, and to clarify that the specific environmental effects of floods depended on their magnitude, duration, and time since a previous flood. The notion that all floods were harmful had been challenged by a committee of the National Research Council charged with reviewing the GCES [*National Research Council,* 1987]. The committee argued that the impact of floods, as measured in the mid-1980s, could not be extrapolated to the entire range of possible floods:

> The evidence is not conclusive as to the long-term effects of a single flood ... In the end, the impression [the GCES Phase I document] leaves is that floods are to be avoided. This statement appears to be based on the idea that steady flows led to the development of the resources as they were in 1982 and on the assumption that these are the levels to emulate. While this recommendation cannot be said to be inherently "wrong," it is

supported by indirect rather than direct evidence. [p. 58]

Environmental measurements showing the effects of a flow regime *without* floods were made in the early 1990s following several years of drought in the Colorado River basin and the adoption of Interim Operating Criteria for Glen Canyon Dam in fall 1991. These interim operating rules were established to protect Colorado River resources until the EIS was completed. This was the first time that the operations of Glen Canyon Dam had been greatly altered to protect downstream river resources. The minimum daily discharge was established at either 141 or 226 m^3/s for night and day periods, respectively, and was intended to protect trout spawning areas upstream from Lees Ferry. The maximum release was set at 566 m^3/s, but was later increased to 622 m^3/s. The range of allowable daily flow fluctuations depended on the monthly release volume, was between 142 and 227 m^{3}s^{-1}day^{-1}, and was well below the range of fluctuations that had occurred between 1966 and 1982. Reduction in the range of fluctuations was intended to insure that the daily change in river stage would be about 1 m in all months in most reaches. The downramp rate was designed to reduce erosion caused by dewatering of alluvial ground water in sand bars and to avoid stranding of trout upstream from Lees Ferry.

3.1. Physical Resources

Despite these well-intentioned changes, erosion of sand bars continued throughout Grand Canyon due to wind erosion, rilling, gullying, and debris flows. The research community came to realize that erosion of streamside sand bars could not be stopped. For example, a debris flow in Eighteen Mile Wash in August 1987 completely covered and destroyed a sand bar that had been a popular campsite and that had been monitored since 1976 [*Melis et al.,* 1994]. Wind deflation and gullying were significant elsewhere. Seepage erosion occurred where the rate of decrease in the elevation of the Colorado River's water surface exceeded the rate of decrease in the alluvial water tables in adjacent sand bars [*Budhu and Gobin,* 1994]. Observations in the relatively unregulated Cataract Canyon of the Colorado River, located immediately upstream from Lake Powell, showed that these erosion processes were typical of sand bars in canyons with abundant debris fans [*Schmidt and Rubin,* 1995]. Researchers realized that the size of eddy sand bars is maintained by annual floods that redeposit sand eroded during the months between floods. A similar mechanism was necessary in Grand Canyon if the area and volume of sand bars was to be maintained. Thus, researchers rediscovered a conclusion that had been reached a decade earlier [*Dolan et al.*, 1974, 1977].

The development of a plan for redistribution of sand from the channel bed to its margins depended on determining how much sand was available for redistribution, where this sand was located, and what were the magnitude and duration of flows necessary to accomplish this purpose. The research community suspected that sand accumulated on the channel bed and at low elevation in some eddies, even though it had not been directly measured, because mass balance calculations indicated that sand accumulated within Grand Canyon after adoption of the Interim Flow Criteria, and accumulation had not occurred along channel margins because erosion had been measured there [*Kaplinski et al.*, 1995]. *Rubin et al.* [1994] cautioned, however, that if sand was preferentially stored in eddies, rather than in the main channel, floods might cause erosion of sand bars in the upstream part of Grand Canyon, because high flows would quickly erode the limited amount of bed sediment in upstream reaches. Thus, it was recognized that direct measurement of sand accumulation on the bed was an essential component of sediment-storage monitoring. However, canyon-wide direct measurement of sand accumulation on the bed, if it actually occurred, was not possible because of the time-consuming nature of precise channel cross-section measurements [*Graf et al.*, 1995a, 1995b]. Seismic techniques had also proved inadequate to measure the thickness of sand in the channel. Thus, annual accounting of sediment inflows and outflows from Grand Canyon was a key river management tool before 1996 (Figure 3).

Howard and Dolan [1981] and *Andrews* [1991a] determined that the Paria and Little Colorado rivers were the primary sources of sand delivered to the Colorado River downstream from the dam. *Randle et al.* [1993] showed that in years of average dam releases, mainstem transport was insufficient to remove the estimated average annual volume of sand delivered to the Colorado River from these tributaries. In response to the need to refine elements of this sediment budget, the U.S. Geological Survey expanded its research program to examine sediment inflow from the Paria River [*Topping*, 1997] and from debris flows in ungaged tributaries [*Webb et al.*, 1989] and to improve estimates of mainstem sediment transport. New techniques were also developed to photograph and map the channel bed.

Geomorphologists know that floods are the only natural mechanism to rebuild alluvial landforms such as high-elevation sand bars, because a river can only create landforms that it inundates some of the time. The rates and locations of potential eddy deposition that might occur

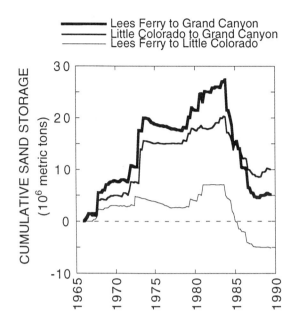

Figure 3. Sediment mass balance for the reach between Lees Ferry and Grand Canyon that was used to develop management recommendations in the Glen Canyon Dam Environmental Impact Statement.

during a flood were determined by a combination of field studies and physical and numerical modeling. After describing the sedimentology of eddy sand bars and proposing a conceptual model of bar deposition during floods [*Rubin et al.*, 1990; *Schmidt*, 1990], *Schmidt et al.* [1993] showed by physical experiment that the rates of deposition in eddies increase as the rates of mainstem transport increase and that the rates of deposition decline as an eddy progressively fills. These results suggested that a restoration flood should be of high magnitude but could be of short duration. Field measurements of eddy circulation showed that there was a high rate of exchange of water between the mainstem and eddies [*Andrews*, 1991b; *Nelson et al.*, 1994]. Three mechanisms cause sediment to enter eddies: turbulent diffusion, secondary flow, and unsteadiness of the reattaching shear layer [*Rubin and McDonald*, 1995].

Other research clarified the ways that sediment redistribution occurs during a flood. Many of the predictions of the response of bars to flooding resulted from the numerical modeling of a Little Colorado River flood that delivered 4.93 Tg of sediment to the Colorado River in 1993 [*Wiele et al.*, 1996]. *Wiele et al.* [1996] also demonstrated that high flows mobilize sand stored on the bed, because the approximate volume of newly-deposited eddy bars greatly exceeded the volume of sand delivered by this flood in the

Little Colorado River. They demonstrated that the distribution of new sand bars under known hydrologic and sediment transport conditions could be predicted.

Other changes associated with high discharge were also predicted. *Melis et al.* [1994] predicted that debris fans would be reworked by much lower discharges than those suggested by *Kieffer* [1985]. Rapid rates of recession from peak discharge were anticipated to cause widespread seepage-induced erosion [*Budhu and Gobin*, 1994].

3.2. Biological Resources

Ecologists also came to realize that floods were essential to the continued recruitment of riparian species if these floods were sufficiently large so that they scoured vegetation in some places [*Webb et al.*, this volume], such as was described by *Stevens and Waring* [1988] during the 1983 flood. Riparian marshes had developed on some sand bars in wide reaches of the canyon, and these marshes had been scoured by floods. However, *Stevens et al.* [1995] suggested that marshes are a resilient habitat, because seedlings had reestablished themselves several years after the high floods of the mid-1980s had ended. Some endangered or threatened species expanded their ranges into the new riparian vegetation that had developed at lower elevations, such as the Kanab ambersnail (*Oxyloma haydeni kanabensis*) and the southwestern willow flycatcher (*Empidonax traillii*). Other species had shifted their behavior to utilize aspects of the non-native fishery, such as bald eagles (*Haliaeetus leucocephalus*) feeding on spawning non-native rainbow trout at Nankoweap Creek [*Brown and Stevens*, 1992].

Recognizing that the native, endemic fishes of the pre-dam Colorado River corridor were adapted to annual flooding and associated high sediment loads, the need to reinstate portions of the historic ecosystem became evident and potentially necessary for recovery of endangered species. However, changes caused by Glen Canyon Dam and other anthropogenic actions were extensive, and researchers realized that floods alone could not save these species. For example, the pre-dam aquatic system was heterotrophic and allochthonous. Lower-elevation riparian areas were barren because of flood scour, and riparian vegetation occurred above the elevation of the stage of the mean annual flood. Fish were adapted to feeding in sediment-laden waters and were not prey to sight feeders. In contrast, the post-dam river is characterized by clear, cold waters with no floods. The energy base is autotrophic with autochthonous production and many exotic predatory and competing fishes.

Clearly, reinstating only one aspect of the historic ecosystem, such as floods, could not restore the structure to which the fish had evolved. However, biologists hypothesized that various aspects of the life history of the native fish could benefit and the life history of non-native fishes could be disrupted by floods. There was, nevertheless, the concern that floods of inappropriate magnitude, duration, and timing either could have little or detrimental effect on endangered fishes. *Valdez and Ryel* [1995] had hypothesized that cold flood releases from the dam that occurred when native fish were still in their larval phase of development could cause the loss of an entire year class, and they recommended that controlled floods occur no later than May. An experimental flood was an opportunity to test such hypotheses about native and non-native species.

Thus, the view developed that the pre-dam biota was at a competitive disadvantage under the post-dam conditions of controlled discharge, clear water, lower summer and fall stream temperatures, and the introduction of non-native organisms. The exotic flora and fauna were thought to have the competitive advantage, because floods, high sediment loads, and summer water temperatures were suppressed [*Blinn and Cole*, 1991; *Minckley* 1991]. As evidence mounted, ecologists working in Grand Canyon came to support the concept that periodic floods were essential in maintaining those elements of the river ecosystem that existed prior to the dam.

4. THE INTERFACE BETWEEN EMERGING SCIENTIFIC UNDERSTANDING AND PUBLIC POLICY BETWEEN 1988 AND 1996

A common concern for the resources of the Colorado River in Grand Canyon was the basis for many associations among scientists, river guides, and agency personnel. These relationships provided an environment that fostered greater informal conversation and constant interchange about scientific understanding and management options. The extent of this communication departed from the historical "arms length" relationship among scientists, managers, and the public. In part, these associations arose from environmentally-oriented concerns of the participants, but also because agency personnel, political appointees, and legislative staff requested scientific opinions about ongoing studies within time frames that made peer review impossible.

Scientists had wide ranging influence. Scientists served as technical advisors to the GCES and to its Senior Scientist, Duncan Patten of Arizona State University, who was charged with supervision of that program. Many scientists were friends of boatmen active in the Grand Canyon

River Guides, and this organization was influential in organizing support among its recreational customers. Other scientists were friends, correspondents, and confidants of agency personnel affiliated with the Bureau of Reclamation, the National Park Service, Western Area Power Administration, and the office of the Secretary of the Interior, as well as with elected officials and their staffs. As the potential use of a controlled flood gained acceptance within the Grand Canyon research community, the concept was already familiar to others and was echoed within some agencies and environmental organizations. Thus, public policy had not waited for scientific consensus. Instead, public policy had developed in concert with the evolving scientific understanding about floods, because there had been an extensive involvement of the public and decision makers from the earliest stages.

5. PRELUDE TO THE FLOOD OF 1996

The principal obstacle preventing the flood was the legal issue of whether the Secretary of the Interior had the authority to release water from Glen Canyon Dam in excess of powerplant capacity, except in a hydrologic emergency such as had occurred between 1983 and 1986. The total annual amount of water transferred from the Upper Colorado River Basin to the Lower Basin is established by the Colorado River Compact of 1922, and a flood of short duration would not directly affect that amount. However, the Colorado River Storage Project Act of 1956 directs the Secretary of the Interior to operate the several dams and hydroelectric powerplants authorized in that act, including Glen Canyon Dam and powerplant, "so as to produce the greatest practicable amount of power and energy that can be sold at firm power and energy rates" (43 USC 620f). The Colorado River Basin Act of 1968 refers to the need to avoid "anticipated spills" that bypass the powerplant. The National Park Service Organic Act of 1916 had directed that national parks be managed so as to remain "unimpaired for future generations."

The rules that guided operations of Glen Canyon Dam had been changed by passage of the Grand Canyon Protection Act of 1992. Within the bounds of other laws, this act requires that the dam also be operated "in such a manner as to protect, mitigate adverse impacts to, and improve the values for which Grand Canyon National Park and Glen Canyon National Recreation Area were established including, but not limited to, natural and cultural resources and visitor use" (P.L. 102-575). In 1993, restoration floods, administratively called "beach/habitat building flows," were incorporated into all alternatives considered by the draft EIS for operations of Glen Canyon

Dam. As described in the EIS, these flows would occur in years of low to moderate runoff when there was a low probability that a restoration flood would be followed by an emergency high release.

One agency where a coordinated plan to conduct experimental measurements during a controlled flood emerged was the U.S. Geological Survey. The Geological Survey held several meetings to plan experiments and field measurements and prepared draft work plans and budgets in 1992 and 1993. In 1993, the GCES Senior Scientist and his advisors expanded upon the Geological Survey plan; a proposal for a controlled high release in spring 1994, along with a set of field measurements, was presented to the GCES Cooperating Agencies in late summer 1993. Concern emerged about insufficient time to design such a flood and to develop a plan for conducting and funding logistical support for scientific measurements. Concerns related to endangered or otherwise valued species also emerged. Some cooperating agencies, such as the Arizona Fish and Game Department, presented concerns about potential "damage" to the tailwater trout fishery. Water and power interests indicated that such a flood was sufficiently different from normal dam operations that additional NEPA compliance was needed. Finally, representatives of state government were concerned that a flood was inconsistent with provisions of the Colorado River Compact. However, these same concerns had earlier been raised when experimental fluctuating-flow releases were proposed in 1989, but no EIS had been required when those "research flows" were conducted in 1990 and 1991.

In summer 1994, scientific support expanded and a more extensive flood research plan was developed by the GCES Senior Scientist and his advisors. This plan called for a flood in spring 1995, timed to occur after spawning of trout in the tailwater and prior to spawning by native fish and nesting by some bird species further downstream. The flood would occur after wintering bald eagles and waterfowl had left the canyon, before nesting by the endangered southwestern willow flycatcher, and prior to seed dispersal by the abundant non-native *Tamarix* sp. The flood was also timed to occur before spawning by humpback chub in the Little Colorado River and descent by larval fishes into the mainstem [*Valdez and Ryel*, 1995]. The recommendation for a flood in spring 1995 was referred to the Secretary of the Interior. The final EIS concerning Glen Canyon Dam operations had not yet been issued, however, and opponents of the flood successfully argued that there still was no documentation of administrative environmental compliance in place to support the flood; they also opposed the magnitude of the proposed expenditure of money to support this scientific program.

Throughout fall and winter 1995, a new lower-budget study plan was developed. This time, the plan was developed by GCES with cooperation from the Geological Survey. Rather than having the plan developed by a small group of experienced scientists, GCES pursued a "stake-holder" approach to planning. GCES convened a meeting attended by representatives of every agency who had research or management responsibility about some aspect of the river resources of Grand Canyon. Research proposals were developed and voted on by this group. Priorities were assigned, and sometimes projects with higher priority were better funded. The final recommendation for distribution of funds among the various projects was made by the Senior Scientist, within the overall constraint that the total research program, including logistics, could not exceed $1.5 M. This approach received widespread support, perhaps because this resulted in research and monitoring funds being more widely distributed. The major focus of scientific measurements was on the physical environment (Table 1).

However, the fundamental policy issue remained of how a flood would be interpreted within the context of the other legislation that guided management of the river. Controlled floods in years of low to moderate runoff represented exactly the "anticipated spills" that the 1968 Colorado River Basin Act prohibited. Randall Peterson, an engineer with the Bureau of Reclamation, proposed that controlled floods occur when Lake Powell was "full or expected to fill." Such a rule would avoid conflict with the 1968 Act, because controlled floods could be interpreted as part of prudent reservoir management. This was the administrative justification for the 1996 flood. Nevertheless, as described by *Harpman* [this volume], reservoir management as early as January 1996, before the status of the snowpack was known with certainty, was altered in order to plan the controlled flood.

These actions resulted in transfer of some of the risk associated with the controlled flood from water users, who feared loss of hydroelectric power revenue and adverse impact on water supplies, to the environment, where the probability of high releases immediately following the flood was increased. The risk in such a plan for the controlled flood was a higher probability that unanticipated high flows might follow the controlled flood and simulate the conditions that had occurred when high releases between 1984 and 1986 eroded some of the deposits created in 1983. Years of large tributary sediment input from the Paria River and Little Colorado River in the early 1990s provided scientists the hope that adequate sediment was available in Grand Canyon if this situation arose after the 1996 controlled flood. As described elsewhere in this book,

unusually high powerplant releases did follow the flood. These high releases confounded the hydrologic regime following the flood, and made interpretation of many changes in physical and biological resources difficult.

The Assistant Secretary for Water and Power waited until the EIS, which recommended that beach/habitat building flows occur, was finalized before she approved the controlled flood. In the months immediately preceding the flood, the environmental organization Trout Unlimited and the Hualapai Tribe attempted to block the flood by threatening lawsuits. In both cases, these suits were withdrawn at the last minute.

6. PURPOSES AND OBJECTIVES OF THE FLOOD

One purpose of the flood was to provide the opportunity to measure geomorphic and ecologic processes during flood passage and test hypotheses about the flood's effects. The other purpose was to determine if a suite of management objectives described in the EIS could be met by the flood. Thus, the former purpose was experimental and the latter purpose was demonstration. The former purpose could be met regardless of the environmental conditions that existed after recession of the flood, but the latter purpose could only be met by determining if the status of specific river corridor resources had been improved by the flood. The public's perception about the "success" of the flood hinged on the achievement of some of the management objectives, especially concerning the abundance and size of sand bars. As described in other chapters, "success" also depended on when the status of river resources was measured, because some flood-caused resources readjusted after the flood receded.

There is some disagreement about the management objectives of the 1996 controlled flood. Many of the administrative memos concerning the proposed flood, written before 1996, refer to only two: redistribution of sand to high elevation and providing a disturbance to the riparian and aquatic ecosystem. These objectives were expanded during stakeholder meetings, discussions, and review, however. GCES' final articulation of the goals of the flood listed 6 management objectives for the flood that were drawn from the final EIS. These objectives were to:

1. remove non-native fishes;
2. rejuvenate backwater habitats for native fishes;
3. redeposit sand bars at higher elevations;
4. preserve and restore camping beaches;
5. reduce near-shore vegetation; and,
6. provide water to the upper riparian zone vegetation that had been well-established prior to the dam.

TABLE 1. Funded research activities during the 1996 controlled flood

PROJECT TITLE	PRINCIPAL INVESTIGATORS		AGENCY
Main channel streamflow, sediment-transport, and sediment storage — collection of critical data	J.B. Graf G.G. Fisk	J.D. Smith	U.S. Geological Survey
Main channel streamflow, sediment transport, and sand storage — development of predictive models	J.D. Smith S.M. Wiele	E.R. Griffin	U.S. Geological Survey
Reworking of aggraded debris fans by the controlled flood	R.H. Webb T.S. Melis	P.G. Griffiths	U.S. Geological Survey
Deposition rate and topographic evolution of sand bars in lateral separation eddies during high flows	E.D. Andrews D. Cacchione J. Nelson	D.M. Rubin J.C. Schmidt	U.S. Geological Survey; Utah State University
Evaluation of the effects of the 1996 controlled flood on Colorado River sand bars in Grand Canyon	R. Parnell L. Dexter	M. Kaplinski J.E. Hazel	Northern Arizona University
Controlled flood terrace deposit monitoring plan, Glen Canyon Dam to Lees Ferry	N. Henderson		Glen Canyon National Recreation Area
The effects of flooding on the vertical, thermal, and chemical structure in Lake Powell and an estimate of the flood effect on primary productivity in the Colorado River: Glen Canyon Dam to Lees Ferry	G.R. Marzolf R.J. Hart D.W. Stephens		U.S. Geological Survey
The effects of the 1996 flood on the aquatic food base in the Colorado River through Grand Canyon	D.W. Blinn J.P. Shannon		Northern Arizona University
Evaluation of the effects of a controlled flood on the aquatic ecosystem of the Colorado River downstream from Glen Canyon Dam	W.R. Persons A.D. Ayers T.L. Hoffnagle R.A. Valdez	W.C. Leibfried C. McIvor N. Henderson	Arizona Game and Fish Department; Hualapai Natural Resources Department; Bio/West, Inc.; University of Arizona; Glen Canyon National Recreation Area
Evaluation of backwater rejuvenation along the Colorado River in Grand Canyon	L.E. Stevens R. Parnell	A. Springer	Glen Canyon Environmental Studies; Northern Arizona University
Assessment, mitigation, and monitoring of the impacts of the controlled flood on the endangered Kanab ambersnail at Vasey's Paradise	L.E. Stevens D.M. Kubly J.R. Patterson	F.R. Protiva V.J. Meretsky	Glen Canyon Environmental Studies; Arizona Game and Fish Department; Grand Canyon National Park; U.S. Fish and Wildlife Service
Effects of the 1996 flood on riparian vegetation in the Colorado River corridor in Glen and Grand canyons	M.J. Kearsley T.J. Ayers		Northern Arizona University
Monitoring songbird use in the Colorado River riparian zones between the Glen Canyon Dam and Lees Ferry	C.A. Pinnock J.R. Spence		Glen Canyon National Recreation Area
Monitoring waterfowl use on the Colorado River between Glen Canyon Dam and Lees Ferry	C.A. Pinnock J.R. Spence		Glen Canyon National Recreation Area
Cultural Resources monitoring proposal	J.R. Balsom A.S. Bulletts T.W. Burchett C.M. Coder	L. Jackson S. Larralde L.M. Leap M. Yeatts	Glen Canyon National Recreation Area; Grand Canyon National Park; Hopi, Hualapai, and Southern Paiute tribes

These objectives were to be accomplished without significant adverse impacts to the tailwater trout fishery, endangered species, cultural resources, or the regional and local economy. The Final Environmental Assessment predicted impacts to other resources as well (Table 2), and these anticipated impacts can be compared with the measured effects described in the remainder of this volume.

These management goals were, by necessity, value-based and impossible to comprehensively meet without conflict. These goals assign objectives to environmental resources that were either formed in the hydrologic environment that existed before the dam or in the post-dam environment. It is difficult, if not impossible, to improve the condition of resources that are flood adapted while also improving the

TABLE 2. Summary of the predicted impacts of the 1996 controlled flood in the Final Environmental Assessment

RESOURCE	IMPACT
Water	The volume of monthly releases from the dam will be adjusted to schedule more water in April and March. No effect on end-of-year water storage in Lakes Powell and Mead.
Sediment	0.3 to 1 m sand deposition on most sandbars followed by erosion over time. Net erosion on some sandbars during the flood. Sand transport upstream from the Little Colorado River estimated at $0.76 \cdot 10^6$ Mg during the flood.
Fish	Temporary reduction in *Cladophora* biomass and increased drift downstream. Backwaters re-formed. Non-native populations temporarily disrupted by high flows; interactions between native and non-native fish rapidly return to conditions before the flood. Some trout eggs, fry, and young lost downstream; adult trout may be affected for a period after the flood.
Vegetation and Habitat	Some woody and emergent marsh vegetation lost through scouring or burial; vegetation recovery to pre-flood levels in months/years following test flow. Some wildlife habitat lost; recovery to pre-flood levels following test flow. No long-term effects on aquatic food base; few wintering waterfowl present during the flood.
Endangered and Other Special Status Species	Habitat improvement for southwestern willow flycatcher and humpback chub. Some Kanab ambersnail and northern leopard frog habitat inundated by test flow; leopard frog population may be lost.
Cultural Resources	High terrace erosion rates may be reduced in short-term. Temporary restoration of natural processes generally beneficial.
Recreation	River-based recreation activities affected to some degree during test flow. Number and size of camping beaches increased.
Hydropower	Two percent less energy generated during test flow. Little or no effect on wholesale or retail power rates. Total financial cost: $3.1 to 4.3 M; economic cost: $0.5 to 2.2 M.

condition of resources that flourish in the absence of floods [*Marzolf et al.*, 1998; *Schmidt et al.*, 1998].

7. CONSTRAINTS ON THE MAGNITUDE AND DURATION OF THE FLOOD

As discussed above, instantaneous discharge is irrelevant to the total annual volume of water that must be delivered from the Upper Colorado Basin to the Lower Basin under the Colorado River Compact. Thus, the Compact presented no constraint in the design of the flood.

Earlier studies and calculations had shown that a larger flood would transport much larger volumes of sand and lead to much larger rates of sand deposition in eddies. *Andrews* [1991b] estimated that deposition rates in eddies vary with the third or fourth power of river discharge, and *Nelson et al.* [1994] predicted even higher rates. Thus, a shorter duration, larger flood would efficiently build sand bars. The Geological Survey recommended a high release of 1584 m^3/s. The Bureau of Reclamation was unwilling, however, to use the emergency spillways [*Collier et al.*, 1997]. Maximum discharge without using these facilities is approximately 1275 m^3/s. Thus, the capacity of the powerplant plus the river outlet works was an engineering constraint.

The combined magnitude of discharge through the powerplant and river outlet works also was the magnitude ultimately allowed in the administrative Biological Opinion of the U.S. Fish and Wildlife Service. The extensive lower riparian zone along the Colorado River is critical habitat for two endangered species — the Kanab ambersnail and the southwestern willow flycatcher. The Service ruled that the zone of potentially-scoured critical habitat for the Kanab ambersnail, located only at a large spring called Vasey's Paradise, could not exceed 10% of the total habitat available prior to the flood. Estimates indicated that a flood magnitude of approximately 1274 m^3/s would inundate 10% of the habitat and would potentially eliminate more than 10% of the population of the Kanab ambersnail. The U.S. Fish and Wildlife Service required that snails in the potentially-scoured zone be transplanted to higher elevations as mitigation and that the flood could not be of higher magnitude. This was a significant administrative constraint.

8. THE FLOOD AS IT OCCURRED

The flood was designed to include a period of low discharge immediately before the hydrograph rise, a period

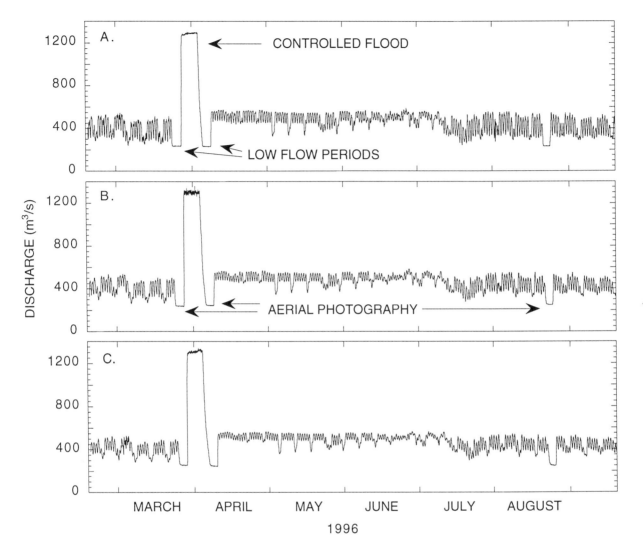

Figure 4. Hydrographs showing instantaneous discharges for the 1996 controlled flood at three locations in Grand Canyon. A. Lees Ferry. B. Grand Canyon. C. Diamond Creek.

of sustained steady high discharge, and a period of low discharge immediately after the flood (Figure 4). Thus, there were periods when the condition of resources could be assessed immediately before and after the flood; the low discharge of 227 m³/s for 4 days also permitted aerial photographs to be taken. The low discharge was initiated at 0100 hrs MST on March 22. Beginning at 0100 hrs MST on March 26, releases were increased by 113 m³s⁻¹hr⁻¹ until a maximum discharge of 1274 m³/s was attained at 1200 hrs on March 26. This high release was maintained for 7 days. At approximately 1100 hrs on April 2, the flow rate was decreased by 42.5 m³s⁻¹hr⁻¹ until the discharge reached 991 m³/s. In an effort to reduce seepage erosion, the rate of recession was reduced to 28 m³s⁻¹hr⁻¹ until a discharge of

566 m³/s was reached. Thereafter, the rate of recession was further reduced to 14 m³s⁻¹hr⁻¹ until a discharge of 227 m³/s was attained at 0800 hrs on April 4. Following 4 days of steady 227 m³/s, dam releases were restored at 0200 hrs on April 8 to those determined by the Interim Operating Criteria.

Discharges measured elsewhere in Grand Canyon slightly differed from those released at the dam, because the flood wave moved faster than the preceding trough of low discharge, and because perennial tributaries and springs contributed baseflow (Figure 4). The maximum measured discharge was 1311, 1297, 1345, 1334, and 1430 m³/s at U.S. Geological Survey cableways at Lees Ferry, above the Little Colorado River, near Grand Canyon, above National

Canyon, and above Diamond Creek, respectively [*Konieczki et al.*, 1997]. Change in stage between the steady 227 m³/s and the peak of the controlled flood ranged between 2 and 5 m, depending on channel geometry [*Konieczki et al.*, 1997]. Between Lees Ferry and Diamond Creek, the reach-average velocity during the flood was 1.8 m/s and ranged between 1.5 and 2.1 m/s [*Konieczki et al.*, 1997]. Thus, the rate of rise of the flood was somewhat greater downstream, and the rate of flood recession was somewhat less than was released at the dam. These deformations in the shape of the controlled flood hydrograph had been predicted by *Wiele and Smith* [1996]. Tributaries had no runoff during the controlled flood [*Konieczki et al.*, 1997].

9. FIELD ACTIVITIES DURING THE FLOOD

Dimock [1996] described the 3 floods that occurred in late March 1996: the flood of water, the flood of the media, and the "rubber armada" of scientists whose goal was to assess the flood's effects. Teams of scientists had begun to collect "pre-flood" data during February 1996, and most scientific teams launched downstream immediately before the low-discharge period. Teams collected data, or installed instrumentation, during low discharge and then waited for the flood to arrive.

While the media and government officials celebrated the opening of the river outlet works, more than 100 scientists waited downstream. The flood began its rise at the Grand Canyon gage at about 2000 hrs on March 26 and at about 1800 hrs on March 27 at Diamond Creek. Thus, peak flows arrived in the middle of the night at some places. There was widespread accumulation of woody debris on the water surface during the first day of the flood. This debris hindered data collection at some sites by entangling fish sampling nets and the fine-mesh nets used to sample drifting invertebrates. At cableways, cable-suspended streamflow and sediment-measuring gear had the potential of getting caught in floating trees and thereby endangering lives. The springtime weather was clear, and there were mild temperatures; working conditions were favorable. Pre- and post-flood aerial photographs were taken successfully, although the post-flood photographs of Marble Canyon were taken before river discharges had receded to 227 m³/s, confounding analysis of those data [*Schmidt et al.*, this volume].

Basic hydrologic and sediment-transport data were collected by the Geological Survey at 5 streamflow-gaging stations and 4 on tributaries. River-stage data were collected at an additional 29 sites and suspended-sediment data were collected at 4 of the 5 mainstem gages. *Konieczki*

et al. [1997] summarized these and other basic measurements of flow, sediment transport, water quality, and bathymetry.

Media attention was substantial during this event. The opening of the first valve of the river outlet works was shown live on the "Today Show" and "Good Morning America," both national morning television shows. Front-page stories appeared in most large newspapers throughout the United States and elsewhere.

10. CONCLUSIONS

Many observers, including the Secretary of the Interior, proclaimed the flood a success in subsequent talks at professional and public meetings. The basis of these comments was the emergence of new sand bars throughout the river corridor when the flood receded [*Collier et al.*, 1997]; one of the primary purposes of the demonstration project had been achieved. Although less attention was given to the experimental objectives of the flood, this event provided a unique opportunity for essential scientific measurements to be made, as described elsewhere in this book. Many people noted, however, that the most important aspect of the 1996 controlled flood in Grand Canyon was that it occurred at all. One of this article's authors wrote:

> In my opinion ... the biggest success of the flood was that it happened at all. In spite of historic conflicts over Colorado River resources, the flood was carried out with the unprecedented cooperation of traditionally contentious interests who chose dialog over litigation. Thanks to direction given by passage of the Grand Canyon Protection Act and to the patient, hard work of the basin states, federal and state agencies, tribes, power and water users, [and] environmental and recreation interests, the flood was accomplished without litigation or bloodshed. The 13-year effort represents a balance driven by our country's changing social values. [*Moody*, 1996, p. 9]

Scientists can take pride that they played an important role in the design and implementation of this event, yet the flood would not have occurred without substantial tenacity and political will from elsewhere.

Acknowledgments. Earlier versions of this draft were reviewed by R.H. Webb and R.A. Valdez.

REFERENCES

Andrews, E.D., Sediment transport in the Colorado River basin, in *Colorado River ecology and dam management*, ed. G.R. Marzolf, pp. 54-74, Natl. Acad. Press, Washington, DC, 1991a.

Andrews, E.D., Deposition rate of sand in lateral separation zones, Colorado River, *Eos, Trans., Amer. Geophys. Union*, 72(44), 219, 1991b.

Blinn, D.W., and G.W. Cole, Algae and invertebrate biota in the Colorado River: comparison of pre- and post-dam conditions, in *Colorado River ecology and dam management*, ed. G.R. Marzolf, pp. 102-123, Natl. Acad. Press, Washington, DC, 1991.

Brian, N.J. and J.R. Thomas, *1983 Colorado River beach campsite inventory, Grand Canyon National Park, Arizona*, Grand Canyon Natl. Park, Div. Res. Manage. rept., Grand Canyon, Arizona, 1984.

Brown, B.T., and L.E. Stevens, Winter abundance, age structure, and distribution of bald eagles along the Colorado River, Arizona, *Southwest. Natur.*, 37, 404-408, 1992.

Budhu, M., and R. Gobin, Instability of sand bars in Grand Canyon, *J. Hydraul. Engr.*, 120, 919-933, 1994.

Bureau of Reclamation, *Glen Canyon Environmental Studies, Final Report*, U.S. Dept. Interior, Bur. Reclam., Salt Lake City, UT, 1988.

Collier, M.P., R.H. Webb, and E.D. Andrews, Experimental flooding in Grand Canyon, *Sci. Amer.*, 276, 82-89, 1997.

Dimock, B., Spring floods, *Boatman's Quarterly Rev.*, 9(2):11, 1996.

Dolan, R., A. Howard, and A. Gallenson, Man's impact on the Colorado River in the Grand Canyon, *Amer. Sci.*, 62, 392-401, 1974.

Dolan, R., B. Hayden, and A. Howard, Environmental management of the Colorado River within the Grand Canyon, *Environ. Manage.*, 1, 391-400, 1977.

Graf, J.B., S.M.D. Jansen, G.G. Fisk, and J.E. Marlow, *Topography and bathymetry of the Colorado River, Grand Canyon National Park, Little Colorado River to Tanner Rapids*, U.S. Geol. Surv. Open-File Rept. 95-725, Tucson, AZ, 1995a.

Graf, J.B., J.E. Marlow, G.G. Fisk, and S.M.D. Jansen, *Sand-storage changes in the Colorado River downstream from the Paria and Little Colorado Rivers, June 1992 to February 1994*, U.S. Geol. Surv. Open-File Rept. 95-446, 61 pp., Tucson, AZ, 1995b.

Howard, A., and R. Dolan, Geomorphology of the Colorado River in the Grand Canyon, *J. Geol.*, 89(3), 269-298, 1981.

Kaplinski, M., J.E. Hazel, Jr., and S.S. Beus, *Monitoring the effects of interim flows from Glen Canyon Dam on sand bars in the Colorado River corridor, Grand Canyon National Park, Arizona*, Final Rept. to Natl. Park Serv., N. Ariz. Univ., Flagstaff, 1995.

Kearsley, L.H., J.C. Schmidt, and K.D. Warren, Effects of Glen Canyon Dam on Colorado River sand deposits used as campsites in Grand Canyon National Park, USA, *Regul. Rivers*, 9, 137-149, 1994.

Konieczki, A.D., J.B. Graf, and M.C. Carpenter, *Streamflow and sediment data collected to determine the effects of a controlled flood in March and April 1996 on the Colorado River between Lees Ferry and Diamond Creek, Arizona*, U.S. Geol. Surv. Open-file Rept. 97-224, 53 pp., 1997.

Marzolf, G.R., R.A. Valdez, J.C. Schmidt, and R.H. Webb, Perspectives on river restoration in the Grand Canyon, *Bull. Ecol. Soc. Amer.*, 79, 250-254, 1998.

Melis, T.S., R.H. Webb, P.G. Griffiths, and T.J. Wise, *Magnitude and frequency data for historic debris flows in Grand Canyon National Park and vicinity, Arizona*, U.S. Geol. Surv. Water Res. Invest. Rept. 94-4214, 285 pp., 1994.

Minckley, W.L., Native fishes of the Grand Canon region: an obituary?, in *Colorado River ecology and dam management*, ed. G.R. Marzolf, pp. 124-177, Natl. Acad. Press, Washington, DC, 1991.

Moody, T., Was the Glen Canyon Dam flood really a success?, *Boatman's Quarterly Rev.*, 9(3), 9, 1996.

National Research Council, *River and dam management: a review of the Bureau of Reclamation's Glen Canyon Environmental Studies*, Natl. Acad. Press, Washington, DC, 1987.

Nelson, J.M., R.R. McDonald, and D.M. Rubin, Computational prediction of flow and sediment transport patterns in lateral separation eddies, *EOS, Trans. Amer. Geophys. Union*, 75, 268, 1994.

Poff, N.L., J.D. Allan, M.B. Bain, J.R. Karr, K.L. Prestegaard, B.D. Richter, R.E. Sparks, and J.C. Stromberg, The natural flow regime: a paradigm for river conservation and restoration, *Bioscience*, 47, 769-784, 1997.

Randle, T.J., R.I. Strand, and A. Streifel, *Engineering and environmental considerations of Grand Canyon sediment management*, U.S. Committee on Large Dams, Proc. Thirteenth Ann. Lect. Ser., pp. 1-13, 1993.

Rubin, D.M., and R.R. McDonald, Nonperiodic eddy pulsations, *Water Res. Res.*, 31, 1595-1605, 1995.

Rubin, D.M., R.A. Anima, and R. Sanders, *Measurements of sand thicknesses in Grand Canyon and a conceptual model for characterizing changes in sandbar volume through time and space*, U.S. Geol. Surv. Open-File Rept. 94-597, 16 pp., 1994.

Rubin, D.M., J.C. Schmidt, and J.N. Moore, Origin, structure, and evolution of a reattachment bar, Colorado River, Grand Canyon, Arizona, *J. Sed. Pet.*, 60, 982-991, 1990.

Schmidt, J.C., Recirculating flow and sedimentation in the Colorado River in Grand Canyon, *J. Geol.*, 98, 709-724, 1990.

Schmidt, J.C., and J.B. Graf, *Aggradation and degradation of alluvial sand deposits, 1965-1986, Colorado River, Grand Canyon National Park*, U.S. Geol. Surv. Prof. Paper 1493, 1990.

Schmidt, J.C., and D.M. Rubin, Regulated streamflow, fine-grained deposits, and effective discharge in canyons with abundant debris fans, in *Natural and anthropogenic influences in fluvial geomorphology*, ed. J.E. Costa, A.J. Miller, K.W. Potter, and P.R. Wilcock, pp. 177-195, Amer. Geophys. Union, Washington, DC, 1995.

Schmidt, J.C., D.M. Rubin, and H. Ikeda, Flume simulation of recirculating flow and sedimentation, *Water Res. Res.*, 29 (8), 2925-2939, 1993.

Schmidt, J.C., R.H. Webb, R.A. Valdez, G.R. Marzolf, and L.E. Stevens, Science and values in river restoration in the Grand Canyon, *Bioscience* 48(9), 735-747, 1998.

Stevens, L.E., and G.L. Waring, Effects of post-dam flooding on riparian substrate, vegetation, and invertebrate populations in the Colorado River corridor in Grand Canyon, in *Glen Canyon Environmental Studies: Executive summaries of technical reports*, pp. 229-243, U.S. Dept. Int., Bur. Reclam., Upper Colo. River Region, Salt Lake City, UT, 1988.

Stevens, L.E., J.C. Schmidt, T.J. Ayers, and B.T. Brown, Flow regulation, geomorphology, and Colorado River marsh development in the Grand Canyon, Arizona, *Ecol. Applic.*, 5, 1025-1039, 1995.

Topping, D.J., *Physics of flow, sediment transport, hydraulic geometry, and channel geomorphic adjustment during flash floods in an ephemeral river, the Paria River, Utah and Arizona*, unpubl. diss., Univ. Washington, Seattle, WA, 1997.

U.S. Department of the Interior, *Operation of Glen Canyon Dam - Final Environmental Impact Statement, Operation of Glen Canyon Dam*, Colorado River Storage Project, Coconino County, AZ, 337 pp. + appendices, Salt Lake City, UT, 1995.

Valdez, R.A., and R.J. Ryel, *Life history and ecology of the humpback chub* (Gila cypha) *in the Colorado River, Grand Canyon, Arizona*, Logan, Utah, BIO/WEST Inc., Rept. No. TR-250-08, Rept. Bur. Recl. Cont. No. 0-CS-40-09110, 1995.

Webb, R.H., P.T. Pringle, and G.R. Rink, *Debris flows from tributaries of the Colorado River, Grand Canyon National Park, Arizona*, U.S. Geol. Surv. Prof. Paper 1492, Washington, DC, 1989.

Wiele, S.M., J.B. Graf, and J.D. Smith, Sand deposition in the Colorado River in the Grand Canyon from flooding of the Little Colorado River, *Water Res. Res.*, 32, 3579-3596, 1996.

Wiele, S.M., and J.D. Smith, A reach-averaged model of diurnal discharge wave propagation down the colorado River through the Grand Canyon, *Water Res. Res.*, 32 (5), 1375-1386, 1996.

John C. Schmidt, Department of Geography and Earth Resources, Utah State University, Logan, UT 84322-5240; email: jschmidt@cc.usu.edu

E.D. Andrews, U.S. Geological Survey, 3215 Marine Street, MS-458, Boulder, CO 80303

David L. Wegner, Ecosystem Management International, P.O. Box 23369, Flagstaff, AZ 86002

Duncan T. Patten, Department of Plant Biology, Box 871601, Arizona State University, Tempe, AZ 85287-1601

G. Richard Marzolf, U.S. Geological Survey, 432 National Center, 12201 Sunrise Valley Drive, Reston, VA 20192

Thomas O. Moody, College of Engineering and Technology, Northern Arizona University, P.O. Box 15600, Flagstaff, AZ 86011

Reworking of Aggraded Debris Fans

Robert H. Webb[1], Theodore S. Melis[2], Peter G. Griffiths[1], and John G. Elliott[3]

Debris flows from 600 tributaries in Grand Canyon periodically deposit poorly sorted sediment on debris fans along the Colorado River. Before regulation, mainstem floods maintained fans and rapids as highly-reworked deposits dominated by large boulders. After regulation, the reduced peak discharges have entrained only particles up to 1 m in diameter, and debris fans have aggraded. We measured the effects of the 1996 controlled flood on 18 recently aggraded debris fans. At most sites, fan area decreased by 2-42%, volume decreased by 3-34%, distal margins became armored with a lag of cobbles and boulders, the width of the reworked zone increased by 4-30 m, and river constrictions decreased slightly. Stream power decreased in most rapids because water-surface fall decreased and rapids widened. The amount of reworking is a function of stream power and the elapsed time between debris flow and flood. The effectiveness of future floods of similar magnitude in reworking debris fans will depend in part on the release history and extent of armoring in the period between debris flow and flood.

1. INTRODUCTION

Debris flows from 600 tributary canyons created and maintain the rapids of the Colorado River in Grand Canyon, Arizona [*Webb et al.*, 1988; *Webb et al.*, 1989; *Melis et al.*, 1994; *Webb et al.*, 1996]. Since 1986, the last year in which Glen Canyon Dam released at least 1270 m³/s, at least 25 debris flows have further constricted the Colorado River in Grand Canyon [*Melis et al.*, 1994], creating 2 new rapids and enlarging debris fans at 9 existing riffles and rapids. These constrictions range in size from small riffles to Lava

[1]U.S. Geological Survey, Tucson, Arizona
[2]Grand Canyon Monitoring and Research Center, Flagstaff, Arizona
[3]U.S. Geological Survey, Lakewood, Colorado

The Controlled Flood in Grand Canyon
Geophysical Monograph 110
This paper not subject to U.S. copyright
Published in 1999 by the American Geophysical Union

Falls, the largest rapid in Grand Canyon [*Webb and Melis*, 1995; *Webb et al.*, 1996]. In general, deposition on debris-fan margins constricts streamflow, thereby increasing flow velocities and water-surface fall through rapids [*Melis et al.*, 1994; *Webb*, 1996], changing navigational severity and affecting sand storage in eddies and upstream pools.

Before closure of Glen Canyon Dam in 1963, the Colorado River removed most of the deposits on aggraded debris fans during early summer floods that averaged 2330 m³/s and were as large as 6230 m³/s. All but the largest particles were swept downstream, and cobbles and small boulders were redeposited on debris bars that constrain the downstream extent of eddies [*Howard and Dolan*, 1981; *Schmidt*, 1990; *Schmidt and Graf*, 1990; *Schmidt and Rubin*, 1995] and control secondary rapids and riffles [*Webb et al.*, 1989]. The residual deposits formed boulder-laden debris fans that create rapids (Figure 1). The interaction between the frequency and magnitude of tributary debris flows and mainstem floods resulted in debris fans and rapids that were relatively stable in the long term. From 1963 through 1982, operations of Glen Canyon Dam

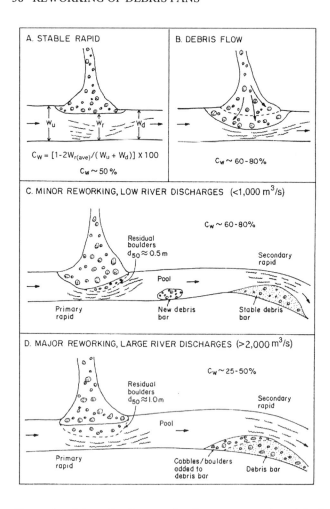

Figure 1. Schematic diagram showing a conceptual model of aggradation and reworking of a typical debris fan by the Colorado River in Grand Canyon [from *Webb et al.*, 1996, 1997]. Stable debris fans usually constrict the river by 50% at major rapids [*Kieffer*, 1985]. Debris-flow deposition generally increases the constriction to 60-80%, raising the bed elevation [*Howard and Dolan*, 1981]. Colorado River discharges <1000 m³/s only erode the distal margins of newly aggraded debris fans and slightly widen the constriction. The amount of widening is dependent on local topographic conditions of the fan and mainstem channel, particularly the stage-discharge relation and water-surface fall, as well as the particle-size distribution of the debris-flow deposit. Colorado River discharges >1500 m³/s rework most of the new deposit, leaving a residual lag of boulders on the debris fan and in the widened rapid.

constrained discharges to an average annual peak of 870 m³/s, which prevented many aggrading debris fans from being eroded [*Howard and Dolan*, 1979, 1981]. The large dam release of 1983, which was of a magnitude similar to pre-dam floods, reworked at least one aggraded debris fan [*Kieffer*, 1985].

In 1987, we began monitoring the reworking of aggraded debris fans [*Melis et al.*, 1994; *Webb et al.*, 1997], which numbered 18 by 1996. Powerplant releases of up to 870 m³/s significantly reworked the distal margins of newly aggraded debris fans and armored older, previously reworked fans (Figure 2). Powerplant releases combined with a flood in the Little Colorado River in January 1993 resulted in boulders as large as 1 m in diameter being entrained [*Melis et al.*, 1994] and increased the armoring of the distal margins of debris fans. *Webb et al.* [1996] documented the complete removal of two historic debris fans at Lava Falls Rapid by dam releases; the first in 1965 by a dam release of 1640 m³/s and the second in 1973 by a flow of 1080 m³/s. Cobbles and boulders entrained from the eroded debris fans appeared to be redeposited in the pool immediately downstream from Lava Falls Rapid instead of on alternating bars farther downstream (Figure 1). This altered pattern of redeposition reflects a change in the geomorphic framework of the Colorado River [*Webb*, 1996].

In this paper, we document the effects of the 1996 controlled flood on 18 debris fans and associated rapids in Grand Canyon National Park, Arizona. Our results demonstrate the effectiveness of a discharge of 1370 m³/s in reworking aggraded debris fans and provide data that will be useful in designing future controlled floods.

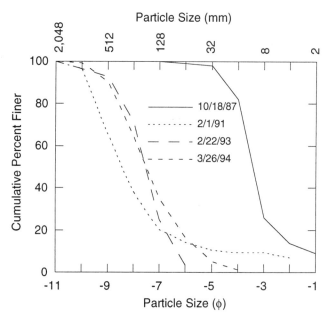

Figure 2. Particle-size distributions showing armoring of the distal margin of the aggraded debris fan at 18-Mile Wash following the 1987 debris flow.

1.1. Previous Studies

Nearly all of the rapids of the Colorado River in Grand Canyon result from the accumulation of large boulders on debris fans at the mouths of side canyons. These fans are deposited during tributary floods and, particularly, debris flows [*Péwé*, 1968; *Hamblin and Rigby*, 1968; *Simmons and Gaskill*, 1969; *Graf*, 1979; *Howard and Dolan*, 1981; *Webb et al.*, 1988, 1989; *Melis et al.*, 1994]. Large debris fans create rapids by constricting the width of the river and raising its bed elevation [*Howard and Dolan*, 1981]. Deposition of unusually large boulders or arrangements of boulders on the bed of the rapid cause spectacular hydraulic features that impede navigation [*Kieffer*, 1985, 1987, 1990]. These boulders are essentially permanent features of the rapids, having withstood historic discharges of up to 8500 m^3/s and possibly even larger prehistoric floods [*O'Connor et al.*, 1994].

Repeated debris-flow deposition alters the configuration of existing rapids and their controlling debris fans [*Webb et al.*, 1989; *Melis et al.*, 1994; *Webb et al.*, 1996]. Not all debris flows increase the constriction of the Colorado River; some aggrade the existing surface area of a debris fan, which may change the stage-discharge relation for higher discharges without constricting the river at low discharges. Because the pattern of flow around the debris fan is altered at higher discharges, navigation through the rapid and sand deposition in downstream eddies may also be changed [*Melis et al.*, 1994]. The low-water control on the rapid — an arrangement of boulders on the bed and channel margin of the river — is not likely to be significantly reworked by dam releases, but higher deposits can be reworked, given sufficient discharges in the Colorado River. The unregulated Colorado River periodically widened constricted rapids by eroding boulders and debris-flow matrix from debris fans [*Kieffer*, 1985]. This reworking left behind residual debris-fan surfaces composed of large, immobile boulders overlain by alluvial sand deposited in eddies [*Howard and Dolan*, 1981].

Many rapids in the Colorado River system have become more severe during the last 30 years because debris-fan constrictions and individual boulders cannot be totally removed by typical dam releases [*Graf*, 1980; *Howard and Dolan*, 1981; *Melis et al.*, 1994; *Webb*, 1996]. *Péwé* [1968] speculated that regulated flows would have minimal effects on debris-fan reworking, and that debris fans would therefore aggrade throughout Grand Canyon. By analyzing 1965 and 1973 aerial photographs, *Howard and Dolan* [1979] estimated that 25% of the debris fans in Grand Canyon had been aggraded by tributary floods. *Graf* [1980] hypothesized that debris fans on the Green River would

aggrade or remain stable in response to operations of Flaming Gorge Dam. *Melis et al.* [1994] also documented the aggradation of debris fans by debris flows and other floods in Grand Canyon as well as the reworking of selected fans by dam releases less than the maximum powerplant capacity of Glen Canyon Dam. Using field observations and 1-dimensional hydraulic modelling, *Hammack and Wohl* [1996] documented the effects of the unregulated Yampa River on the reworking of Warm Springs Rapid, which was aggraded by a debris flow in 1966.

Kieffer [1985] concluded that stable debris fans in Grand Canyon constrict the channel by an average of 50% and that floods of about 11,300 m^3/s would be required to rework Crystal Rapid to a stable constriction. *Hammack and Wohl* [1996] report that the duration of a flood is as important as its peak discharge in reworking aggraded debris fans. *Webb et al.* [1996] documented several debris flows at Lava Falls Rapid that created constrictions similar to the one at Crystal Rapid and were reworked by flows <3530 m^3/s. *Webb et al.* [1996] modified Kieffer's conceptual model and showed that dam releases above powerplant capacity but less than the annual pre-dam flood, such as the 1996 controlled flood, may partially or completely remove aggraded debris fans.

2. METHODS

We assessed the amount of reworking by the 1996 controlled flood on 16 debris fans recently aggraded by debris flows and 2 other debris fans (Table 1). These fans are distributed along the river corridor from Badger Creek to Lava Falls rapids (river miles 7.9-179.4) and represent a variety of geomorphic and hydraulic configurations. Debris flows at several sites, particularly 62.5 Mile, Tanner Canyon, 127.6 Mile, 157.6 Mile, and Prospect Canyon, initially constricted the river significantly. At 18 Mile, the 1987 debris flow covered both an existing debris fan and a separation sandbar and significantly constricted the river. Some of the debris fans, particularly at Palisades Creek and 160.8 Mile, represent debris flows that covered existing debris fans without significantly increasing the constriction of the river (Table 1).

Before the controlled flood, most of these debris fans were partially reworked by dam releases combined with tributary floods [*Melis et al.*, 1994]. Two debris fans — at 71.2 Mile and 72.1 Mile — were included despite reworking by discharges up to 1510 m^3/s in 1986 and were not expected to be eroded by the 1996 controlled flood. The debris fan at Tanner Canyon, which aggraded in August 1993, was mostly reworked by a dam release enhanced by a small flood in August 1994. The highest discharge that

TABLE 1. Changes in area and volume of debris fans aggraded after 1983 and associated rapids that were monitored during the 1996 controlled flood in Grand Canyon

Debris fan (Rapid)	River mile-side	Year of debris flow	Type[a] of aggrad-ation	Previous[b] peak discharge (m^3/s)	Pre-flood area (m^2)	Post-flood area (m^2)	Area change (m^2)	Pre-flood volume (m^3)	Post-flood volume (m^3)	Volume change (m^3)
Jackass Canyon[c] (Badger Creek)	7.9-L	1994	C	630	1,900	1,100	-800	5,800	3,800	2,000
18 Mile	18.0-L	1987	S	830	4,200	4,000	-200	16,800	14,900	1,900
62.5 Mile[d]	62.5-R	1990	C	960	1,600[e]	1,500[e]	-100	i	i	i
63.3 Mile[d]	63.3-R	1990	C	960	1,400[e]	1,300[e]	-100	i	i	i
Palisades Creek[d] (Lava Canyon)	65.5-L	1987 1990	D	960	2,300[e]	2,300[e]	0	i	i	i
Tanner Canyon[f] (Tanner)	68.5-L	1993	C	620	2,000	1,900	-100	7,200	6,700	500
Cardenas Creek	70.9-L	1993	D	620	7,800[e]	7,900[e]	+100	i	i	i
71.2 Mile	71.2-R	1983	C	1,510	4,400[e]	4,400[e]	0	i	i	i
72.1 Mile	72.1-R	1983	C	1,510	1,500[e]	1,400[e]	-100	i	i	i
Crystal Creek[h] (Crystal)	98.3-R	1995	C	720	5,300[e]	5,200[e]	-100	i	i	i
126.9 Mile	126.9-L	1989	C	960	3,200	3,100	-100	15,700	15,700	0
127.3 Mile	127.3-L	1989	C	960	2,100	2,100	0	8,800	8,500	300
127.6 Mile	127.6-L	1989	C	960	3,000	2,700	-300	13,100	11,500	1,600
Specter Chasm (Specter)	129.0-L	1989	C	960	3,800[e]	3,400[e]	-400	i	i	i
Bedrock Canyon (Bedrock)	130.0-R	1989	C	960	3,000[e]	3,200[e]	+200	i	i	i
157.6 Mile	157.6-R	1993	C	670	3,600	3,100	-500	9,800	7,900	1,900
160.8 Mile	160.8-R	1993	D	670	4,200	4,200	0	13,500	13,500	0
Prospect Canyon[h] (Lava Falls)	179.4-L	1995	C	670	5,300	4,200	-900	32,900	27,000	5,900

[a]C, the debris flow constricted the Colorado River; S, the debris fan covered an existing debris fan and a sandbar, constricting the Colorado River; D, the debris flow covered an existing debris fan but did not constrict the river.

[b]Largest peak discharge between the date of the debris flow and the 1996 controlled flood.

[c]Only the area and volume of 1994 deposition [Melis and others, 1994].

[d]Only the area and volume of 1990 deposition [Melis and others, 1994].

[e]Only determined by rectification of registered aerial photographs.

[f]Only the area and volume of 1993 deposition [Melis and others, 1994].

[g]This debris fan was not expected to change during the controlled flood (see Methods section).

[h]Only the area and volume of the lowest part of the debris fan, which was covered by the 1995 deposition [Webb et al., 1996].

[i]Not measured.

reworked 8 of the 18 debris fans was 960 m^3/s, which was a combination of a flood on the Little Colorado River and dam releases on January 13, 1993. Other debris fans had been reworked by dam releases that ranged from 620-830 m^3/s (Table 1). Debris fans at Crystal and Lava Falls rapids were aggraded in March 1995 and, because of their young age and poor sorting, were expected to change significantly during the 1996 controlled flood.

2.1. Discharges of the Controlled Flood

We calculated pre- and post-flood discharges at all the monitored debris fans using known travel times between gaging stations at about 250 m^3/s; for the peak of the flood, we report the highest discharge at the nearest gaging station. Flood frequency for unregulated and regulated periods of the record (1921-1962 and 1963-1996) was estimated at the gaging station on the Colorado River near Grand Canyon, Arizona [Garrett and Gellenbeck, 1991], using a log-Pearson type III distribution [U.S. Water Resources Council, 1981].

2.2. Areas and Volumes of Debris Fans

We surveyed 10 debris fans before and after the controlled flood to calculate changes in fan area, fan volume, and water-surface fall through the rapid [Webb et al., 1997]. We calculated fan volume above the lowest elevation measured at the fan-water edge. Although the accuracy of ground points is on the order of ±0.1 m horizontally and ±0.01 m vertically, the irregular surface and shape of debris fans is difficult to survey. Despite the fact that the accuracy of surveying data is higher than that of image-processing data, for consistency we rounded all area measurements to the nearest 100 m^2 (implied accuracy of ±50 m^2), all volume measurements to the nearest 100 m^3 (implied accuracy of ±50 m^3), and all distance measurements to the nearest 1 m (implied accuracy of ±0.5 m).

We used photogrammetric analysis of aerial photographs, taken at steady river discharges of 230-260 m^3/s before and after the controlled flood, to measure debris-fan area, changes in river constriction, and the width of the reworked zone [Webb et al., 1997]. Using image-processing software, we measured the area of 8 debris fans that were not surveyed and the change in width of the reworked distal edge of the debris fan and the channel widths necessary to determine the constriction of the Colorado River for all 18 debris fans. The accuracy of image processing varied among the debris fans depending on the amount of

distortion in the aerial photographs, the accuracy of surveyed or orthophotograph controls, the clarity of the image, and the choice of borders between debris fans, sandbars, and water stage in the Colorado River. Although the pixel resolution for the rectified images ranged from 0.1-0.5 m, rectification involved the alteration of pixel locations by as many as several meters. For consistency, we rounded all area measurements to the nearest 100 m^2 (implied accuracy of ±50 m^2) and all distance measurements to the nearest 1 m (implied accuracy of ±0.5 m). We attained broad agreement in areas calculated from survey data and image processing [Webb et al., 1997].

2.3. Particle-Size Distributions, Constrictions, and Stream Power

Point counts [Wolman, 1954] on the distal margins of debris fans were used to document changes in particle-size distribution and armoring [Melis et al., 1994]. The change in river constriction at a debris fan caused by reworking during the controlled flood was calculated as the percentage of channel width constricted by the debris fan [Webb et al., 1996]. We define percent constriction, C_w, as

$$C_w = \left[1 - \frac{2W_r}{(W_u + W_d)} \right] \times 100 \qquad (1)$$

where W_r = the average constricted channel width (m), W_u = the width (m) upstream from the rapid, and W_d = the width (m) downstream from the rapid below the expansion zone. All widths were measured at discharges between 230-260 m^3/s, and we considered the accuracy of C_w to be ±0.5%.

Following Baker and Costa [1987], we calculated stream power per unit width of channel, ω, before and after the flood at each rapid as

$$\omega = \gamma Q S W_r^{-1} \qquad (2)$$

where γ = specific weight of water, Q = discharge in the Colorado River, and S = the maximum water-surface slope through the rapid. S was surveyed before and after the controlled flood along the margin of the debris fan from the top of the rapid to the top of the recirculation zone. This method may overestimate the actual slope through a given rapid, but it does provide a relative measure for comparison among rapids. In addition, the slope is obtained along the margin of the debris fan where particles were entrained during the controlled flood. We used γ = 9810 N/m^3 in our calculations.

Figure 3. Photographs that document reworking of the debris fan at 127.6 Mile. A. (March 1996). The combination of a debris flow in 1990 and small streamflow floods, the last of which occurred in 1993, aggraded the debris fan at 127.6 Mile, covering an existing sandbar. The large volume of driftwood was deposited following the 1993 flood in the Little Colorado River. The discharge is about 245 m^3/s. B. (April 1996). Reworking by the 1996 controlled flood reduced the area and volume of the debris fan by 300 m^2 and 1600 m^3, respectively. The width of the reworked zone increased 14 m, and the particle-size distribution on the distal margin of the debris fan coarsened from D_{85} = 440 mm before the flood to D_{85} = 620 mm afterwards. The discharge is about 248 m^3/s.

Figure 3 (continued)

3. RESULTS

3.1. Areas and Volumes of Debris Fans

The areas of recently aggraded debris fans in Grand Canyon generally decreased during the controlled flood (Table 1; Figure 3). The largest changes occurred at Jackass Canyon (Badger Creek Rapid) and Prospect Canyon (Lava Falls Rapid). In both cases, the most recent debris flow had occurred <2 years before the controlled flood, and previous river discharges were too low to significantly rework these deposits. Decreases in fan area were observed at most of the rapids, but the area of the debris fan at Bedrock Canyon increased slightly because reworked sediment was deposited at the downstream margin of the debris fan. As expected, the debris fan at 71.2 Mile, which had been reworked at higher discharges, was unchanged. The debris fan at 72.1 Mile had also been partially reworked previously, but lost 100 m^2 of fan area during the controlled flood.

In general, changes in debris-fan volumes mirrored changes in areas (Table 1). The debris fan at Lava Falls Rapid lost the largest volume of sediment (5900 m^3), whereas the new deposits on the Jackass Canyon debris fan had the largest percentage change (-34%). In most cases, the changes in area and volume accurately depict flood-reworked conditions on the debris fans except for 160.8 Mile, where a considerable volume of fine sediment, mostly debris-flow matrix of particles <16 mm in diameter, was eroded from between boulders that were present before the 1993 debris flows. This type of erosion cannot readily be accounted for using surveying or image-processing techniques. We had few inconsistencies between the changes in area and volume except at 126.9 Mile, for which the survey data showed no change in volume yet a 200 m^2 decrease in area. Deposits that are poorly sorted, contain large numbers of boulders, and have irregular boundaries are difficult to describe using standard techniques.

3.2. Armoring of the Distal Margins of Debris Fans

Armoring of the distal margin was evaluated at 9 debris fans (Table 2). Most debris fans showed significant increases in the sizes of particles in the reworked zones (Figures 3 and 4). The D$_{50}$ of particles decreased, remained unchanged, or increased only slightly for some debris fans, such as the debris fans at 18-Mile Wash, 126.9 Mile, 157.6 Mile and 160.8 Mile. But D$_{50}$ increased by up to a factor of 2.5 at others, including Jackass Canyon and Prospect Canyon. The increases in D$_{85}$ were generally greater; the

TABLE 2. Change in particle-size distributions on the distal margins of debris fans reworked by the 1996 controlled flood in Grand Canyon

Debris fan	Date of debris flow (yr)	Particle diameter	Debris flow deposit (mm)	Before flood (mm)	After flood (mm)
Jackass Canyon	1994	D$_{50}$	110	140	220
		D$_{85}$	480	470	540
18 Mile	1987	D$_{50}$	11	180	170
		D$_{85}$	18	410	440
Tanner Canyon	1993	D$_{50}$	71	89	160
		D$_{85}$	290	550	540
126.9 Mile	1989	D$_{50}$	31	81	82
		D$_{85}$	110	290	310
127.3 Mile	1989	D$_{50}$	54[a]	54	89
		D$_{85}$	170[a]	170	240
127.6 Mile	1989	D$_{50}$	10	170	260
		D$_{85}$	86	440	620
157.6 Mile	1993	D$_{50}$	120	180	180
		D$_{85}$	360	470	410
160.8 Mile	1993	D$_{50}$	150	140	190
		D$_{85}$	480	410	530
Prospect Canyon	1995	D$_{50}$	140	210	530
		D$_{85}$	370	650	1170

[a] Particle size measured in March 1994 because a 1993 tributary flood altered the particle-size distribution of the distal margin of the debris fan.

typical D$_{85}$ after the controlled flood was about 0.5 m but was as high as 1.2 m at Prospect Canyon (Table 2).

At Lava Falls Rapid, D$_{50}$ on the Prospect Canyon debris fan increased from 0.21-0.53 m during the controlled flood (Table 2). As a result of reworking, the particles on the distal margin of the debris fan became larger and became better sorted. Reworking also preferentially removed limestones, leaving a deposit dominated by higher density, basalt particles. These results are in agreement with *Melis et al.* [1994], who report similar preferential lithologic removal during reworking.

The width of the reworked zone on the debris fan was one of the most sensitive measurements of the effects of the 1996 controlled flood (Table 3). For 12 of the debris fans, the width of the reworked zone increased by at least 4-14 m (the increase of 30 m at Crystal Rapid occurred over a relatively small and low-relief debris fan deposited in

TABLE 3. Changes in the width of the reworked zone on the distal margin of debris fans reworked by the 1996 controlled flood in Grand Canyon

Debris fan	Pre-flood width (m)	Post-flood width (m)	Change (%)	Maximum fan[a] height (m)	Mean channel width (m)	Area of fan innundated (%)
Jackass Canyon[c]	23	14	-39	5.4	61	100
18 Mile	7	15	114	8.8	38	79
62.5 Mile	3	13	333	b	10	b
63.3 Mile[d]	7	9	29	b	51	b
Palisades Creek[d]	11	11	0	b	55	b
Tanner Canyon[e]	5	8	60	8.2	106	52
Cardenas Creek[e]	10	16	60	b	69	b
71.2 Mile	8	9	13	b	152	b
72.1 Mile	10	14	40	b	b	b
Crystal Creek[f]	3	33	1000	b	87	b
126.9 Mile	6	10	67	8.3	35	66
127.3 Mile	7	7	0	6.1	38	69
127.6 Mile	10	24	140	7.1	31	68
Specter Chasm	5	9	80	b	30	b
Bedrock Canyon	8	12	50	b	34	b
157.6 Mile	5	12	140	5.7	36	100
160.8 Mile	6	12	100	6.8	37	96
Prospect Canyon[f]	6	11	83	8.6	43	58

The reworked zone is the width of the debris fan between the 277 and 1,320 m³/s stages.
[a] Difference between the lowest and highest debris-fan elevations at 210 m³/s.
[b] No data.
[c] Only the reworked width of the 1994 deposition (Melis and others, 1994).
[d] Only the reworked width of the 1990 deposition (Melis and others, 1994).
[e] Only the reworked width of the 1993 deposition (Melis and others, 1994).
[f] Only the reworked width of the 1995 deposition (Webb and others, 1996).

1995). For most of the debris fans, the reworked zone after the flood was at least twice as wide as it was before the flood (Table 3). The width of the reworked zone at Jackass Canyon (Badger Rapid) decreased because a third of the aggraded debris fan was removed by the controlled flood. We measured no change in the width of the reworked zone at Palisades Creek and 127.3 Mile. Changes in the size of the reworked zone are likely dependent on local stage-discharge relations during the controlled flood (Table 3).

3.3. Constriction, Water-Surface Fall, and Stream Power

Of the 18 debris fans we monitored, the percent constriction, C_w, decreased at 12 sites (Table 4), indicating widening of the rapid and (or) changes in the stage-discharge relation. The largest amount of widening

occurred at Prospect Canyon and 62.5 Mile, where changes in constriction were >10% of the initial value. The constriction increased at Bedrock Canyon because eroded sediment was redeposited on the downstream side of the debris fan across from the island in the center of the rapid [see *Stevens*, 1990]. Surprisingly, the constriction decreased at 71.2 Mile despite its history of reworking by larger dam releases in the 1980s. This widening may reflect decreases in sand bars on both sides of the river instead of real change in the debris fan.

The water-surface fall through 7 of 10 rapids was changed less than ±0.1 m by the controlled flood (Table 5). The largest changes were in Tanner and Lava Falls rapids, where the water-surface fall decreased by 0.27 m and increased by 0.21 m, respectively. The locations of changes in water-surface profile varied among the rapids and debris fans (Figure 5). At Tanner Rapid, most of the change

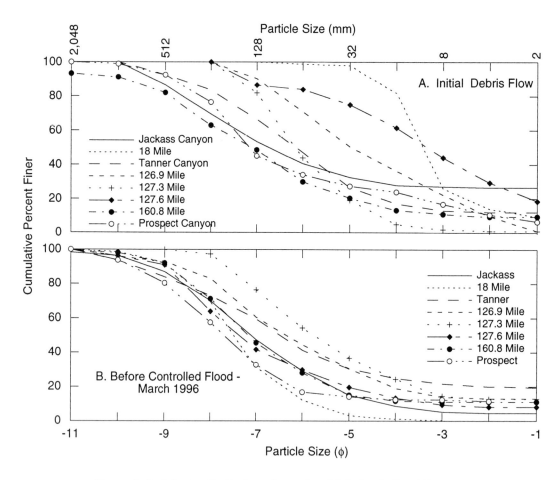

Figure 4. Particle-size distributions of reworked debris fans in Grand Canyon.

occurred at the top of the rapid, whereas at Lava Falls, the largest changes occurred at the bottom of the rapid. The change at Lava Falls Rapid probably is related to the partial removal of a small, submerged debris bar that had formed downstream from the rapid after the 1995 debris flow [*Webb and Melis*, 1995; *Webb et al.*, 1996, Figure 5].

Changes in the percent constriction reflect complex interactions between velocity, stage-discharge relation, width of the rapid, and changes in bed roughness. Stream power is one measure that reflects all of these variables. Unit stream power decreased at 9 of 10 debris fans by 1-16% (Table 5). The exception was the debris fan at Tanner Rapid, where a slight increase was calculated. The largest decreases (-7 to -16%) were for the debris fans at 126.9, 127.3, and 127.6 Mile.

3.4. Changes in Navigability of Rapids

The navigational severity of several rapids changed as a result of the 1996 controlled flood. Although velocities

decreased through Lava Falls Rapid, less water moved down the left side after the flood, exposing rocks and reducing that navigational possibility at low discharges. The height of waves appeared to be lower for similar discharges in Badger Creek and Specter rapids. Because the constriction at Bedrock Rapid narrowed, that rapid is now more difficult to navigate at lower discharges.

The navigational severity of Crystal Rapid, the second largest rapid in Grand Canyon [*Stevens*, 1990; *Webb*, 1996], significantly decreased during the 1996 controlled flood. On the basis of slight changes in area and percent constriction (Tables 1 and 4), one might conclude that Crystal Rapid changed very little. However, several large boulders in the right side of the rapid were removed, decreasing the intensity of the lateral waves on the right side [Webb et al., 1996]. The 1996 controlled flood caused significant changes to the two largest rapids in Grand Canyon, although in the case of Lava Falls Rapid, the changes were easily detected with our measurements.

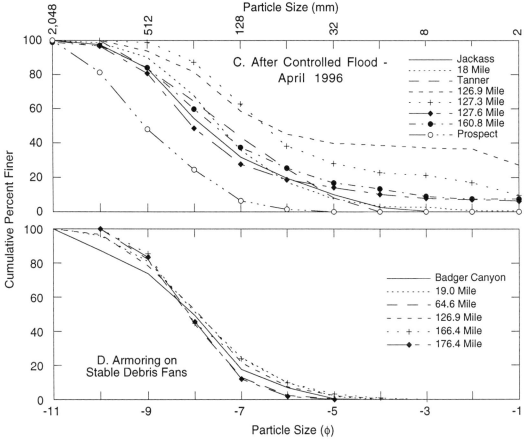

Figure 4 (continued)

4. DISCUSSION AND CONCLUSIONS

The 1996 controlled flood partially reworked many of the debris fans aggraded since 1983 that we monitored in Grand Canyon. Using several measures of debris-fan reworking, including changes in fan area, fan volume, amount of river constriction, and sediment armoring, we found that the controlled flood caused substantial changes. The peak discharge of 1370 m³/s was a 10-year flood in the regulated Colorado River, but was smaller than the mean annual flood in the pre-dam river. Our results do not agree with those of *Kieffer* [1985], who concluded that extremely high discharges are required for significant reworking of debris fans. Instead, our general results are in accord with the observations of *Hammack and Wohl* [1996], who documented reworking of Warm Springs Rapid on the Yampa River, and *Webb et al.* [1996], who documented the removal of 2 historic debris fans at Lava Falls Rapid, by moderate flows.

We measured significant reductions in the area and volume of recently aggraded debris fans at Prospect Canyon, Jackass Canyon, 157.6 Mile, and 127.6 Mile. The areas of other aggraded debris fans changed only slightly, and we measured no change in area at only 2 debris fans (Table 1). Large volumes of sediment were removed from debris fans at Prospect Canyon, Jackass Canyon, 157.6 Mile, 18 Mile, and 127.6 Mile (Table 1). Differences in stream power explain many of the differences in area and volume change among the debris fans (Figure 7). Debris fans exposed to high stream power before the controlled flood, such as the Prospect Canyon and 127.6 Mile debris fans (Lava Falls and 127.6 Mile rapids), had large changes in area and volume, whereas sites exposed to low stream power, such as at Tanner Canyon and the aggraded fans at 126.9 Mile and 127.3 Mile, changed relatively little (Tables 1 and 5). The fan at Mile 160.8 lost a large volume of fine sediment from between boulders, but we were not able to document this type of change with ground surveys. Recent aggradation increases the amount of reworking despite lower stream power (Figure 6). This effect is evident at the Jackass Canyon debris fan, aggraded in 1994, where moderate stream power markedly decreased its area and volume.

TABLE 4. Changes in percent constrictions (C_w) of debris fans reworked by the 1996 controlled flood in Grand Canyon

Debris fan (rapid)	Pre-flood C_w (%)	Post-flood C_w (%)	Change (%)
Jackass Canyon	22	20	-10
18 Mile	54	50	-7
62.5 Mile	55	45	-18
63.3 Mile	70	69	-1
Palisades Creek	55	56	+2
Tanner Canyon	31	33	+6
Cardenas Creek	56	55	-2
71.2 Mile	13	12	-8
72.1 Mile	10	10	0
Crystal Creek	55	52	-5
126.9 Mile	54	53	-2
127.3 Mile	24	24	0
127.6 Mile	47	42	-11
Specter Chasm	32	27	-16
Bedrock Canyon	64	66	+3
157.6 Mile	33	35	+6
160.8 Mile	47	48	+2
Prospect Canyon	42	34	-19

The constrictions at many rapids, particularly 62.5 Mile and Lava Falls Rapid, increased considerably, whereas other constrictions changed only slightly due to the stability of boulders at the distal edges of older fans (Table 4, Figure 7a). Initial and reworked constrictions for the debris fans reported here fall within the range of constrictions measured throughout the river system at confluences where fans are present (Figure 7). Changes in percent channel constriction, which is a function of debris-fan area, water-surface fall, and the stage-discharge relation, do not appear to be related to the elapsed time between the most-recent debris flow and the controlled flood. For some debris fans, we measured relatively small changes in percent constrictions, despite the fact that large particles were removed from the fan surface. On the basis of the amount of reworking by the controlled flood, we conclude that the constriction ratio, which has been used previously to depict the reworking status of rapids [*Kieffer*, 1985, 1987, 1990], is not a particularly sensitive measure of changes in debris fans and rapids caused by relatively small floods.

Because of reworking by the 1996 controlled flood, navigation of Lava Falls Rapid is essentially the same as it was before the 1995 debris flow [*Webb et al.*, 1996]. Although the total water-surface fall through Lava Falls Rapid increased by 0.3 m (Table 5), the stage-discharge

relation at the top of the rapid decreased by 0.4 m (Figure 5), exposing several large boulders on the left side at low discharges. At Crystal Rapid, removal of only a few boulders, which were lodged in positions that significantly increased the severity of the rapid, decreased the navigational severity despite only small changes in the debris fan.

We attribute the relation between amount of reworking, as measured by changes in area and volume, and the elapsed time between the debris flow and the controlled flood to the history of reworking and armoring by lower discharges in the Colorado River. The low-water controls at several debris fans that aggraded before 1993 and that were reworked by the January 1993 flood were unchanged by the controlled flood. We observed that closest-packing imbrication and even suturing [*Webb*, 1996] had occurred on parts of debris fans that had been submerged for long

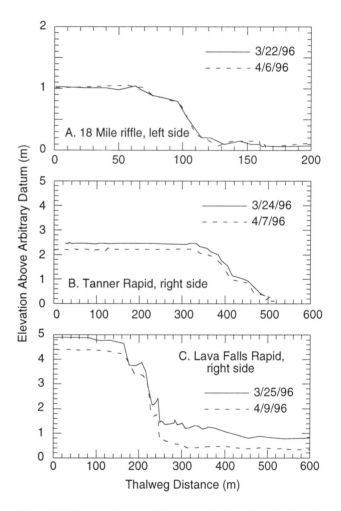

Figure 5. Water-surface profiles through selected rapids showing the effects of the controlled flood of 1996. All water-surface profiles were surveyed at discharges between 230 and 248 m^3/s.

TABLE 5. Changes in water-surface profiles and streampower around selected debris fans at discharges of 230-254 m³/s before and after the 1996 controlled flood in Grand Canyon

Debris fan (Rapid)	WATER-SURFACE FALL			LENGTH OF RAPID (m)	WATER-SURFACE SLOPE		WIDTH OF RAPID		UNIT STREAM POWER		
	Before flood (m)	After flood (m)	Differ-ence (m)		Before flood	After flood	Before flood (m)	After flood (m)	Before flood (w/m²)	After flood (w/m²)	Differ-ence (%)
Jackass Canyon (Badger Creek)	4.39	4.41	0.02	230	0.024	0.024	59	63	940	890	-5
18 Mile	0.94	0.99	0.05	75	0.014	0.014	38	38	840	830	-1
Palisades Creek (Lava Canyon)	1.75	1.63	-0.12	115	0.021	0.019	55	54	880	800	-9
Tanner Canyon (Tanner)	2.23	1.96	-0.27	145	0.017	0.017	107	104	360	370	+3
126.9 Mile[a]	0.57	0.57	0.00	110	0.014	0.015	35	35	1120	1040	-7
127.3 Mile	0.16	0.16	0.00	50	0.005	0.004	38	38	310	260	-16
127.6 Mile	0.95	1.01	0.06	70	0.024	0.024	29	32	2040	1820	-11
157.6 Mile	0.98	1.01	0.03	60	0.016	0.016	35	36	1090	1070	-2
160.8 Mile	1.37	1.32	-0.05	75	0.018	0.017	36	37	1190	1110	-7
Prospect Canyon (Lava Falls)	3.12	3.42	0.21	85	0.038	0.041	40	46	2270	2150	-5

[a] Pre-flood data collected in May 1994 at a discharge of 986 m³/s.

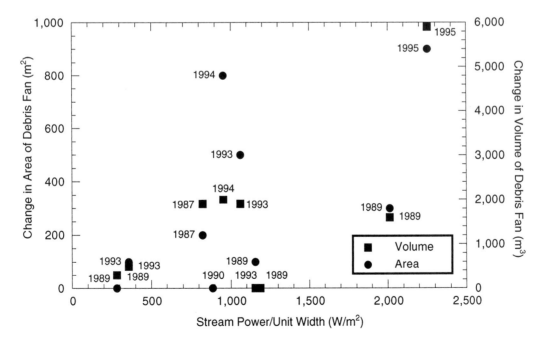

Figure 6. Relations between stream power before the 1996 controlled flood (calculated at discharges of 245-254 m³/s), the date of the debris flow that aggraded the debris fan, and changes in area and volume of selected debris fans on the Colorado River in Grand Canyon.

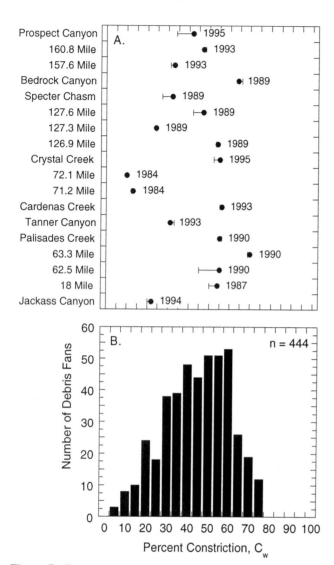

Figure 7. Percent constriction of the Colorado River by debris fans in Grand Canyon. A. Changes in aggraded debris fans caused by the 1996 controlled flood. Solid circle, percentage of channel constricted before the flood. Vertical bar, percentage of the channel constricted after the flood. B. Percent constrictions (C_w) for 444 debris fans between Lees Ferry and Diamond Creek measured from March 1996 aerial photographs [*Melis*, 1997].

periods, typically longer than 3 years. For example, the 1993 debris flow that constricted Tanner Rapid [*Melis et al.*, 1994] increased the stage-discharge relation by 1 m in the pool above the rapid. This change increased the storage of sand in the upper pool between October 1993 and March 1996. We expected that this condition would be reversed by reworking during the controlled flood, but large boulders deposited by the debris flow that control the elevation of the pool were unchanged by the flood, and the upper-pool

elevation was only reduced by 0.25 m (a 25% decrease).

Debris fans aggraded after 1993 were more easily reworked. For example, the low-water control added by the 1995 debris flow at Lava Falls was almost completely removed (1 boulder remains), but the low-water controls at Bedrock and Specter rapids, created by debris flows in 1989, remain. As the interval between the aggrading debris flow and the reworking flood increases, a flood will be less effective in clearing out the constriction, and larger peak discharges are required. If more significant reworking of aggraded debris fans is a desirable characteristic of future deliberate floods in Grand Canyon, the amount of elapsed time since the last debris flow is one of the most significant criteria.

Because most of the reworking occurred on the rise in the flood hydrograph, we conclude that the 7-day duration of the controlled flood was unnecessary for reworking of debris fans. If management of aggraded debris fans is one of the criteria for design of future controlled floods, the duration of such a flood could be much shorter. However, the design discharge of 1270 m³/s, which actually was as high as 1370 m³/s in Grand Canyon, could be increased to have a greater effect on reworking debris fans. Although some recently aggraded debris fans such as at Lava Falls Rapid were partially removed by the 1996 controlled flood, others, such as at Tanner Canyon, were not greatly affected because the stage of the flood was not high enough to erode the side of the aggraded fan or overtop it.

Very short-duration and high-magnitude controlled floods would be highly effective in reworking aggraded debris fans. Such a flood could have a discharge as high as 2800 m³/s, or twice the discharge of the 1996 controlled flood (less than a 5-year flood on the unregulated river), but be designed for a duration of only minutes of peak discharge at sites downstream from the Little Colorado River. The recession of the hydrograph of such a flood, which would mimic some of the pre-dam flash floods through Grand Canyon, could be designed to include steady, beach-building discharges. One potential drawback of such a flood may be that boulders deposited in the downstream pool might require more sustained, high discharges to be flushed downstream.

Acknowledgments. The authors gratefully acknowledge the contributions the many individuals who helped with the field and office work that made this report possible. We thank Travis McGrath, Holly Metz, and Mia Wise, who were volunteers; Cassie Fenton of the University of Utah; Lauren Hammack, John Rihs, and Tamara Wiggins of the National Park Service; Joe Hazel of Northern Arizona University; Dominic Oldershaw, an independent contractor; Jim Pizzuto of the University of Delaware; Tim Randle of the Bureau of Reclamation; Robert Ruffner of the University of Minnesota; Steve Tharnstrom of the Defense Nuclear Agency; and Mimi Murov and Steve Wiele of the U.S. Geological Survey. We especially thank the professional river guides who made the difficult logistics possible: Tony Anderson, Steve Bledsoe, Bob Grusy, Matt Herman, Pete Reznick, Rachael Running, Kelly Smith, and Greg Williams. The Defense Nuclear Agency generously provided Tharnstrom's salary for the duration of the flood experiment. Special thanks to Jim Pizzuto and Tim Randle for their review of our manuscript.

REFERENCES

Baker, V.R., and J.E. Costa, Flood power, in *Catastrophic Flooding*, ed. L. Mayer and D. Nash, pp. 1-21, Allen Unwin, London, 1987.

Garrett, J.M., and D.J. Gellenbeck, *Basin characteristics and streamflow statistics in Arizona as of 1989*, U.S. Geol. Surv. Water-Resour. Invest. Rept. 91-4041, 612 pp., 1991.

Graf, W.L., Rapids in canyon rivers, *J. Geol.*, 87, 533-551, 1979.

Graf, W.L., The effect of dam closure on downstream rapids, *Water Res. Res.*, 16, 129-136, 1980.

Hamblin, W.K., and J.K. Rigby, *Guidebook to the Colorado River, Part 1: Lee's Ferry to Phantom Ranch in Grand Canyon National Park*, 84 pp., Brigham Young Univ., Geol. Studies, 15, pt. 5, Provo, UT, 1968.

Hammack, L., and E. Wohl, Debris-fan formation and rapid modification at Warm Springs Rapid, Yampa River, Colorado, *J. Geol.*, 104, 729-740, 1996.

Howard, A.D., and R. Dolan, Changes in the fluvial deposits of the Colorado River in the Grand Canyon caused by Glen Canyon Dam, in *Proc. First Conf. Sci. Res. Natl. Parks*, Vol. II, ed. R.M. Linn, pp. 845-851, Washington, D.C., 1979.

Howard, A., and R. Dolan, Geomorphology of the Colorado River in Grand Canyon, *J. Geol.*, 89, 269-297, 1981.

Kieffer, S.W., 1985, The 1983 hydraulic jump in Crystal Rapid: Implications for river-running and geomorphic evolution in the Grand Canyon, *J. Geol.*, 93, 385-406.

Kieffer, S.W., *The rapids and waves of the Colorado River, Grand Canyon, Arizona*, U.S. Geol. Surv. Open-File Rept. 87-096, 69 pp., 1987.

Kieffer, S.W., Hydraulics and geomorphology of the Colorado River in the Grand Canyon, in *Grand Canyon Geology*, ed. S.S. Beus and M. Morales, pp. 333-383, Oxford Univ. Press, New York, 1990.

Melis, T.S., 1997, *Geomorphology of debris flows and alluvial fans in Grand Canyon National Park and their influences on the Colorado River below Glen Canyon Dam, Arizona*, [Ph.D dissertation], Tucson, Univ. Ariz., 1997.

Melis, T.S., R.H. Webb, P.G. Griffiths, and T.J. Wise, *Magnitude and frequency data for historic debris flows in Grand Canyon National Park and vicinity, Arizona*, U.S. Geol. Surv. Water Res. Invest. Rept. 94-4214, 285 pp., 1994.

O'Connor, J.E., L.L. Ely, E.E. Wohl, L.E. Stevens, T.S. Melis, V.S. Kale, and V.R. Baker, A 4,500 year record of large floods on the Colorado River in Grand Canyon, Arizona, *J. Geol.*, 102, 1-9, 1994.

Péwé, T.L., *Colorado River Guidebook, Lees Ferry to Phantom Ranch*, 78 pp., privately published, Tempe, AZ, 1968.

Schmidt, J.C., Recirculating flow and sedimentation in the Colorado River in Grand Canyon, Arizona, *J. Geol.*, 98, 709-724, 1990.

Schmidt, J.C., and J.B. Graf, *Aggradation and degradation of alluvial sand deposits, 1965-1986, Colorado River, Grand Canyon National Park, Arizona*, U.S. Geol. Surv. Prof. Paper 1493, 74 pp., 1990.

Schmidt, J.C., and Rubin, D.M., Regulated streamflow, fine-grained deposits, and effective discharge in canyons with abundant debris fans, in *Natural and anthropogenic influences in fluvial geomorphology*, ed. J.E. Costa, A.J. Miller, K.W. Potter, and P.R. Wilcock, pp. 177-195, Amer. Geophys. Union Mono. 89, 1995.

Simmons, G.C., and D.L. Gaskill, *River Runner's Guide to the Canyons of the Green and Colorado Rivers, with Emphasis on Geologic Features, Volume III, Marble Gorge and Grand Canyon*, 132 pp., Northland Press, Flagstaff, AZ, 1969.

Stevens, L., *The Colorado River in Grand Canyon, A guide*, 115 pp., Red Lake Books, Flagstaff, Arizona, 1990.

U.S. Water Resources Council, *Guidelines for determining flood flow frequency*, Hydrology Subcommittee Bulletin 17B, U.S. Govt. Print. Off., Washington, D.C., 1981.

Webb, R.H., *Grand Canyon, a Century of Change*, 290 pp., Univ. Ariz. Press, Tucson, AZ, 1996.

Webb, R.H., and T.S. Melis, The 1995 debris flow at Lava Falls Rapid, *Grand Canyon Nature Notes*, 11, 1-4, 1995.

Webb, R.H., P.T. Pringle, S.L. Reneau, and G.R. Rink, Monument Creek debris flow, 1984, Implications for formation of rapids on the Colorado River in Grand Canyon National Park, *Geol.* 16, 50-54, 1988.

Webb, R.H., P.T. Pringle, and G.R. Rink, *Debris flows from tributaries of the Colorado River, Grand Canyon National Park, Arizona*, U.S. Geol. Surv. Prof. Paper 1492, 39 pp., 1989.

Webb, R.H., T.S. Melis, T.W. Wise, and J.G. Elliott, *"The great cataract," Effects of late Holocene debris flows on Lava Falls Rapid, Grand Canyon National Park and Hualapai Indian Reservation, Arizona*, U.S. Geol. Surv. Open-file Rept. 96-460, 91 pp., 1996.

Webb, R.H., T.S. Melis, P.G. Griffiths, and J.G. Elliott, *Reworking of aggraded debris fans by the 1996 controlled flood on the Colorado River in Grand Canyon National Park, Arizona*, U.S. Geol. Surv. Open-file Rept. 97-16, 96 pp., 1997.

Wolman, M.G., A method of sampling coarse-river-bed material, *Trans. Amer. Geophys. Union*, 35(6), 951-956, 1954.

Robert H. Webb, U.S. Geological Survey, 1675 W. Anklam Road, Tucson, AZ 85745, email: rhwebb@usgs.gov

John G. Elliott, U.S. Geological Survey, Denver Federal Center, Lakewood, CO 80205

Peter G. Griffiths, U.S. Geological Survey, 1675 W. Anklam Road, Tucson, AZ 85745

Theodore S. Melis, Grand Canyon Monitoring and Research Center, 2255 N. Gemini Drive, Flagstaff, AZ 86001

Entrainment and Transport of Cobbles and Boulders from Debris Fans

James E. Pizzuto[1], Robert H. Webb[2], Peter G. Griffiths[2], John G. Elliott[3], and Theodore S. Melis[4]

Observations of marked cobbles and boulders during the 1996 controlled flood document the flood's effectiveness in reworking debris fans along the Colorado River in Grand Canyon. In addition, the hydraulics associated with initial sediment motion were defined on debris fans at 127.6 Mile and Lava Falls Rapid. At 127.6 Mile, peak Shields stresses, scaled by the mean particle size, varied from 0.06 to 0.24 and exceeded empirical entrainment threshold values by a factor of two. At Lava Falls Rapid, peak Shields stresses exceeded 0.3. Rocks tagged with radio transmitters at Lava Falls moved at Shields stresses that varied by more than an order of magnitude because of the unique shape, orientation, and topographic setting of each particle. Critical Shields stresses for the tagged particles are quantified using a new method for computing the ratio of threshold drag force to submerged weight that accounts for varying particle orientation, geometry, and packing. Theoretical computations of critical Shields stresses are well-correlated with critical Shields stresses estimated from field measurements of the local depth-slope product. Particles with a b-axis diameter as large as 2 m were displaced during the flood, and the tagged particles at Lava Falls Rapid moved an average distance of 230 m.

1. INTRODUCTION

What discharges are capable of moving boulders of a given size and shape? To understand and predict the stability of many bedrock and gravel-bed rivers, it is important to answer this question. For example, geologists often infer the hydraulics of catastrophic floods by examining large particles in flood deposits [Baker, 1974; O'Connor, 1993]. Engineers need to assess the stability of

channels and engineered structures during extreme flows [Chang, 1988] and ecologists may be interested in the stability of riverine habitats under a variety of flow regimes [Stevens et al., 1996]. In Grand Canyon, understanding and predicting the stability of fan-eddy systems requires an accurate assessment of the mobility of large particles [Howard and Dolan, 1981; Schmidt and Rubin, 1995]. Because fan-eddy systems are an important geomorphic element of the Colorado River in Grand Canyon, the design of future controlled floods will require accurate methods for determining when cobbles and boulders are entrained.

The 1996 controlled flood in Grand Canyon provided an important opportunity to document conditions of entrainment for large particles on recently aggraded debris fans [Webb et al., 1997]. This paper, with its companion [Webb et al., this volume], presents the results of our observations and analyses of the effects of this flood. Our results suggest that it is possible to predict the initial motion of irregularly shaped angular particles in Grand Canyon and that accurate predictions of debris-fan reworking require important modifications to existing methods. In addition,

[1]Department of Geology, University of Delaware at Newark
[2]U.S. Geological Survey, Tucson, Arizona
[3]U.S. Geological Survey, Lakewood, Colorado
[4]Grand Canyon Monitoring and Research Center, Flagstaff, Arizona

The Controlled Flood in Grand Canyon
Geophysical Monograph 110
Copyright 1999 by the American Geophysical Union

most of the large particles entrained from debris-flow fans that we observed were deposited in pools at the bottom of rapids. Before the construction of Glen Canyon Dam, however, particles entrained from debris-flow fans during floods were likely transported through these pools and were deposited in secondary rapids and bars farther downstream [*Webb et al.*, 1997], suggesting that the 1996 flood may not have been large enough to effectively maintain the existing fan-eddy system.

2. PREDICTIONS OF PARTICLE ENTRAINMENT

2.1. Empirical Methods for Predicting Initial Motion

The threshold of entrainment is usually expressed in terms of the dimensionless Shields stress, τ_{*c}, required to initiate motion:

$$\tau_{*c} = \frac{\tau_c}{(\rho_s - \rho) g D_s}, \qquad (1)$$

where τ_c is the shear stress exerted by the flow at initial motion (N/m^2); ρ_s and ρ are the densities of particle and water (kg/m^3); g is the acceleration of gravity (9.8 m/s^2); and D_s is the diameter of the particle (m), usually represented by the intermediate (b) axis.

Empirical estimates of τ_{*c} are available for a flat bed, here denoted as τ_{*c0}. For a well-sorted, well-rounded sediment with regular packing, and for fully rough flows with relative submergence (defined by the ratio of depth to particle diameter) greater than about 15, τ_{*c0} is a constant equal to approximately 0.047 [*Buffington and Montgomery*, 1997]. However, for sediment mixtures typical of many coarse-grained riverbeds, τ_{*c0} is quite variable. In a recent comprehensive review, *Buffington and Montgomery* [1997] report τ_{*c0} values ranging from 0.03 to 0.073 for previous field studies of initial motion. In this paper, this range will be used to represent a reasonable empirical range for τ_{*c0}.

Empirical estimates of τ_{*c0} must be modified if the bed is not flat. We have used the method of *Kovacs and Parker* [1994] to account for the influence of a sloping bed. Because the downstream slope has a negligible influence, even for the steep rapids typical of Grand Canyon, only the cross-channel slope β (defined "down" the fan perpendicular to the channel) is accounted for:

$$\tau_{*c} = \tau_{*c0} \cos(\beta) \sqrt{1 - \frac{[\tan(\beta)]^2}{\mu^2}}, \qquad (2)$$

where μ is a friction coefficient.

Equations (1) and (2) can be used to estimate τ_c if τ_{*c0}, β, μ, ρ_s, and ρ are known values. The fluid shear stress at the threshold of motion, τ_c, can also be related to more readily measured hydraulic variables using a variety of methods. Here, we relate τ_c to the depth and slope on inclined debris fan surfaces using the area method [*Lundgren and Jonsson*, 1964; *Parker*, 1978]:

$$\tau_c = \rho g D S \cos(\beta), \qquad (3)$$

where D is the depth and S is the slope of the water surface.

2.2. A Theoretical Method For Estimating τ_{*c}

Wiberg and Smith [1987] derived an equation for τ_{*c} from basic fluid mechanical principles:

$$\tau_{*c} = \frac{F_d}{W_s} \cdot \frac{C_d}{2} \cdot \frac{V}{A_x D_s} \cdot \frac{1}{<f^2(y/y_0)>}, \qquad (4)$$

where F_d is the drag force (N); W_s is the submerged weight of the particle (N); C_d is the drag coefficient; V is the particle volume (m^3); A_x is the cross-sectional area of the particle exposed to the flow (m^2); and $f(y/y_0)$ is a velocity profile function (m/s) where y is the height above the bed and y_0 is the 'roughness height' where the velocity nominally is zero. The notation <> indicates vertical averaging from the bottom to the top of the particle.

2.3. A New Method For Estimating F_d/W_s

Wiberg and Smith [1987] derived an expression for F_d/W_s by considering a spherical particle resting on a bed of other spherical particles in two-dimensional space. The resulting expression determines F_d/W_s in terms of the downstream slope of the bed, the ratio of lift to drag, and a pivot angle ϕ_0. *Wiberg and Smith's* [1987] method may be extended to three dimensions, but an estimate of the pivot angle ϕ_0 appropriate for three-dimensional particles is required. Because we could not use standard methods for directly measuring ϕ_0 [*Li and Komar*, 1986, *Buffington et al.*, 1992], and because simple visual estimates lacked precision, a new theory was developed to estimate F_d/W_s for the irregular boulders and cobbles of Grand Canyon debris fans.

Our method for estimating F_d/W_s is based on the coordinate system and forces illustrated in Figure 1. The origin is the center of mass of the particle, the x-axis is oriented parallel with the flow direction (and hence the drag force), and the y-axis is vertical. The z-axis is defined

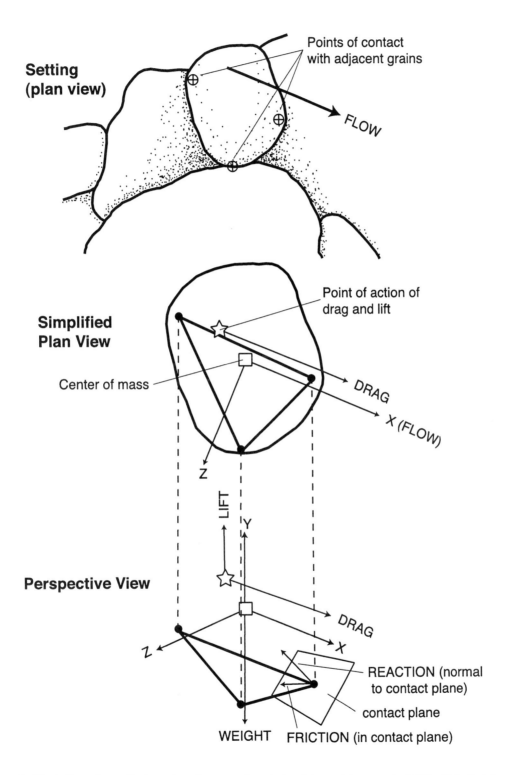

Figure 1. Illustration of coordinate system, forces, and contact points used to compute F_d/W_s. Note that the contact points are located beneath the particle in the upper diagram.

orthogonal to the other two axes to create a right-handed coordinate system. Lift and drag forces act at a specified point of action, and the submerged weight vector points in the negative y direction. [The derivation of the equations used to assess conditions of initial particle motion are too involved to present in detail, and only a brief summary is described below. A more complete derivation for one mode of initial motion, sliding along a planar surface, is provided in Appendix I.]

The grain is supported at "contact points" where the particle touches neighboring particles. The contact points serve as points of action for reaction and frictional forces. The reaction forces act normal to contact "planes," while the frictional forces act within the contact planes. To remain at rest, a three-dimensional particle must have at least three contact points, although all of the points may not be relevant to the entrainment of the particle. At the threshold of motion, two equations are available to define the equilibrium of a particle. These equations state that the vector sums of the forces and moments on the particle are zero. The forces include drag, lift, submerged weight, and the reaction and frictional forces acting at each contact point. The moments are taken about the center of mass.

The magnitudes of the friction and lift vectors may be eliminated from the equations by introducing a friction coefficient, μ, and the ratio of lift to drag, L_d:

$$\mu = \frac{F_{ri}}{R_i} \tag{5}$$

and

$$L_d = \frac{F_L}{F_d}, \tag{6}$$

where F_{ri} and R_i are the magnitudes of the friction and reaction vectors associated with the contact point i; and F_L and F_d are the magnitudes of the lift-force and drag-force vectors.

Even after equations (5) and (6) are introduced, the moment and force balance equations are difficult to solve. However, for specific modes of particle motion the equations may be greatly simplified and readily solved (e.g., the example presented in Appendix I). Four possible modes of initial motion are identified in this study, including 1) rolling about 2 contact points, 2) sliding on a planar surface without rotation where the direction of friction is unknown (treated in detail in Appendix I), 3) sliding on a planar surface without rotation where the direction of friction is specified by the geometry of the contact points, and 4) sliding with rotation where the direction of friction is specified. The equations are applied

in the following way. For each particle, F_d/W_s is computed for each possible mode of motion. The mode of motion that requires the smallest fluid forces for entrainment then controls the initial motion of the particle. The value of F_d/W_s for this mode of motion is then used in equation (4) to compute the critical Shields stress.

2.4. Velocities and Drag Coefficients

According to *Wiberg and Smith* [1987], the drag force and submerged weight may be defined by

$$F_d = \frac{1}{2}\rho A_x C_d <u^2> \tag{7}$$

where $<u^2>$ is the average of the squared velocity across the exposed portion of the particle. By dividing (7) by (8),

$$W_s = \langle \rho_s - \rho \rangle gV, \tag{8}$$

and solving for $<u^2>$, the following is obtained:

$$\sqrt{<u^2>} = \sqrt{\frac{F_d}{W_s} \cdot \frac{2}{C_d} \cdot \frac{(\rho_s - \rho)}{\rho} \cdot \frac{gV}{A_x}}. \tag{9}$$

Once F_d/W_s has been estimated, equation (9) can be used to compute an average velocity near the particle at the threshold of motion.

Wiberg and Smith [1987] use a logarithmic velocity profile function to represent $f(y/y_o)$, an approach we adapt. Because the debris fans of Grand Canyon slope towards the river, the cross-stream coordinate z (Figure 1) is adjusted by dividing by $\cos(\beta)$. This is equivalent to assuming that the velocity profile normal to the sloping bed surface is logarithmic. The roughness length y_o is specified by 0.1 D_{85} [*Whiting and Dietrich*, 1990, *Wilcock et al.*, 1996].

The location of the bed surface, where y = 0, is often difficult to specify. The following approach, consistent with the methods used by *Wiberg and Smith* [1987], is used here. A local, average bed slope (oriented in the downstream direction) was determined by fitting a straight line through a detailed profile surveyed to define bed roughness. A line with this slope was then fitted through the lowest elevations along the surveyed profile to define bed elevation. The location of the bed was then projected into three-dimensions using the mean bank slope. Thus, the bed surface was defined as a planar surface that slopes downstream at the mean downstream slope and cross-stream at the mean cross-stream slope.

We used MAPLEV symbolic mathematics software to average $<f^2(y/y_o)>$ across the particle. MAPLEV calculated a solution to the integral required to average the velocity-profile function over the particle. We used this solution to compute values of $<f^2(y/y_0)>$ for the appropriate bottom and top elevations of particles.

Wiberg and Smith [1987] used experimental data for spheres to estimate values for C_d. However, the rocks of the Grand Canyon debris fans we studied are typically angular, and are not well-represented by spheres. Other commonly-used methods for determining C_d were also unconvincing. Therefore, we have selected the constant value of 1.36 obtained by *Nelson et al.* [1997] for smooth cubes on a rough bed. The correct values for Grand Canyon rapids could be either higher or lower than this value by as much as 50%, and they could vary systematically with particle geometry. We consider possible variability in the drag coefficient as a source of error that can only be reduced with additional study.

The ratio of lift to drag, L_d, is also difficult to specify accurately. *James* [1990] notes experimental observations of both positive and negative lift forces on particles near the bed, with values of L_d ranging from -5 to 2.5. Recognizing the current uncertainty in estimating values of L_d, we have used values from 0.5 to 1.5, a range that includes the majority of reported experimental measurements.

3. METHODS

3.1. Study Area

The entrainment of boulders and cobbles was monitored on six recently aggraded debris fans during the 1996 controlled flood [*Webb et al.*, 1997]. In addition to gross changes in morphology of debris fans [*Webb et al.*, this volume], we estimated the conditions of particle entrainment for the debris fan on river left at 127.6 Mile and the Prospect Canyon debris fan at Lava Falls Rapid (mile 179.3), where a 1995 debris flow [*Webb et al.*, 1996] constricted the river by about 50%. Debris flows had deposited sediment on these debris fans relatively recently, and none of the debris fans had been previously reworked by discharges greater than 1000 m³/s (Table 1). The age of debris-flow deposition is relevant, because clasts are wedged together by turbulence during high flows in Grand Canyon, making them increasingly difficult to rework [*Webb et al.*, 1997]. Before the 1996 controlled flood, the debris fan at 127.6 Mile had a D_{50} of 170 mm and a D_{85} of 440 mm, while the Prospect Canyon debris fan had a D_{50} of 210 mm and a D_{85} of 650 mm. D_{50} for all debris fans before the flood ranged from 90 to 210 mm, and D_{85} similarly ranged from 410 to 650 (Table 1).

3.2. General Field and Laboratory Measurements

A detailed summary of the field measurements obtained during the 1996 controlled flood is provided by *Webb et al.* [1997]. The measurements that are relevant to entrainment studies defined the topography of the debris fans, the hydraulics of the flow, the characteristics of the sediment on the debris fan, and the timing and nature of particle entrainment during the flood. Both aerial photography (Figure 2) and traditional instrument surveys were analyzed for changes. In addition to documenting the overall debris-fan morphology, we surveyed detailed profiles for topographic roughness and mean downstream bed slope at the 250 m³/s stage before and after the flood at both 127.6 and Lava Falls Rapid. Before and after the flood at all debris fans, the distributions of particle sizes and lithologies were measured using point counts based on the *Wolman* [1954] method. Individual cobbles and boulders were marked with numbered bolts inserted into drilled holes and located by surveys before and after the flood.

The densities of several samples of sandstone and basalt from the Prospect Canyon debris fan were determined by weighing the rocks in both air and water. Friction coefficients were measured using a tilting board. The samples all had well-developed planar joint faces on at least one side. Tilting experiments with the planar surfaces in contact were used to measure the coefficients of friction for basalt in contact with basalt, for sandstone in contact with sandstone, and for basalt in contact with sandstone. Replicate measurements using smooth aluminum spheres on the tilting board demonstrated that angles could be measured to within approximately 0.5°.

3.3. Field Measurements at 127.6 mile and Lava Falls Rapid

Hydraulic data available at both sites include measurements of stage, surface velocities, and water-surface elevation. These data were collected continuously at Lava Falls Rapid during the controlled flood, but the equipment at 127.6 Mile failed early in the flood. Stage data were obtained in the pools above the rapids. Surface velocities were measured using tetherballs before and after the 1996 controlled flood at both sites, and during the flood at Lava Falls Rapid [*Webb et al.*, 1997]. Water-surface profiles were also surveyed before and after the flood at both sites. At Lava Falls Rapid, water-surface profiles were surveyed on river right during the flood at 0910, 1050, 1112, 1129, 1134, 1159, 1218, 1249, 1345, 1533, and 1632 hrs on March 27, 1996. These data were projected across the rapid to river left and interpolated in space and time to determine local

TABLE 1. Geomorphic and sedimentologic characteristics of the debris fans studied in this paper

| | DEBRIS FAN | | | | | |
	18.0 Mile	Tanner Canyon	127.6 Mile	157.6 Mile	160.8 Mile	Prospect Canyon
Date of last debris flow	1987	1993	1989	1993	1993	1995
Maximum discharge experienced by most recent debris flow (m^3/s)	830	620	960	670	670	670
Water-surface fall before 1996 flood (m)[a]	0.94	2.23	0.95	0.98	1.37	3.12
Water-surface slope before 1996 flood (m/m)[a]	0.014	0.017	0.024	0.016	0.018	0.038
Unit stream power, τ, before the flood (W/m^2)[a,b]	840	360	2040	1090	1190	2270
Mean debris-fan slope near water's edge before the flood (degrees)	nd	nd	13.3	nd	nd	16.7
D_{50} before the 1996 flood (mm)	180	89	170	180	140	210
D_{85} before the 1996 flood (mm)	410	550	440	470	410	650

[a] Water-surface characteristics were measured at a discharge of about 250 m^3/s.

[b] $\tau = \gamma QS/W$, where Q is the discharge, S is the water-surface slope, and W is the mean width.

water-surface slopes and water depths at the times and locations where radio-tagged rocks were entrained by the controlled flood.

Using water surface elevation data from one side of Lava Falls Rapid to infer elevations on the other side is likely to be quite inaccurate. For example, *Kieffer* [1998] used photogrammetric methods to map the water surface topography of Lava Falls Rapid. Her maps show that water surface elevations may vary considerably from one side of the rapid to another. As a result, water surface slopes surveyed on one side of the rapid may deviate significantly from local water surface slopes on the other side. Our method was adopted for safety: surveying water surface elevations near an eroding debris-fan margin during a flood is dangerous. Our estimates of local water surface slopes and depths at Lava Falls Rapid, therefore, are likely to be imprecise.

At 127.6 Mile, 41 of 83 marked rocks were located after the flood and resurveyed. Only rocks that were located after the flood are used to document entrainment conditions at

127.6 Mile because particles that were not located could have been buried in place rather than transported. Only a profile of high-water marks is available to determine the water-surface elevation and slope during the flood. A straight line was fit to the surveyed high-water marks; the slope and intercept of the fitted line were used to determine the peak shear stresses at the locations of marked particles during the 1996 controlled flood. By using the 95% confidence intervals for the predicted slopes and depths, a range of values for the peak stress at each particle location was obtained.

3.4. Particle tracking at Lava Falls Rapid

At Lava Falls Rapid, radio transmitters (tags) were used to determine the timing and conditions of entrainment of individual rocks. Radio tags were inserted into 9 rocks; a 10th tag was attached to the outside of one particle. Three of the 10 particles were cobble-sized (less than 256 mm);

 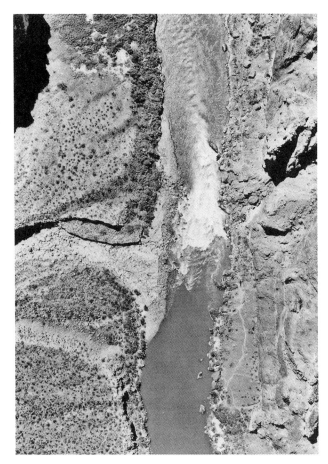

Figure 2. Aerial photographs that document reworking of the Prospect Canyon debris fan at Lava Falls Rapid. A. (March 24, 1996). Lava Falls Rapid was constricted by a 1995 debris flow from Prospect Canyon, shown on the left. The discharge is about 254 m³/s. B. (April 9, 1996). Reworking during the rising limb of the 1996 controlled flood removed 5900 m³ of the edge of the debris fan, increasing the width of the rapid by an average of 6 m. The discharge is about 246 m³/s.

the rest were boulders (greater than 256 mm) (Table 2). We used specially built radio tags transmitting in the 40-Mhz wavelength with a battery life of 30 days for maximum transmission power (Lee Carstenson, Smith-Root Incorporated, pers. commun., 1995). The tags were inserted into holes drilled into the rocks, glued in place using polyester body filler, and then the rocks were placed along the steeply sloping edge of the debris fan.

Before the flood, we determined the characteristics of all of the tagged particles. We surveyed their locations and measured the sizes and shapes of the particles, their orientations, lithology, exposure to the flow, and topographic pocket angles. The sizes and shapes of the particles were determined using a shape classification. Particles were classified as being one of the following shapes: 1) a rectangular solid, 2) an ellipsoid, 3) a cylindrical solid, 4) an ellipsoidal solid, and 5) a triangular solid. The size of each

particle was defined using appropriate A, B, and C axes, and the volume of each particle was determined using mensuration formulae [*Melis et al.*, 1994]. The orientation of each particle was determined by measuring the azimuth of the A axis and the strike and dip of the A-B plane using a Brunton compass. The lithology of each particle was determined visually. The area exposed to the flow was estimated using a tape measure. Detailed measurements were also used to define where each of the tagged rocks was supported by contact with surrounding rocks. Contact points were located relative to any useful reference point on each rock. Measurements were made using a Brunton Compass and a tape measure. An estimate of the direction of flow around each rock was also made.

During the flood, the time of initial motion was documented using both a scanning receiver attached to a data logger and an audible receiver. Radio signals from

TABLE 2. Characteristics of the 10 rocks tagged with radio transmitters on the Prospect Canyon debris fan at Lava Falls Rapid

Particle Number	Length of axes (m)			Shape[a]	Lithology	Weight[b] (N)
	a	b	c			
1	0.26	0.22	0.11	RS	Sandstone	154
2	0.31	0.23	0.21	RS	Sandstone	367
3	0.27	0.21	0.11	ES	Basalt	183
4	0.35	0.34	0.11	ES	Sandstone	230
5	0.55	0.49	0.46	EL	Basalt	1970
6	0.48	0.31	0.28	TS	Sandstone	510
7	0.35	0.28	0.14	ES	Basalt	327
8	0.53	0.38	0.35	ES	Sandstone	1356
9	0.86	0.66	0.52	ES	Sandstone	5679
10	0.94	0.70	0.44	EL	Sandstone	3714

[a] RS – rectangular solid, ES – ellipsoidal solid, EL – ellipsoid, TS – triangular solid.

[b] computed from the shape classification, measured axes, and measured density for each lithology.

dislodged particles were faint during the peak of the controlled flood because of high signal attenuation in the deep water. Weak to strong signals were detected during the steady low discharges after the flood. By triangulation with manual receivers, we located 8 of 10 radio-tagged particles after the flood. Particles were located to within an accuracy of about 5 m.

4. RESULTS

4.1. Physical Properties of Selected Lithologies

The densities of three samples of sandstone were all 2.5 g/cm^3. The densities of 4 samples of basalt ranged from 2.9 to 3.3 g/cm^3, with an average of 3.1 g/cm^3. The 69 friction angles (note that μ as defined in equation (5) is tan(ϕ)) range from 26 to 39°, with a mean and standard deviation of 32° and 3°, respectively (Figure 3). Friction angles for basalt and sandstone were not significantly different.

4.2. Entrainment of Marked Particles at 127.6 Mile

A line fit to the surveyed high-water marks at 127.6 Mile has a coefficient of determination (r^2) of 0.65. The 95% confidence interval for the slope of the water surface ranges from 0.0069 to 0.022, values that are lower than the water-surface slope of 0.024 surveyed before and after the flood (Table 1). The 41 marked particles located after the flood

moved an average distance of 8 m, with a minimum distance of 3 m and a maximum of 20 m to the downstream edge of the debris fan. The marked particles that we did not find after the flood were either buried or deposited farther downstream.

Peak Shields stresses for the 24 marked particles for which B-axis measurements were available are illustrated in Figure 4. Peak Shields stresses were computed using equations (3) and (1). The range of peak shear stresses scaled by the mean sediment diameter and the range of empirical critical Shields stresses reported by *Buffington and Montgomery* [1997] are also illustrated (both ranges have been adjusted for the lateral slope of 13.3° using equation (3)). Peak Shields stresses defined using mean sediment size exceed threshold values by at least a factor of two. The computed peak Shields stresses for the marked particles generally fall within the range of empirical critical Shields stresses.

4.3. Entrainment of Particles at Lava Falls Rapid

At Lava Falls Rapid, the leading edge of the flood wave of the 1996 controlled flood reached the site at about 1010 hrs on March 27, 1996 (Figure 5). At 1102 hrs, the top of rock 1 was just submerged, and we observed it immediately being carried downstream by the flow. Fortunately, a TV crew at the site was filming rock 1 as it departed; the film shows the rock vibrating vigorously in place just before it moved. Radio transmitters indicated that rock 5 was moved at 1105 hrs. At the time of movement, we observed a pronounced wave over the top of rock 5, indicating shallow submergence and a significant water surface deformation at the time of initial motion. By 1214 hrs, all of the tagged rocks had been transported downstream (Table 3).

We also observed other sediment transport processes as the stage rose on the morning of March 27. By 1140 hrs, the steep bank at the edge of the debris fan began to erode by slab failures. By 1200 hrs, slab failures were occurring continuously (Figure 5). We also heard crashing noises louder than the background noise of the rapid, which we interpreted as the sounds of boulders colliding. These observations suggest that the entrainment of the tagged rocks may have been influenced by falling bank debris or by collisions with rocks already in transport, although most of the radio-tagged particles were entrained well before the bank began to fail continuously at 1200 hrs. By 1300 hrs, the rate of bank collapse had greatly decreased, and by 1400 hrs both the rate of bank erosion and the noise of crashing boulders had nearly ceased (Figure 5), even though the fluid stresses clearly remained high.

Computations of F_d/W_s for 3 potential modes of motion for rock 5 are illustrated in Figure 6. Computed values of F_d/W_s were obtained by solving equations (A1) and (A2) for different modes of motion: for each mode of motion the actual equations solved are simplified forms of equations (A1) and (A2) similar to equations (A13) and (A14). The horizontal axis of Figure 6 represents assumed values of the ratio of lift to drag (L_d). The three modes of motion considered include rolling about 2 contact points and sliding without rotation in two prescribed directions. (Note that the morphology of the particle and the locations of the contact points are highly schematic in Figure 6 and do not accurately reflect field measurements.) Possible variations in the potential directions of sliding are created by taking different trial locations for the point of action of drag and lift. Variations in L_d and μ also create a wide range in computed values of F_d/W_s. Rolling requires the lowest fluid forces to initiate motion, and therefore the highest and lowest values of F_d/W_s for this mode of motion are chosen to compute the range of critical Shields stresses for this particle (Table 3).

Quantitative estimates of the hydraulic conditions required to entrain the 10 tagged particles are presented in Table 3 and Figures 7 and 8. Critical Shields stresses computed from simplified versions of equations (A1) and (A2) vary by more than an order of magnitude, falling well outside the empirical range presented by *Buffington and Montgomery* [1997]. Although some large particles moved at small critical Shields stresses (e.g., particle 5) and some small particles moved at large critical Shields stresses (e.g., particle 7), there is no correlation between the critical Shields stress and either absolute or relative particle size.

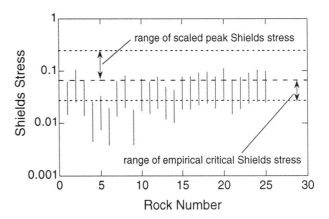

Figure 4. Peak critical Shields stresses at river mile 127.6. The ranges of peak Shields stresses scaled by D_{50} and of critical Shields stresses reported by *Buffington and Montgomery* [1997] are illustrated by horizontal lines (both have been corrected for the cross-channel debris fan slope). Peak Shields stresses for 24 marked particles are also illustrated. The ranges of predicted values for the 24 particles are a result of 95% confidence intervals for the local depth and slope at each particle determined from high-water marks.

Although the data scatter widely about the line of perfect agreement in Figure 7, critical Shields stresses computed from equations (A1) and (A2) are significantly correlated ($r^2 = 0.75$) with the critical Shields stresses estimated from the depth-slope product. The mean ratio of computed to estimated Shields stress is 0.8, with a standard error of 0.16, indicating a tendency to slightly underpredict the critical Shields stress.

In Figure 8, the predicted critical Shields stresses are plotted as a function of the time when the particles were moved. The solid line in Figure 8 represents the Shields stress determined for an arbitrarily-selected surveyed location on the bank of the fan; the curve indicates the depth-slope product at this point, scaled by the mean particle size before the flood and a submerged density appropriate for a sandstone clast. The solid line increases with time as the water depth increases during the flood (the survey data indicate that the water-surface slope also increases with time at this location, but the increased depth has the greater influence on the depth-slope product). A line drawn through the points for the 10 tagged particles represents a predicted trend of increasing stress with time exerted by the flow during the rising stage of the flood. This line has a trend broadly similar to that of the line defined by the field data, providing additional evidence that the theoretical computations are reasonable estimates of the relative erodibility of the 10 tagged rocks.

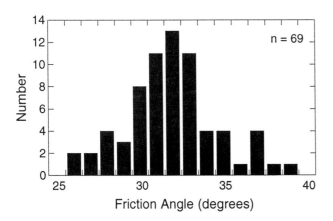

Figure 3. Histogram of 69 measurements of the friction angle μ determined in the laboratory on samples of sandstone and basalt from the Prospect Canyon debris fan. Mean = 32; standard deviation = 3.

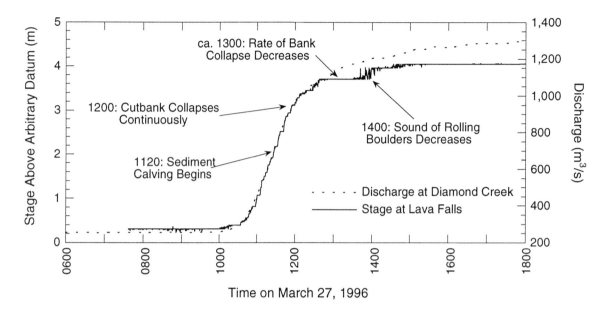

Figure 5. The relation between stage, discharge, and reworking of the debris fan at Lava Falls Rapid. The stage record at Lava Falls Rapid is for March 27, 1996, and the hydrograph for the Colorado River above Diamond Creek is adjusted for a discharge-independent travel time of 7.46 hrs [*Webb et al.*, 1997].

Average velocities computed using equation (9) range from 0.9 to 3.9 m/s (Table 3). These values are consistent with the mean surface velocities of 5.1 and 6.6 m/s measured on the left side of Lava Falls Rapid before the flood and during the peak stage of the flood, respectively [*Webb et al.*, 1997]. However, the data are not available to evaluate the computed velocities more rigorously.

Entrainment of particles from the Prospect Canyon debris fan significantly coarsened its face (Figure 9). As a result of the 1996 flood, most fine sediment (>-4 φ) was removed, but sand was deposited among the larger remaining particles. The largest change was in particles between 16 and 112 mm (-4 to -7 φ). Although many large particles were removed, the reworked particle-size distribution does not approach the particle-size distribution of the "fully reworked" state, which is the state of the debris fan as a result of the 1983 flood of 2720 m³/s.

4.4. Deposition of Radio-Tagged Particles at Lava Falls Rapid

At Lava Falls Rapid, 8 of the 10 radio-tagged particles dislodged from the debris fan were relocated after the flood (Figure 10). The smallest particle, a cobble, travelled 420 m to another debris bar that forms the secondary rapid (Lower Lava Rapid). The 6 particles were deposited in the pool immediately downstream from the main rapid and one,

particle 6, remained lodged at the bottom of the rapid. The average travel distance for the 8 relocated particles was 230 m (Table 3).

4.5. Entrainment and Transport of Particles on Other Debris Fans

The 1996 controlled flood removed large boulders from some debris fans, but did not noticeably shift the larger particles on other fans. The number of marked cobbles and boulders that were transported downstream varied among 5 debris fans (Figure 11). Fewer large particles were transported from previously-reworked debris fans than at recently aggraded fans where debris-flow matrix remained a major component of the deposit.

After the flood, we found between 24 and 89% of the marked particles on five debris fans (Table 4), suggesting that between 11 and 76% of the marked particles either were transported off the debris fan or could not be found owing to burial. Of the relocated particles, the entrained particles and the surveyed travel distance were a function of b-axis diameter, as expected (Figure 11, Table 4). At 157.6 Mile and 160.8 Mile, 76 and 40 percent of the marked particles, respectively, were transported downstream (Figure 11, Table 4); boulders with b-axis diameters up to 0.5 m were removed (Figure 11). In contrast, only 11 percent of the marked particles, with b-axis diameters up to

TABLE 3. Conditions of entrainment and transport distances for 10 radio-tagged particles from the Prospect Canyon debris fan during the controlled flood on March 27, 1996

Particle number	Time of initial motion	F_d/W_s	A_x (m^2)	Mode of motion[a]	Range of computed τ_{*c}	Measured τ_{*c}	Range of velocity[b] (m/s)	Transport distance (m)
1	1102	0.27–0.45	0.024	2	0.010–0.017	0.045	1.2–1.6	nd
2	1112	0.30–0.52	0.048	2	0.010–0.018	0.023	1.4–1.9	110
3	1115–1141	0.28–0.88	0.023	1	0.020–0.061	0.175	1.3–2.3	420
4	1142	0.15–0.38	0.041	1	0.005–0.012	0.008	0.9–1.4	230
5	1105	0.11–0.24	0.213	1	0.002–0.005	0.004	1.0–1.5	nd
6	1202–1214	0.44–0.78	0.134	3	0.014–0.025	0.011	1.2–1.6	50
7	1152–1202	0.39–0.71	0.012	2	0.060–0.110	0.125	3.1–4.2	310
8	1119–1145	0.07–0.17	0.053	1	0.009–0.023	0.013	1.3–2.0	250
9	1156–1207	0.48–0.92	0.321	1	0.023–0.045	0.098	2.8–3.9	220
10	1208–1212	0.17–0.39	0.132	1	0.012–0.028	0.015	2.0–3.1	240

nd, no data.

[a] 1 – rolling, 2 – sliding without rotation (direction of friction not specified), 3 – sliding without rotation (direction of friction specified).

[b] Computed using equation (9).

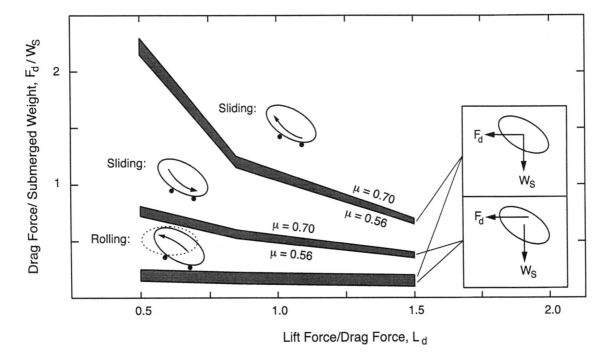

Figure 6. Computations of F_d/W_s for rock 5 for three modes of initial motion. A range of values for each mode of motion is computed by varying L_d, μ (relevant for sliding only), and the location of action of the drag and lift forces (illustrated in the two boxes to the right). Locations of contact points and the shape of rock 5 are highly schematic and do not accurately reflect field measurements.

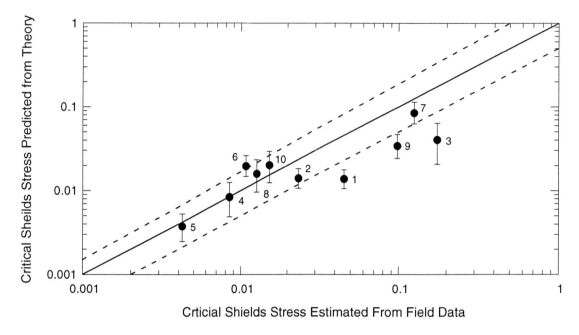

Figure 7. Critical Shields stresses estimated from the local depth-slope product compared with critical Shields stresses predicted using equations (A1) and (A2) for the 10 tagged rocks from the Prospect Canyon debris fan. Dashed lines are 95% confidence limits about the best-fit regression line.

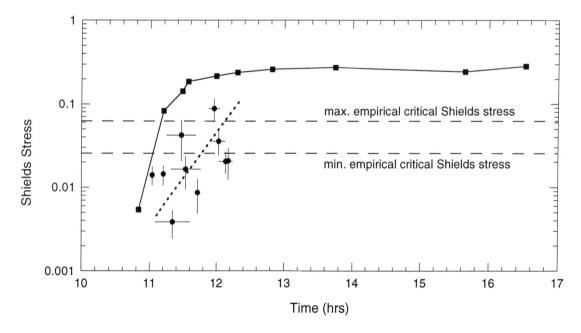

Figure 8. Shields stresses as a function of time on March 27, 1996, on the Prospect Canyon debris fan. The solid line represents shear stresses computed for an arbitrarily selected point on the edge of the fan, scaled to a sandstone clast with a b-axis equal to the mean particle diameter measured before the 1996 controlled flood.

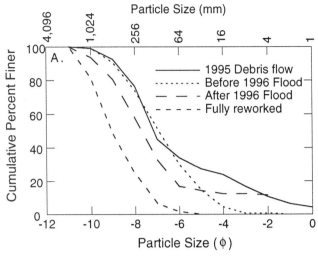

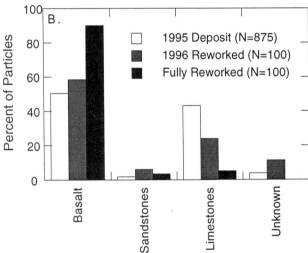

Figure 9. Effects of the 1996 controlled flood on particles on the distal margin of the Prospect Canyon debris fan at Lava Falls Rapid [modified from *Webb et al.*, 1996], which was reworked by the 1983 flood. A. Particle-size distributions of the distal margins of the debris fan. The "fully reworked" curve is for the debris fan before the 1995 debris flow [*Webb et al.*, 1996]. B. Lithologies of particles at various stages of reworking.

only 0.26 m, were removed from the 18 Mile debris fan. The number of particles entrained on each debris fan is related in a complex way to the average stream power across the debris fan (Table 1; *Webb et al.*, 1997). Most of the large particles either were tilted or rolled a short distance; some of the smaller particles were found wedged a short distance downstream parallel to the flow direction over the debris fan.

An important factor in large particle transport was the initial position of particles on the debris fan relative to the

inundation stage [*Webb et al.*, 1997]. For some debris fans, such as at Tanner Rapid, many large particles remained above the stage of the controlled flood and therefore could not be entrained. Particle position, combined with the previous reworking of the debris fan by lower discharges in August 1994, resulted in only one marked boulder being removed by the 1996 controlled flood (Figure 11b). In contrast, despite the fact that the debris fan at 18.0 Mile was completely inundated, only a few marked cobbles and small boulders were removed (Figure 11a), because the debris fan had been previously reworked to a limited extent by power-plant releases [*Melis et al.*, 1994].

5. DISCUSSION AND CONCLUSIONS

The 1996 controlled flood was of sufficient magnitude to entrain boulders as large as 2 m in diameter and to transport particles with diameters of less than 0.5 m off of the margins of debris fans. At Lava Falls Rapid, we observed significant reworking at rising discharges between 1000 and 1200 m^3/s. We observed two types of reworking: (1) failure of unconsolidated debris-flow deposits by lateral erosion, and (2) the entrainment of individual particles from the bed of the river. Most of the reworking of the Prospect Canyon debris fan, which had not been subjected to river discharges greater than 670 m^3/s before the 1996 controlled flood, resulted from slab failures of unconsolidated debris-flow deposits that were laterally eroded during the flood rise; most of the debris fan was not overtopped during the flood. These slab failures provided initial motion for many large particles at discharges less than what would normally be required to entrain these particle sizes from a previously reworked debris fan, such as the fan at Palisades Creek. Other cobbles and boulders, particularly the ones embedded with radio transmitters, were entrained from the bed as individual particles.

Our results indicate that the Shield's stresses exerted by the 1996 controlled flood exceeded empirical estimates of the critical Shield's stress at both 127.6 Mile and Lava Falls Rapid. At Lava Falls, the peak dimensionless Shield's stress on the edge of the Prospect Canyon debris fan was approximately 0.3 (Figure 8), an extremely high value for flow over boulders, and nearly an order of magnitude larger than that required to initiate sediment motion in gravel-bed streams. At 127.6 Mile, peak Shield's stresses defined for the mean particle size varied from 0.06 to 0.24, exceeding critical values by at least a factor of 2 (Figure 4).

The timing and magnitude of critical Shield's stresses for the 10 tagged rocks at Lava Falls Rapid are predicted remarkably well by a modified version of the theory of

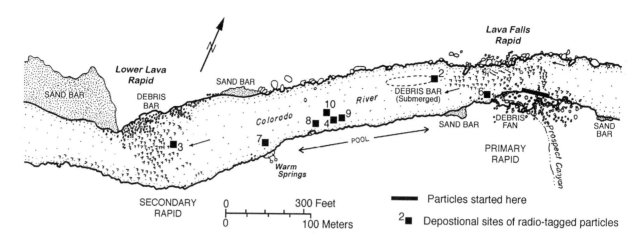

Figure 10. Map showing the depositional sites of 8 radio-tagged cobbles and boulders downstream from Lava Falls Rapids.

TABLE 4. Movement of marked boulders from four debris fans during the 1996 controlled flood

	DEBRIS FAN			
	18 Mile	127.6 Mile	157.6 Mile	160.8 Mile
Length of fan (m)	110	86	96	140
Total number of marked particles	27	83	50	50
Percentage of particles found on fan after flood	89%	55%	24%	60%
Percentage of particles presumed removed from debris fan	11%	45%	76%	40%

Average travel distance (m) of particles on debris fan by b-axis diameter (number of particles)				
φ	mm			
-10	1024	0.1 (1)	0.3 (2)	np	np
-9	512	2.6 (8)	3.9 (16)	8.2 (8)	3.3 (8)
-8	256	2.2 (6)	5.1 (13)	7.8 (4)	3.3 (19)
-7	128	0.1 (1)	np	np	19.5 (1)

np, no particles marked in this size class.

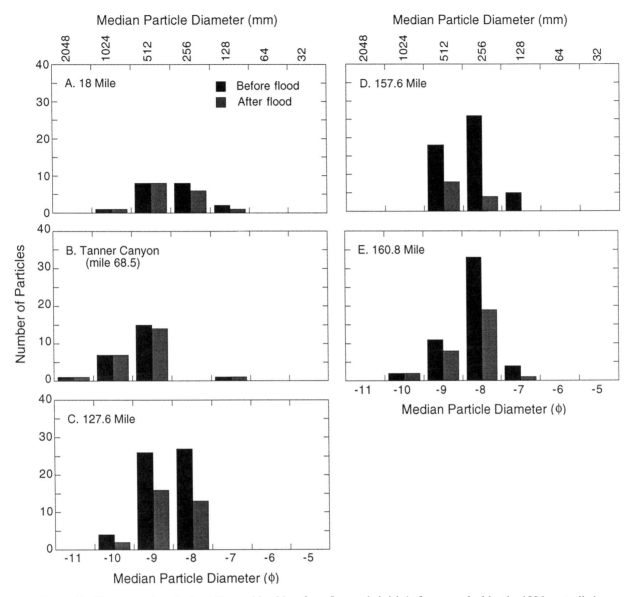

Figure 11. Transport of marked cobbles and boulders from 5 aggraded debris fans reworked by the 1996 controlled flood.

Wiberg and Smith [1987]. Our new method for computing F_d/W_s accounts for variations in particle shape, orientation, and packing, and it also predicts whether initial motion will occur by sliding or rolling. Predicted critical Shield's stresses vary by more than an order of magnitude due to the unique shape, orientation, and topographic setting of each particle. Critical Shield's stresses are not correlated with either absolute or relative particle size, variables that are often used to represent differences in protrusion and pivot angles in poorly sorted sediment mixtures.

Reworking of the Prospect Canyon debris fan ended after approximately 4 hours when large boulders armored the unconsolidated bank, preventing further bank failures. In this case, duration of the flood appeared to be unimportant to reworking, contrary to the predictions of *Hammack and Wohl* [1996] at Warm Springs Rapid on the Yampa River in Colorado. The two mechanisms of coarse-particle entrainment documented in this study — slab failures by bank undercutting as well as individual particle entrainment — have important implications for understanding the mobility and evolution of channel features, such as islands and channel bars downstream from large fans and rapids, during both regulated and unregulated flows in the Colorado River.

We documented movement of cobbles and boulders, some with diameters as large as 2 m, during the 1996 controlled flood. At 3 of 4 debris fans, 40 percent or more of the marked particles were removed from the debris fan (Table 4), and others were transported as far as 20 m to the downstream ends of debris fans. As demonstrated with the radio-tagged particles at Lava Falls Rapid, most of the particles were deposited in the pool downstream from the rapid and not on the alternating debris bars farther downstream (Figure 10). The average travel distance of 230 m demonstrates the effectiveness of discharges as small as 1300 m³/s in rearranging boulders on some debris fans in the Colorado River. These results indicate that large cobbles and small boulders can be readily transported from previously reworked debris fans by floods half the size of the pre-dam annual peak discharge. These size classes are mobile at low discharges (relative to pre-dam conditions) in settings where particle-to-particle interactions and pocket topography are critical in controlling entrainment. One potential drawback of floods of similar size to the 1996 flood is that boulders deposited in the downstream pool might require more sustained, high discharges to be flushed further downstream.

APPENDIX I - EQUATIONS USED TO CALCULATE INCIPIENT MOTION FOR SLIDING WITHOUT ROTATION ON ONE ARBITRARILY ORIENTED PLANAR SURFACE

In this appendix, we present the detailed derivation of the equations used to calculate critical stresses for initial motion by one process: sliding along an arbitrarily oriented planar surface. Other modes of motion require different equations, which cannot be presented in detail here. However, the approach used to derive all of the equations is similar, as is the method of solution, so hopefully the details presented here will help the reader better understand the computations presented in the main body of the paper.

At the threshold of motion, all the forces and moments on a solid sum to zero:

$$\sum \text{Forces on the particle} = 0 \qquad (A1)$$

$$\sum \text{Moments on the particle} = 0 \qquad (A2)$$

where the forces are reactions and frictional forces at contact points, drag and lift forces acting at the center of lift, and the gravitational force acting at the center of mass (Figure 1).

For sliding along a single planar surface, only motion by translation is considered, and therefore only equation (A1) is needed. Furthermore, because only one plane of sliding is involved, reaction forces at individual contact points are all parallel to one another, and therefore it is not necessary to consider these forces separately; they may be added together to obtain the total reaction force vector $\mathbf{R}$ acting normal to the sliding surface. A similar situation holds for the frictional forces at individual contact points: they are all parallel, and may be added together to obtain one frictional force vector, $\mathbf{F_r}$, acting within the sliding plane. Under these conditions, equation (A1) becomes:

$$\mathbf{R} + \mathbf{F_r} + \mathbf{F_D} + \mathbf{F_L} + \mathbf{W_S} = 0 \qquad (A3)$$

where $\mathbf{F_D}$, $\mathbf{F_L}$, and $\mathbf{W_s}$ are the drag, lift, and weight vectors.

The drag, lift, and weight vectors act in the x, y, and -y directions, respectively (Figure 1). As a result, these vectors in equation (A3) can all be written in terms of unit vectors $\mathbf{i}$ and $\mathbf{j}$ in the x and y directions, respectively, and the magnitudes of the vectors themselves:

$$\mathbf{F_D} = F_D\mathbf{i} \qquad (A4)$$

$$\mathbf{F_L} = F_L\mathbf{j} \qquad (A5)$$

$$\mathbf{W_s} = -W_s\mathbf{j} \qquad (A6)$$

The lift force may be eliminated by introducing L_d, the ratio of lift to drag (i.e., equation (6) of the main body of the paper), leading to:

$$\mathbf{F_L} = L_dF_D\mathbf{j} \qquad (A7)$$

It is convenient to introduce unit vectors $\mathbf{n}$ and $\mathbf{q}$. The unit vector $\mathbf{n}$ is normal to the plane of sliding and it 'points' towards the interior of the grain: it is therefore parallel to $\mathbf{R}$. The unit vector $\mathbf{q}$ is tangent to the plane of sliding, and it points in the direction of the frictional force $\mathbf{F_r}$. Thus:

$$\mathbf{R} = R\mathbf{n} \qquad (A8)$$

$$\mathbf{F_r} = F_r\mathbf{q} \qquad (A9)$$

where R is the magnitude of $\mathbf{R}$ and F_r is the magnitude of $\mathbf{F_r}$. Because the orientation of the plane of sliding is known, the unit vector $\mathbf{n}$ is specified, leaving only R to be determined in equation (A8).

At the threshold of sliding, when all possible friction resists motion, the frictional forces may be represented by a Coulomb sliding law (equation (5)):

$$F_r = \mu R \qquad (A10)$$

Equation (A10) allows the magnitudes of the friction vectors to be eliminated from equation (A9):

$$\mathbf{F_r} = \mu R \mathbf{q} \qquad (A11)$$

In order to further clarify which variables are known and which are unknown, the unit vector $\mathbf{q}$ is decomposed into two orthogonal components within the plane of sliding, one parallel to the strike of the plane of sliding and the other parallel to its dip direction:

$$\mathbf{q} = q_r \mathbf{r_d} + q_s \mathbf{s_t} \qquad (A12)$$

where q_r and q_s are components of $\mathbf{q}$ in the directions of the dip and strike respectively, and $\mathbf{r_d}$ and $\mathbf{s_t}$ are unit vectors parallel to the dip and strike directions of the sliding plane. The strike vector $\mathbf{s_t}$ may be in either of two directions; the orientation of $\mathbf{s_t}$ is chosen so $\mathbf{r_d}$, $\mathbf{s_t}$, and $\mathbf{n}$ define a right-handed coordinate system. Note also that because $\mathbf{q}$ is a unit vector, the magnitudes of q_r and q_s are not independent:

$$q_s = \sqrt{1 - q_r^2} \qquad (A13)$$

With the modifications indicated by equations (A4)-(A12), equation (A3) may now be reduced and solved. The following variables are assumed to be known: $\mathbf{i}$, $\mathbf{j}$, $\mathbf{n}$, L_d, W_s, L_d, μ, $\mathbf{r_d}$, and $\mathbf{s_t}$. This leaves 3 unknowns: R, q_r, and F_D. Written in component form, however, the vector equation (A3) represents 3 algebraic equations, and therefore the equations can be solved. The equations are developed as follows:

1. Equations (A4) to (A12) are incorporated into the vector equation (A3).
2. The resulting vector equation is written in components in the $\mathbf{r_d}$, $\mathbf{s_t}$, and $\mathbf{n}$ coordinate system, to create 3 equations in 3 unknowns.
3. Written in $\mathbf{r_d}$, $\mathbf{s_t}$, and $\mathbf{n}$, one equation is readily solved for R, and the result is used to eliminate R from the remaining two equations, leaving the following two equations:

$$(W_s \mathbf{j} \cdot \mathbf{n} - F_D [\mathbf{i} \cdot \mathbf{n} + L_d \mathbf{j} \cdot \mathbf{n}]) \mu q_r$$
$$- W_s \mathbf{j} \cdot \mathbf{r_d} + F_D (\mathbf{i} \cdot \mathbf{r_d} + L_d \mathbf{j} \cdot \mathbf{r_d}) = 0 \qquad (A14)$$

$$(W_s \mathbf{j} \cdot \mathbf{n} - F_D [\mathbf{i} \cdot \mathbf{n} + L_d \mathbf{j} \cdot \mathbf{n}]) \mu \sqrt{1 - q_r^2} + F_D \mathbf{i} \cdot \mathbf{s_t} = 0 \qquad (A15)$$

Equations (A13) and (A14) may be readily solved for F_D and q_r using standard methods. Note that by definition the dot products are all cosines of known angles. For example, the dot product of $\mathbf{i}$ and $\mathbf{n}$ is the cosine of the angle between the x axis (the direction of flow) and the normal to the plane of sliding.

Acknowledgments. We acknowledge the assistance of many individuals who helped make this report possible. We thank Travis McGrath, Holly Metz, and Mia Wise, who were volunteers; Lauren Hammack and John Rihs of the National Park Service; Tim Randle of the Bureau of Reclamation; Robert Ruffner of the University of Minnesota; and Mimi Murov of the U.S. Geological Survey. We especially thank professional river guides Tony Anderson, Steve Bledsoe, Bob Grusy, Matt Herman, Pete Reznick, Rachael Running, Kelly Smith, and Greg Williams for their help with our research.

REFERENCES

Baker, V.R., Paleohydraulic interpretation of Quaternary alluvium near Golden, Colorado, *Quat. Res.*, 4, 94-112, 1974.

Buffington, J.M., W.E. Dietrich, and J.W. Kirchner, Friction angle measurements on a naturally formed gravel streambed: implications for critical boundary shear stress, *Water Res. Res.*, 28, 411-425, 1992.

Buffington, J.M., and D.R. Montgomery, A systematic analysis of eight decades of incipient motion studies, with special reference to gravel-bedded rivers, *Water Res. Res.*, 33, 1993-2039, 1997.

Chang, H.H., *Fluvial processes in river engineering*, John Wiley Sons, New York, 432 pp., 1988.

Hammack, L., and E. Wohl, Debris-fan formation and rapid modification at Warm Springs Rapid, Yampa River, Colorado, *J. Geol.*, 104, 729-740, 1996.

Howard, A., and R. Dolan, Geomorphology of the Colorado River in Grand Canyon, *J. Geol.*, 89, 269-297, 1981.

James, C.S., Prediction of entrainment conditions for nonuniform, noncohesive sediments, *J. Hydraul. Res.*, 28, 25-41, 1990.

Kieffer, S.W., *Hydraulic map of Lava Falls Rapids, Grand Canyon, Arizona*, U.S. Geol. Surv. Misc. Invest. Series, 1998.

Kovacs, A., and G. Parker, A new vectorial bedload formulation and its application to the time evolution of straight river channels, *J. Fluid Mech.*, 267, 153-183, 1994.

Li, Z., and P.D. Komar, Laboratory measurements of pivoting angles for applications to selective entrainment of gravel in a current, *Sediment.*, 33, 413-424, 1986.

Lundgren, H., and I.G. Jonsson, Shear and velocity distribution in shallow channels, *J. Hydraul. Engin.*, 90, 1-21, 1964.

Melis, T.S., R.H. Webb, P. G. Griffiths, and T. J. Wise, *Magnitude and frequency data for historic debris flows in Grand Canyon National Park and vicinity*, U.S. Geol. Surv. Water-Res. Invest. Rept. 94-4214, 285 pp., 1994.

Nelson, J.M., M.W. Schmeeckle, R.L. Shreve, and W.M. Bruner, High-frequency measurements of lift and drag on sediment grains, *EOS* (Fall Meeting Supplement), 1997.

O'Connor, J.E., *Hydrology, hydraulics, and geomorphology of the Bonneville Flood*, Geol. Soc. Amer. Special Paper 274, 83 pp., 1993.

Parker, G., Self-formed straight rivers with equilibrium banks and mobile bed: Pt. 2, The gravel river, *J. Fluid Mech.*, 89, 127-146, 1978.

Schmidt, J.C., and Rubin, D.M., Regulated streamflow, fine-grained deposits, and effective discharge in canyons with abundant debris fans, in *Natural and anthropogenic influences in fluvial geomorphology*, ed. J.E. Costa, A.J. Miller, K.W. Potter, and P.R. Wilcock, pp. 177-195, Amer. Geophys. Union Mono. 89, 1995.

Stevens, L.E., J.C. Schmidt, T.J. Ayers, and B.T. Brown, Flow regulation, geomorphology, and Colorado River marsh development in the Grand Canyon, Arizona, *Ecol. Appl.*, 5, 1035-1039, 1996.

Webb, R.H., T.S. Melis, T.W. Wise, and J.G. Elliott, *"The great cataract," Effects of late Holocene debris flows on Lava Falls Rapid, Grand Canyon National Park, Arizona*, U.S. Geol. Surv. Open-File Rept. 96-460, 96 pp., 1996.

Webb, R.H., T.S. Melis, P.G. Griffiths, and J.G. Elliott, *Reworking of aggraded debris fans by the 1996 controlled flood on the Colorado River in Grand Canyon National Park, Arizona*, U.S. Geol. Surv. Open-File Rept. 97-16, 96 pp., 1997.

Whiting, P.J., and W.E. Dietrich, Boundary shear stress and roughness over mobile alluvial beds, *J. Hydr. Engin.*, 116, 1495-1511, 1990.

Wiberg, P.L., and J.D. Smith, Calculations of the critical shear stress for motion of uniform and heterogeneous sediments, *Water Res. Res.*, 23, 1471-1480, 1987.

Wilcock, P.R., A.F. Barta, C.C. Shea, G.M. Kondolf, W.V. Matthews, and J. Pitlick, Observations of flow and sediment entrainment on a large gravel-bed river, *Water Res. Res.*, 32, 2897-2909, 1996.

Wolman, M.G., A method of sampling coarse riverbed material, *Trans. Amer. Geophys. Union*, 35, 951-956, 1954.

James E. Pizzuto, Department of Geology, University of Delaware, Newark, DE 19716; email: pizzuto@udel.edu

Robert H. Webb and Peter G. Griffiths, U.S. Geological Survey, 1675 W. Anklam Road, Tucson, AZ 85745

John G. Elliott, U.S. Geological Survey, Denver Federal Center, Lakewood, CO 80205

Theodore S. Melis, Grand Canyon Monitoring and Research Center, 2255 N. Gemini Drive, Flagstaff, AZ 86001

Linkage Between Grain-Size Evolution and Sediment Depletion During Colorado River Floods

David J. Topping

U.S. Geological Survey, Denver, Colorado

David M. Rubin

U.S. Geological Survey, Menlo Park, California

Jonathan M. Nelson, Paul J. Kinzel III, and James P. Bennett

U.S. Geological Survey, Denver, Colorado

Suspended-sediment concentrations decrease and suspended sediment and bed sediment coarsen significantly during floods or seasonal flood periods in the Colorado River in Grand Canyon. We present evidence of these processes for the 1996 controlled flood, examine their role in producing the sedimentologic signatures of both modern and historical flood deposits, and suggest how these processes may be exploited to optimize flood releases to achieve management objectives. In general, the processes that cause grain-size evolution in the Colorado River arise because of sediment supply limitation and a mismatch between the timing of tributary sediment supply to the Colorado River and high sediment-transporting events in the mainstem. The system fines immediately following large tributary sediment inputs, and the system coarsens as the bed sediment is winnowed during subsequent mainstem flows.

1. INTRODUCTION

The Colorado River in Grand Canyon falls somewhere in the middle of the spectrum between classical self-formed alluvial rivers and bedrock rivers. Like purely alluvial rivers, the morphology of the bed and bars is somewhat adjustable and is coupled to flow in the river; on the other hand, this reach also displays the effects of sediment-supply limitation and bedrock control. This mixture of behaviors

makes predicting the geometric rearrangement of sediment in the system during floods particularly difficult, because any such prediction must be based on both a detailed understanding of how the flow interacts with the available sediment to produce the observed bed and bar morphology and also on what volume and size classes of sediment are available on the bed for transport by the flow. *Wiele et al.* [this volume] have shown that the response of the channel bed and bar morphology during floods is highly sensitive to the supply of sediment. Moreover, it is difficult and time consuming to acquire the detailed local information about sediment volumes and sizes on the bed that is required to make accurate predictions of the topographic response of the system to high flows. Furthermore, even if that information is known for a particular flow event, the results may have little generality, as the pre-existing sediment volume and size-class conditions in the channel are likely to be different for subsequent high-flow events. Therefore, it would be advantageous to develop a different methodology for determining the state of the sediment supply.

In this paper, we describe the basis for a different methodology for extracting information about the state of the supply of fine sediment in Marble and Grand Canyons. Rather than trying to assess the actual volumes and sizes of sediment present on the bed of the channel, our approach views the problem indirectly. Specifically, by tracking the temporal evolution of the concentration and grain-size distribution of sediment in suspension, it is possible to invert the problem and determine the state of the supply of fine sediment in Marble and Grand Canyons. As a first step in solving this problem, this paper documents the linkage between the grain-size distribution of the bed, the concentration of each size class of suspended sediment in the main channel and in a lateral recirculating eddy, and the grain-size distribution in flood deposits in eddies.

One of the primary strengths of this perspective on sediment-supply limitation is that it allows amalgamation of both modern and historical suspended-sediment data with the observed sedimentology of both pre-dam and recent channel deposits. This combination of seemingly disparate data produces an excellent context for evaluating the behavior of the river with respect to different supplies of sediment. In addition, this view of the sediment-supply problem has implications for river management. For example, this perspective illustrates how future controlled floods released from Glen Canyon Dam [*U.S. Department of the Interior*, 1995] could be designed to compensate for very different supplies of sediment.

2. PRE- AND POST-DAM SEDIMENT SUPPLY

Sediment transport in the Colorado River is characterized by two key features that are similar in both the pre- and post-dam systems in Marble and Grand Canyons. This similarity allows these systems to be linked in the data analysis presented in this paper. First, both pre- and post-dam systems are characterized by annual supply limitation with respect to fine sediment (i.e., sand, silt, and clay), meaning that, over the course of a year, the river has the capacity to transport more fine sediment than is typically supplied to it by its tributaries. Second, both the pre- and post-dam systems display a mismatch between the timing of high sediment supply to the channel and the timing of high sediment-transport events in the channel.

For the pre-dam river, annual supply limitation and the mismatch in timing of sediment supply and mainstem transport are closely related. Before completion of the dam, most of the water that passed through Grand Canyon originated in the mountains of Colorado, Utah, and Wyoming, whereas most of the sediment was supplied by Colorado Plateau tributaries that drained directly into the canyons of the Colorado River downstream of the confluence of the Green and Colorado Rivers [*Smith et al.*, 1960; *Andrews*, 1990, 1991]. High discharges in the Colorado were primarily associated with the annual snowmelt floods that typically occurred from late March through early July. In contrast, the highest sediment inputs to the Colorado River occurred during floods on the Colorado Plateau tributaries during the summer thunderstorm season (July through October) and also during late-winter and early-spring floods on two of the largest tributaries, the San Juan and Little Colorado Rivers. Mainstem discharges during times of high sediment supply were typically low, so some portion of the sediment supplied during this period was stored in the channel and eddies of the Colorado River until the next snowmelt season. Over the course of the snowmelt flood, sediment stored in the river during the previous months gradually became depleted, such that, at the same discharge of water, sediment concentrations decreased over time. This hysteresis in suspended-sediment concentration, which is a typical characteristic of a supply-limited system, was first recognized and investigated at the Grand Canyon gage by *Leopold and Maddock* [1953]. Hysteresis in sediment concentration in the Colorado River has been related both to seasonal scour of the bed [*Leopold and Maddock*, 1953; *Brooks*, 1958; *Burkham*, 1986] and seasonal coarsening of the bed [*Colby*, 1964; *Burkham*, 1986; *Rubin et al.*, 1998].

After completion of Glen Canyon Dam in 1963, a similar situation arose in the Colorado River below the dam, but for somewhat different reasons. Although the dam dramatically reduced the frequency and magnitude of relatively high sediment-transporting flows, it also virtually eliminated the sediment supply to the portion of Glen Canyon below the dam and greatly reduced the supply of fine sediment to Marble and Grand Canyons. Furthermore, the dam introduced wide-ranging daily discharge fluctuations that sometimes exceeded 820 m^3/s in a single day, while greatly diminishing the seasonal discharge variation relative to pre-dam conditions. The dam effectively replaced relatively high and low flows with a greater frequency of moderate flows that range from 300 and 600 m^3/s; these moderate discharges have significant sediment-transporting capabilities [*U.S. Department of the Interior*, 1995]. The net result of these alterations in the hydrograph is that the sediment-transport capacity of the Colorado

River decreased, but by a smaller fraction than the decrease in sediment supply.

In the post-dam system, the majority of sediment supplied to the Colorado River in Marble and Grand Canyons comes from the Paria and Little Colorado rivers. Large inputs of sediment from the Paria River occur during high-discharge, short-duration floods that typically occur in July through October, whereas large inputs of sediment from the Little Colorado River occur during floods that typically occur both in July through October and in January through early April. High mainstem flows in the post-dam system, however, do not necessarily coincide with these tributary sediment-supplying events, because high mainstem flows are produced by the dam operators to generate power and maintain reservoir levels. Therefore, as with pre-dam conditions, there may be still a mismatch in the timing of large sediment inputs to the system relative to the timing of high flows in the mainstem. Thus, during high flows in the post-dam river, we expect suspended-sediment concentrations to decrease with time, and grain sizes, in both suspension and on the bed, to coarsen with time in response to depletion of fine sediment in the system [*Rubin et al.*, 1998], just as during pre-dam snowmelt floods.

During the 1996 controlled flood [*Schmidt et al.*, this volume], as part of a larger USGS data collection effort, we set out to improve our understanding of the intricate linkage between flow, sediment supply, sediment transport, and bed morphology. To accomplish this goal, we focused on two major objectives. First, we monitored bathymetry during the flood at both a main channel location and in a typical lateral separation eddy to monitor the effect of the flood on the bed morphology. Second, and the subject of this paper, we documented the linkages between the grain-size distribution on the bed of the channel, the concentration and grain-size distribution of suspended sediment, and the grain-size distribution of deposits produced in depositional regions (primarily in eddies). In doing so, we focused on measuring both concentrations and grain-size distributions of suspended sediment with sufficient temporal resolution to provide insights into the issue of supply limitation.

3. FIELD OBSERVATIONS DURING THE 1996 CONTROLLED FLOOD

To study the linkage between the suspended sediment and bed of the Colorado River during a flood, we designed a 9-day measurement program to monitor the suspended sediment, bed sediment, flow velocity, water-surface slope, and bed topography at both a main channel location and in a typical lateral-recirculation eddy during the 1996 flood experiment. This chapter presents mainly the depth-

integrated suspended-sediment and bed-sediment data that we collected at the 2 sites during the 1996 controlled flood. The main-channel site chosen for this work was the reach at the USGS gage at river mile 87.4. This gage is officially designated as "Colorado River near Grand Canyon, Arizona, station number 09402500" and is referred to herein as the Grand Canyon gage. The Grand Canyon gage reach was chosen both because of its key location in upper Grand Canyon and also because of its wealth of USGS flow, suspended-sediment, bed-sediment, and bed-topographic data that span the period from November 1923 to the present. The eddy site chosen for this work was 56-km downstream of the Grand Canyon gage at river mile 122 at the mouth of Hundred Twenty-two Mile Creek (herein referred to as the 122-Mile eddy site). The 122-Mile eddy site was chosen both because it had a large eddy in which topographic changes in the bed during the flood could be readily detected and also because it had a history of prior sedimentologic investigations.

Other suspended-sediment data presented in this chapter were also collected by the USGS at 2 other gages on the Colorado River. The first of these gages is at river mile 61.0, 42 km upstream of the Grand Canyon gage and at the downstream end of Marble Canyon; this gage is officially designated as the "Colorado River above the mouth of the Little Colorado River near Desert View, Arizona, station number 0938100" and is herein referred to as the above LCR gage. The second of these gages is at river mile 166.1, downstream of the 122-Mile eddy site; this gage is officially designated as "Colorado River above National Canyon near Supai, Arizona, station number 09404120" and is referred to herein as the National Canyon gage.

As measured at the Grand Canyon gage, the controlled-flood experiment consisted of three days of a steady 238 m^3/s discharge followed by an increase over 5.75 hrs to a discharge of 1290 m^3/s. This high discharge was constant for 7 days and then decreased over 3.2 days back to a steady discharge of 238 m^3/s. In this paper, the day prior to the arrival of the flood is referred to as "day -1", the seven days of 1290 m^3/s discharge are referred to as "days 1-7", and the first day of the recessional limb of the flood is referred to as "day +1".

4. METHODS

4.1. Grand Canyon Gage Reach

To monitor the suspended sediment, we collected samples from the USGS cableway at the Grand Canyon gage (Figure 1) using both a P-61 sampler and a bag sampler equipped with a D-77 sampler head (*Edwards and*

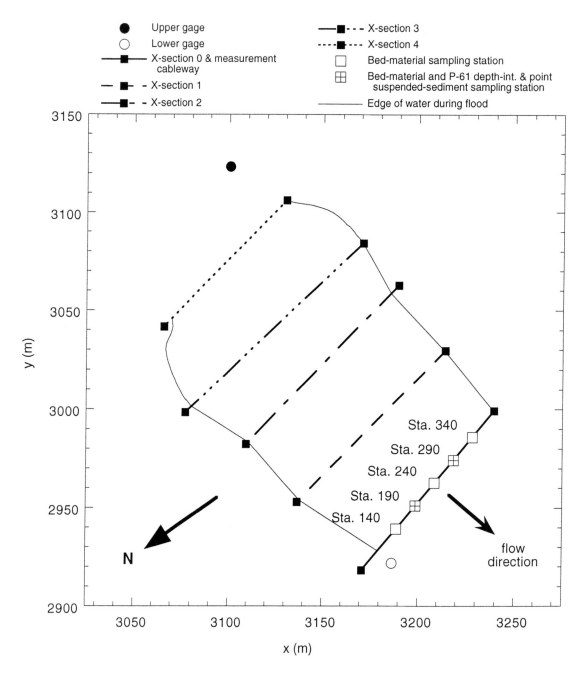

Figure 1. Map of the study area at the Grand Canyon gage showing the locations of the upper and lower gages, the measurement cableway, the locations of the bed-material and P-61 depth-integrated and point suspended-sediment measurement stations on the cableway, and the cross-sections surveyed during the 1996 controlled flood.

Glysson [1988] provide a detailed description of these samplers). The P-61 sampler was deployed at two verticals located 1/3 (Station 290; sampling stations are measured in feet from the right-bank endpoint of the cable) and 2/3 (Station 190) of the distance across the channel in both a depth-integrating and a point-integrating mode. In the

depth-integrating mode, 2 to 4 samples were collected at each vertical on days 1-7 and +1 by opening the P-61 sampler at the bottom and raising it to the surface at a uniform rate. In the point-integrating mode, 3 samples were collected with the P-61 sampler at 6 points in each vertical on days 2, 4, and 7. On days -1, 2, 4, 6 and 7, the D-77 bag sampler was used to collect cross-sectionally integrated suspended-sediment samples using the equal-discharge-increment methodology (described by *Edwards and Glysson* [1988]). When used properly, the D-77 and P-61 suspended-sediment samplers sample different portions of the flow. Therefore, the concentrations and grain-size distributions measured with these 2 types of samplers are systematically different. Because a D-77 samples to no closer than 18 cm of the bed, whereas a P-61 samples to within 12 cm of the bed [*Edwards and Gysson*, 1988], suspended-sand concentrations measured with a D-77 sampler are somewhat lower than those measured with a P-61 sampler, and suspended-sand grain-size distributions measured with a D-77 sampler are somewhat finer than those measured with a P-61 sampler.

To monitor the bed sediment at the Grand Canyon gage, we collected samples of the bed material at 3 to 5 locations under the cableway (Stations 140, 190, 240, 290, and 340) on days -1, 1-3, 5-7, and +1 (Figure 1). With the exception of 1 sample, these bed samples were collected with a BM-54 sampler (one sample on day 5 was collected in the center of the channel with a pipe dredge). Because of mechanical problems with the BM-54 sampler, reliable samples of the bed sediment were not obtained on day 4.

4.2. 122-Mile Eddy

In order to monitor alteration in the shape and volume of the eddy bar, we measured bathymetry on a daily basis in a large lateral-recirculation eddy on river right, immediately downstream from the mouth of Hundred Twenty-two Mile Creek. These topographic measurements were made using a combination of ground surveys and coupled tracking theodolite/echo sounder (Hydro) measurements. In addition, suspended-sediment concentration and grain-size information were collected using a D-74 depth-integrating sampler deployed from a boat at least twice daily at 4 different sites spanning the width of the eddy. Continuous indirect measurements of sediment concentration were obtained at discrete locations in the eddy using optical-backscatter sensors (OBS).

Suspended-sediment concentrations and grain-size distributions at the two sites were determined from the samples using the same methodologies. Concentrations of suspended sediment were determined using standard USGS techniques [see *Guy*, 1969]. Grain-size distributions of the suspended sand in P-61 and D-74 samples were measured at 1/4-ϕ intervals in the USGS Colorado District laboratory using a visual-accumulation tube specifically calibrated for sediment from the Colorado River in Grand Canyon. Typical sediment from this site has a mean Powers index of

3.0 and a mean Corey shape factor of 0.7 [*Topping*, 1997]. Grain-size distributions of the suspended sand in the D-77 samples were measured at 1-ϕ intervals in the sediment laboratory at the USGS Cascade Volcano Observatory using wet sieving. Grain-size distributions of the bed material were measured at 1/2-ϕ intervals using dry sieving.

5. RESULTS

5.1. Grand Canyon Gage Reach

Our observations during the 1996 controlled flood suggest that many sediment-related processes are common to both the pre- and post-dam systems. During the 1996 controlled flood, bed topography at the Grand Canyon gage measurement cableway, grain-size distributions of both the bed and suspended sediment, and suspended-sediment concentrations all evolved in manners somewhat similar to their evolution during pre-dam floods.

Except for one 24-hr period (day 2 to day 3) during the 1996 controlled flood, magnitudes of bed aggradation or degradation at the measurement cableway were in approximate balance with opposing magnitudes of erosion or deposition in 4 cross-sections within 158 m upstream, such that little change in sediment volume occurred in the 158-m-long reach upstream of the cableway. The most erosion in this reach occurred during the first day of the recessional limb of the flood, not during the 7 days of high discharge. Furthermore, less sediment was eroded from the reach during the 7 days of high discharge than during the 3 weeks prior to the flood. Finally, during the 1996 controlled flood, as in the pre-dam system [*Leopold and Maddock*, 1953; *Colby*, 1964; *Howard and Dolan*, 1981; *Burkham*, 1986], the bed at the measurement cableway aggraded during the rising limb of the flood and then scoured during the latter part of the flood. Most of this scour occurred during the recessional limb of the 1996 flood.

During the 1996 controlled flood, bed grain-size distributions measured at the Grand Canyon gage were similar to those measured near the peak of the only pre-dam snowmelt flood for which bed sediment data are available (i.e., 1956 snowmelt flood), and as during the 1956 snowmelt flood, the grain-size distribution of the bed sediment coarsened (Figure 2). Measured bed-sediment grain-size distributions (Figure 2) during the 1996 controlled flood were coarser than the grain-size distribution of sediment supplied during tributary (e.g., Paria River) floods, similar to that measured near the peak discharge of the pre-dam 1956 snowmelt flood, coarser than that measured during the rising limb of the pre-dam 1956 snowmelt flood, and finer than that measured near the

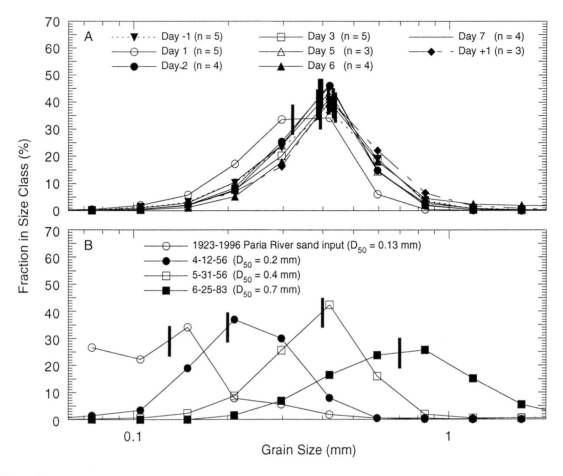

Figure 2. (A) Daily mean grain-size distributions of the bed sediment at the Grand Canyon gage during the 1996 controlled flood. During the rising limb of the flood, the median size of the bed sediment fined; during the 7 days of high discharge, the median size of the bed sediment coarsened, mainly during the first 2 days of high discharge. The number of samples in each spatially averaged measurement is indicated by n in the legend; the thick vertical lines indicate the median grain sizes for each day. (B) Grain-size distributions of: the calculated 1923-1996 Paria River input of sand to the Colorado River [after *Topping*, 1997]; the bed sediment at the Grand Canyon gage on 4-12-56, during the rising limb of the 1956 snowmelt flood [*U.S. Geological Survey*, 1961]; the bed sediment at the Grand Canyon gage on 5-31-56, at near-peak discharge of the 1956 snowmelt flood [*U.S. Geological Survey*, 1961]; and, the bed sediment at the Grand Canyon gage on 6-25-83, at near-peak discharge of the 1983 flood [*Garrett and others*, 1993]. The thick vertical lines indicate the median grain sizes in each distribution. The grain-size distributions of the 1956 and 1983 samples were calculated at 1/2-ϕ increments by weighted cubic-spline interpolation so that they could be more easily compared to the grain-size distributions in (A).

peak discharge of 2720 m³/s during the 1983 flood. Prior to coarsening during the 1996 controlled flood, the bed under the measurement cableway at the Grand Canyon gage first fined as it aggraded with the initial increase in water-surface stage (from day -1 to day 1 in Figure 3). The sources of this fine sand were presumably the upstream channel bed, bars, and banks that eroded during the rising limb of the flood. Following this initial aggradation and fining, the bed under the measurement cableway at the

Grand Canyon gage coarsened (Figures 2 and 3) prior to any significant scour of the bed.

Bed coarsening at the Grand Canyon gage during the 1996 controlled flood was accompanied by both a coarsening of the suspended sand and a decrease in suspended-sand concentration (Figures 3 and 4, Tables 1 and 2). Analyses of pre-dam sediment data at the Grand Canyon gage indicate that, as during the 1996 controlled flood, decreasing suspended-sand concentrations during the annual snowmelt flood were also associated with coarsening of both the suspended sand and the bed sediment. Coarsening of the suspended sand at the Grand Canyon gage during the 1996 flood generally occurred by both a decrease in the concentration of the finest fractions and an increase in the concentration of the coarsest sizes. This style of concentration change associated with coarsening of the suspended sand occurred at sites located along at least 170 km of the river, and was observed at the above LCR gage [42 km upstream from the Grand Canyon gage], the 122-Mile eddy site [56 km downstream from the Grand Canyon gage], and the National Canyon gage [127 km downstream from the Grand Canyon gage] (Figures 3 and 4, Table 2).

5.2. 122-Mile Eddy

During the 1996 controlled flood, the eddy bar at the 122-Mile eddy site changed dramatically. Initially, a significant amount of erosion occurred near the upstream end of the eddy on day 1, followed by over 4 m of deposition near the middle and downstream end of the eddy during the following 6 days. This pattern of erosion and deposition within the eddy was similar to the pattern of erosion and deposition in other eddies during the flood [Andrews et al., this volume; Hazel et al., this volume; Schmidt et al., this volume; Wiele et al., this volume], and occurred primarily in response to enlargement of the eddy both in width and length. At high flow, the reattachment point of the flow moved downstream such that the length of the eddy almost doubled relative to the lower flow preceding the flood. Data collected previously at this site [Schmidt and Graf, 1990] suggest that this style of geometric response to the increase in stage during the 1996 controlled flood was the typical geometric response of the 122-Mile eddy during previous floods. In Figure 5, photographs of the eddy are shown before, during, and immediately after the 1996 flood. Before the flood, the existing eddy deposit consisted of a relatively flat platform about 70 cm above the 238 m^3/s water surface. Our bathymetric measurements show that much of this deposit eroded on the first day of the flood in response to the change in eddy geometry. When the flood receded and the bar was exposed at 238 m^3/s on day +1 (Figure 5c), much of the former upstream extent of the pre-flood bar had been eroded, but there had been a great deal of deposition farther downstream.

Comparison of the volume of this bar on day -1 and day +1 shows that sand was redistributed from lower to higher elevations, but that the actual increase in sediment stored in the eddy was modest. The volume of the eddy deposit above the 142 m^3/s water-surface elevation increased by only 38% during the 1996 controlled flood, while the

volume of the eddy deposit above the 566 m^3/s water-surface elevation increased by 106% [Hazel et al., 1997]. Net erosion on the channelward and upstream end of the pre-flood bar was nearly balanced by the very large amount of deposition at the middle and downstream end of the same eddy. This observed style of geometric rearrangement of eddy deposits is in agreement with that observed at 34 other sites in Marble and Grand Canyons during the 1996 controlled flood [Hazel et al., 1997]. As shown in Figure 5d, the channelward margin of the flood deposit evolved into a steep cutbank during the first day of the recessional limb of the flood, and a significant portion of this deposit was eroded and redeposited in the main channel during the low flows following the flood. However, even after the erosive effects of the low flows of days +1 through +3, thick flood deposits remained for a period of many months [Kaplinski et al., 1998].

As in the main channel at the Grand Canyon gage, the suspended-sand concentration in the eddy decreased during the controlled flood, and the grain sizes in suspension coarsened significantly, as expected from the supply-limitation effect discussed above. The decrease in concentration was corroborated by the continuous OBS measurements, which displayed a near monotonic decay between the first and last days of the flood event. In Figures 3 and 4, the spatially averaged concentrations of the suspended sediment sampled in the eddy are shown along with the same quantities from the main channel sites upstream at the Grand Canyon gage and downstream at the National Canyon gage. It is surprising that the suspended sediment at the 2 main-channel sites and the 122-Mile eddy site evolved so similarly during the 7 days of high discharge, given the respective 56- and 71-km separations between these two main channel sites and the eddy site, and the distinct difference between main-channel and eddy environments. Furthermore, coarsening of the suspended sand associated with decreasing suspended-sand concentration was also observed at the above LCR gage, 98 km upstream from the eddy site (Figure 3).

These observations suggest that the response we observed in the suspended sediment was representative of the reach in Marble and Grand Canyons, and that the longitudinal variations in suspended-sediment concentration were relatively small throughout at least the 127-km-long reach from the Grand Canyon gage to the National Canyon gage. In contrast to changes in eddy-bar volume — that exhibited irregular variability both from site to site [Hazel et al., this volume] and from day to day at a given site [Andrews et al., this volume] during the 1996 controlled flood — suspended-sediment concentration and grain size showed little longitudinal variability and showed relatively

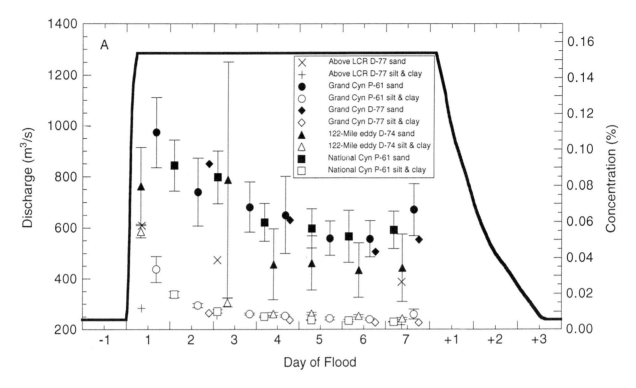

Figure 3. (A) Hydrograph of the 1996 controlled flood (as measured at the Grand Canyon gage) and spatially averaged, depth-integrated sand and silt & clay concentrations measured at the Grand Canyon gage, 122-mile eddy site, above LCR gage, and the National Canyon gage. Error bars are one standard deviation. The travel-time of the flood between the various main-channel and eddy sites has been removed in this figure such that the beginning of day one at each site corresponds to the time of the first arrival of the flood wave at each site. Mean values and error bars for the 122-mile eddy data shown in this plot, and in Figure 4b, supercede those in *Rubin and others* [1998]. The above LCR gage and National Canyon gage data are from *Konieczki and others* [1997]. (B) Hydrograph of the 1996 controlled flood (as measured at the Grand Canyon gage); the spatially averaged, depth-integrated median grain sizes of suspended sand measured at the Grand Canyon gage, 122-mile eddy site, above LCR gage, and the National Canyon gage; and the spatially averaged, median grain sizes of the sand, silt, and clay on the bed measured at the Grand Canyon gage. The samples collected with the D-77 samplers are finer than those collected with the P-61 samplers for the reason described in the text.

smooth trends through time. This reduced spatial and temporal variability suggests that tracking the temporal evolution of the concentration and grain-size distribution of sediment in suspension could be an effective tool for monitoring the state of the sediment supply in Marble and Grand Canyons.

5.3. Mechanics of Suspended-Sediment Concentration Decrease and Coarsening

Rubin et al. [1998] interpreted the decrease in suspended-sand concentration and the associated coarsening of the suspended sand during the 1996 controlled flood to result from winnowing of the finest sizes of sand from the channel bed and preferential deposition of these

finest sizes of sand in eddies. In their interpretation, as the finest sizes of sand were winnowed from the bed, the bed coarsened, which caused the median size of sediment supplied to deposits in the eddies to coarsen. On the reach scale in any river, because of their lower settling velocities, the finer sizes of suspended sediment travel downstream at progressively higher velocities than the coarser grain sizes. Winnowing of the finest grain sizes from the bed occurs during mainstem floods in a system like the Colorado River in Marble and Grand Canyons because of both the upstream supply limitation of sediment and the mismatch in the timing of maximum tributary sediment supply and mainstem sediment transport. Because there is a limited upstream supply of sediment during each mainstem flood event, and the finer grain sizes travel downstream at

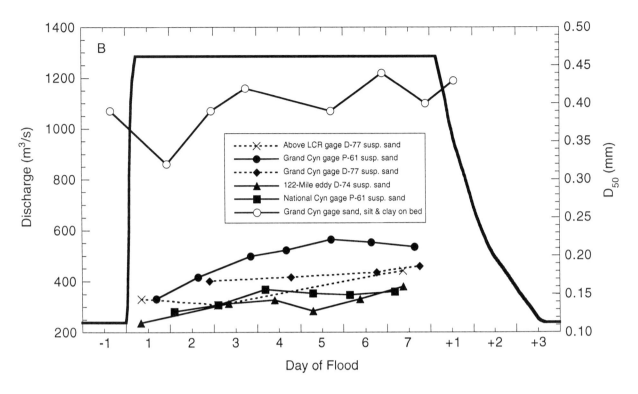

Figure 3 (continued)

progressively higher velocities (and are deposited first in eddies), the concentration of the finest grain sizes will change the fastest during mainstem flood events. Furthermore, as the concentration of the finer grain sizes in suspension decreases below the concentration of these sizes that can be supported by the grain-size distribution of the bed sediment, a mass transfer of these sizes from the bed to the suspended sediment will occur (i.e., the finest sizes will be winnowed from the bed), and the bed will coarsen.

The physical coupling between the coarsening of the bed and the suspended sediment in the channel can be explained in the following manner. For a sandy bed, the overall equilibrium concentration of suspended sediment near the bed scales approximately with the shear stress in excess of the critical value for initiation of motion for the median size comprising the bed, whereas the equilibrium concentration of each size fraction scales with the fraction of the bed composed of that size class [e.g., *Smith and McLean,* 1977; *McLean,* 1992; *Topping,* 1997]. Thus, the near-bed total concentration of suspended sediment will decrease as the median grain size of the bed increases, but, as observed in both the suspended sediment and in the deposits of the 1996 controlled flood, the concentration of the coarser fractions will increase as a result of higher representation in the bed

material. The mathematical representation of this hypothesized physical balance is:

$$(c_m)_{za} = A_s\, i_m\, c_b\, \gamma\, [(\tau_{sf} - \tau_{cr})/\tau_{cr}] \qquad (1)$$

where $(c_m)_{za}$ = the near-bed, time-averaged concentration of suspended sediment in size-class m, A_s = the fractional area of the bed that is covered by fine sediment (i.e., sand, silt, and clay), i_m = the volume fraction of sediment size-class m in that portion of the bed covered by fine sediment, $c_b = 0.65$ is the volumetric concentration of fine sediment in that portion of the bed covered by fine sediment, $\gamma = 0.0045$ [*P. Wiberg,* University of Virginia, pers. comm, 1989], τ_{sf} is the skin-friction component of the boundary shear stress, and τ_{cr} is the critical shear stress of the median size (D_{50}) of the fine sediment on the bed. *Topping* [1997] showed that equation 1, used in combination with the suspended-sediment theory of *Smith and McLean* [1977], is a good predictor of both the measured depth-integrated concentration and grain-size distribution of suspended sediment in the flume experiments of *Kennedy* [1961] and *Guy et al.* [1966], and is also a good predictor of both the measured near-bed concentration and grain-size distribution of suspended sediment in the Rio Puerco data of *Nordin* [1963].

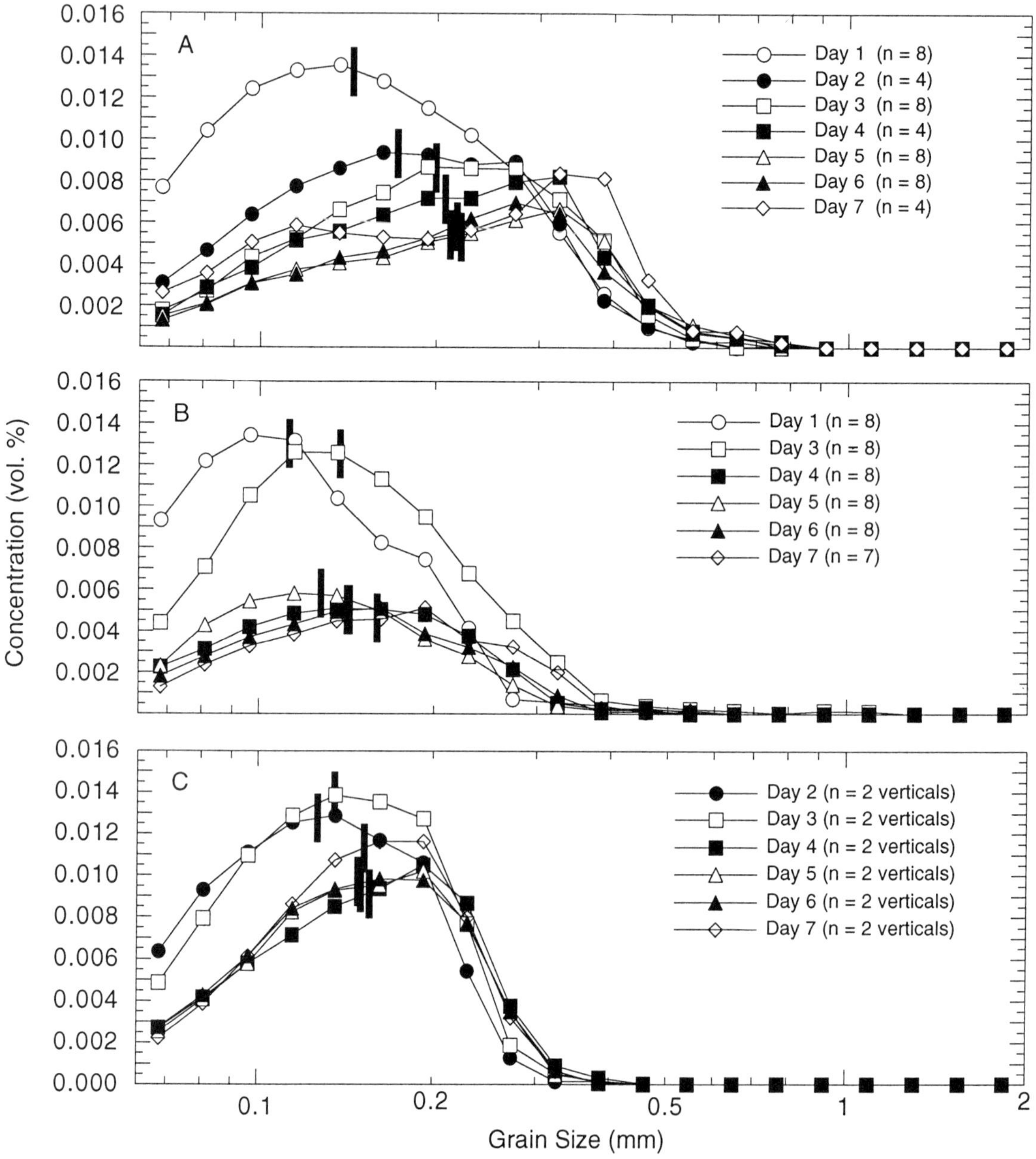

Figure 4. (A) Daily mean concentrations by size class of the suspended sand depicted in Figure 3a measured with the P-61 sampler at the Grand Canyon gage. The number of samples in each spatially averaged measurement is indicated by n in the legend. The thick vertical lines indicate the median grain sizes for each day; during the 7 days of high discharge, the median size of the suspended sand increased from 0.14 to 0.21 mm. (B) Daily mean concentrations by size class of the suspended sand depicted in Figure 3a measured with the D-74 sampler at the 122-mile eddy site. During the 7 days of high discharge, the median size of the suspended sand in the eddy increased from 0.11 to 0.16 mm. (C) Daily mean concentrations by size class of the suspended sand measured with the P-61 sampler at stations 245 and 325 at the National Canyon gage; see Table 2 for a description of the data analysis. The number of verticals in each spatially averaged measurement is indicated by n in the legend. During the days 2 through 7 of the 7 days of high discharge, the median size of the suspended sand increased from 0.13 to 0.16 mm; samples were not collected on day 1 of the flood at the National Canyon gage.

TABLE 1. Bed-sediment data from the Grand Canyon gage during the 1996 controlled flood

DAY		DAY -1 (3-26-96)		DAY 1 (3-27-96)		DAY 2 (3-28-96)		DAY 3 (3-29-96)		DAY 5 (3-31-96)		DAY 6 (4-1-96)		DAY 7 (4-2-96)		DAY +1 (4-3-96)	
SAMPLER TYPE & NUMBER OF SAMPLES		BM-54 (n = 5)		BM-54 (n = 5)		BM-54 (n = 4)		BM-54 (n = 5)		BM-54 (n = 2) PIPE DREDGE (n = 1)		BM-54 (n = 4)		BM-54 (n = 4)		BM-54 (n = 3)	
D50 (mm)		0.39		0.32		0.39		0.42		0.39		0.44		0.40		0.43	
SIZE CLASS mm	φ	MEAN FRACTION IN SIZE CLASS (%)	σ FRACTION IN SIZE CLASS (%)	MEAN FRACTION IN SIZE CLASS (%)	σ FRACTION IN SIZE CLASS (%)	MEAN FRACTION IN SIZE CLASS (%)	σ FRACTION IN SIZE CLASS (%)	MEAN FRACTION IN SIZE CLASS (%)	σ FRACTION IN SIZE CLASS (%)	MEAN FRACTION IN SIZE CLASS (%)	σ FRACTION IN SIZE CLASS (%)	MEAN FRACTION IN SIZE CLASS (%)	σ FRACTION IN SIZE CLASS (%)	MEAN FRACTION IN SIZE CLASS (%)	σ FRACTION IN SIZE CLASS (%)	MEAN FRACTION IN SIZE CLASS (%)	σ FRACTION IN SIZE CLASS (%)
>0.0625 mm	>4.0	0.11	0.10	0.14	0.07	0.05	0.02	0.06	0.03	0.07	0.06	0.04	0.03	0.03	0.02	0.09	0.09
0.0625 to 0.088	4.0 to 3.5	0.24	0.36	0.44	0.25	0.10	0.08	0.11	0.07	0.28	0.38	0.08	0.05	0.11	0.06	0.12	0.09
0.088 to 0.125	3.5 to 3.0	1.42	2.12	2.01	1.09	0.54	0.36	0.56	0.41	1.07	0.98	0.42	0.25	0.66	0.39	0.59	0.38
0.125 to 0.177	3.0 to 2.5	3.21	3.96	5.81	2.40	2.08	1.22	1.95	0.77	2.85	2.08	1.26	0.69	2.00	0.96	2.20	1.43
0.177 to 0.25	2.5 to 2.0	10.24	8.96	17.19	4.67	8.46	2.74	7.69	1.67	10.62	6.40	5.26	2.45	7.39	2.25	7.04	4.30
0.25 to 0.35	2.0 to 1.5	23.44	12.63	33.53	4.98	25.41	5.10	20.17	2.72	24.08	6.12	17.97	7.76	24.17	6.89	16.22	9.34
0.35 to 0.50	1.5 to 1.0	36.99	10.62	34.15	7.51	46.03	4.64	41.42	8.26	43.56	9.33	40.13	15.88	46.74	4.70	41.09	11.71
0.50 to 0.71	1.0 to 0.5	18.92	15.91	6.02	3.56	14.88	7.46	19.08	6.55	14.48	7.00	18.21	4.47	14.47	6.62	22.15	15.27
0.71 to 1.00	0.5 to 0.0	3.79	4.12	0.50	0.41	1.85	2.22	3.30	2.22	2.12	1.36	4.53	4.17	2.73	2.24	6.54	7.58
1.00 to 1.41	0.0 to -0.5	0.95	1.19	0.12	0.12	0.35	0.58	0.75	0.50	0.39	0.21	2.36	3.78	0.85	0.72	1.67	1.77
1.41 to 2.00	-0.5 to -1.0	0.35	0.47	0.04	0.03	0.18	0.34	0.22	0.17	0.21	0.13	1.85	3.49	0.39	0.32	0.81	0.71
2.00 to 2.8	-1.0 to -1.5	0.18	0.29	0.03	0.05	0.03	0.03	0.11	0.10	0.11	0.02	1.62	3.04	0.18	0.13	0.16	0.17
2.8 to 4.0	-1.5 to -2.0	0.12	0.14	0.01	0.02	0.03	0.06	0.12	0.12	0.10	0.10	1.39	2.76	0.14	0.12	0.12	0.16
4.0 to 5.6	-2.0 to -2.5	0.02	0.04	0.01	0.02	0	0	0.11	0.19	0.03	0.05	1.23	2.43	0.12	0.09	0.18	0.20
5.6 to 8.0	-2.5 to -3.0	0.02	0.04	0	0	0	0	0.38	0.85	0.06	0.10	0.95	1.90	0.05	0.09	0.52	0.45
8.0 to 11.3	-3.0 to -3.5	0	0	0	0	0	0	0.25	0.55	0	0	0.64	1.27	0	0	0.33	0.57
11.3 to 16.0	-3.5 to -4.0	0	0	0	0	0	0	0.66	1.48	0	0	2.06	4.11	0	0	0.17	0.29
16.0 to 22.6	-4.0 to -4.5	0	0	0	0	0	0	3.08	6.88	0	0	0	0	0	0	0	0
22.6 to 32.0	-4.5 to -5.0	0	0	0	0	0	0	0	0	0	0	0	0	0	0	0	0

TABLE 2. Suspended-sediment data from the 1996 controlled flood

LOCATION		ABOVE LCR GAGE	GRAND CANYON GAGE			DAY 1 (3-27-96) 122-MILE EDDY		NATIONAL CANYON GAGE			
		D-77*	P-61 (n = 8)		D-77*	D-74 (n = 8)		P-61 (n = 0)§		P-61 SUBSET† (n = 0)§	
SAMPLER TYPE & NUMBER OF SAMPLES (n)		CONC. (vol. %)	MEAN CONC. (vol. %)	σ CONC. (vol. %)	CONC. (vol. %)	MEAN CONC. (vol. %)	σ CONC. (vol. %)	MEAN CONC. (vol. %)	σ CONC. (vol. %)	MEAN CONC. (vol. %)	σ CONC. (vol. %)
SUSPENDED-SAND D$_{50}$ (mm)		0.143	0.144		---	0.112		---		------	
SILT & CLAY		0.012	0.034	7.3e-3	---	0.055	3.4e-3	---	---	---	---
TOTAL SAND (0.0625-2.0 mm)		0.058	0.11	0.020	---	0.080	0.021	---	---	---	---
SAND SIZE CLASS mm	φ										
0.0625 to 0.074	4.00 to 3.75		7.7e-3	1.8e-3		9.3e-3	1.5e-3	---	---	---	---
0.074 to 0.088	3.75 to 3.50		0.010	1.9e-3		0.012	2.3e-3	---	---	---	---
0.088 to 0.105	3.50 to 3.25	0.024	0.012	1.6e-3		0.013	3.3e-3	---	---	---	---
0.105 to 0.125	3.25 to 3.00		0.013	2.2e-3	---	0.013	4.0e-3	---	---	---	---
0.125 to 0.149	3.00 to 2.75		0.014	2.2e-3		0.010	3.0e-3	---	---	---	---
0.149 to 0.177	2.75 to 2.50		0.013	2.6e-3		8.2e-3	2.3e-3	---	---	---	---
0.177 to 0.210	2.50 to 2.25	0.023	0.011	3.3e-3		7.4e-3	2.9e-3	---	---	---	---
0.210 to 0.25	2.25 to 2.00		0.010	2.8e-3		4.1e-3	2.5e-3	---	---	---	---
0.25 to 0.30	2.00 to 1.75		8.4e-3	2.4e-3	---	6.9e-4	4.2e-4	---	---	---	---
0.30 to 0.35	1.75 to 1.50		5.5e-3	3.6e-3		5.4e-4	3.0e-4	---	---	---	---
0.35 to 0.42	1.50 to 1.25	9.8e-3	2.6e-3	1.7e-3		4.1e-4	5.5e-4	---	---	---	---
0.42 to 0.50	1.25 to 1.00		9.5e-4	6.0e-4	---	7.6e-5	1.2e-4	---	---	---	---
0.50 to 0.59	1.00 to 0.75		3.8e-4	2.3e-4		0	0	---	---	---	---
0.59 to 0.71	0.75 to 0.50		2.1e-6	5.8e-6		0	0	---	---	---	---
0.71 to 0.84	0.50 to 0.25	7.0e-4	0	0		0	0	---	---	---	---
0.84 to 1.00	0.25 to 0.00		0	0		0	0	---	---	---	---
1.00 to 1.19	0.00 to -0.25		0	0		0	0	---	---	---	---
1.19 to 1.41	-0.25 to -0.50		0	0		0	0	---	---	---	---
1.41 to 1.68	-0.50 to -0.75		0	0		0	0	---	---	---	---
1.68 to 2.00	-0.75 to -1.00	0	0	0	---	0	0	---	---	---	---

TABLE 2. Suspended-sediment data from the 1996 controlled flood—continued

		ABOVE LCR GAGE	GRAND CANYON GAGE			122-MILE EDDY		NATIONAL CANYON GAGE			
DAY 2 (3-28-96)											
LOCATION		D-77*	P-61 (n = 4)		D-77*	D-74 (n = 0)		P-61 (n = 10)		P-61 SUBSET† (n = 2)	
SAMPLER TYPE & NUMBER OF SAMPLES (n)		---				---					
SUSPENDED-SAND D_{50} (mm)		---	0.172		0.167	---		---		0.127	
		CONC. (vol. %)	MEAN CONC. (vol. %)	σ CONC. (vol. %)	CONC. (vol. %)	MEAN CONC. (vol. %)	σ CONC. (vol. %)	MEAN CONC. (vol. %)	σ CONC. (vol. %)	MEAN CONC. (vol. %)	σ CONC. (vol. %)
SILT & CLAY			0.013	1.2e-3	9.1e-3			0.020	1.9e-3	0.027	1.9e-3
TOTAL SAND (0.0625-2.0 mm)			0.077	0.019	0.092			0.091	0.014	0.082	3.9e-3
SAND SIZE CLASS mm	φ										
0.0625 to 0.074	4.00 to 3.75		3.1e-3	4.1e-4		---	---	---	---	6.4e-3	8.9e-4
0.074 to 0.088	3.75 to 3.50		4.6e-3	4.6e-4		---	---	---	---	9.3e-3	8.9e-4
0.088 to 0.105	3.50 to 3.25		6.4e-3	6.8e-4		---	---	---	---	0.011	3.6e-4
0.105 to 0.125	3.25 to 3.00	---	7.7e-3	1.0e-3	0.025	---	---	---	---	0.013	2.9e-4
0.125 to 0.149	3.00 to 2.75		8.6e-3	1.6e-3		---	---	---	---	0.013	8.3e-4
0.149 to 0.177	2.75 to 2.50		9.4e-3	2.6e-3		---	---	---	---	0.012	2.6e-4
0.177 to 0.210	2.50 to 2.25		9.2e-3	2.5e-3		---	---	---	---	0.011	1.4e-4
0.210 to 0.25	2.25 to 2.00	---	8.8e-3	2.3e-3	0.050	---	---	---	---	5.4e-3	1.9e-4
0.25 to 0.30	2.00 to 1.75		8.9e-3	2.7e-3		---	---	---	---	1.3e-3	5.8e-5
0.30 to 0.35	1.75 to 1.50		5.9e-3	2.8e-3		---	---	---	---	1.6e-4	1.6e-4
0.35 to 0.42	1.50 to 1.25		2.3e-3	1.7e-3		---	---	---	---	1.6e-4	2.9e-5
0.42 to 0.50	1.25 to 1.00	---	1.1e-3	7.9e-4	0.016	---	---	---	---	3.0e-5	4.3e-5
0.50 to 0.59	1.00 to 0.75		2.7e-4	1.1e-4		---	---	---	---	0	0
0.59 to 0.71	0.75 to 0.50		3.0e-4	3.6e-4		---	---	---	---	0	0
0.71 to 0.84	0.50 to 0.25		0	0		---	---	---	---	0	0
0.84 to 1.00	0.25 to 0.00	---	0	0	1.0e-3	---	---	---	---	0	0
1.00 to 1.19	0.00 to -0.25		0	0		---	---	---	---	0	0
1.19 to 1.41	-0.25 to -0.50		0	0		---	---	---	---	0	0
1.41 to 1.68	-0.50 to -0.75		0	0		---	---	---	---	0	0
1.68 to 2.00	-0.75 to -1.00	---	0	0	0	---	---	---	---	0	0

TABLE 2. Suspended-sediment data from the 1996 controlled flood—continued

SAND SIZE CLASS mm	SAND SIZE CLASS φ	ABOVE LCR GAGE D-77* CONC. (vol. %)	GRAND CANYON GAGE P-61 (n=8) MEAN CONC. (vol. %)	GRAND CANYON GAGE P-61 (n=8) σ CONC. (vol. %)	GRAND CANYON GAGE D-77* CONC. (vol. %)	DAY 3 (3-29-96) 122-MILE EDDY D-74 (n=8) MEAN CONC. (vol. %)	DAY 3 (3-29-96) 122-MILE EDDY D-74 (n=8) σ CONC. (vol. %)	NATIONAL CANYON GAGE P-61 (n=12) MEAN CONC. (vol. %)	NATIONAL CANYON GAGE P-61 (n=12) σ CONC. (vol. %)	NATIONAL CANYON GAGE P-61 SUBSET† (n=2) MEAN CONC. (vol. %)	NATIONAL CANYON GAGE P-61 SUBSET† (n=2) σ CONC. (vol. %)
AMPLER TYPE & NUMBER OF SAMPLES (n)		D-77*	P-61 (n=8)		D-77*	D-74 (n=8)		P-61 (n=12)		P-61 SUBSET† (n=2)	
SUSPENDED-SAND D_{50} (mm)		0.136	0.200		---	0.138		---		0.136	
SILT & CLAY		8.0e-3	8.6e-3	5.0e-4		0.015	2.5e-3	0.010	9.9e-4	0.016	2.0e-3
TOTAL SAND (0.0625-2.0 mm)		0.039	0.068	0.014		0.083	0.066	0.084	0.015	0.087	0.022
0.0625 to 0.074	4.00 to 3.75		1.8e-3	3.1e-4		4.4e-3	1.9e-3	---	---	4.9e-3	3.1e-4
0.074 to 0.088	3.75 to 3.50		2.7e-3	5.9e-4		7.1e-3	4.0e-3	---	---	7.9e-3	1.6e-3
0.088 to 0.105	3.50 to 3.25	0.018	4.3e-3	7.3e-4		0.011	8.2e-3	---	---	0.011	3.0e-3
0.105 to 0.125	3.25 to 3.00		5.2e-3	1.2e-3	---	0.013	0.011	---	---	0.013	3.9e-3
0.125 to 0.149	3.00 to 2.75		6.6e-3	1.3e-3		0.013	0.011	---	---	0.014	3.7e-3
0.149 to 0.177	2.75 to 2.50		7.4e-3	1.8e-3		0.011	8.7e-3	---	---	0.014	3.4e-3
0.177 to 0.210	2.50 to 2.25		8.7e-3	2.6e-3		9.5e-3	7.4e-3	---	---	0.013	3.6e-3
0.210 to 0.25	2.25 to 2.00	0.014	8.6e-3	2.3e-3	---	6.7e-3	6.6e-3	---	---	7.7e-3	1.6e-3
0.25 to 0.30	2.00 to 1.75		8.6e-3	2.0e-3		4.5e-3	5.0e-3	---	---	1.9e-3	2.1e-4
0.30 to 0.35	1.75 to 1.50		7.1e-3	1.5e-3		2.5e-3	3.2e-3	---	---	5.2e-4	3.3e-4
0.35 to 0.42	1.50 to 1.25		5.1e-3	2.2e-3		6.4e-4	9.2e-4	---	---	1.2e-4	3.3e-5
0.42 to 0.50	1.25 to 1.00	6.6e-3	1.6e-3	4.4e-4	---	3.8e-4	5.2e-4	---	---	2.0e-5	2.9e-5
0.50 to 0.59	1.00 to 0.75		4.3e-4	3.4e-4		2.4e-4	3.5e-4	---	---	0	0
0.59 to 0.71	0.75 to 0.50		4.4e-5	9.1e-5		1.5e-4	3.4e-4	---	---	0	0
0.71 to 0.84	0.50 to 0.25		8.6e-5	2.4e-4		1.9e-5	5.3e-5	---	---	0	0
0.84 to 1.00	0.25 to 0.00	4.7e-5	0	0	---	1.4e-4	3.9e-4	---	---	0	0
1.00 to 1.19	0.00 to -0.25		0	0		1.0e-4	2.9e-4	---	---	0	0
1.19 to 1.41	-0.25 to -0.50		0	0		0	0	---	---	0	0
1.41 to 1.68	-0.50 to -0.75		0	0		0	0	---	---	0	0
1.68 to 2.00	-0.75 to -1.00	0	0	0	---	0	0	---	---	0	0

TABLE 2. Suspended-sediment data from the 1996 controlled flood—continued

		DAY 4 (3-30-96)									
LOCATION		ABOVE LCR GAGE	GRAND CANYON GAGE			122-MILE EDDY		NATIONAL CANYON GAGE			
AMPLER TYPE & NUMBER OF SAMPLES (n)		D-77*	P-61 (n =4)		D-77*	D-74 (n = 8)		P-61 (n = 13)		P-61 SUBSET† (n = 2)	
SUSPENDED-SAND D_{50} (mm)		---	0.207		0.172	0.142		---		0.156	
		CONC. (vol. %)	MEAN CONC. (vol. %)	σ CONC. (vol. %)	CONC. (vol. %)	MEAN CONC. (vol. %)	σ CONC. (vol. %)	MEAN CONC. (vol. %)	σ CONC. (vol. %)	MEAN CONC. (vol. %)	σ CONC. (vol. %)
SILT & CLAY			7.4e-3	6.1e-4	5.3e-3	8.6e-3	8.9e-4	7.2e-3	4.4e-4	0.012	4.9e-4
TOTAL SAND (0.0625-2.0 mm)			0.064	0.021	0.061	0.036	0.020	0.060	0.011	0.062	6.2e-3
SAND SIZE CLASS mm	φ										
0.0625 to 0.074	4.00 to 3.75	---	1.5e-3	6.3e-4		2.2e-3	8.7e-4	---	---	2.7e-3	2.1e-4
0.074 to 0.088	3.75 to 3.50		2.9e-3	1.0e-3		3.1e-3	6.1e-4	---	---	4.1e-3	5.8e-4
0.088 to 0.105	3.50 to 3.25		3.8e-3	1.3e-3	0.019	4.2e-3	1.2e-3	---	---	5.8e-3	5.8e-4
0.105 to 0.125	3.25 to 3.00	---	5.2e-3	2.4e-3		4.9e-3	1.3e-3	---	---	7.1e-3	6.9e-4
0.125 to 0.149	3.00 to 2.75		5.6e-3	2.4e-3		5.1e-3	1.5e-3	---	---	8.5e-3	3.6e-4
0.149 to 0.177	2.75 to 2.50		6.4e-3	3.2e-3		5.0e-3	3.2e-3	---	---	9.4e-3	4.2e-4
0.177 to 0.210	2.50 to 2.25		7.2e-3	3.3e-3	0.026	4.8e-3	3.9e-3	---	---	0.011	9.0e-4
0.210 to 0.25	2.25 to 2.00	---	7.2e-3	3.2e-3		3.8e-3	4.4e-3	---	---	8.7e-3	1.0e-3
0.25 to 0.30	2.00 to 1.75		7.9e-3	2.8e-3		2.1e-3	3.2e-3	---	---	3.8e-3	6.6e-4
0.30 to 0.35	1.75 to 1.50		8.2e-3	2.2e-3		5.2e-4	6.4e-4	---	---	9.2e-4	5.0e-4
0.35 to 0.42	1.50 to 1.25		4.3e-3	6.1e-4	0.015	2.4e-4	3.3e-4	---	---	3.3e-4	2.6e-4
0.42 to 0.50	1.25 to 1.00	---	2.0e-3	6.1e-4		3.0e-4	3.6e-4	---	---	3.8e-5	4.1e-5
0.50 to 0.59	1.00 to 0.75		8.0e-4	5.9e-4		8.8e-5	2.5e-5	---	---	6.2e-6	8.8e-6
0.59 to 0.71	0.75 to 0.50	---	5.2e-4	3.5e-4		0	0	---	---	3.8e-6	5.4e-6
0.71 to 0.84	0.50 to 0.25		2.9e-4	5.2e-4		0	0	---	---	0	0
0.84 to 1.00	0.25 to 0.00	---	0	0	1.3e-3	0	0	---	---	0	0
1.00 to 1.19	0.00 to -0.25		0	0		0	0	---	---	0	0
1.19 to 1.41	-0.25 to -0.50		0	0		0	0	---	---	0	0
1.41 to 1.68	-0.50 to -0.75		0	0		0	0	---	---	0	0
1.68 to 2.00	-0.75 to -1.00	---	0	0	0	0	0	---	---	0	0

TABLE 2. Suspended-sediment data from the 1996 controlled flood—continued

		ABOVE LCR GAGE	DAY 5 (3-31-96) GRAND CANYON GAGE			122-MILE EDDY		NATIONAL CANYON GAGE			
LOCATION			GRAND CANYON GAGE			122-MILE EDDY		NATIONAL CANYON GAGE			
AMPLER TYPE & NUMBER OF SAMPLES (n)		D-77*	P-61 (n = 8)		D-77*	D-74 (n = 8)		P-61 (n = 13)		P-61 SUBSET† (n = 2)	
SUSPENDED-SAND D50 (mm)		---	0.221		---	0.128		---		0.151	
		CONC. (vol. %)	MEAN CONC. (vol. %)	σ CONC. (vol. %)	CONC. (vol. %)	MEAN CONC. (vol. %)	σ CONC. (vol. %)	MEAN CONC. (vol. %)	σ CONC. (vol. %)	MEAN CONC. (vol. %)	σ CONC. (vol. %)
SILT & CLAY			6.1e-3	7.8e-4		8.9e-3	6.9e-4	5.3e-3	2.1e-4	9.6e-3	2.0e-4
TOTAL SAND (0.0625-2.0 mm)			0.051	9.6e-3		0.037	0.015	0.056	0.011	0.062	5.1e-3
SAND SIZE CLASS mm	φ										
0.0625 to 0.074	4.00 to 3.75		1.5e-3	3.5e-4		2.3e-3	5.9e-4	---	---	2.6e-3	3.1e-4
0.074 to 0.088	3.75 to 3.50		2.1e-3	4.4e-4		4.3e-3	9.7e-4	---	---	4.1e-3	5.0e-4
0.088 to 0.105	3.50 to 3.25	---	3.1e-3	5.9e-4	---	5.4e-3	1.4e-3	---	---	5.8e-3	2.2e-4
0.105 to 0.125	3.25 to 3.00		3.7e-3	6.4e-4		5.8e-3	1.7e-3	---	---	8.3e-3	1.7e-4
0.125 to 0.149	3.00 to 2.75		4.0e-3	5.1e-4		5.7e-3	1.6e-3	---	---	9.3e-3	6.1e-4
0.149 to 0.177	2.75 to 2.50		4.3e-3	4.7e-4		5.0e-3	2.3e-3	---	---	9.6e-3	8.6e-4
0.177 to 0.210	2.50 to 2.25		5.1e-3	8.8e-4		3.6e-3	2.8e-3	---	---	0.010	6.3e-4
0.210 to 0.25	2.25 to 2.00	---	5.5e-3	9.1e-4	---	2.8e-3	2.9e-3	---	---	7.7e-3	5.8e-4
0.25 to 0.30	2.00 to 1.75		6.1e-3	1.1e-3		1.4e-3	2.1e-3	---	---	3.5e-3	8.0e-4
0.30 to 0.35	1.75 to 1.50		6.6e-3	2.3e-3		3.7e-4	3.7e-4	---	---	4.6e-4	3.8e-4
0.35 to 0.42	1.50 to 1.25		5.1e-3	2.3e-3		1.9e-4	1.5e-4	---	---	1.8e-4	3.9e-4
0.42 to 0.50	1.25 to 1.00	---	2.0e-3	1.0e-3	---	2.1e-4	3.6e-4	---	---	1.4e-5	1.2e-5
0.50 to 0.59	1.00 to 0.75		1.1e-3	3.0e-4		1.8e-5	4.0e-5	---	---	0	0
0.59 to 0.71	0.75 to 0.50		5.8e-4	5.5e-4		1.1e-5	3.1e-5	---	---	0	0
0.71 to 0.84	0.50 to 0.25		0	0		5.7e-6	1.6e-5	---	---	0	0
0.84 to 1.00	0.25 to 0.00	---	0	0	---	0	0	---	---	0	0
1.00 to 1.19	0.00 to -0.25		0	0		0	0	---	---	0	0
1.19 to 1.41	-0.25 to -0.50		0	0		0	0	---	---	0	0
1.41 to 1.68	-0.50 to -0.75		0	0		0	0	---	---	0	0
1.68 to 2.00	-0.75 to -1.00	---	0	0	---	0	0	---	---	0	0

TABLE 2. Suspended-sediment data from the 1996 controlled flood—continued

LOCATION	ABOVE LCR GAGE	GRAND CANYON GAGE			DAY 6 (4-1-96) 122-MILE EDDY		NATIONAL CANYON GAGE			
AMPLER TYPE & NUMBER OF SAMPLES (n)	D-77*	P-61 (n = 8)		D-77*	D-74 (n = 8)		P-61 (n = 13)		P-61 SUBSET† (n = 2)	
	CONC. (vol. %)	MEAN CONC. (vol. %)	σ CONC. (vol. %)	CONC. (vol. %)	MEAN CONC. (vol. %)	σ CONC. (vol. %)	MEAN CONC. (vol. %)	σ CONC. (vol. %)	MEAN CONC. (vol. %)	σ CONC. (vol. %)
SUSPENDED-SAND D_{50} (mm)	---	0.217		0.178	0.143		---		0.149	
SILT & CLAY		5.5e-3	6.4e-4	3.8e-3	7.6e-3	6.7e-4	4.9e-3	5.9e-4	9.4e-3	2.5e-3
TOTAL SAND (0.0625-2.0 mm)		0.051	0.010	0.043	0.033	0.015	0.052	0.014	0.063	0.016
SAND SIZE CLASS										
mm φ										
0.0625 to 0.074 4.00 to 3.75		1.3e-3	2.9e-4		1.8e-3	3.3e-4	---	---	2.7e-3	9.4e-4
0.074 to 0.088 3.75 to 3.50		2.1e-3	4.5e-4		2.8e-3	5.6e-4	---	---	4.3e-3	1.7e-3
0.088 to 0.105 3.50 to 3.25		3.1e-3	5.2e-4		3.7e-3	1.0e-3	---	---	6.1e-3	2.4e-3
0.105 to 0.125 3.25 to 3.00	---	3.5e-3	5.1e-4	0.012	4.3e-3	1.5e-3	---	---	8.4e-3	3.6e-3
0.125 to 0.149 3.00 to 2.75		4.3e-3	1.1e-3		4.9e-3	1.7e-3	---	---	9.4e-3	3.3e-3
0.149 to 0.177 2.75 to 2.50		4.6e-3	1.1e-3		5.1e-3	1.7e-3	---	---	9.8e-3	2.2e-3
0.177 to 0.210 2.50 to 2.25		5.2e-3	1.3e-3	0.018	3.9e-3	2.4e-3	---	---	9.8e-3	1.7e-3
0.210 to 0.25 2.25 to 2.00	---	6.2e-3	1.5e-3		3.2e-3	2.7e-3	---	---	7.7e-3	1.1e-4
0.25 to 0.30 2.00 to 1.75		7.0e-3	1.7e-3		2.3e-3	3.1e-3	---	---	3.5e-3	1.5e-4
0.30 to 0.35 1.75 to 1.50		6.4e-3	2.2e-3	0.011	8.8e-4	1.4e-3	---	---	6.8e-4	8.5e-6
0.35 to 0.42 1.50 to 1.25		3.6e-3	2.7e-3		9.5e-5	2.4e-4	---	---	6.7e-5	3.3e-5
0.42 to 0.50 1.25 to 1.00	---	2.0e-3	1.9e-3		1.2e-4	3.2e-4	---	---	2.2e-5	2.4e-5
0.50 to 0.59 1.00 to 0.75		7.1e-4	4.0e-4	1.9e-3	4.7e-5	1.3e-4	---	---	0	0
0.59 to 0.71 0.75 to 0.50		4.8e-4	4.7e-4		0	0	---	---	0	0
0.71 to 0.84 0.50 to 0.25		1.6e-4	3.8e-4		0	0	---	---	0	0
0.84 to 1.00 0.25 to 0.00	---	0	0	0	0	0	---	---	0	0
1.00 to 1.19 0.00 to -0.25		0	0		0	0	---	---	0	0
1.19 to 1.41 -0.25 to -0.50		0	0		0	0	---	---	0	0
1.41 to 1.68 -0.50 to -0.75		0	0		0	0	---	---	0	0
1.68 to 2.00 -0.75 to -1.00	---	0	0	0	0	0	---	---	0	0

TABLE 2. Suspended-sediment data from the 1996 controlled flood—continued

		DAY 7 (4-2-96)									
		ABOVE LCR GAGE	GRAND CANYON GAGE			122-MILE EDDY		NATIONAL CANYON GAGE			
LOCATION											
AMPLER TYPE & NUMBER OF SAMPLES (n)		D-77*	P-61 (n=4)		D-77*	D-74 (n=7)		P-61 (n=13)		P-61 SUBSET† (n=2)	
SUSPENDED-SAND D_{50} (mm)		0.181	0.212		0.186	0.160		---		0.153	
		CONC. (vol. %)	MEAN CONC. (vol. %)	σ CONC. (vol. %)	CONC. (vol. %)	MEAN CONC. (vol. %)	σ CONC. (vol. %)	MEAN CONC. (vol. %)	σ CONC. (vol. %)	MEAN CONC. (vol. %)	σ CONC. (vol. %)
SILT & CLAY		2.6e-3	8.6e-3	2.8e-3	3.8e-3	6.1e-3	6.3e-4	4.2e-3	1.9e-4	7.5e-3	4.3e-4
TOTAL SAND (0.0625-2.0 mm)		0.027	0.067	0.014	0.050	0.034	0.019	0.055	0.010	0.067	5.6e-3
SAND SIZE CLASS mm	φ										
0.0625 to 0.074	4.00 to 3.75		2.6e-3	6.9e-4		1.3e-3	3.3e-4	---	---	2.2e-3	9.6e-5
0.074 to 0.088	3.75 to 3.50		3.6e-3	1.2e-3		2.4e-3	4.8e-4	---	---	3.9e-3	8.7e-5
0.088 to 0.105	3.50 to 3.25	8.3e-3	5.1e-3	1.4e-3	0.013	3.3e-3	8.6e-4	---	---	6.2e-3	1.5e-4
0.105 to 0.125	3.25 to 3.00		5.9e-3	6.6e-4		3.8e-3	1.2e-3	---	---	8.6e-3	4.9e-5
0.125 to 0.149	3.00 to 2.75		5.5e-3	3.6e-4		4.5e-3	1.6e-3	---	---	0.011	8.7e-4
0.149 to 0.177	2.75 to 2.50		5.3e-3	6.5e-4		4.6e-3	1.9e-3	---	---	0.012	2.6e-3
0.177 to 0.210	2.50 to 2.25		5.2e-3	1.2e-3		5.1e-3	3.2e-3	---	---	0.012	2.4e-3
0.210 to 0.25	2.25 to 2.00	9.8e-3	5.6e-3	4.9e-4	0.021	3.6e-3	3.4e-3	---	---	8.0e-3	2.8e-4
0.25 to 0.30	2.00 to 1.75		6.4e-3	1.1e-3		3.2e-3	4.1e-3	---	---	3.2e-3	5.3e-4
0.30 to 0.35	1.75 to 1.50		8.3e-3	3.3e-3		2.0e-3	3.0e-3	---	---	6.5e-4	1.8e-4
0.35 to 0.42	1.50 to 1.25		8.1e-3	5.7e-3		3.0e-4	3.0e-4	---	---	6.6e-5	6.6e-5
0.42 to 0.50	1.25 to 1.00	7.1e-3	3.3e-3	2.3e-3	0.014	2.2e-4	4.6e-4	---	---	1.4e-6	2.0e-6
0.50 to 0.59	1.00 to 0.75		8.2e-4	1.3e-4		1.6e-4	4.2e-4	---	---	3.1e-7	4.4e-7
0.59 to 0.71	0.75 to 0.50		7.6e-4	5.0e-4		0	0	---	---	0	0
0.71 to 0.84	0.50 to 0.25		2.4e-4	2.8e-4		0	0	---	---	0	0
0.84 to 1.00	0.25 to 0.00	1.5e-3	0	0	2.2e-3	0	0	---	---	0	0
1.00 to 1.19	0.00 to -0.25		0	0		0	0	---	---	0	0
1.19 to 1.41	-0.25 to -0.50		0	0		0	0	---	---	0	0
1.41 to 1.68	-0.50 to -0.75		0	0		0	0	---	---	0	0
1.68 to 2.00	-0.75 to -1.00	3.0e-4	0	0	0	0	0	---	---	0	0

Notes: In this table, a number with an exponent (for example, A x10-b) is expressed in the form Ae-b. Because of differences in laboratory procedures, some differences exist among the data presented in this table. For the Grand Canyon gage P-61 samples, 122-mile eddy D-74 samples, and National Canyon gage D-74 samples, the sand fraction is defined as all sediment with a settling velocity greater than that for 0.0625 mm sediment with a Corey shape factor of 0.7 and Powers index of 3.0; platy material larger than 0.0625 mm with lower settling velocities is grouped with the silt and clay. For the above LCR gage D-77 samples, Grand Canyon gage D-77 samples, and National Canyon P-61 samples, the sand fraction is defined as all sediment retained on a 0.0625 mm sieve, and the silt and clay fraction is defined as all sediment that passes through a 0.0625 mm sieve.

* Because of differences sampler design, the concentrations of suspended sand measured with a D-77 sampler are expected to be lower than those measured with either a P-61 or a D-74 sampler, and the grain-size distributions of suspended sand measured with a D-77 sampler are expected to be finer than those measured with either a P-61 or a D-74 sampler.

§ n in this case refers to not the number of depth-integrated samples, but rather to the number of verticals included in the spatial average. The depth-integrated mean concentration of each size class in each vertical at National Canyon was determined by depth-integrating the point-samples at each vertical and dividing by the local flow depth.

† The samples in the P-61 subset are the only samples from the National Canyon gage that were analyzed for sand grain size; they were collected at stations 245 (the mid-channel station) and 345 (a near-bank station). These grain-size distributions were analyzed in the same manner as those for the Grand Canyon gage P-61 and 122-mile eddy D-74 samples.

Figure 5. (A) View of the eddy and sub-aerial eddy bar at the 122-mile eddy site during the low flow immediately preceding the controlled flood. Flow is from left to right, and the 122-mile eddy is just upstream from this reach. The large open sand bar shown here is about 70 cm above the low flow water surface. (B) Same view as in (A) during the flood. (C) Same view as in (A) and (B) immediately following the artificial flood. The large eddy bar in the center of the photo has maximum elevations more than 4 meters above the low flow water surface, as shown in the view from the channelward side in (D).

c

d

Figure 5 (continued)

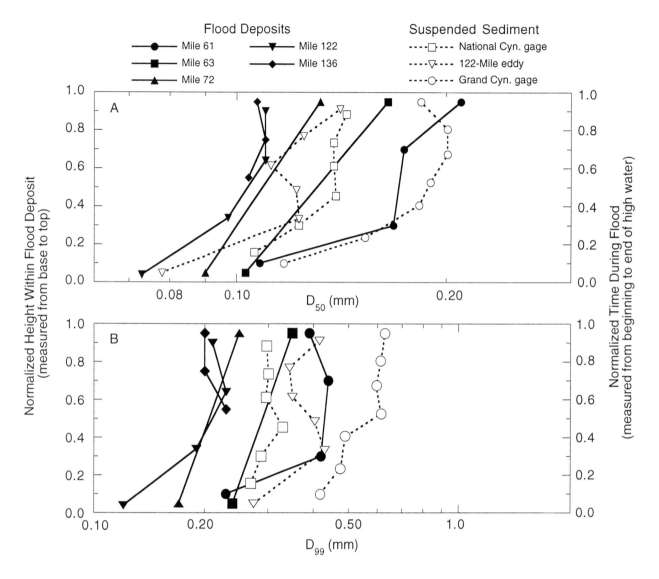

Figure 6. (A) Plot of median grain size (D50) as function of normalized height within the 1996 flood deposits, and plot of the median grain size of the suspended sediment in the main channel at the Grand Canyon (P-61 samples only) and National Canyon gages, and in the 122-mile eddy as a normalized function of time during the 1996 controlled flood. (B) Plot of D99 as function of normalized height within the 1996 flood deposits, and plot of D99 of the suspended sediment in the main channel at the Grand Canyon (P-61 samples only) and National Canyon gages, and in the 122-mile eddy as a normalized function of time during the 1996 controlled flood. (C) Plot of median grain size (D50) and D99 as function of normalized height within the pre-dam flood deposits.

6. VERTICAL TRENDS IN THE GRAIN-SIZE DISTRIBUTION OF COLORADO RIVER FLOOD DEPOSITS

6.1. Deposits of the 1996 Flood

The sedimentology of fine-grained deposits in Grand Canyon was previously investigated by *McKee* [1938], *Howard and Dolan* [1981], *Schmidt and Graf* [1990], and *Rubin et al.* [1990]. To build on this earlier work and to couple the sedimentology of deposits produced during the 1996 controlled flood to the grain-size distributions of suspended sediment measured during the 1996 controlled flood, we sampled deposits of the 1996 flood in trenches on 5 eddy bars between Lees Ferry and Diamond Creek. At 3 of these sites, identification of the base of the flood deposit was aided by scour chains that were emplaced prior to the

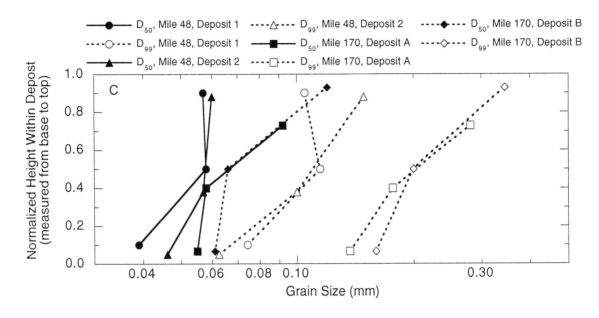

Figure 6 (continued)

flood [*Schmidt et al.*, this volume]. At each site, we collected samples at multiple elevations between the base and top of the deposit. Just as the suspended sediment and bed sediment coarsened during the flood, the sediment deposited during the flood also coarsened. Inverse grading of flood deposits has been described by *Osterkamp and Costa* [1987] and studied by *Iseya* [1989], who attributed formation of inversely graded flood deposits in Japanese rivers to changing sediment supply during floods. In the 1996 flood deposits that we studied at the 5 eddy bars, the median grain size increased from 0.073-0.11 mm at the base to 0.11-0.21 mm at the top (Figure 6); on average, the median grain size coarsened upward by a factor of 1.6. As with the suspended load (Figure 4), the increase in median grain size generally occurred not merely by the removal of fines, but also by an increase in the modal size and an increase in size of the coarsest fraction (Figure 6).

The most common sedimentary structure in the 1996 flood deposits was climbing-ripple cross-stratification, at some sites overlain by trough cross-stratification deposited by migrating fluvial dunes (Figure 7). This locally observed change from ripples to dunes was not caused by a change in flow conditions, but by the increase in grain size of sediment supplied to the eddies. The transition from ripples to dunes, driven only by a change in grain size, with no change in flow conditions, has been empirically documented through bedform phase diagrams [*Southard*, 1971; *Rubin and McCulloch*, 1980; *Southard and Boguchwal*, 1990]. In the rock record, observations of upward coarsening and change in bed configuration from ripples to dunes are typically interpreted to indicate stronger flows, but this is not necessarily the case. In the 1996 flood, peak discharge was held constant; winnowing caused the bed to coarsen, and coarsening of the bed caused the change in bed configuration. This was reflected in the sedimentology of the newly deposited bars throughout the river below Glen Canyon Dam.

6.2. Deposits of Pre-Dam Floods and Other Unsteady High Flows

The hydrograph of the 1996 controlled flood was unusual with respect to both pre-dam snowmelt and post-dam tributary-driven floods, not because of the magnitude or duration of high flow, but because peak discharge was constant. Natural floods are typically more unsteady, making it more difficult to interpret vertical trends in grain size. The difficulty arises not only because individual flood deposits become more difficult to recognize, but also because fluctuations in stage can cause changes in depositional processes at any one site (and such changes might influence local grain-size sorting).

Changes in depositional processes are particularly pronounced at higher-elevation sites on the bank, where a small change in stage can produce a large change in the local flow regime. Such a situation occurred in January 1993, for example, when a flood of 510 m^3/s on the Little Colorado River combined with a mainstem discharge of

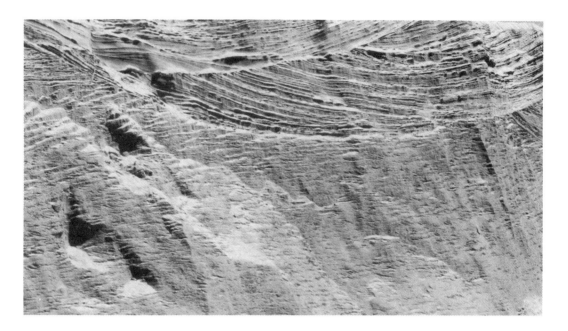

Figure 7. Photo of the upper 1.5 m of a 5-m-thick deposit of the 1996 flood. The lower left shows finer-grained climbing-ripple structures (ripple foresets dip toward the right), whereas the top 0.5 m shows coarser cross-bedded strata.

270-470 m³/s (fluctuating daily for powerplant needs). Deposits of this 1993 combined flow commonly exhibit a vertical sequence of 3 sedimentary structures produced by the daily flow fluctuations (Figure 8a): re-deposition of clasts of older sand (formed as waves undercut banks and clasts of older cohesive sand fell off the bank), deposition of flat-sand beds (within the wave-swash zone adjacent to the cut bank), and deposition of climbing ripples (produced by waves and currents in slightly deeper water farther from the bank). This sequence can be thought of as a transgressive sequence consisting of basal conglomerate, beach, and offshore deposits. The vertical succession arises because the 3 environments (cut bank, swash zone, and offshore ripples) occur adjacent to each other in successively deeper water (Figure 8b). As the river stage fluctuates, as during the 1993 example, the depositional environments shift laterally, producing a vertical succession of these deposits at any one point. Because a single flood may have multiple fluctuations in discharge (and therefore multiple local inundations), identifying the deposit of an individual flood can be difficult. Moreover, grain-size trends may be influenced by differences in sorting between bank undercutting, beach swash, and offshore ripples.

In some beds produced during pre-dam floods, however, the base and top of the deposit are clearly identifiable (e.g., by thin soil horizons with concentrations of rootlets), and the entire deposit consists of a single depositional facies. We located 4 such pre-dam flood beds, and sampled them

vertically for grain size. The results (Figure 6) show an upward coarsening from 0.039-0.061 mm to 0.057-0.12 mm.

Although these pre-dam deposits are finer than the 1996 deposits, the relative upward coarsening is the same (a factor of 1.6 from base to top). These similar rates of coarsening are probably fortuitous; a longer or shorter flood may have achieved a greater or lesser degree of upward coarsening, respectively. The 1996 flood might have been expected to produce a more rapid grain-size coarsening, because the 1996 flood supplied clear water released from Glen Canyon Dam, rather than the sediment-laden water in the pre-dam regime. However, the 1996 flood was of shorter duration and lower peak discharge relative to pre-dam floods, and these factors would tend to reduce the rate of coarsening. Evidently these opposing changes from pre-dam to 1996 balanced out in such a way as to produce a similar degree of upward coarsening between the 1996 and pre-dam flood deposits.

The fact that the individual flood deposits coarsen upward might seem to conflict with the fining-upward structure that is characteristic of other fluvial deposits. The apparent conflict is due to a difference in scale between individual flood deposits and the overall depositional sequence, which encompasses many flood deposits. Despite the upward coarsening of individual flood beds, the generalized fluvial sequence in Grand Canyon still fines upward on a larger scale, from boulders, gravel, and sand on the

Figure 8. (A) Three common depositional facies in the Little-Colorado-River-driven unsteady 1993 flood. Vertical sequence showing basal conglomerate (sand clasts, some with vertical bedding), beach swash (flat beds), and climbing ripples. Daily flow fluctuations from the dam caused daily transgressions and regressions of these facies. (B) Photograph showing the processes that produce the 3 facies in (A).

subaqueous channel bed, to sandy eddy and channel-margin bars, and then to silty and muddy marsh or floodplain-like deposits that occur along channel margins or atop some eddy bars [*Rubin et al.*, 1990].

7. CONCLUSIONS

During the 1996 Grand Canyon controlled flood, we documented the coupled evolution of the concentration of suspended sediment, the grain-size distribution of suspended sediment, the grain-size distribution of the sediment on the channel bed, and the grain-size distribution of the flood deposits on eddy bars. At all sites where suspended-sediment measurements were made in Marble and Grand Canyons during the 7 high-discharge days of the 1996 controlled flood (i.e., the above LCR gage, the Grand Canyon gage, the 122-Mile eddy site, and the National Canyon gage), the concentrations of sediment in suspension decreased as the suspended sediment coarsened. Thus, decreasing suspended-sediment concentrations associated with coarsening of the suspended sediment were observed at sites along 170 km of the river. At the Grand Canyon

a

b

gage (the only place where bed-sediment grain-size distributions were measured during the 1996 controlled flood), coarsening of the suspended sediment was observed to be coupled to coarsening of the bed sediment. Deposits produced during both the 1996 controlled flood and pre-dam floods in the Colorado River in Marble and Grand Canyons are inversely graded. This style of system-wide evolution of sediment grain size coupled to the evolution of suspended-sediment concentration during floods is characteristic of a supply-limited system like the Colorado River in Marble and Grand Canyons, in which a mismatch in the timing of maximum tributary sediment supply and maximum mainstem sediment transport exists.

8. IMPLICATIONS FOR RIVER MANAGEMENT

Before the recent controlled flood, we thought that dam operations had only two major impacts on sand in the canyon: the exchange of sand between bars and the channel bed, and transport down the channel [*Rubin et al.*, 1994]. The approach to building or maintaining bars was to optimize the balance and timing of high flows to transfer sand from the channel bed to the bars and low flows to minimize both erosion of bars and sediment transport out of the canyon [*U.S. Department of the Interior*, 1995]. The results of the controlled flood show that the situation is more complicated because dam operations also affect the grain sizes of sediment retained within the canyon following tributary sediment-input events. The preferred alternative outlined in the Glen Canyon Dam Environmental Impact Statement [*U.S. Department of the Interior*, 1995], based on calculations using stable sediment-rating curves, was designed on the assumption that significant quantities of sediment would be stored in Marble and Grand Canyons during normal powerplant releases from Glen Canyon Dam. Our work, however, suggests that, because sediment rating curves shift over time as a function of the grain sizes of sediment present on the bed and that the grain size of the bed changes significantly over time (Figure 2b), the prediction of net sediment accumulation in the reach between Lees Ferry and the Grand Canyon gage under normal powerplant releases [*U.S. Department of the Interior*, 1995] may not be valid.

Because the grain-size distribution of fine sediment evolves as functions of both recent tributary sediment input and dam operation, stable sediment-rating curves are precluded in supply-limited rivers such as the Colorado River in Marble and Grand Canyons. For the range of bed grain-size distributions that are possible in the post-dam Colorado River (Figure 2b), suspended-sand concentrations at a given discharge of water will vary by over a factor of

10 [*Topping*, in prep.].Therefore, sediment budgets cannot be constructed for reaches of the river using stable sediment rating curves. Because sediment rating curves shift with time in the Colorado River, curves fit to suspended-sand data from time periods when the system is coarse (e.g., 1983-1986) will likely underpredict sediment transport (export) and overpredict sediment storage in the mainstem [*U.S. Department of the Interior*, 1995], whereas curves fit to suspended-sand data from time periods when the system is fine (e.g., immediately after a significant flood on the Paria or Little Colorado Rivers) will likely overpredict sediment transport and underpredict sediment storage in the mainstem.

The key to managing the fine-sediment resource of the Colorado River below Glen Canyon Dam is to develop a physically based understanding of the processes that control both the short-term fining of the system following large tributary sediment inputs, and the subsequent coarsening of the system as fines are winnowed from the bed and deposited in the eddies or transported downstream. This understanding will allow for the design of future controlled floods to obtain similar management objectives in the presence of very different supplies of sediment. For example, because sediment-deposition rates scale approximately with the concentration of sediment in suspension, and the concentration of sediment in suspension will be higher when the system is dominated by finer sediment (e.g., immediately after large tributary sediment inputs), sediment-deposition rates in eddies will be much higher when the system is finer. Thus, the response of the system to a shorter-duration controlled flood, in the presence of finer sediment, will be similar to the response of the system to a longer duration controlled flood, in the presence of coarser sediment.

The link between floods and grain-size winnowing allows for several additional management capabilities. First, the winnowing process may be used advantageously by managers to manipulate dam releases to keep the main channel relatively coarse, thereby reducing the rate of net downstream sediment transport. If the channel can be maintained in a coarsened state, sediment transport down the channel will be reduced. Floods can, therefore, be used to perform a dual function by transferring sand from channel to bars: (1) causing aggradation on the bars, and (2) coarsening the bed and driving the system toward a "sediment-rating curve" in which sand discharge is reduced (for any given water discharge). Second, dam operations can be used to influence grain size of surficial sediment on bars. If floods are relatively short and infrequent, the bed will undergo less coarsening during each flood, and the resulting surficial deposits will be finer grained (and contain more nutrients and moisture to support vegetation). In contrast, longer and less frequent floods will cause more coarsening, depositing coarser sediment at the bar surface.

Acknowledgments. This work was funded by the Bureau of Reclamation's Glen Canyon Environmental Studies. We thank the National Park Service for providing housing, and thank the Bureau of Reclamation for providing sediment-sampling equipment. R. Anima (USGS), K. Brown (USGS), C. Crouch (USGS), B. Dierker (OARS), T. Hopson (Univ. of CO), F. Iseya (Jobu Univ., Japan), S. Jansen (USGS), J. Lyons (USBR), T. Reiss (USGS), M. Rubin, J.P. Running (OARS), J. Schmidt (Utah State), D. Schoellhamer (USGS), R.L. Shreve (UCLA), M. Smelser (Utah State), R. Stanley (USGS), G. Tate (USGS), M. Walker (OARS), and G. Williams (OARS) helped collect data. F. Iseya, J. D. Smith, and S. Wiele contributed stimulating scientific discussions on the work presented in this paper. Constructive reviews that improved the quality of this manuscript were provided by M.S. Hendrix, T.S. Melis, and J. C. Schmidt.

REFERENCES

Andrews, E.D., The Colorado River; A perspective from Lees Ferry, Arizona, in *Surface water hydrology,* ed. M.G. Wolman, and H.C. Riggs, The Geology of North America, v. O-1, Geol. Soc. Amer., Boulder, CO, pp. 281-328, 1990.

Andrews, E.D., Sediment transport in the Colorado River basin, in *Colorado River ecology and dam management,* ed. G.R. Marzolf, Natl. Acad. Press, Washington, D.C., pp. 54-74, 1991.

Brooks, N.H., Mechanics of streams with movable beds of fine sand, *Amer. Soc. Civil Engin. Trans.,* 123, 526-594, 1958.

Burkham, D.E., *Trends in selected hydraulic variables for the Colorado River at Lees Ferry and near Grand Canyon, Arizona: 1922-1984,* Glen Canyon Environ. Studies Rept. No. PB88-216098, 58 pp., 1986.

Colby, B.R., *Scour and fill in sand-bed streams,* U.S. Geol. Surv. Prof. Paper 462-D, 32 pp., 1964.

Edwards, T.K., and G.D. Glysson, *Field methods for measurement of fluvial sediment,* U.S. Geol. Surv. Open-File Rept. 86-531, 118 pp., 1988.

Garrett, W.B., E.K. Van De Vanter, and J.B. Graf, *Streamflow and sediment-transport data, Colorado River and three tributaries in Grand Canyon, Arizona, 1983 and 1985-86,* U.S. Geol. Surv. Open-File Rept. 93-174, 624 pp., 1993.

Guy, H.P., *Laboratory theory and methods for sediment analysis,* U.S. Geol. Surv. Tech. Water-Res. Invest., Book 5, Chapter C1, 58 pp., 1969.

Guy, H.P., D.B. Simons, and E.V. Richardson, *Summary of alluvial channel data from flume experiments, 1956-61,* U.S. Geol. Surv. Prof. Paper 462-I, 96 pp., 1966.

Hazel, J.E., Jr., M. Kaplinski, R.A. Parnell, M.F. Manone, and A.R. Dale, *The effects of the 1996 Glen Canyon Dam Beach/Habitat-building test flow on Colorado River sand bars in Grand Canyon,* Final Rept. to Grand Can. Monitor. Res. Center, 57 pp., 1997.

Howard, A., and R. Dolan, Geomorphology of the Colorado River in the Grand Canyon, *J. Geol.,* 89, 269-298, 1981.

Iseya, F., Mechanism of inverse grading of suspended load deposits, in *Sedimentary facies in the active plate margin,* ed. A. Taira and F. Masuda, Terra Sci. Publ. Co., Tokyo, Japan, pp. 113-129, 1989.

Kaplinski, M., J.E. Hazel, Jr., M.F. Manone, R.A. Parnell, and A.R. Dale, *Colorado River sediment storage in Grand Canyon during calendar year 1997,* Final Rept. to Grand Canyon Monitor. Res. Center, 28 pp., 1998.

Kennedy, J.F., *Stationary waves and antidunes in alluvial channels,* Calif. Inst. Tech., Div. Engin., W.M. Keck Lab. Hydr. Water Res. Rept. No. KH-R-2, 146 pp., 1961.

Konieczki, A.D., J.B. Graf, and M.C. Carpenter, *Streamflow and sediment data collected to determine the effects of a controlled flood in March and April 1996 on the Colorado River between Lees Ferry and Diamond Creek, Arizona,* U.S. Geol. Surv. Open-File Rept. 97-224, 55 pp., 1997.

Leopold, L.B., and T. Maddock, Jr., *The hydraulic geometry of stream channels and some physiographic implications,* U.S. Geol. Surv. Prof. Paper 252, 57 pp., 1953.

McKee, E.D., Original structures in Colorado River flood deposits of Grand Canyon, *J. Sed. Pet.,* 8, 77-83, 1938.

McLean, S.R., On the calculation of suspended load for noncohesive sediments, *J. Geophys. Res.,* 97, 5759-5770, 1992.

Nordin, C.F., Jr., *A preliminary study of sediment transport parameters, Rio Puerco near Bernardo, New Mexico,* U.S. Geol. Surv. Prof. Paper 462-C, 21 pp., 1963.

Osterkamp, W.R., and J.E. Costa, Changes accompanying an extraordinary flood on a sand-bed stream, in *Catastrophic flooding,* ed. L. Mayer and D. Nash, Allen Unwin, Boston, MA, pp. 201-224, 1987.

Rubin, D.M., J.C. Schmidt, and J.N. Moore, Origin, structure, and evolution of a reattachment bar, Colorado River, Grand Canyon, Arizona, *J. Sed. Pet.,* 60, 982-991, 1990.

Rubin, D.M., J.C. Schmidt, R.A. Anima, K.M. Brown, R.E. Hunter, H. Ikeda, B.E. Jaffe, R. McDonald, J.M. Nelson, T.E. Reiss, R. Sanders, and R.G. Stanley, *Internal structure of bars in Grand Canyon and evaluation of proposed flow alternatives for Glen Canyon Dam,* U.S. Geol. Surv. Open-File Rept. 94-594, 56 pp., 1994.

Rubin, D.M., J.M. Nelson, and D.J. Topping, Relation of inversely graded deposits to suspended-sediment grain-size evolution during the 1996 flood experiment in Grand Canyon, *Geol.,* 26, 99-102, 1998.

Rubin, D.M., and D.S. McCulloch, Single and superimposed bedforms: A synthesis of San Francisco Bay and flume observations, in Shallow marine processes and products, ed. A.H. Bouma, D.S. Gorsline, G. Allen, and C. Monty, *Sed. Geol.,* 26, 207-231, 1980.

Schmidt, J.C., and J.B. Graf, *Aggradation and degradation of alluvial sand deposits, 1965-1986, Colorado River, Grand Canyon National Park, Arizona,* U.S. Geol. Surv. Prof. Paper 1493, 100 pp., 1990.

Smith, J.D., and S.R. McLean, Spatially averaged flow over a wavy surface, *J. Geophys. Res. (Oceans),* 82, 1735-1746, 1977.

Smith, W.O., C.P. Vetter, and G.B. Cummings, and others, *Comprehensive survey of sedimentation in Lake Mead, 1948-49,* U.S. Geol. Surv. Prof. Paper 295, 254 pp., 1960.

Southard, J.B., Representation of bed configurations in depth-velocity-size diagrams, *J. Sed. Pet.,* 41, 903-915, 1971.

Southard, J. B., and L.A. Boguchwal, Bed configurations in steady unidirectional water flows, Part 2, Synthesis of flume data, *J. Sed. Pet.,* 60, 658-679, 1990.

Topping, D.J., *Physics of flow, sediment transport, hydraulic geometry, and channel geomorphic adjustment during flash floods in an ephemeral river, the Paria River, Utah and Arizona,* unpubl. Ph.D. thesis, Univ. Wash., Seattle, WA, 406 pp., 1997.

U.S. Department of the Interior, *Operation of Glen Canyon Dam,* *Final Environmental Impact Statement,* Bur. Reclam., Salt Lake City, UT, 337 pp., 1995.

U.S. Geological Survey, *Quality of surface waters of the United States, 1957,* U.S. Geol. Surv. Water-Supply Paper 1523, part 9-14, 1961.

David J. Topping, U.S. Geological Survey, MS-413, Denver Federal Center, Lakewood, CO 80225; email: dtopping@usgs.gov

David M. Rubin, U.S. Geological Survey, MS-999, 345 Middlefield Road, Menlo Park, CA 94025

Jonathan M. Nelson, Paul J. Kinzel III, and James P. Bennett, U.S. Geological Survey, MS-413, Denver Federal Center, Lakewood, CO 80225

Flow and Suspended-Sediment Transport in the Colorado River Near National Canyon

J. Dungan Smith

U.S. Geological Survey, Boulder, Colorado

Point measurements of flow speed and suspended-sand concentration were made from a cableway 293-km downstream from Glen Canyon Dam during the 1996 controlled flood. The data demonstrate a systematic fining of the suspended load in the Colorado River, a reduction in near-bed sand concentration with time, and a strong secondary circulation that very effectively transported suspended sand toward the channel margins. In the center of the river, the primary flow was well represented by steady, horizontally uniform flow theory, with a shear velocity of 0.081 m/s and a sand grain related roughness parameter of $4.5 \cdot 10^{-6}$ m; at the channel margins the primary flow exhibited a distinct internal boundary layer with a shear velocity of approximately 0.081 m/s and an outer boundary layer with a shear velocity of approximately twice that value. The secondary circulation was caused by long wavelength irregularities in the rockfall-produced sloping banks of the approximately trapezoidal channel. The primary flow was forced upward and toward the river center by these topographic features causing a fully 3-dimensional circulation. The upward forced vertical velocities apparently interacted with turbulence in the primary flow to produce boils. Consequently, the upwelling zone degraded to an irregular, bank-parallel boil line. Downwelling occurred over a broad region in the center of the river, but also was concentrated along well-defined convergence zones over which woody debris concentrated. This secondary circulation was very effective in transporting suspended sand toward the channel margins at the bottom, then lifting it in the boils and depositing it inshore of the boil line on the riverbanks.

1. INTRODUCTION

In the Grand Canyon, the flux of fine sediment from the main flow of the Colorado River to open depositional sites along the banks and to eddies is controlled by the transport rate of suspended sediment during high flow. To understand this process better, suspended-load discharge and some ancillary flow and sediment-transport variables were measured at 4 gaging stations during the controlled flood [*Konieczki et al.*, 1997]. As part of this effort, dense sets of flow-speed and suspended-sediment (silt-plus-clay and sand) concentration measurements were collected at a cableway across the Colorado River just upstream from National Canyon [for location, see *Webb et al.*, this volume]. The purpose of this paper is to summarize

The Controlled Flood in Grand Canyon
Geophysical Monograph 110

measurements of mainstem suspended-sediment transport at this site. National Canyon is located 70% of the distance between Glen Canyon Dam and Lake Mead reservoir and is well downstream from the major tributaries contributing sand to the Colorado below the dam [*Schmidt et al.*, this volume]. The results reported here, therefore, characterize mainstem flow and sediment-transport processes in the downstream half of the river. The goals of the research above National Canyon were to (1) determine the cross-sectional structure of the downstream-velocity component using flow-speed measurements and suspended-sand concentration fields during the flood, (2) characterize the mechanisms by which suspended sand is transported to the sides of the river at high discharge, and (3) evaluate the rates of downstream and lateral sand transport.

Reach-averaged morphology and flow are well-known and well-defined for those parts of the river in which limestone crops out at river level [*Griffin*, 1997], making it easy to relate local controlled flood results to reach-averaged flow and suspended-sand-transport models [*Wiele and Smith*, 1996; *Smith and Griffin*, in press]. Muav Limestone crops out continuously at river level for 42 km upstream and 5 km downstream from the National Canyon cableway, and the closest large rapids are 26 km upstream and 21 km downstream. The river is nearly straight upstream from the cableway. National Canyon Rapid is several river widths downstream and the river is approximately straight below the rapid for another few river widths. Upstream from the measurement site is a well-defined, approximately symmetrical pool with a bed composed primarily of sand [*Christensen*, 1993]. This sand bed maintains the sand flux past the cableway. From its prior history of scour and fill, the layer of sand in the pool appears to be thin [*Christensen*, 1993]. Consequently, this pool is likely to be able to adjust to changes in fine sediment input rapidly, making measurements at the National Canyon cableway potentially representative of transport through the Muav reach.

Choice of an appropriate site is critical to successful generalization of results in fluvial geomorphic and sediment-transport investigations. For this controlled-flood investigation, a consistent river-level geologic setting was considered desirable. First, it allowed an equilibrium channel and bed morphology to have been established over the centuries before the dam and the decades before the controlled flood for a significant distance (several tens of kilometers) upstream from the measurement site. Second, for a specific input of fine sediment (sand, silt, and clay) and for specific time-average flow conditions, the long upstream reach with a simple channel geometry permitted an equilibrium coverage of the river bed with fine sediment and a constant discharge of this material. As a consequence of these 2 geomorphic features, implications of empirical flow and suspended-sediment transport results can be extrapolated more easily and more accurately to other locations along the river corridor. These 2 geomorphic features also make possible accurate local sand-transport measurements and their interpretation in a reach-averaged context.

This paper describes some aspects of the measurement program undertaken at National Canyon during the 1996 controlled flood. Of primary concern are the velocity, suspended sediment, and bed erosion /deposition conditions during the steady high flow part of the event [*Schmidt et al.*, this volume]. The riverbed topography, velocity field and suspended-sediment profiles are presented and compared and contrasted to steady, horizontally uniform flow theory. The relations between the measured fields are discussed in detail and the implications of their interplay on erosion of the river bed and deposition on the river margins is described.

2. MEASUREMENT PROGRAM

The measurements made at and near the National Canyon cableway during the 1996 controlled flood that are analyzed in this paper are listed in Table 1. Similar measurements had been made in 1991 at this same site [*Christensen*, 1993]. The coordinate system used to depict the locations of some of these measurements is shown in Figure 1, along with the approximately trapezoidal geometry of the cross section. Station 75 (at 75 m from the left cable abutment) is approximately in the center of the river, and the bottom of the trapezoid extends from Station 50 to 100. The left edge of the river was at 27 m and the right edge was 97 m away at 124 m. During the steady high flow period of the controlled flood, the average depth of the cross-section was 7.1 m and the average depth of the approximately flat central 50 m was 8.9 m. The cross-sectional area during this period was 708 m^2. These and other features of the cross-section during the steady high flow are presented in Table 2.

Flow-speed measurements for use in estimating profiles of the downstream-velocity component were procured at 11 stations in the central 50 m of the river. At each station measurements were made at 7 positions in the vertical direction as shown in Figure 1. Suspended-sediment samples were procured in the same array, except that the bottom 2 samples were collected 0.15 m closer to the river bed (Table 1). An attempt was made to obtain 1 set of flow-

TABLE 1. Measurements made upstream from National Canyon during the 1996 controlled flood

Date[1]	Water discharge[2]	Flow-speed profiles[3,4]	Suspended-sediment concentration profiles[3,5] and water temperature[6]	Echo-sounding surveys[7]
March 5	--	--	--	0-4, 7
March 26	X	X	--	--
March 27	--	--	--	0-1
March 28	X	--	X	1-7
March 29	X	--	X	0-1
March 30	--	X	X	0-4, 7
March 31	--	X	X	0-1
April 1	--	X, X[8]	X	0-7
April 2	--	X	X	0-2

[1] The steady high flow at National Canyon occurred from March 26 - April 2, 1996.
[2] Calculations based on these numbers as well as similar data from other gaging stations yields a best estimate of 1310 m³/s for the steady high flow at the National Canyon cableway
[3] Flow-speed and suspended-sediment concentration profiles were sampled every 5 m from stations 50 to 100 in the deep part of the channel in reference to the cross-stream coordinate system shown in Figure 1.
[4] Point measurements were made 0.30, 0.61, 0.91, 1.83, 3.66, 5.49 m above the riverbed and 0.61 m below the river surface.
[5] Point samples were collected 0.15, 0.46, 0.91, 1.83, 3.66, 5.49 m above the riverbed and 0.61 m below the river surface.
[6] Values ranged from 10.5°C to 11.0°C and averaged 10.8°C.
[7] Different cross-sections were measured each day, and the ones that were measured are indicated by the numbers listed in this column. The section under the cableway is denoted as 0 and the rest of the sections are located sequentially every 122 m upstream.
[8] Two sets of flow-speed profiles were measured on April 1.

speed measurements and 1 set of suspended-sediment samples each day during the steady high flow, but no suspended-sediment samples were obtained on the first day (March 27) and no flow-speed measurements were made until March 30. Ultimately, 6 sets of suspended-sediment samples and 5 sets of flow-speed measurements were collected at the points in the array of Figure 1 during the steady high flow (Table 1). In addition, 1 set of flow-speed measurements was obtained before the controlled flood. Water temperature, measured during the suspended-sediment sampling, ranged from 10.5°C to 11.0°C and averaged 10.8°C; thus, the average kinematic viscosity was $1.28 \cdot 10^{-6}$ m²/s (Table 1). The basic flow speed, suspended-sediment concentration, water-discharge and water-temperature data for this and other gaging stations are presented by Konieczki et al. [1997].

The cross-sectionally averaged velocity at the National Canyon cableway during the steady high flow was determined from an accurate estimate of river discharge. Local measurements of discharge ranged between 1296 m³/s and 1323 m³/s and averaged 1309 m³/s. Velocity and discharge measurements made during the period of steady low flow before the flood yielded a net tributary and groundwater input to the Colorado River between Lees Ferry and National Canyon of 35 m³/s. Adding this supplemental discharge to that of the flood peak at Lees Ferry produces a net discharge at the National Canyon cableway of 1310 m³/s, which is consistent with the estimate determined from the local discharge measurements. The cross-sectionally averaged velocity, calculated by dividing the average discharge by the cross-sectional area, is 1.83 m/s, which is similar to the reach-averaged velocity determined from dye-advection measurements for the 70 km river segment between 122 Mile Creek and National Canyon of 1.89 m/s [Konieczki et al., 1997]. This agreement shows how closely the cross-sectional area of the National Canyon measurement site matched the average cross-sectional area of the upstream 70 km of the river.

To record variations in the river-bed topography upstream from the cableway during the flood, echo-sounding surveys were made along a set of previously established cross-sections. The cross-sections used were among a set that had been established in 1991, a subset of which had been remeasured frequently since that time [Graf et al., 1997], including on March 5, 1996. Just before the 1996 controlled flood, 7 of the original sections were

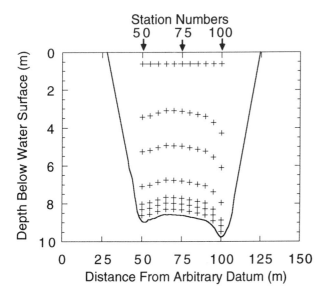

Figure 1. Topographic cross-section beneath the cableway upstream from National Canyon on April 2. The view is downstream. Only the topography of the middle 50 m of the river at this section varied measurably with time between March 5 and April 2. The plus signs (+) denote the locations of the flow-speed measurements.

located and the anchor points were moved up so they would not be submerged. Echo-sounding surveys were conducted under the cableway 6 times during the steady high flow, and the next cross-section upstream was measured 7 times during this period. Those further upstream were measured less frequently (Table 1).

For each survey, a boat with an echo sounder made 10 or more runs back and forth across the river beneath a kevlar cable stretched between two anchors. The cable was marked every 6.1 m and the positions along the cable were noted by an observer as the boat passed beneath and marked on the echo-sounding record. The record for each of the runs was later digitized and those for each set of runs were averaged. The boat operator attempted to stay directly beneath the kevlar cable but, in fact, followed an irregular path because of the interaction of the boat and the macroturbulence in the flow. In addition, there were timing errors associated with reading the marks on the kevlar cable as the boat passed beneath. When averaged, these 2 statistically independent deviations combined with the spread of the acoustic beam from the echo sounder to smooth out small-scale irregularities in the river bed, permitting good reproducibility of the mean profile obtained from averaging the digitized traces. Standard deviations were calculated for each set of runs to provide a measure of accuracy of the associated mean profile and to determine whether or not the

daily variations were statistically significant. Prior experience by the author with this method at the National Canyon site indicated that for a sediment bed devoid of large boulders, 10 runs provided a mean that was accurate to 0.05 m or better in the central part of the river at this site. Errors in estimating the mean bed elevation at the edges of the channel were larger because of the interplay between the rapid change in depth and the error in determining the along-cable position. Neglecting channel-margin observations, typical daily variations associated with local scour and fill in the sand pool ranged from 0.1 to 0.2 m, which are sufficiently larger than the standard deviations; thus, scour and fill are well resolved. The velocity, suspended-sediment, and bed-topography measurements obtained during the high-flow part of the 1996 controlled flood provide sufficient information to evaluate the dominant flow and suspended-sediment dynamic processes associated with above power-plant capacity flows at this site.

3. RESULTS OF THE MEASUREMENT PROGRAM

3.1. Topography

The river upstream from the National Canyon cableway is approximately symmetrical, trapezoidal, and straight for more than 5 channel widths (485 m); further upstream, the river curves slightly to the north. Channel geometry is similar at low and high discharge. The steep margins of the channel (e.g., Figure 1) are nearly linear with top-to-bottom slopes ranging from 20° to 30°; the average bank slope is 24°. The channel cross-section 485 m upstream from the National Canyon cableway is the mirror image of the one beneath the cableway but is narrower (90 m) and deeper in the central third of the river. The area of this upstream cross section is 96% of the cableway cross section. Between these cross sections, the bed in the central half of the river is distorted by very long (over 800 m), low (less than 2 m in height), first-mode alternating bars, which also occur at lower discharge [*Christensen*, 1993]. The thalweg is on the right side of the river for at least 250 m upstream from the cableway. During the flood, the cross-sectional area of the flow decreased systematically from the section 485 m upstream to the one 122 m upstream from the cableway, then it increased again beneath the cableway.

The bed in the center of the river under the cableway sloped gently from left to right before the 1996 controlled flood, as shown by the March 5 profile in Figure 2. The relative intensity and irregularity of the acoustic returns displayed on the echo-sounding records for March 5 suggest that the bed surface at this time was neither entirely sand nor entirely gravel, but rather a mixture of these

TABLE 2. Summary of geometric and some flow parameters for the Colorado River at the National Canyon cableway during the 1996 controlled flood

Discharge at cableway	1310 m³/s
Cross-sectional area of flow	708 m²
Cross-sectionally-averaged downstream velocity component	1.83 m/s
River surface width	97 m
Thalweg depth	9.5-9.8 m
Depth at channel center	8.5-8.9 m
Average depth of central 50 m of channel	8.9 m
Average depth of entire channel	7.1 m
Hydraulic radius	7.1 m
Water temperature	10.8°C
Kinematic viscosity	$1.28 \cdot 10^{-6}$ m²/s

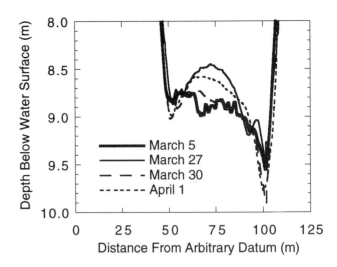

Figure 2. Measured topographic cross-sections showing changes in the mid-channel bar beneath the cableway with time. The March 5 cross-section shows the channel shape before the 1996 controlled flood. The cross-section for March 27 demonstrates that there was deposition of a 0.35 m high mid-channel bar on this surface during the first day of high flow. The March 30 profile shows that the initial deposition was followed by erosion of the bar top and both lateral channels.

materials. During the first day of the flood, a mid-channel bar formed at this cross-section. This is shown by the March 27 profile on Figure 2. In the next few days, the channels on both sides of the bar and the bar top eroded, leading to a deepening of the entire form by 0.25 m on March 30. After March 30, there was no further erosion of the lateral channels and there was only a small amount of scour and fill on the bar top, resulting in some additional deposition there by the end of the steady high-flow period. The cross sections immediately upstream responded in the same manner. Thus, there was a net transfer of mass from the margins of the deep channel to its center that was most rapid at the beginning of the flood, but that continued at a diminishing rate until the end of the steady high-flow period. The relative intensity of the acoustic returns and the surface smoothness displayed by the individual echo-sounding records are similar to those in marine environments where a bed of fine sand is deposited from suspension because the bed surface was smooth and the acoustic reflection was weak.

3.2. Velocity Field

The Price AA current meters measure flow speed, which under steady, horizontally uniform conditions is equal to the down-stream component of the velocity. This assumption is appropriate for the National Canyon cableway. When the lateral- and vertical-velocity components both are 10% of the downstream component, they still contribute only 2% to the downstream flow speed. Even if one of the other velocity components is 32% of the downstream component, its contribution to the flow speed is only 5%. Consequently,

lateral- and vertical-velocity components rarely are large enough relative to the downstream component to contribute in a resolvable way to the flow speed. Therefore, in most field investigations, including the one undertaken of the National Canyon cableway, flow speed was used as a surrogate for the downstream velocity component, and even under the most complex flow conditions, it usually can be used as an initial estimate for the dominant velocity component.

Reasonably accurate, empirical profiles of the downstream-velocity component and suspended-sand concentration profiles, especially for the bottom half of the flow, are necessary to define key flow and suspended-sediment parameters. Point measurements of velocity and suspended-sediment concentration, made with standard U.S. Geological Survey sampling devices and methods in large rivers, however, can be highly variable because of the short sampling time relative to the periods of the peaks in the fluctuating velocity and suspended-sediment concentration spectra that are caused by turbulence and boils. To reduce this variability, as required to resolve physically important trends, the data were averaged in various ways. Flow speeds were treated as downstream-velocity components, then the 5 measured speeds made during high steady discharge (Table 1) were averaged. This could be done because, except for the changing density stratification effects on the turbulent flow [*Smith and McLean*, 1977a; *Gelfenbaum and Smith*, 1986], the velocity field was steady on an hourly or longer time scale, and because the density stratification caused by suspended sediment was small and varied slowly by the time the first set of flow-speed measurements was made on March 30.

Time averaging, in contrast, was not appropriate for the suspended-sand concentration data because the concentration field was evolving. Nevertheless, an averaged concentration field had to be calculated for the analysis of the velocity field because of the density stratification produced by the vertical gradient of the suspended-sand concentration field. To accomplish this smoothing, the suspended-sand concentration data for the center of the river were averaged spatially for the 4 stations for which both the flow-speed and suspended-sand concentration profiles were similar (stations 70, 75, 80, and 85). The parameters obtained from this average profile were used to make the density stratification correction for these stations, and also for the lower parts of the profiles for the stations between 65 and 90. For stations 55, 60, and 95, which are located on the flanks of the mid-channel bar, the suspended-sand concentration data were averaged for the 3 stations for pairs of days and the parameters obtained from these average profiles were used to make the stratification correction for the bottom parts of the velocity profiles for stations 50-60 and 95-100. No density stratification correction could be made for the upper parts of any velocity profiles (or the associated suspended-sand concentration profiles) for which the velocity did not increase monotonically with distance from the river bed.

Individual profiles of the downstream-velocity component for stations 55 to 95 have a systematic cross-stream structure, but show no temporal trends, as displayed for 3 stations in Figure 3. Station 60 is typical of flow over the far left flank of the mid-channel bar, station 75 is typical of flow in the center of the river, and station 95 is typical of flow over the far right flank of the mid-channel bar. Owing to the absence of any resolvable temporal trend in the measured flow-speed profiles, the variations in values at each point in the measurement array can be considered statistically independent and averaged to produce a mean value at that measurement site. These calculated mean flow-speed values are included, connected by line segments, for each of the 3 stations in Figure 3.

3.2.1. Temporal Variation. Standard deviations for the flow-speed measurements vary considerably among the individual sites of the array, but when calculated for all of the measurements at a fixed distance above the river bed, or all measurements at a station, systematic spatial trends emerge. The decreasing standard deviation with distance

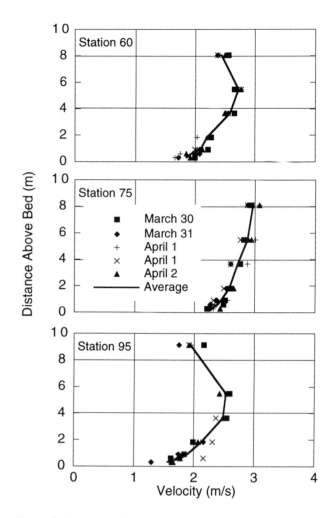

Figure 3. Profiles of flow speed at stations on the left flank of the mid-channel bar (station 60), in the center of the river (station 75), and on the right flank of the bar (station 95). Individual values and time averages connected by line segments are shown. No temporal trend in profile magnitude or shape is evident.

from the river bed (Figure 4a) reflects the normal vertical structure of the intensity of turbulence in open channel flow, which is calculated as the square root of the turbulent kinetic energy [*Schlichting*, 1979]. The magnitude of this variable is largest near the river bed where the shear and shear stress, and hence production of turbulent kinetic energy, are highest, and it falls off as the shear and shear stress decrease toward the river surface. In the center of the channel, this estimate of the intensity of turbulence is low by about a factor of 2.8 relative to laboratory measurements [*Schlichting*, 1988], because the full turbulent kinetic energy spectrum was not resolved.

The cross-stream structure of the standard deviation (Figure 4b) shows that there is substantially enhanced

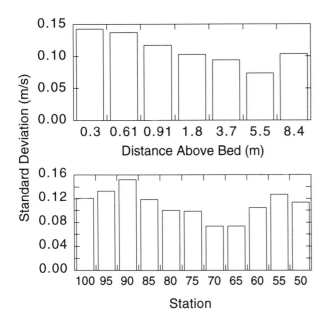

Figure 4. Standard deviations for flow speed for each (A) measurement level at all stations, and (B) station at all levels, showing systematic trends both in the vertical (A) and lateral (B) directions. These standard deviations mirror the cross-sectional structure of the intensity of turbulence.

turbulence near the channel margins. The high value near the surface (z = 8.35 in Figure 4a) is a consequence of additional turbulence production in the vicinity of the free surface, and large boil-caused variations in both the downstream and the cross-stream components of velocity near the surface at the margins of the flow (Figure 4b). This effect is more pronounced on the right side of the flow. Separate analyses of the 3 stations on the right side of the river (90, 95, and 100) and the other 8 suggest that the standard deviation increases 11% from 5.49 to 8.35 m above the river bottom as a result of the normal increase in turbulence production, and 30% due to lateral velocity variations on the right side of the river. The increase due to boil-caused macroturbulence on the left side of the river is 18%. The overall standard deviation for stations 90, 95, and 100 of 0.130 m/s is 31% higher than that for stations 50-85 (0.099 m/s), and the standard deviation more than doubles between 5.49 m and 8.35 m (0.069 to 0.142 m/s) for these 3 stations. These data, therefore, indicate a normal turbulent structure for the central and near-bottom parts of the cross-section, but also show regions of substantially elevated turbulence (30-100%, Figure 4b) near the river surface, and especially toward the edges.

3.2.2. Velocity Profiles. In channel flows and turbulent boundary layers, profiles of the downstream-velocity component have a predictable vertical structure in turbulent flows that are steady and horizontally uniform in the mean [*Schlichting*, 1988; *Smith and McLean*, 1977; *Long et al.*, 1993]. In this situation, the downstream-velocity component for a flow with a homogeneous density field varies logarithmically with distance above the bottom in the lower 20% of the flow, and approximately parabolically with distance above the bottom in the upper 80% of the flow [*Rattray and Mitsuda*, 1974; *Gelfenbaum and Smith*, 1986; *Christensen*, 1993]. In a suspended-sediment trans-porting flow, such as the one of concern here, the suspended sediment produces a stable density stratification that reduces the local turbulent diffusivities of momentum and mass in accordance with the local gradient Richardson number [*Smith and McLean*, 1977; *Gelfenbaum and Smith*, 1986; *Kachel and Smith*, 1989; *McLean*, 1992] and causes deviations from these relations.

Mean velocity-profile data for 2 segments of the cross-section (stations 55-60 and 70-85) are graphed in a semi-logarithmic manner with respect to distance above the bed (z) in Figure 5. The data for each point in the vertical have been averaged for 2 and 4 stations respectively, in addition to the time averaging, to produce the greatest possible accuracy in the empirical profile. This accuracy is necessary to obtain good definition of the key flow parameters, namely the shear velocity u_* and the roughness parameter z_0 [*Schlichting*, 1988]. The theoretical curves that have been fit to the empirical velocity profiles of Figure 5, and the suspended-sand concentration profiles presented in subsequent figures, include the correction of *Gelfenbaum and Smith* [1986] for density stratification, but with the empirical constants of *Wieringa* [1980]. For the river center (stations 70-85), the fit was to all but the highest data point, which was omitted for reasons discussed below. This fit relation yields a shear velocity of 0.081 m/s and a roughness parameter of $4.5 \cdot 10^{-6}$ m. For the near-bed portion of the profile, in the lower 20% of the flow depth over the far left flank of the mid-channel bar (55-60), the shear velocity is also 0.081 m/s, but the roughness parameter is an order of magnitude higher ($z_0 = 4.4 \cdot 10^{-5}$ m). Using the method of *Nikuradse* [1932, 1933a, 1933b; see *Schlichting*, 1979] with the fit to his data for hydraulically transitional flow suggested by *Smith* [1977], a z_0 of $4.5 \cdot 10^{-6}$ m and a u_{*SF} of 0.081 m/s gives a median grain diameter of $D_{50} = 0.21$ mm (stations 70-85, averaged for March 30 to April 2).

Often z_0 is not solely a consequence of grain roughness. Sediment for which the settling velocity is significantly greater than the shear velocity moves as bed load, not in suspension, and if this occurs the saltating sand will produce bedforms, such as under the conditions that existed

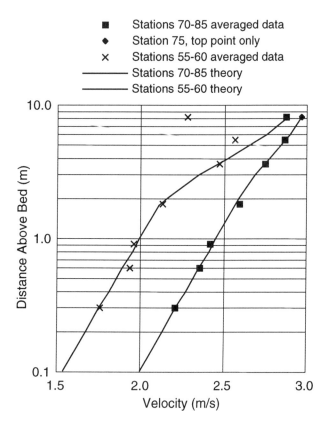

Figure 5. Velocity profiles composed of values for each level in the vertical averaged temporally and then spatially for 4 (70-85) and 2 (55-60) stations, respectively, graphed semi-logarithmically with distance above the river bed. The empirical profiles are fitted in each case to a steady, horizontally uniform flow theory that includes the effects of suspended sediment caused density stratification.

during the controlled flood at stations 55, 60, and 95. At a shear velocity of 0.081 m/s, bedforms can develop for sand with a median diameter greater than 0.74 mm; therefore, unless the D_{50} predicted from z_0 is less than 0.74 mm, an analysis that assumes the bed to be roughened only by the sand grains is not valid at this shear velocity. If the z_0 of $4.4 \cdot 10^{-5}$ m (stations 55-60) were a result of grain roughness, at a shear velocity of 0.081 m/s the median grain diameter would have been 1.4 mm, which does not satisfy this criterion. The limiting value of z_0 for incipient suspended-load transport of sand with a median grain diameter of 0.74 mm is $2.3 \cdot 10^{-5}$ m.

The method of *Nikuradse* requires the sediment bed to be geometrically smooth. To remove this constraint, *Smith and McLean* [1977a] present a method for treating the effects of periodic topographic elements such as bedforms, and they use the method to calculate the effective roughness of subaqueous dunes. *Wiberg and Smith* [1984] have shown

that the method of *Smith and McLean* [1977a] also works well with ripples. It also can be extended for application to undulations on sloping river banks, although for the present problem not enough is known about the geometry of the topographic elements to pursue an accurate calculation of this type.

The foundation for the method of *Smith and McLean* [1977a] is (1) averaging the flow over a characteristic downstream scale λ (such as the wavelength of a bedform), (2) separation of the spatially-averaged boundary shear stress (τ_{bT}) into a skin-friction shear stress (τ_{bSF}) and a form drag per unit area of river bed (τ_{bFD}), and (3) treatment of the flow as being two parts composed of an internal boundary layer which produces the velocity field that is averaged to obtain the drag force, and an outer region that is affected by both the skin friction and the form drag. The internal boundary layer has a velocity field that is proportional to the skin-friction (SF) shear velocity

$$u_{*SF} = (\tau_{bSF}/\rho)^{1/2}, \qquad (1)$$

where ρ = water density, and it has a z_0 that can be calculated using the method of *Nikuradse* [1933b] described above. The outer region has a velocity profile that is proportional to

$$u_{*T} = (\tau_{bT}/\rho)^{1/2}, \qquad (2)$$

where the subscript T denotes the outer region, and a z_0 that is found by matching the 2 profiles at the top of the spatially averaged internal boundary layer ($z = \delta$).

The profile for stations 55-60 was fit in 2 parts, an internal boundary layer with its top at the break in slope of the velocity profile shown in Figure 5, and an outer region calculated using (2). For the outer layer, the shear velocity is 0.16 m and the roughness parameter is $8.3 \cdot 10^{-3}$ m. From this analysis the top of the internal boundary layer could not be determined accurately. Slightly better resolution is obtained when the data from station 95 are averaged with those from 55 and 60, and this analysis gives an internal boundary layer height of 1.6 m. An empirical estimate of the spacing between roughness elements can be obtained using the measured values of δ and z_{0SF} (1.6 m and $4.4 \cdot 10^{-5}$ m, respectively) and solving

$$\delta/z_{0SF} = 0.10(\lambda/z_{0SF})^{4/5} \qquad (3)$$

to get $\lambda = 393$ m.

Resolution of the velocity profiles using the flow-speed measurements is insufficient for cases in which 2 layers are required to determine either the internal boundary layer or

TABLE 3. Values of flow parameters from velocity-profile analysis

Station[1]	Skin friction shear velocity, u_{*SF} (m/s)	Roughness parameter for internal boundary layer, z_{0SF} (m)	Shear velocity for outer boundary[2] layer (from total boundary shear stress), u_{*T} (m/s)	Roughness parameter for[2] outer boundary layer, z_{0T} (m)
50	0.058	$4.4 \cdot 10^{-5}$	0.12	$1.1 \cdot 10^{-2}$
55	0.081	$4.4 \cdot 10^{-5}$	0.16	$8.3 \cdot 10^{-3}$
60	0.081	$4.4 \cdot 10^{-5}$	0.16	$8.3 \cdot 10^{-3}$
65	0.081	$2.3 \cdot 10^{-5}$	--	--
70	0.081	$4.5 \cdot 10^{-6}$	--	--
75	0.081	$4.5 \cdot 10^{-6}$	--	--
80	0.081	$4.5 \cdot 10^{-6}$	--	--
85	0.081	$4.5 \cdot 10^{-6}$	--	--
90	0.079	$4.5 \cdot 10^{-6}$	--	--
95	0.14	$3.4 \cdot 10^{-3}$	0.14	$3.4 \cdot 10^{-3}$

[1] Reliable values could not be obtained for station 100.

[2] Values are given only for profiles that reflect form drag, thus requiring a two-part curve.

the outer layer parameters accurately. Nevertheless, this 2-layer analysis can be applied to the time-averaged velocity-profile data for each station to determine whether or not an outer layer yields an improved fit. If there is no apparent need for 2 layers, then a more accurate analysis can be done using a single layer (and calculating single values of $u_* = u_{*SF}$ and $z_0 = z_{0SF}$). The fit shown in Figure 5 for stations 55-60 is typical of a situation in which 2 layers are required and that for station 70-85 is typical of a situation where only a single layer is warranted. In situations where the fit was made using 0.081 m/s and $4.5 \cdot 10^{-6}$ m, or one of these parameters was nearly as good as the best fit, it was selected because these were the parameters obtained from fitting the most accurate empirical profile for the center of the river (composed of values averaged in time and for stations 70-85 in Figure 5). In all analyses, including those for data from individual stations, the correction for density stratification was determined from theory using the parameters obtained from the sand concentration profiles.

3.2.3. Secondary Circulation. Results of the analyses of the time-averaged velocity-profile measurements for each station are presented in Table 3. Only the profiles at stations 50, 55, 60, 95 and 100 show any effects of form drag, and in these cases the best estimate of u_{*T} ranges from 0.12 m/s (50) to 0.16 m/s (55 and 60). The measured velocity profile for station 100 was too complex to be fit reliably. Locations of the stations with 2-layer velocity profiles suggests that the form drag causing the enhanced outer-layer shear velocities is associated with the banks of the river. As shown in Table 3, from stations 55 to 90 u_{*SF} is about 0.081 m/s, and

from stations 65-90 there is no outer layer. For all of the stations between 55 and 95, the velocity field in the lower half of the water column is well predicted using steady, horizontally uniform flow theory with a density-stratification correction.

Although u_{*SF} is essentially constant with respect to cross-stream position between stations 55 and 90, the structure of the velocity field in the upper half or more of the water column varies considerably, systematically, and unusually across the river. This is clearly shown in the time-averaged profiles of the downstream-velocity component profiles for the right half of the river (stations 70-95 in Figure 6). The profiles for stations 75 and 95 are the same as the ones connected with lines on Figure 3. Profiles covering the center 20 m of the river all increase monotonically to the surface, but those toward the right bank show a large, near-surface momentum deficit that increases systematically toward the side of the river. The same situation occurs on the left (c.f., Station 60 in Figure 3). It is more likely that this deficit is a surface momentum phenomenon and not a mid-level velocity maximum because the bottom several meters of the flow behaved systematically and predictably with cross-stream direction (Table 3). Figure 3 demonstrates that these structures are part of the mean velocity field at stations 60 and 95 and are not an artifact of the averaging procedure.

In an unbounded, steady, horizontally uniform flow, the fluid at the top of the river rides on that below and would be moving at the same speed except that there is an added gravitational force on the upper layer causing it to move

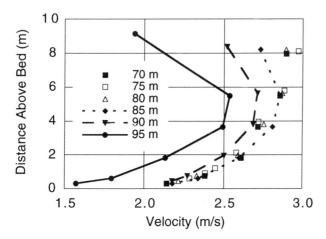

Figure 6. Time-averaged velocity profiles for a series of stations ranging from the central part of the river (70) toward its right edge (95), showing a distinct zone with a low near-surface downstream-velocity component. This zone is largest near the edge of the river and decreases toward its center.

slightly faster. There are only 2 ways that a near-surface low-velocity zone such as the one evident in the near-bank velocity profiles at the National Canyon cableway can be produced in a turbulent flow with a free surface. There has to be either (1) a flow-opposing, near-surface pressure gradient that is higher near the edges of the river, or (2) fluid with a low downstream component of momentum being advected from the margins of the river toward its center by a secondary circulation. The near-surface, low-momentum layer penetrates too far toward the center of the river to be supported by a bank-generated, flow-opposing near-surface pressure gradient alone, such as is associated with near-shore eddies. Away from the banks, there is no mechanism to produce such a pressure gradient. Thus, it is likely that a secondary circulation was generated by the downstream-velocity component interacting in some manner with the banks. The direction of the circulation was toward the center of the river near the surface and toward the banks near the bottom. Owing to the simple structure of the turbulent flow near the bottom in the central part of the river, the circulation had to be driven from the near bank region by an interaction between the downstream flow and the topography of the sloping banks. Moreover, it was likely related to the high river-edge turbulence levels and the high river-edge total shear stresses related to form drag.

Other evidence supports this hypothesis. During the steady high flow, intense boils bringing sand to the river surface were observed along both margins of the river from the banks, the boat, and the cableway. In addition, floating woody debris, which posed a substantial hazard to the

instrumentation, moved in a somewhat organized pattern near the middle of the river. The debris typically collected along irregular lines in surface-flow convergence zones. These convergence zones were systematically but irregularly connected, and below them, moderate to strong vertical-velocity components were indicated by downward advecting leaves and twigs. The sinking flow was fed in a pulsating manner by the boils on the river margins. As a consequence, the convergence zones (clearly delineated by woody debris) moved back and forth in response to the highly variable, boil-produced cross-stream velocity components. The lines of woody debris were being pushed to one side of the river in some places and to the other side in other places. The river also was lighter in color near its margins, where sand was upwelling, and darker over the downwelling region near the center of the channel. Video tapes made during the high flow show these same optical features in other parts of the system, suggesting a similar hydrodynamic situation.

3.2.4. Discussion of Velocity Field. The cause of the secondary circulation (Figure 7) can be inferred by consideration of the nature of the channel banks and the salient features of the velocity and turbulence fields. The margins of the approximately trapezoidal channel upstream from the National Canyon cableway are not straight in the streamwise direction, owing to talus that forms the banks. These streamwise undulations in the sloping river banks (Figure 1) act as topographic elements on which the downstream flow produces form drag (Figure 5; Table 3). The consequence is a substantially enhanced total boundary shear stress near the river margins and a reduced skin-friction shear stress on the actual bed of the river. The streamwise undulations also force the downstream flow to rise on their upsloping sides.

When the topographic elements are steep enough, the flow separates at the crest, and inertia carries the uplifted fluid both downstream and toward the river surface. The upward acceleration produces a low-pressure region just downstream from the crest and causes water from the river bottom to flow into and up the separation zone toward the bank. In this manner, the rising fluid is replaced on the downstream side of the undulation. In a 2-dimensional flow, there would be no source from which to replace uprising fluid and that fluid, instead, would be forced back toward the river bed on the downstream side of the obstruction by the low pressure. In contrast, this is not the case in the fully 3-dimensional situation over sloping channel margins near the bases of upsloping banks. Rather, the low-pressure zone on the downstream side of the undulations causes lateral flow along the river bed. The small-scale, 3-dimensional aspects of this secondary flow cause it to interact nonlin-

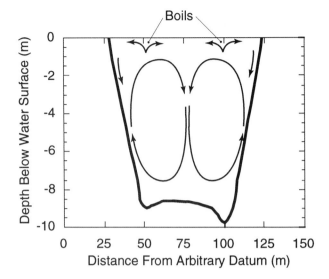

Figure 7. Diagram showing direction of secondary circulation and general location of boils.

early with channel-depth scale vorticity in the turbulence field, producing boils, rather than a systematic upward vertical velocity component field.

When the upward-traveling cores of the boils reach the surface of the rive, they decelerate to produce a bulge in the surface that was estimated to be as much as 0.3 m high. This small-scale, approximately symmetric bulge in the river surface accelerates the flow radially over a distance on the order of a local river depth. The streamwise components of these perturbations in the velocity field add to the turbulent kinetic energy but do not increase the mean downstream velocity component. In contrast, the cross-river components of velocity produced by the boil-caused humps in the river surface combine to produce a net cross-stream flow centered at the upwelling zone. The shoreward component of this flow decelerates at the bank, causing the river surface to rise and the fluid to sink. This, in turn, counteracts any tendency for the upwelling zone to expand toward the bank, which would allow it to increase in size and make possible a less chaotic upwelling. The boil-production zone cannot expand in the other direction, because the irregular channel margins are required to produce the initial upward velocity.

The pulsating cross-stream velocities from the 2 sides of the river transport low-momentum fluid back and forth past the center of the channel producing lower downstream near-surface velocities that are less than otherwise predicted at all stations including the one in the middle of the river (Figures 5, 6). Suspended sand is carried upward in the boils and transported both toward the banks and toward the center of the river. The sand carried toward the banks is deposited rapidly at the channel margins. Sand carried toward the channel center is deposited over a broad zone, and perhaps, as at the National Canyon site, maintaining a mid-channel bar. Near the National Canyon cableway, the upwelling during the controlled flood was nearly symmetrical, but at other locations during this event it appeared to be confined to one side of the river or the other.

In summary, the velocity field at the National Canyon cableway during the steady high-flow period of the 1996 controlled flood displayed a set of consistent features. These features delineate a horizontally uniform downstream velocity with a relatively constant shear velocity and a z_0 controlled by grain roughness in the central 40 m of the channel, interacting with a complex near-bank flow through an inertially driven secondary circulation. Closer to the banks, the primary flow also is horizontally uniform, but it is divided into an internal boundary layer scaled by the skin-friction shear velocity, and an outer region scaled by a total shear velocity that is dominated by form drag on rockfall-produced topographic elements. The z_0 for the internal boundary layer is related to the grain roughness of the sloping river banks, and z_0 for the outer layer is enhanced by form drag on the topographic elements. Superimposed on the 2-part, horizontally uniform downstream flow is a secondary circulation driven by topographically forced upwelling near the banks (Figure 7). In addition to producing additional resistance through form drag, topographic elements in the form of undulations in the plane of the sloping bank with crest lines perpendicular to the direction of primary flow cause strong upward vertical-velocity components. This upward flow is the same scale (namely the order of the flow depth) as the energy-bearing eddies in the primary flow and interacts with them to generate boils. In this manner, the systematic upward mean-velocity component is converted to an intense pulsating one with the same spatially averaged value.

The upwelling, manifested as boils, is concentrated over the base of the upsloping banks (Figure 7). As the boils rise to the surface, they cause the air-water interface to dome upward and the time-dependent sloping free-surface to produce bankward flow inshore from the boil line, strong pulsating cross-stream velocities on the riverward side, and high variability in all components of the near-surface velocity field in the general vicinity of the upwelling zone. Owing to the small diameter of the dome, the effects of the boils are concentrated near the river surface. The circulation is fully 3 dimensional and the local near-surface flow acts as if independent of the near-bottom velocity, although they are coupled by the topography-produced, non-hydrostatic pressure field. At the base of the upsloping banks, water is drawn toward and up the banks on the downstream sides of the crests of the undulations by this pressure gradient and replaces the fluid carried upward by the boils. The circulation is closed by sinking of the fluid in the central part of the river, both over broad regions and along convergence lines. The measured suspended-sediment concentration field was imbedded in this velocity field.

3.3. Suspended Sediment

3.3.1. Background. The suspended-sediment concentration field in a steady, horizontally uniform flow arises from a balance between upward diffusion of sediment induced by flow turbulence and the downward advective transport of the sedimentary particles due primarily to the settling of the grains [*Graf*, 1971; *Yalin*, 1972; *Middleton and Southard*, 1984]. The scale for upward turbulent diffusion of sediment grains is the shear velocity (u_*), the scale for downward advection of grains of diameter D_n is the still-water settling velocity (w_n) of this material, and the scale for upward penetration of a sedimentary particle into the water column as a result of turbulent diffusion is the ratio of these parameters times von Karman's constant k [*Graf*, 1971; *Yalin*, 1972; *Smith*, 1977]. The inverse of this ratio is called the Rouse number,

$$p_n = w_n/(k \cdot u_*). \qquad (4)$$

The Rouse number depends on the size D_n of the particles through the settling velocity. Near a sediment bed in a steady horizontally uniform flow, suspended-sediment concentrations vary approximately as the negative power of distance above the bed. The exponent in this relation is the Rouse number. This variation is modified by the effects of stable density stratification produced by suspended sediment [*Smith and McLean*, 1977a; *Gelfenbaum and Smith*, 1986]. Stable density stratification inhibits turbulent fluctuations and reduces the upward diffusion of suspended particles in a calculable manner that depends on the vertical component of the concentration gradient [*Smith and McLean*, 1977b; *Gelfenbaum and Smith*, 1986; *Kachel and Smith*, 1989; *McLean*, 1992].

There must be a source of available sediment just above the river bed that feeds an upward turbulent diffusion of sediment to the turbulence-related fluctuating velocity field in the fluid and the associated random-pressure gradient field. That source is the top of a layer next to the river bed in which the fluid and collisional forces on the sediment grains cause some of these grains to move by hopping, called saltation. When the fluctuating pressure gradients that cause water parcels to move randomly in the turbulent

fluid are sufficient in magnitude to cause a saltating sediment grain to move upward from the top of its trajectory, in spite of its weight, then this grain can go into suspension. Near the bed, the root-mean-squared fluctuating vertical velocity in a turbulent flow is very nearly equal to the shear velocity. Therefore a criterion for estimating whether or not a sediment grain will go into suspension is $w_n < u_*$ [*Wiberg*, 1987].

The suspended load in a river is supported by sediment grains of the same size in the saltation layer and, hence, at the sediment-water interface. The composition of the bed surface controls the composition of the suspended load. In the case of fine sediment in the presence of significant amounts of coarser material, the concentration of the fine sediment in the saltation layer and on the bed can be extremely small. The concentration is very sensitive to the mixed-grain-size saltation-layer processes [*Hunt*, 1954; *Smith*, 1977; *Smith and McLean*, 1977b]. In general, the near-bed concentration of suspended sediment of a particular size (or type) depends on the fraction of the bed covered by saltating sediment and the size distribution of the saltating material. Sediment that is rolling (e.g., granules and pebbles on a sand bed) rather than saltating rarely contributes to the suspended load, except to reduce the bed-area covered by saltating material. It is extremely difficult to sample the top layer of the sand bed on the river bottom accurately, especially if the bed is distorted by bedforms or the sand is lodged in the interstices of pebbles and cobbles, as is common on the bed of the Colorado River, including at the National Canyon cableway. For these reasons, in suspended-sediment transport problems it often is more accurate to sample the suspended load near the sediment-water interface, but above the saltation layer, which is only a few median grain diameters in height, than to try to sample the bed surface and calculate the suspended-sediment concentration at the top of the saltation layer. In the National Canyon investigation, emphasis was placed on measurements of both silt-plus-clay and sand concentrations in the water column, the lowest samples having been collected at 0.15 m above the river bed. Collection of accurate bed-surface samples was considered impossible, and collection of bed samples had previously proven unreliable at this site.

3.3.2. Suspended Silt and Clay. The suspended-sediment samples collected at the National Canyon cableway were separated into silt-plus-clay and sand [*Konieczki et al.*, 1997]. Silt and clay often are distributed uniformly from bank-to-bank and in the vertical in rivers that transport fine sediment (sand, silt, and clay), because of the low settling velocities of these grain sizes [*McNown and Malaika*, 1950; *Dietrich*, 1982]. The near-bed (z = 0.15 m) silt-plus-clay concentrations for the central portion of the Colorado River (averaged for stations 70-85) are higher than those for the entire deep part of the cross-section (averaged for stations 50-100) and display cross-stream uniformity in sediment concentration only after day 2 (March 29) of the flood. These data, normalized by their values on day 6 (April 2) of the steady, high-flow period, are shown in Figure 8, along with similarly normalized data for the near-bed suspended-sand concentration (averaged for stations 70-85). The daily profiles of normalized concentration (averaged for stations 70-85) also display vertical uniformity (Figure 9) after the second day of steady high flow.

The close correspondence of the normalized near-bed concentrations of silt-plus-clay and sand after day 2 indicates that the ratio of silt-plus-clay to sand in the river bed was essentially constant by that time. Once this equilibrium had been established, the 2 suspended-sediment fields evolved in parallel, and accurate measurement of the time-dependent suspended-silt concentration at any point in the cross-section would have provided an extremely accurate estimate of the time evolution of the near-bed suspended-sand concentration in the center of the river. If, in addition, the asymptotic ratio of the silt-to-sand concentrations and the size of the sand being transported by the flow were known, then the temporally changing suspended-load transport could have been calculated from the silt-concentration measurements.

During the first day of the 1996 controlled flood, the silt-plus-clay concentration was higher in the center of the channel than near the margins, and it was higher than the normalized near-bed sand concentration (Figure 8). During this initial transient period, silt apparently was winnowed out of the bed of the river and was transported or diffused from the center of the river to the sides where it was deposited along with sand on the banks and in the eddies. Silt, like sand, must be supported by material of that size in the bed when in suspension. If mixed with significant amounts of sand in the suspended load, the concentration of silt in the bed can be in a lower ratio to the other suspendable material than for the flow [*Hunt*, 1954; *Smith*, 1977]; nevertheless, the silt concentration in the river bed had to be much higher during the first 2 days of the controlled flood than after that. Silt and clay is likely to have accumulated on the channel bottom during the prolonged period of relatively low flow in the years before the flood.

Close examination of the normalized silt-plus-clay profile in Figure 9 shows that it declined slightly with height above the bed. Measurement of the slope of this profile gives a Rouse number of 0.051, which indicates that the silt-plus-clay is mostly coarse silt. The deviation of the

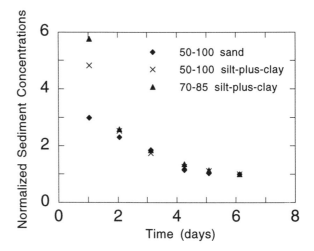

Figure 8. Normalized near-bed silt-plus-clay concentrations averaged for stations 70-85 and 50-100 respectively, and sand concentrations averaged for stations 70-85. These data were normalized by the respective values on the last day of steady high flow during the controlled flood (day 6).

bottom set of measurements from the downward projection of this line is a consequence of settling sand on the suspended silt that reduces the silt concentration in the river bed required to support the silt flux.

3.3.3. Suspended Sand. The suspended-sand measurements indicate that the bed rapidly evolved during the first several days of the high flow and that approximate equilibrium was achieved by day 2. The cross-sectional and temporal structures of the suspended-sand concentration field are more complex than the silt field, because the Rouse number of the sand sizes is strongly dependent on the changing settling velocity of the mass of sand. Moreover, the suspended-sand concentration profiles cannot be normalized as could be done with the suspended silt-plus-clay profiles because their shapes depend sensitively on the Rouse number. The temporal evolution of the suspended-sand field for the central part of the river (70-85) is shown in Figure 10.

The theoretical curves that have been fit to the suspended-sand concentration data (Figure 10) include the density-stratification correction and have been fit to both an outer and an inner region. When the sediment bed is composed of a wide variety of grain sizes, concentration profiles for each size must be calculated and then added together [*Smith*, 1977; *Smith and McLean*, 1977b]. *Smith* [1977] shows that this produces a concentration-weighted average settling velocity that varies with distance from the bed. In an empirical analysis of suspended-sediment concentration data, therefore, the analysis must be broken

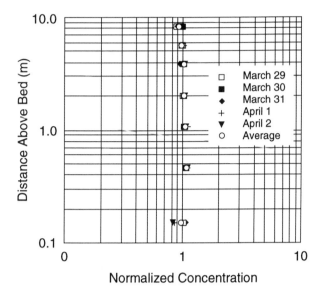

Figure 9. Normalized daily silt-plus-clay concentration profiles for stations 70-85 compared with time-averaged values. These profiles have been normalized by the vertical average rather than the value at the lowest level because of the higher variability of the value at the bottom. The slight slope of the top 6 levels of the profile indicates that the dominant size is coarse silt.

into several layers, the number of which depends on the density of the measurements and the complexity of the profiles. In the case of the data discussed here, 2 layers were used. There was no significant improvement when more were added to the analyses. These curves and the associated suspended-sand concentration data display very clear decreases in (1) Rouse number (equivalent to an increase in profile slope), and (2) near-bed sediment concentration, both of which persist throughout the entire measurement period.

The decreasing settling velocities for both layers and the decreasing near-bed sand concentration from these profile analyses are shown in Figure 11. In the upper layer, the settling velocity decreases by only 38% from 9.0 to 6.5 m/s. In the lower layer, however, the settling velocity drops by almost a factor of 3 from nearly 18 to 6.5 m/s. The associated sand sizes in the upper layer range around 0.125 mm, and in the lower layer they decrease from 0.22 to 0.11 mm. All of the profile-determined settling velocities are within the range of error of the settling tube measurements. A decreasing grain size for a fixed near-bed concentration would result in an increased suspended-load transport. In the case of the controlled flood, the decreasing near-bed concentration caused a decrease in suspended-load transport in spite of the decrease in settling velocity. This decrease most likely was the result of a decrease in the area

of the bed covered with suspendable sand, perhaps due to the encroachment of pebbles from the eroding channels lateral to the mid-channel bar (shown in Figure 2).

The question arises as to whether or not the settling velocity decreased similarly at the channel margins. An analogous but smaller decrease in settling velocity is shown near the bed at station 55-60 in Figure 12. These profiles are affected by an upward-velocity component of significant but unknown magnitude. Comparisons of 2 profiles from the channel margin and 2 from the center are shown in Figure 13. These indicate that the changes from March 28-29 to April 1-2 are essentially the same at both locations. Consequently, it appears that the suspended sediment became finer everywhere in the cross-section, while at the same time the suspended load transport decreased. The most likely explanation for this result is a reduction in bed-area covered by suspendable material upstream from the National Canyon cableway. This result is different from that found by *Rubin et al.* [1998] at the Grand Canyon cableway. In their situation, the entire perimeter of the river was covered with sand, so a reduction in bed-area coverage was not possible and the adjustment in suspended-load

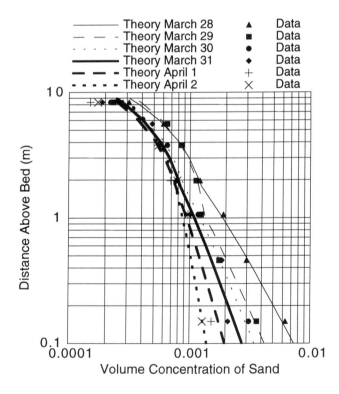

Figure 10. Channel center (70-85), daily sand concentration profiles. These data show a clear temporal trend, but they cannot be normalized like the silt-clay profiles because the profile shape changed as the Rouse number decreased with time.

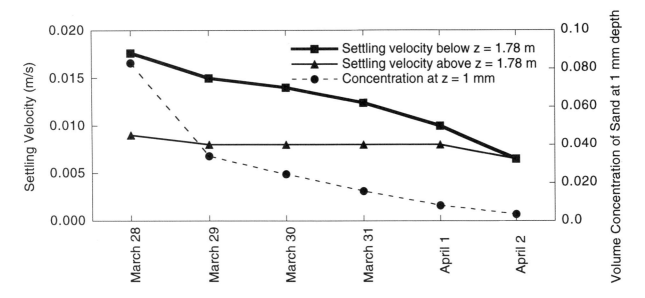

Figure 11. Inner- and outer-layer settling velocities and near-bed suspended-sand concentrations in the central part of the river (70 through 85) as functions of time during the controlled flood.

transport had to occur through grain-size increase. River beds adjust in many ways to maintain a constant sediment flux, but the one described here and the one described by *Rubin et al.* [1998] are probably the most important ones in the Colorado River through Grand Canyon.

In summary, the suspended-sand concentration field during the steady high flow part of the controlled flood became finer and the near-bed concentration decreased at the National Canyon cableway. This evolution was most obvious in the central part of the river, but the same trend also occurred toward the margins. Data from stations in the center of the river (70-85) display the characteristics of a slowly evolving suspended-sand concentration field in a steady horizontally uniform flow. Toward the channel margins, modifications in the suspended-sand concentration field caused by the secondary circulation reflected the net upward velocity associated with the boil produced upwelling.

3.3.4. Discussion of Suspended-Sediment Results. The velocity, suspended-sand concentration and bed-elevation fields are interconnected. Flow over the irregular margins of the channel appear to have caused a secondary circulation that, in turn, modified the suspended-sand concentration field. The resulting patterns of erosion and deposition altered the height and shape of the mid-channel bar. If the sand being deposited on the bar had been coarsening rather than becoming finer, then dunes eventually would have formed on the bar and they would have altered the roughness of the near bed flow. In addition, if sufficient deposition had occurred on the channel margins, they might have become smooth enough to eliminate the secondary circulation. Fortunately, the patterns were simple enough and the measurements were dense enough to resolve the essential features of the flow and suspended-sediment transport and to model some aspects of them. The flow speed and suspended-sediment concentration data collected at the National Canyon cableway during the controlled flood demonstrated that steady, horizontally uniform flow theory could be used to calculate a primary flow field on which a secondary circulation could be superimposed.

The major consequence of the secondary circulation is to transport suspended sand to the margins of the channel extremely effectively, thereby promoting rapid deposition of channel-margin deposits. This mechanism may also effectively route sand to downstream recirculation zones, owing to the relatively small lateral diffusion that occurs in most rapids. Channel-margin deposits and eddy bars accumulated rapidly downstream from the National Canyon cableway during the 1996 controlled flood. In the first day of high flow, an old distributary channel of National Creek filled with sand. It is likely that similar regions of secondary circulation occurred elsewhere along the river during this event. *Smith and Griffin* [in press] argue that most of the open depositional sites along the Colorado River at the beginning of the flood were nearly full after the first 13 hrs of high flow. If this is true, there must have been some very effective suspended-sediment extraction mechanisms active along the river during the early stages of high flow. This

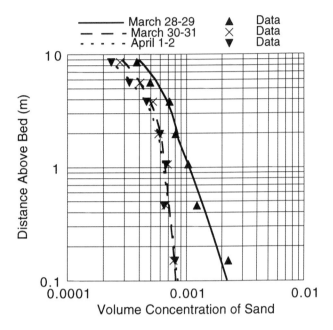

Figure 12. Suspended sand concentration profiles for stations 55-60 for March 28-29, March 30-31, and April 1-2.

type of topographically driven, channel-margin secondary circulation could have been one of them.

4. SUMMARY AND CONCLUSIONS

Measured current-speed fields display a distinct low velocity zone just below the river surface at the channel margins, one that encroaches on the channel from the sides. In addition, measured suspended-sand concentration fields show higher than calculated sand concentrations just below the river surface near the channel margins. These measurements are consistent with a river-margin, boil-driven secondary circulation that resulted from the high channel-edge, downstream-directed velocity being forced upward over rock-fall-produced irregularities in the channel margins. Owing to the channel-depth scale of the upward forced flow, the upward flow was able to interact with the turbulence in the primary flow and become manifested as boils. The irregular, but generally bank parallel, boil line was observed by the field crew from the talus, the cableway, and a survey boat. Also, mid-river surface convergence zones were observed from these locations. The secondary circulation formed on both sides of the river and produced mean lateral velocity components that efficiently carried suspended sand to the channel margins where it was either deposited on the river bank or carried downstream into nearby recirculation zones.

Near-bed suspended silt-plus-clay concentrations and near-bed suspended-sand concentrations, when normalized using their respective values on the last day of the controlled high flow, trace the same curve with time during the event after the second day. This demonstrates that the structure, percent coverage by sand, and sedimentology of the river bed, which together control the suspended-load transport, had come to equilibrium with the flow by the second day. After that, there was a slow evolution toward a simpler cross-stream structure, a lower percent coverage of the river bed with sand, and a finer suspended load at the National Canyon cableway.

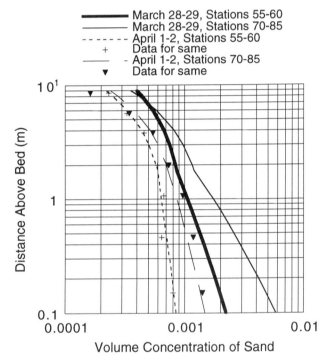

Figure 13. Suspended sand concentration profiles for stations 55-60 and 70-85 for March 28-29 and April 1-2.

Acknowledgments. The field work that formed the foundation this paper represented a huge effort on the part of a large group of devoted USGS employees. Julia B. Graf, the Project Chief in the Arizona District Office, took care of all of the logistics for the experiment and provided the author with a truly outstanding field crew. It was her effort that enabled a first rate scientific investigation during the controlled flood, not only just at the National Canyon cableway, but throughout the Canyon. In the field, Jeff Phillips, Kent Sherman, and Rick Siedeman each took charge of a major facet of the study and each insured that all data procured was of the highest quality. Also, Rod Roeske and Christry O'Day performed above and beyond the call of duty. Various Geological Survey employees of the Flagstaff Field Office also assisted the field program in critical ways. Good boat operators always are essential to the success of experiments on large rivers and in the coastal ocean. Both Mike Geneous and John Toner tolerated the author's numerous requests for difficult maneuvers with clumsy vessels in turbulent waters and then made the maneuvers look easy. Finally, I want to thank Ellie Griffin, Mary Hill, and Jack Schmidt for numerous very productive scientific discussions on issues related to flow and sediment transport in the Colorado River.

REFERENCES

Christensen, T., *Suspended sediment transport in the Colorado River*, 113 pp., unpubl. M.S. thesis, Univ. Wash., Seattle, WA, 1993.

Dietrich, W.E., Settling velocity of natural particles, *Water Res. Res.*, 18(6), 1615-1626, 1982.

Einstein, H.A., and N. Chien, Effects of heavy sediment concentration near the bed on velocity and sediment distribution, *Institute of Engineering Research, Rept. no. 8*, Univ. Calif., Berkeley, CA, 1955.

Gelfenbaum, G.R., and J.D. Smith, Experimental evaluation of a generalized suspended sediment transport theory, in *Sedimentology of shelf sands and sandstones*, ed. R.J. Knight and J.R. McLean, Can. Soc. Petrol. Geol., Memoir II, 133-144, 1986.

Graf, J.B, J.E. Marlow, P.D. Rigas, and S.M.D. Jansen, *Sand-storage changes in the Colorado River downstream from the Paria and Little Colorado Rivers, April 1994 to August 1995*, U.S. Geol. Surv. Open-File Rept. 97-206, 41 pp., 1997.

Graf, W.H, *Hydraulics of sediment transport*, 513 pp., McGraw-Hill, New York, 1971.

Griffin, E.R., *Use of geographic information system to extract topography for modeling flow in the Colorado River through Marble and Grand Canyons*, 113 pp., unpubl. M.S. thesis, Univ. Colorado, Boulder, CO, 1997.

Hunt, J.N., The turbulent transport of suspended sediment in open channels, *Proc. Royal Soc. London, Series A, 224*, 332-335, 1954.

Kachel, N.B., and J.D. Smith, Sediment transport and deposition of the Washington continental shelf, in *Coastal oceanography of Washington and Oregon*, ed. M.R. Landry and B.M. Hickey, pp. 287-348, Elsevier Oceanography Series, 47, Amsterdam, 1989.

Konieczki, A.D., J.B. Graf, and M.C. Carpenter, *Streamflow and sediment data collected to determine the effects of a controlled flood in March and April 1996 on the Colorado River between Lees Ferry and Diamond Creek, Arizona*, U.S. Geol. Surv. Open-File Rept. 97-224, 55 pp., 1997.

Long, C.E., P.L. Wiberg, and A.R.M. Nowell, 1993, Evaluation of von Karman's constant from integral flow parameters, *J. Hydraul. Engin.*, 119, 1182-1190, 1993.

McKnown, J.S., and J. Malaika, Effects of particle shape on settling velocity at low Reynolds number, *Trans. Amer. Geophys. Union*, 31, 74-81, 1950.

McLean, S.R., On the calculation of suspended load for noncohesive sediments, *J. Geophys. Res.*, 97, 5759-5770, 1992.

Middleton, G.V., and J.B. Southard, *Mechanics of sediment movement*, Soc. Econ. Paleon. Min., short course no. 3, 401 pp., Tulsa, OK, 1984.

Nikuradse, J., Getsetzmassigkeit der turbulenten Strommungen in glatten Rohren, *Forschungsheft 356*, 36 pp., Kaiser Wilhelm Inst. Fur Strommungsfrosch., Gotteingen, Germany, 1932.

Nikuradse, J., Strommungsgesetze in raughen Rohren, *Forschungsheft 361*, 22 pp., Kaiser Wilhelm Inst. Fur Strommungsfrosch., Gotteingen, Germany, 1933a.

Nikuradse, J., *Laws of flows in rough pipes*, National Advisory Committee for Aeronautics, Tech. Memo. 1292, trans. A.A. Briclmaier, 1933b.

Rattray, M., Jr., and E. Mitsuda, Theoretical analysis of conditions in a salt wedge, *Estuar. Coast. Marine Sci.*, 2, 373-394, 1974.

Rubin, D.M., J.M. Nelson, and D.J. Topping, Relation of inversely graded deposits to suspended-sediment grain-size evolution during the 1996 flood experiment in Grand Canyon, *Geol.*, 26, 99-102, 1998.

Schlichting, H., *Boundary layer theory*, 817 pp., McGraw-Hill, New York, 1979.

Smith, J.D., Modeling of sediment transport on continental shelves, in *The Sea: Ideas and observations on progress in the study of the seas*, ed. E.D. Goldberg, pp. 538-577, John Wiley Sons, New York, 1977.

Smith, J.D., and E.R. Griffin, Flow, suspended sediment transport, and bank deposition in the Colorado River during the 1996 test flood, *Ecol. Applic.*, in press, 1999.

Smith, J.D., and S.R. McLean, Spatially averaged flow over a wavy surface, *J. Geophys. Res.*, 82(12), 1735-1746, 1977a.

Smith, J.D., and S.R. McLean, Boundary layer adjustments to bottom topography and suspended sediment, in *Proceedings of the 8th International Liege Colloquium on Ocean Hydrodynamics*, ed. J.C.J. Nihoul, pp. 123-151, Elsevier Publishing Company, New York, 1977b.

Topping, D.J., *Physics of flow, sediment transport, hydraulic geometry, and channel geomorphic adjustment during flash floods in an ephemeral river, the Paria River, Utah and Arizona*, 405 pp., unpubl. Ph.D. diss., Univ. Washington, Seattle, WA, 1997.

Vanoni, V.A., Transportation of suspended sediment by water, *Trans. Amer. Soc. Civil Engin.*, 111, 67-133, 1946.

Wiberg, P.L., *Mechanics of bedload transport*, 132 pp., unpubl. Ph.D. diss., Univ. Wash, Seattle, WA, 1987.

Wiele, S.M., and J.D. Smith, A reach-averaged model of diurnal discharge wave propagation down the Colorado River through the Grand Canyon, *Water Res. Res.*, 32, 5, 1375-1386, 1996.

Yalin, M.S., *Mechanics of sediment transport*, 290 pp., Pergamon Press, Oxford, 1972.

J. Dungan Smith, U.S. Geological Survey, 3215 Marine Street, Boulder, Colorado 80303; email: jdsmith@usgs.gov

Topographic Evolution of Sand Bars

E.D. Andrews, Christopher E. Johnston

U.S. Geological Survey, Boulder, Colorado

John C. Schmidt

Department of Geography and Earth Resources, Utah State University, Logan, Utah

Mark Gonzales

Grand Canyon Monitoring and Research Center, Flagstaff, Arizona

Sand bars deposited in lateral separation eddies are an essential biological and recreational resource of the Colorado River downstream from Glen Canyon Dam. Since 1986, however, sustained discharges substantially in excess of the power-plant capacity have not occurred, and approximately half of the sand bars that existed in 1986 have been degraded by erosion and encroachment of vegetation. A primary purpose of the 1996 controlled flood release from Glen Canyon Dam was to measure the rate of sand deposition and erosion as well as to observe the adjustment of sand bar topography during a period of sustained high flow. Repeated, detailed bathymetric surveys of 5 eddies were made before, during, and after a flood of 1275 m^3/s for 7 days to determine the topographic evolution of sand bars. Two of the eddies are located upstream of the Little Colorado River, the primary source of sand to the Colorado River through Grand Canyon, and 3 eddies are located downstream of the Little Colorado River. The topography of sand bars in all 5 eddies adjusted rapidly during the first several hours of the flood. Sand bars aggraded and degraded by as much as 3.5 m within less than 24 hrs. The general pattern of deposition and erosion observed during the first day persisted to varying degrees throughout the flood, even though a few to several thousand cubic meters of sand were eroded and deposited from one day to the next. The area of sand exposed above the 565 m^3/s water-surface elevation and available for camping, however, increased in all the eddies studied. Subaqueous mass failures of over steepened portions of the sand bar appeared to occur in all eddies.

1. INTRODUCTION

The Colorado River flows through Grand Canyon in a deeply incised channel confined by bedrock walls and coarse talus slopes. Although the river carries several million tons of sand in an average year, flow velocities are relatively high and extensive accumulation of fine-grained fluvial sediment are limited. Abrupt changes in channel width and depth, however, create numerous local areas for deposition. Typically, sand accumulates in discontinuous strips along the channel margin in zones of recirculating flow called eddies. In these eddies, the near-bank current is in the up-river direction, and the prolonged retention of sediment-laden water entering them from the main channel results in rapid deposition of the suspended sand and silt. Sand bars deposited within recirculation zones during periods of high flow and exposed when the river stage falls can be quite large, typically several hundred to a few thousand square meters in area.

These sand bars are important because, although they are quite numerous, they occupy only a few percent of the river banks. Bedrock and steep talus slopes form the vast majority of the river banks. Sand bars are an essential recreational and ecological resource in Grand Canyon. Thus, sand bars provide nearly all of the campsites used by the more than 22,000 river runners that pass through Grand Canyon annually. The relatively low velocity, and abundant food supply available in eddies make them frequently utilized habitat by the native fish, especially the juveniles. Sand bars also form an important substrate for riparian vegetation.

Within a few years following the completion of Glen Canyon Dam, river runners, hydrologists, and other long-time observers of the Colorado River in Grand Canyon began to report the gradual depletion and loss of sand bars along the channel margin. The basis for those reports were largely anecdotal and involved favorite or noteworthy campsites. Like most physical and biological features of the river corridor, sand bars had not been surveyed or studied in any way prior to the construction of Glen Canyon Dam.

The extent and causes of sand-bar degradation in Grand Canyon have been considered by numerous investigations since the mid-1970s [e.g., *Howard and Dolan*, 1979, 1981, *Beus et al.*, 1985, *Schmidt and Graf*, 1990, *Schmidt*, 1992, *Kearsley et al.*, 1994, *Kaplinski et al.*, 1995, *Cluer*, 1995]. A detailed summary of these studies is provided in *Webb et al.* [this volume] and *Schmidt et al.* [this volume]. Understanding and insight gained from these studies resulted in significant modification to the normal operating regime of the Glen Canyon powerplant [*Bureau of Reclamation*, 1996]. The new operating regime was designed primarily to slow the rate of sand-bar erosion, while maintaining the flexibility to follow the regional demand for electrical energy. The scientists and resource managers who devised the new operating regime recognized that simply slowing the rate of sand-bar erosion would be insufficient to maintain them as an important river resource. Accordingly the new operating regime for the Glen Canyon powerplant

provided for the possibility of flow releases in excess of the powerplant capacity to rebuild and maintain sand bars along the Colorado River within the Grand Canyon.

The limited supply of sand available in the Colorado River downstream from Glen Canyon Dam [*Andrews*, 1991] was a major constraint on the goal to replenish and rebuild sand bars. It was widely believed that the gradual erosion and loss of sand bars was due to an insufficient supply of sand in the Colorado River to maintain bars. Sediment budgets calculated for various reaches, however, determined that the sand limitation was not as great as had been believed [*Randle et al.*, 1993]. A January 1993 flood on the Little Colorado River provided support for this conclusion. The flood increased the discharge of the Colorado River below the confluence with the Little Colorado River to approximately 954 m^3/s [*Wiele et al.*, 1966], or almost 35% greater than the normal daily peak powerplant release. Significant aggradation and reconstruction of sand bars, exceeding 1.5 m, occurred as far as 260 km downstream.

Most of the sand contributed to the Colorado River during the flood (~ 4 · 10^6 ton), however, was deposited with the first 20 km downstream of the confluences. Sand deposited on bars far downstream was material stored in the river channel when the flood began. Therefore, the 1993 flood demonstrated that the sand bars could be restored rapidly to an elevation well above the water surface stage of powerplant operations using the existing supply of sand. Recognition of the beneficial effects of the 1993 flood on sand bars prompted the consideration of whether sand bars could be rebuilt throughout Grand Canyon by deliberate releases from Glen Canyon Dam in excess of powerplant capacity. Rebuilding a sand bar, such that the elevation of the bar crest and the area of sand exposed above the water-surface elevation of normal powerplant releases are increased requires the deposition of sand across the top of the bar. Despite the observed effects of the 1993 flood, the efficiency and practicality of a deliberate flood as a river management tool remained unknown.

The effects of a relatively high flow on sand bars upstream of the Little Colorado River was a subject of much uncertainty. This reach receives sediment only from the Paria River and smaller tributaries and has a smaller concentration of suspended sand at a given discharge than the reach downstream of the Little Colorado River. Accordingly, a primary objective of the controlled flood was to determine the deposition rate through time on selected sand bars located upstream and downstream of the Little Colorado River. A second objective was to determine the net sediment budget for several eddies during a period of sustained, high flow. Sand deposited on the crest of a bar may be material that was already stored in the eddy at the beginning of the flood and redistributed within the eddy by the high flow. Alternatively, sand deposited over the bar top may be material that was stored in the channel some distance upstream, suspended by the flood, transported downstream, and carried into the eddy as water was exchanged between the eddy and the main channel current. Although both of these processes are important, redistribution of sand already stored in an eddy when a flood began would aggrade and build the bar top more rapidly than would be possible if most or all of the required materials had to be carried into the eddy by water exchange with the main channel.

As described by *Schmidt et al.* [this volume], eddies along the Colorado River in Grand Canyon are very effective sand traps. The flux of river water into an eddy is quite large. For a typical eddy during the controlled flood, the rate of water exchange between the river and the eddy was estimated to be on the order of 50 m³/s, and the mean residence time of water in an eddy is only about 10 min. The mean residence time of water, however, is 3-5 times longer than is required for a medium-sized sand grain to settle through the water column to the eddy bottom.

Accumulation of sand in an eddy, however, is clearly limited. A bar cannot be built above the water surface or beyond the confines of the eddy. Therefore, the third objective of this investigation was to determine what processes eventually limit the accumulation of sand in an eddy. Is growth of the sand bar eventually limited when the topography of the sand bar suppresses and inhibits the exchange of water between the eddy and the channel? Alternatively, growth of the sand bar may accelerate the flux of sediment out of the eddy, either continuously or episodically, so that a time-averaged balance between sediment accumulation and loss is achieved.

1.1. Morphology of Fan-Eddy Complex

The Carbon Canyon fan-eddy complex is shown in Figure 1. Common geomorphic features of the fan-eddy complex are identified in the legend. The debris fan (A) consists of relatively coarse material transported down Carbon Creek and deposited in and constricting the channel of the Colorado River. Flow accelerates through the constriction and into a rapid downstream. Where the channel widens abruptly, the flow separates, such that flow along the shore is upstream relative to the river. The point of separation, B, typically moves upstream as discharge increases. Vortices shed from the point of separation are advected downstream along the boundary between the channel flow and the eddy to the point of reattachment.

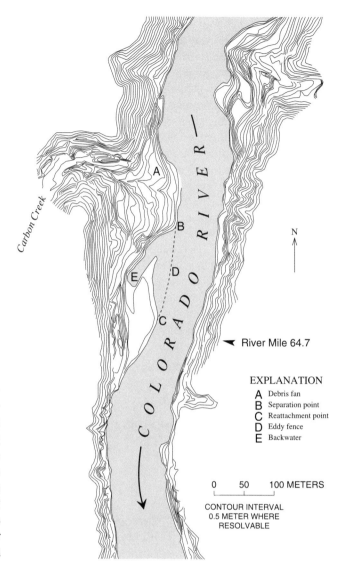

Figure 1. Carbon Canyon eddy-fan complex, river mile 64:7.

Further downstream, streamwise flow along the shore is reestablished. As discharge increases, the point of reattachment generally moves downstream and the eddy becomes larger. At the relatively low discharge shown in Figure 1, approximately 140 m³/s, the point of flow reattachment is at C.

The boundary between the eddy and the channel, D, is a von Karman vortex street [*Schlichting*, 1968]. Grand Canyon scientists and river runners, however, universally refer to the boundary as an "eddy fence." The familiar term will be used herein. A sand bar occupies most of the Carbon eddy in Figure 1. This bar is commonly termed the "reattachment bar" [*Schmidt*, 1992]. A similar deposit of sand overlies the downstream lobe of the debris fan, extending from B part way to E, and is the "separation bar." Areas within the eddy adjacent to the point of flow reattachment and separation are commonly locations of sand accumulation and retention.

At higher flows, sufficient to submerge the reattachment bar, eddy-wide circulation develops including a well-defined upstream current immediately along the shore. This flow scours the "return-current-channel" between the reattachment bar and the bank. When the reattachment bar is emergent, flow in the still inundated part of the return-current channel is nearly stagnant and backwater (E) exists. Water temperature increases and organic material accumulates within the backwater area at low discharges. Under these conditions, the backwater becomes an important ecological habitat, which is otherwise rather uncommon in the Colorado River. Over a period of several months, however, fine sediment and organic material will accumulate in a backwater. Unless the river discharge is periodically sufficient to establish flow in the return-current channel and scour the accumulated material, the backwater will become a marsh and the aquatic habitat lost [*Stevens et al.*, 1995].

2. MEASUREMENT OF SAND BAR EROSION AND DEPOSITION

2.1. Selection of Study Eddies

The study sites described here were selected to complement a multi-scale measurement program that included surveys of 33 eddies located throughout Grand Canyon [*Hazel et al.*, this volume] and an aerial photo analysis of every persistent eddy in two study reaches totaling 31 km [*Schmidt et al.*, this volume]. The Eminence Break eddy (Figure 2) and the Saddle Canyon eddy (Figure 3) are located upstream from the Little Colorado River near Point Hansbrough. The Crash Canyon eddy (Figure 4), Salt Mine eddy (Figure 5), and Carbon Canyon eddy are located in Tapeats Gorge reach downstream from the Little Colorado River [*Schmidt et al.*, this volume].

Two of the study sites are located upstream from the Little Colorado River, the largest source of sand supply to the Colorado River, and 3 sites are located downstream from this source. The 5 eddies selected for this study are geomorphically similar. Each is a fan-eddy complex with a persistent recirculating zone in the lee of a debris fan. There are 24 persistent eddies larger than 2500 m^2 in the Point Hansbrough reach. *Schmidt et al.* [this volume] concluded

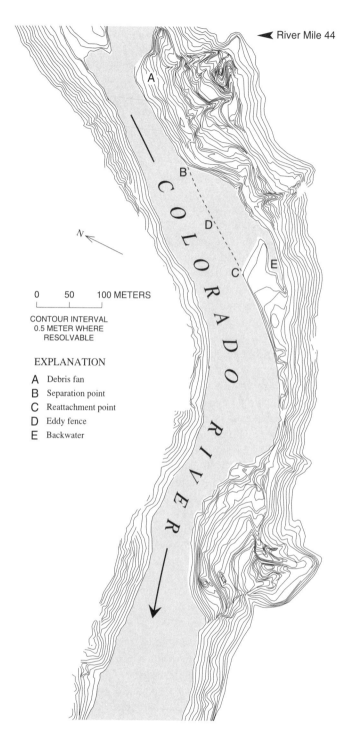

Figure 2. Eminence Break eddy-fan complex, river mile 44.5.

that the area of newly deposited sand as well as the relative extent of erosion and deposition in these 5 eddies were consistent with observed pre-to-post changes at other sand bars in the respective reaches.

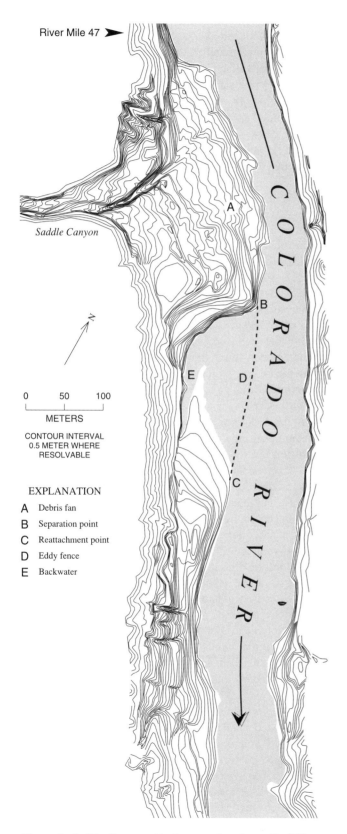

River Mile 47 ➤

Saddle Canyon

N

| 0 | 50 | 100 |

METERS

CONTOUR INTERVAL
0.5 METER WHERE
RESOLVABLE

EXPLANATION

A   Debris fan
B   Separation point
C   Reattachment point
D   Eddy fence
E   Backwater

Figure 3. Saddle Canyon eddy-fan complex, river mile 47.0.

2.2. Bathymetric Surveys

Investigating the topographic evolution of sand bars during the controlled flood required a highly efficient technique for making detailed bathymetric surveys. Over a number of years, surveyors and river guides have developed the equipment and methods needed to make bathymetric surveys of the Colorado River in Grand Canyon. An acoustic depth sounder is mounted on a relatively small inflatable boat powered by an outboard engine. With skillful handling, these boats can be operated along substantial sections of the river, including running upstream through small rapids even at the relatively high discharge of the controlled flood. The position of the boat is tracked by an electronic theodolite and pulsed laser tracking station, located on the river bank. A reflective target is mounted on a mast directly above the transducer. Bench marks referenced to the Arizona State plane coordinates were established at each of the study sites prior to the flood. The tracking station was located over one bench mark with an unobstructed view of the entire eddy, and the other bench marks were used as back sights. The distance, vertical angle, and horizontal angle of the target relative to the tracking station were measured about 4 times per second, and transmitted by radio to a receiver on the boat. The position of the transducer, and depth were recorded in an onboard laptop computer.

Bathymetry of a reach, including the channel and eddy, between local hydraulic controls was surveyed along longitudinal and cross-sectional transects spaced every 10 m. Transects for each study site were determined prior to the flood in the state plane coordinates. To navigate, the boatman relied on a video monitor displaying the transect being surveyed and the actual position of boat. Bathymetric surveys typically required approximately 2-3 hrs to measure the bed elevation at about $1 \cdot 10^4$ points.

In addition to the bathymetric surveys during the flood, each eddy was surveyed immediately before and after the controlled flood at a discharge of about 225 m³/s. Those parts of an eddy emergent at the low pre- and post-flood discharge were surveyed using the conventional total station target-rod method. A contour map and grid of bed elevations were computed from the bathymetric surveys. The gridded eddy bed topography determined from each survey is available and can be retrieved from http://wwwbrr.cr.usgs.gov/projects/GEOMORPH-RIVER/.

Changes in sand bar area and volumes of sand scoured and filled between bathymetric surveys were computed by comparing the gridded bed surfaces within an area bounded by the edge of water and the eddy fence. As will be shown below, substantial depths of scour and fill occurred locally

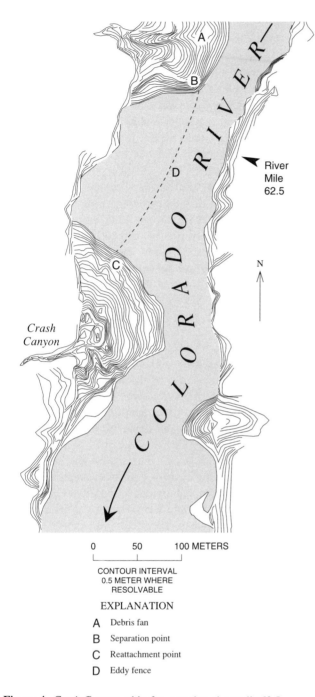

Figure 4. Crash Canyon eddy-fan complex, river mile 62.5.

in all eddies, including points near the eddy boundary. Consequently, the computed eddy-wide volumes of scour and fill depend on the identified eddy boundaries. Occasionally, a transect was repeated to obtain an estimate of the uncertainty in the measurement of bed elevation at the given location. Along a transect, the standard deviation

of bed elevation computed at 1-m intervals from successive surveys is approximately ±0.10 m.

An array of load cells, also, was placed in the Salt Mine eddy to measure sand deposition and erosion during the controlled flood [*Konieczki et al*, 1997]. The array covered only a limited portion of the eddy, but recorded pressure changes every 15 minute, and, thus, provided detailed temporal resolution. [*Konieczki* 1997] describe the load cell measurements and results.

4. TOPOGRAPHIC EVOLUTION OF EDDY SAND DEPOSITS

The volumes of material scoured and filled in each eddy between successive bathymetric surveys are summarized in 1-5. The volumes of sand eroded and deposited were computed by summing the negative and positive changes in bed elevation at nodes on a 2 m x am grid of the eddy bathymetry. The net accumulation or depletion of sand in an eddy between successive surveys is the sum of the volume eroded and the volume deposited. The net volumetric change represents the quantity of sand added to or removed from an eddy during the interval between successive surveys.

The flood, approximately 1275 m^3/s, arrived at the upstream most study site, Eminence Break, at 10:24 PM on March 26, and reached the downstream most study site, Carbon Canyon at 1:12 AM on March 27. Bathymetric surveys of all eddies were completed by late afternoon on March 27, and then repeated daily throughout the flood, as possible. The topography of sand bars in all 5 eddies adjusted rapidly during the first several hours of the flood and large volumes of the sand were redistributed in each eddy. As will be described below, the erosion of sand from some parts of the eddies and deposition in other parts followed a consistent and general pattern. In addition to the redistribution of sand, relatively large net changes, both accumulations and depletion, occurred in every eddy during the initial hours of the flood. At 3 eddies — Eminence Break, Saddle Canyon, and Carbon Canyon — the largest net daily change in sand volume occurred on the first day.

Throughout the flood and in each eddy, the volume of material eroded and deposited in a relatively short time was remarkable. Typically, a few to several thousand cubic meters of material were eroded and deposited from one day to the next. Frequently, more than 10,000 m^3 of material were eroded or deposited between successive bathymetric surveys. Between March 30 and 31, 55,000 m^3 were scoured from the Salt Mine eddy. For comparison the reach-averaged daily flux of sand was approximately

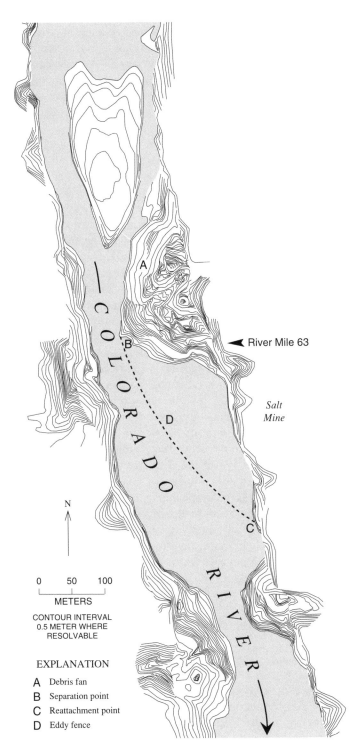

Figure 5. Salt Mine eddy-fan complex, river mile 63.2.

N

0   50   100

METERS

CONTOUR INTERVAL
0.5 METER WHERE
RESOLVABLE

EXPLANATION

A   Debris fan
B   Separation point
C   Reattachment point
D   Eddy fence

River Mile 63

Salt
Mine

TABLE 1.   Volume of bed material scoured and filled in the Eminence Break eddy, March 25 to April 6, 1996

Flood Day		Scour	Fill	Net Change
Begin	End	(m³)	(m³)	(m³)
Pre-Flood	Day 1	22,900	4800	-18,100
Day 1	Day 2	8750	6350	-2400
Day 2	Day 3	8720	4280	-4440
Day 3	Day 4	5090	6130	1040
Day 4	Day 5	6020	3970	-2050
Day 5	Day 6	5140	3370	-1770
Day 6	Day 7	5490	3830	-1660
Day 7	Post-Flood	2020	8340	6320
Pre-Flood	Post-Flood	37,300	12,500	-24,800

bars were typically a significant fraction of the daily volume of sand transported in the river past an eddy.

The volumetric changes reflect both the redistribution of material within the eddy as well as net accumulation or depletion of material in the eddy. Both processes are quite significant. Of the 36 volumetric comparisons between successive surveys shown in Tables 1-5, the volume of material scoured and filled within the eddy is greater than the net accumulation or depletion 19 times. Redistribution of sand in an eddy, including erosion in the center and along the outer portion of the eddy and deposition in the vicinity of the separation and reattachment area, as described above, was very substantial. Accumulations and depletions between successive surveys also were relatively large. In 17 of the 36 comparisons summarized in Tables 1-5, the net accumulation or depletion exceeded the volume of material redistributed. Thus, the flux of sand into or out of an eddy between successive survey was large.

Net accumulation of sand in an eddy was observed in 16 of the 36 comparisons. Among the 16 with net accumulation, 7 were in eddies upstream of the Little Colorado River and 9 of 20 comparisons in were eddies downstream from the Little Colorado River. Net depletion of sand in an eddy was observed in 20 of the 36 comparisons, with 9 of the 16 comparisons upstream of the Little Colorado River and 11 of the 20 comparisons downstream from the Little Colorado River.

4.1. Spatial Pattern of Scour and Fill

A similar pattern of scour and fill was evident in all eddies throughout the flood. Sand was deposited and

50,000 m³/day upstream of the Little Colorado River, and 100,000 m³/day downstream of the Little Colorado River. Thus, the observed daily volumetric changes in eddy sand

TABLE 2. Volume of bed material scoured and filled in the Saddle Canyon eddy, March 25 to April 6, 1996

Flood Day		Scour	Fill	Net Change
Begin	End	(m^3)	(m^3)	(m^3)
Pre-Flood	Day 1	36,000	8620	-27,400
Day 1	Day 2	8010	8970	960
Day 2	Day 3	4150	8940	4790
Day 3	Day 4	8300	6590	-1710
Day 4	Day 5	4820	5850	1030
Day 5	Day 6	3540	6020	2480
Day 6	Day 7	3260	5120	1860
Day 7	Post-Flood	7820	7170	-650
Pre-Flood	Post-Flood	31,500	18,300	-13,200

TABLE 3. Volume of bed material scoured and filled in the Crash Canyon eddy, March 25 to April 6, 1996

Flood Day		Scour	Fill	Net Change
Begin	End	(m^3)	(m^3)	(m^3)
Pre-Flood	Day 1	21,000	18,400	-2600
Day 1	Day 2	15,000	4400	-10,600
Day 2	Day 3	8740	13,200	4460
Day 3	Day 4	13,800	4330	-9470
Day 4	Post-Flood	25,900	6160	-19,700
Pre-Flood	Post-Flood	41,000	5600	-35,400

Table 4. Volume of bed material scoured and filled in the Salt Mine eddy, March 25 to April 6, 1996

Flood Day		Scour	Fill	Net Change
Begin	End	(m^3)	(m^3)	(m^3)
Pre-Flood	Day 1	12,600	14,600	2000
Day 1	Day 2	5550	10,400	4850
Day 2	Day 3	2760	8900	6140
Day 3	Day 4	5790	8980	3190
Day 4	Day 5	55,200	4210	-51,000
Day 5	Day 6	13,100	7060	-6040
Day 6	Post-Flood	905	11,000	10,100
Pre-Flood	Post-Flood	45,000	20,900	-24,100

retained primarily on the separation and reattachment bars, and scoured from the center and outer portion of the eddy. This pattern persisted independently of the net accumulation or depletion. Thus, even when a large net volume of sand was scoured from an eddy, the separation and reattachment bars retained and frequently accumulated sand.

The observations of the Carbon Canyon eddy will be used to illustrate the common pattern of scour and fill, and the resulting topographic evolution of the sand bar. The pre-flood topography of the Carbon Canyon eddy is shown in Figure 6. Changes in the sand bar topography during the first 18 hrs are shown in Figure 7. The maximum deposition exceeded 3.5 m and more than 2 m of sand accumulation was common. Comparing Figures 6 and 7, it can be seen that sand accumulated primarily on the separation and reattachment bars. Immediately prior to the flood, these areas were mostly exposed, or only slightly submerged, at a discharge of 225 m^3/s. Flood deposition near the reattachment point raised elevation of the sand bar crest to within 0.15 m of the water surface. Simultaneously with the deposition of sand on the separation and reattachment bars, sand was scoured over a substantial portion of the Carbon Canyon eddy, primarily from the center of the eddy toward the outer margin of the eddy along the eddy fence. There was, however, very little or no erosion of the river bed immediately outside the eddy. Those areas scoured during the first several hours of the flood were submerged at a discharge of 225 m^3/s before the flood.

The crest of the reattachment bars in both the Eminence Break and Salt Mine eddies, also, aggraded more than 3 m and were only several centimeters under the surface water on March 27. Significantly, substantial deposition of sand occurred on the Eminence Break reattachment bar, while a net volume of nearly 23,000 m^3 was scoured from the eddy. These observations suggest that the maximum thickness of accumulation in an eddy were limited by the water-surface elevation rather than the transport rate or available supply of sand.

The general pattern of deposition on the separation and reattachment bars and erosion in the center and outer portion of the eddy occurred in all 5 eddies and persisted to varying degrees throughout the flood. In addition, sand was scoured from the return-current channel between the reattachment bar and the bank. The return-current channel became wider and deeper. As much as 2.5 m of sand were scoured from the return-current channel during the first 18 hrs of the controlled flood. Those parts of an eddy where material accumulated during the first day of the flood tended to be areas of net accumulation throughout the flood. Similarly, those parts of an eddy where material was eroded

TABLE 5. Volume of bed material scoured and filled in the Carbon Canyon eddy, March 25 to April 6, 1996

Flood Day		Scour	Fill	Net Change
Begin	End	(m^3)	(m^3)	(m^3)
Pre-Flood	Day 1	3460	16,300	12,800
Day 1	Day 2	12,500	4510	-8000
Day 2	Day 3	4550	4870	320
Day 3	Day 4	5110	1830	-3280
Day 4	Day 5	635	9570	8930
Day 5	Day 6	13,900	2760	-11,100
Day 6	Day 7	10,500	2820	-7680
Day 7	Post-Flood	3200	6300	3100
Pre-Flood	Post-Flood	10,100	6540	-3560

during the first day of the flood tended to be areas of net erosion throughout the flood.

The general pattern of erosion and deposition in eddies during the flood, as well as considerable local variations, are illustrated by the comparison of transects constructed from bathymetric surveys on successive days. The variation in bed elevation along two transects of the Carbon Canyon eddy, A-A' and B-B' (Figure 6) are shown in Figure 8. Transect A-A' passes near the point of flow separation, whereas transect B-B' passes near the point of flow reattachment. Transects A-A' and B-B' are aligned approximately perpendicular to the flow direction in the eddy. During the first day, a thick deposit of sand accumulated along the entire length of transect A-A'. The newly deposited material exceeded 3 m in places. Subsequently, sand was eroded along the entire transect. The depth of erosion, however, varied significantly along the transect. Adjacent to the shore, approximately half of the sand thickness deposited on the initial flood day was later removed. Towards the middle of the river, relatively greater amounts of sand were eroded. By the second day, all of the material deposited along the outer half of the transect (channelward) during the first day and more had been scoured. After the flood subsided, the scour depth along the outer portion of the transect averaged slightly more than 1 m and exceeded 2 m in places.

The pattern of erosion and deposition along transect B-B' was generally similar, both spatially and temporally, as that observed along transect A-A'. In specific details, however, there were differences, especially during the first day of the flood. The reattachment bar in the Carbon

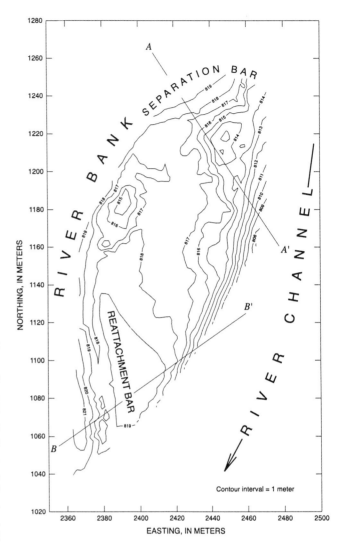

Figure 6. Topography of the Carbon Canyon eddy before the 1996 controlled flood.

Canyon eddy is separated from the shore by the return-current channel. Before the flood began, the reattachment bar had relatively steep-sides and a high, narrow crest. During the initial hours of the flood, sand was eroded along most, though not all, of the transect B-B'. The existing reattachment bar was extensively scoured and the return-current channel enlarged. The outer or channelward half of the reattachment bar along transect B-B', also, was extensively scoured. Maximum depth of erosion exceeded 3 m, whereas, on average less than 2 m of material was removed along most of the transect. The only significant deposition along the transect during the first day was the construction of a new reattachment bar to the channel side of the pre-flood bar. Between days 1 and 2, sand accumulated along

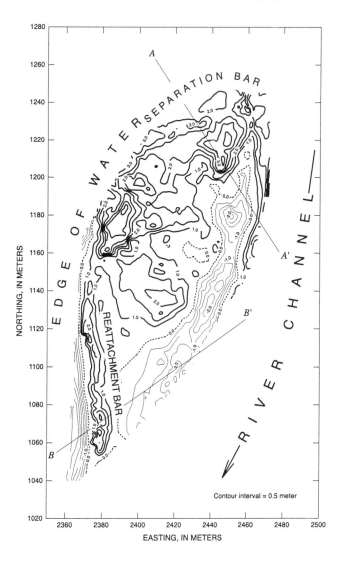

Figure 7. Topographic changes in the Carbon Canyon eddy during the first day of the 1996 controlled flood. Areas of deposition are shown by the thick contour lines and areas of erosion are shown by the thin contour lines.

nearly the entire length of transect B-B'. The crest of the reattachment bar broadened considerably as more than 2 m of material accumulated on both sides of the bar top. Following the bathymetric survey on day 2, the Carbon Canyon reattachment bar aggraded progressively for several days. By the last day of the flood, more than 3 m of sand had accumulated across the bar, and the bar crest was built to within several centimeters of the water surface. Aggradation of the Carbon Canyon reattachment bar during the latter stages of the flood encroached upon the return-current channel somewhat. Even so, the return-current channel remained significantly wider and deeper than it had been prior to the flood.

A primary objective of the controlled flood was to determine whether the area and volume of sand exposed above the normal range of river stage would be increased by a sustained flood substantially greater than the maximum powerplant capacity, about 930 m^3/s. For a couple of days immediately before and after the controlled flood, releases from Glen Canyon Dam were held constant at about 225 m^3/s. Although this discharge is at the lower end of the normal operating range, the sub-aerial volume of the sand bar before and after the flood can be measured quite precisely. Therefore, volumes of sand eroded and deposited above the 225 m^3/s water-surface elevation during of the flood in the 5 eddies studied were computed (Table 6).

The comparison of net volumetric changes between the pre- and post-flood bathymetric surveys includes substantial areas of erosion that are submerged even at the minimum required flow release (140 m^3/s). As described above, the greatest observed scour depths were typically near the center of eddies and outward towards the eddy fence. The volume of sub-aerial deposition exceeded subaerial erosion in 4 of the 5 eddies studied. Erosion exceeded deposition over the sub-aerial part of the eddy deposits only in the Carbon Canyon eddy, where net erosion was less than 1000 m^3. Despite more than 2 m of deposition in the vicinity of the separation and reattachment points, erosion along the shore between these two areas was large.

Changes in the area of exposed sand between the pre- and post-flood surveys is an alternative comparison by which to assess the effect of the controlled flood. Changes in sand bar area above the water-surface elevation of 565 m^3/s, the current maximum expected powerplant release under normal conditions, are shown in Table 7. The changes in sand area are given both in absolute and percentage increase over the duration of the flood. In all of the 5 eddies, the area of sand exposed above the 565 m^3/s water-surface elevation increased. Downstream from the Little Colorado River, the increases varied from 250 m^2 in the Carbon Canyon eddy to 2290 m^2 in the Salt Mine eddy. The percentage increase in area of exposed sand varied from 12% in the Carbon Canyon eddy to 194% in the Crash Canyon eddies. Similarly, sand bars in both of the eddies located upstream of the Little Colorado River, Eminence and Saddle Canyons, increased by approximately 900 m^2 or 25% greater than the bars which had existed prior to the flood. Thus, there is no apparent difference in the response of eddy sand bars during the flood to the increased availability of sand downstream of the Little Colorado River.

The increase of exposed sand-bar area observed during the controlled flood at the 5 eddies studied for this investigation was typical throughout the 400 km downstream from Glen Canyon Dam. *Thompson et al* [1997] evaluated pre-

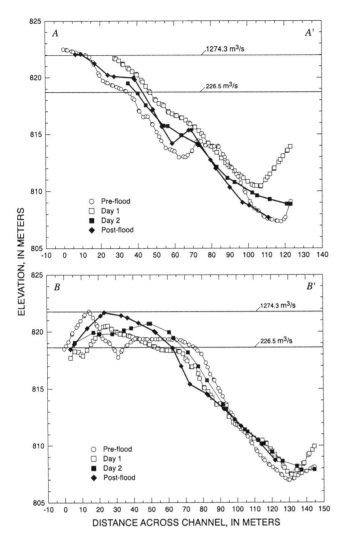

Figure 8. Variation in bed elevation during the flood along transects passing near point of flow separation (A-A′) and flow reattachment (B-B′).

and post-flood photographs of 44 sand bars and found that 82% of the sand bars (36 of 44) were appreciably larger in area following the controlled flood. *Kearsley et al* [this volume] mapped 53 sand bars commonly used as campsites by river runners 2 weeks before and after the controlled flood. They found that sand-bar area available for camping increased substantially at 26 eddies, was unchanged at 21 eddies, and decreased at 11 eddies. *Hazel et al.* [this volume] surveyed the topography of 33 sand bars a few weeks before and after the controlled flood. They found that the volume of sand-bar deposits had increased by 48%, although, the total area of sand bars had increased by only 7%.

TABLE 6. Net volumes of bed material scoured and filled above the 225 m³/s water-surface elevation at 5 study sites during the 1996 controlled flood

Eddy	Sub-aerial Deposit	
	Scour (m³)	Fill (m³)
Eminence Break	5330	7090
Saddle Canyon	2420	4360
Crash Canyon	106	5800
Salt Mine	343	664
Carbon Canyon	4520	3610

4.2. Mass Movements of Sand Bar Deposits

Net daily erosion, exceeding 10,000 m³ of sediment, occurred in each of the 5 eddies studied (Tables 1-5). In the Crash Canyon and Carbon Canyon eddies, large net daily erosion occurred more than once. In the Eminence, Saddle Canyon, and Crash Canyon eddies, large net volumes of sand were eroded either during the first or last day of the flood, and may have been related to the rapid increase or decrease in discharge. In contrast, large daily net volumes of sand were eroded from the Crash Canyon, Salt Mine Canyon, and Carbon Canyon eddies during a period of essentially constant flood discharge.

Rapid and nearly complete erosion of sand bars in some Colorado River eddies has been observed previously [*Cluer and Dexter*, 1994; *Cluer*, 1995]. The circumstances leading up to the erosion of these sand bars, however, were substantially different than existed during the controlled flood. The erosion occurred during periods of very large daily fluctuations in stage and discharge. Furthermore, little or no sand was deposited over the bar top during the days prior to the erosion. *Cluer* [1995] hypothesized that a gradual accumulation of sediment in the channel adjacent to the eddy may have caused a shift in the channel-eddy flow field resulting in the scour of a portion, perhaps most, of the sand bar. Alternatively, it was suggested that fluvial erosion may have removed structural support for a portion of the sand bar and caused additional erosion.

During the controlled flood no discernible change in eddy circulation or size was observed. Furthermore, flow velocities in the eddies were insufficient to erode very large volumes of sand rapidly. The erosion of large volumes of sand from within an eddy in a matter of several hours, at most, must involve a mass movement. Topographic changes in the Salt Mine eddy between day 4 (April 1) and day 5 (April 2) of the controlled flood are shown in Figure 9. A net volume of 51,000 m³ of sand was eroded from the Salt

TABLE 7. Changes in sand bar area above the 565 m³/s water-surface elevation at 5 study sites during the 1996 controlled flood

Eddy	Area Change	
	m²	%
Eminence Break	+875	+25
Saddle Canyon	+900	+28
Crash Canyon	+290	+194
Salt Mine	+2,290	+115
Carbon Canyon	+250	+12

Mine eddy. The maximum depth of scour exceeded 4.5 m. In spite of the very substantial net loss of material, the general pattern of erosion and deposition on sand bars, as described above, is evident. As much as 2 m of sand was deposited over the separation and reattachment bars, while sand was eroded from the center of the eddy outward to the eddy fence.

Some information concerning the stability and possible mass failure of the eddy sand deposit can be determined from the bathymetric surveys. Relatively steep sand slopes are common within Colorado River eddies. The general pattern of topographic change (Figure 8) consisting of deposition near the separation and reattachment points and erosion in the eddy center and along the eddy fence significantly increases the slope of the sand bar side facing toward the eddy fence. For example, maximum slope angles along transects A-A' and B-B' in the Carbon Canyon eddy exceeded 25°. Sand slope of nearly 30° existed elsewhere in the Carbon Canyon, as well as the other four eddies during most of the flood. These slope are only slightly less than the angle of repose for a sand deposit.

As described above, rapid deposition of sand occurred in all of the eddies studied. Such a deposit would initially have very high water content and porosity. A relatively high water content greatly reduces the strength of the deposit and resistance to gravitational forces. Subaqueous mass movements have been initiated on slopes as flat as 10 degrees without additional forces [e.g., an earthquake, *Hampton et al.*, 1978; *Edwards et al.*, 1980; *Schwab and Lee*, 1988]. Failure and movement by a relatively small part of a saturated sand deposit can cause pore-pressure fluctuations sufficient to liquefy a much larger volume of material [*Poulos et al.*, 1985]. Mass movements appear to be the principal process removing large volumes of sand from an eddy and returning it to the river channel. The occurrence and mechanics of these mass movements need to be further investigated, especially as they relate to depositional history and variations in river discharge.

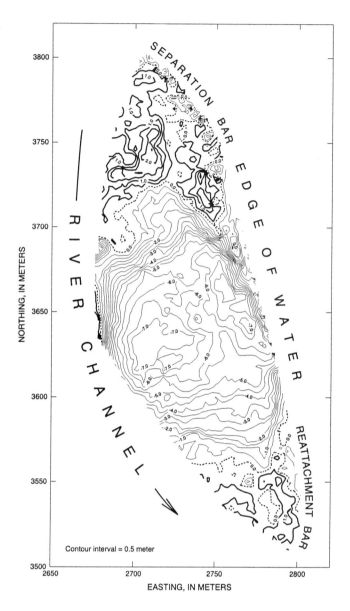

Figure 9. Topographic changes in the Salt Mine eddy between day 4 (April 1) and day 5 (April 2) of the 1996 controlled flood. Areas of deposition are shown by the thick contour lines and areas of erosion are shown by the thin contour lines.

5. CONCLUSION

A primary purpose of the controlled flood release from Glen Canyon during the Spring of 1996 was to measure the rate of sand deposition and erosion as well as the adjustment of sand-bar topography during a period of sustained high flow along the Colorado River through Grand Canyon National Park. Repeated, detailed bathymetric surveys of 5 eddies were made before, during, and after a seven-day flood of 1275 m³/s to determine the topographic evolution of sand bars. Two of the eddies are located upstream of the Little Colorado River, the primary source of sand to the Colorado River through Grand Canyon, and 3 eddies are located downstream of the Little Colorado River.

The topography of sand bars in the 5 eddies adjusted rapidly during the first several hours of the flood. Subsequently, throughout the flood and in every eddy, the volume of sand scoured and filled in a relatively short time was remarkable. Typically, a few to several thousand cubic meters of sand were eroded and deposited within an eddy from one day to the next. These volumetric changes resulted from both the redistribution of material within an eddy as well as the net accumulation or depletion of sand in the eddy. Both processes were significant in every eddy at various times during the flood.

A similar pattern of scour and fill was evident in all eddies throughout the flood. Sand was deposited and retained primarily on the separation and reattachment bars, and scoured from the center and outer parts of the eddy. This pattern persisted independently of the magnitude of net sand accumulation or depletion in the eddy.

All eddies lost material by the end of the flood. Net erosion between the pre- and post-flood bathymetric surveys varied from 3600 m³ in the Carbon Canyon eddy to 35,400 m³ in the Crash Canyon eddies. The greatest depths of scour, however, occurred in those parts of the eddies that are inundated by all flows greater than the minimum sustained dam release under normal conditions (~225 m³/s). When only those parts of the sand bars exposed above the 225 m³/s water-surface elevation are considered, the volume of deposition exceeded erosion in four of the 5 eddies studied. The areal extent exposed above a discharge of 565 m³/s of all 5 eddy sand bars studied in detail during the flood increased from 25 to 90%. Similar increases in the area of sand bars available for camping were observed at eddies throughout Grand Canyon [*Kearsley et al.*, this volume].

Acknowledgments: Many individuals contributed significantly to this investigation. Over a number of years, we have had many fruitful discussions concerning the dynamics of Grand Canyon sand bars and lateral separation eddies with J. Dungan Smith, David Rubin, Jonathan Nelson and Thomas Moody. During the controlled flood, bathymetric surveys were made by a field team that included Tyler Allred, Michelle Brink, Leslie Pizzi, John Nagy, Evan Wallman, Mike Leschin, and Bert Lewis. The bathymetric surveys would not have been possible without the expertise and skill of boatmen, Kenton Gura and Lars Niemi. The draft manuscript was reviewed by Robert Webb, Brian Cluer and Joe Hazel. They made many worthwhile suggestions that have improved the manuscript materially. The support and encouragement provided by David Wegner, Glen Canyon Environmental Studies Office, U.S. Bureau of Reclamation over more than a decade is greatly appreciated.

REFERENCES

Andrews, E.D., Sediment transport in the Colorado River basin, in *Colorado River ecology and dam management*, ed. G.R. Marzolf, G.R., Natl. Academy Press, Washington, D.C., pp. 54-74, 1991.

Beus, S.S., S.W. Carothers, and C.C. Avery, Topographic changes in fluvial terrace deposits used as campsite beaches along the Colorado River in Grand Canyon, *J. Arizona-Nevada Acad. Sci.*, 20, 111-120, 1985.

Cluer, B.L., and L.R. Dexter, *Daily dynamics of Grand Canyon sandbars: Monitoring with terrestrial photogrammetry*, unpublished final report, U.S. Department of the Interior, National Park Service, Flagstaff, AZ, 200 pp., 1994.

Cluer, B.L., Cyclic fluvial processes and bias in environmental monitoring, Colorado River in Grand Canyon, *J. Geol.*, 103, 422-432, 1995.

Edwards, B.D., M.E. Field, and E.C. Clukey, *Geological and geotechnical analysis of a submarine slump, California borderland*, Proc. 12th Ann. Offshore Tech. Conf., Houston, TX, pp. 399-410, 1980.

Hampton, M.A., A.R. Bouma, P.R. Carlson, B.F. Molnia, E.C. Clukey, and D.A. Sangrey, *Quantitative study of slope instability in the Gulf of Alaska*, Proc. 10th Ann. Offshore Tech. Conf., Houston, TX, pp. 2307-2318, 1978.

Howard, A.D., and Dolan, R., Changes in fluvial deposits of the Colorado River in the Grand Canyon caused by Glen Canyon Dam, in *Proceedings of the First Conference on Scientific Research in the National Parks*, ed. R.M. Lin, Trans. Procs. no. 5, Natl. Park Serv., pp. 845-851, 1979.

Howard, A., and R. Dolan, Geomorphology of the Colorado River in the Grand Canyon, *J. Geol.*, 89(3), 269-298, 1981.

Kaplinski, M., J.E. Hazel, and S.S. Beus, *Monitoring the effects of interim flows from Glen Canyon Dam on sand bars in the Colorado River corridor, Grand Canyon National Park, Arizona*, unpubl. final rept. to Natl. Park Service, Flagstaff, AZ, 62 p., 1995.

Kearsley, L.H., J.C. Schmidt, and K.W. Warran, Effects of Glen Canyon Dam on Colorado River sand deposits used as campsites in Grand Canyon National Park, USA, *Regul. Rivers: Res. Manage.*, 9, 137-149, 1994.

Konieczki, A.D., J.B. Graf, and M.C. Carpenter, *Streamflow and sediment data collected to determine the effects of a controlled flood in March and April 1996, on the Colorado River between Lees Ferry and Diamond Creek, Arizona*, U.S. Geol. Surv. Open-File Rept. 97-224, 55 p., 1997.

Poulos, S.J., G. Castro, and J.W. France, Liquefaction evaluation procedure, *J. Geotech. Engin.*, 111, 772-791, 1985.

Randle, T.J., R.I. Strand, and A. Streifel, Engineering and environmental considerations of Grand Canyon sediment management, in *U.S. Committee of Large Dams, Proceedings of Thirteenth Annual Lecture Series*, p. 1-12, 1993.

Schlichting, H., *Boundary-layer theory*, 6th ed., McGraw-Hill Co., New York, 747 p., 1968.

Schmidt, J.C., Recirculating flow and sedimentation in the Colorado River in Grand Canyon, Arizona, *J. Geol.*, 98, 709-724, 1992.

Schmidt, J.C., and J.B. Graf, *Aggradation and degradation of alluvial sand deposits, 1965-1986, Colorado River, Grand Canyon National Park*, U.S. Geol. Surv. Prof. Paper 1493, 74 p., 1990.

Schwab, W.C., and H.J. Lee, Causes of two slope-failure types in continental-shelf sediment, northeastern Gulf of Alaska, *J. Sed. Pet.*, 58, p. 1-11.

Stevens, L.E., J.S. Schmidt, T.J. Ayers, and B.T. Brown, Flow regulation, geomorphology, and Colorado River marsh development in the Grand Canyon, Arizona, *Ecol. Appl.*, 5, 1035-1039.

Thompson, K., K. Burke, and A. Potochnik, *Effects of the beach-habitat building flow and subsequent interim flow from Glen Canyon Dam on Grand Canyon camping beaches, 1996: A repeat photography study*, unpubl. rept., Grand Canyon River Guides, Flagstaff, AZ, 11 pp., 1997.

U.S. Department of the Interior, *Operation of Glen Canyon Dam, Final Environmental Impact Statement*, Bur. Recl., Salt Lake City, UT, 337 p., 1995.

Wiele, S.M., J.B. Graf, and J.D. Smith, Sand depositions in the Colorado River in the Grand Canyon from flooding of the Little Colorado River, *Water Res. Res.*, 32(12), 3579-3596, 1996.

E.D. Andrews, Christopher E. Johnston, U.S. Geological Survey, 3215 Marine Street, Boulder, CO 80303; email: eandrews@usgs.gov

John C. Schmidt, Department of Geography and Earth Resources, Utah State University, Logan, UT 84332

Mark Gonzales, Grand Canyon Monitoring and Research Center, 121 East Birch Street, Flagstaff, AZ 86001

The Effect of Sand Concentration on Depositional Rate, Magnitude, and Location in the Colorado River Below the Little Colorado River

S.M. Wiele

U.S. Geological Survey, Lakewood, Colorado

E.D. Andrews, and E.R. Griffin

U.S. Geological Survey, Boulder, Colorado

A model of flow, sand transport, and bed evolution is used to examine processes during the 1996 controlled flood and to examine the effect of low and high sand concentrations on depositional rate, magnitude, and location. During a flood on the Little Colorado River in January 1993, water discharge in the main stem was increased to about 950 m^3/s and a sand volume sufficient to increase sand concentrations to about twice the sand concentration measured on the first day of the controlled flood was delivered to the main stem and redistributed downstream. This influx of sand led to large deposits in depressions in main channel pools and deposition in recirculation zones that the model predicts originated as separation deposits and spread downstream and into the recirculation zone. In contrast, during the controlled flood at a discharge of 1270 m^3/s sand was eroded from the main channel and the delivery of suspended sand from the main channel to the recirculation zones was focused at the reattachment point. Similar depositional patterns were predicted by the model using sand inputs at the upper and lower bounds of measured sand transport at the controlled flood water discharge. Rates of deposition and deposit volume were proportional to the sand input to the reach.

1. INTRODUCTION

The delivery of sand to the Colorado River in the Grand Canyon has always been variable, depending on the magnitude and duration of seasonal floods in the Colorado River basin, including tributaries local to the Grand Canyon. The closure of Glen Canyon Dam reduced the total sand supply, making tributaries within the Grand Canyon the sole source with which to replenish sand deposits along

the main stem. Thus, the delivery of sand to the Colorado River downstream from the dam now depends on tributary flows, primarily the Paria and Little Colorado Rivers, and the magnitude and duration of dam releases, which transport sand downstream. Consequently, the adjustment of sand deposits to controlled floods [*Webb et al.,* this volume; *Schmidt et al.,* this volume] varies as a function of the sand supply as well as the magnitude and duration of dam releases. In this paper, a model of flow, sand transport, and bed evolution [*Wiele et al.,* 1996] is used to examine the effect of varying sand supplies on depositional rates, patterns, and magnitude in two pools below the confluence with the Little Colorado River (LCR). The Salt reach, named after the salt that forms on the riverside bedrock, is

The Controlled Flood in Grand Canyon
Geophysical Monograph 110
Published in 1999 by the American Geophysical Union

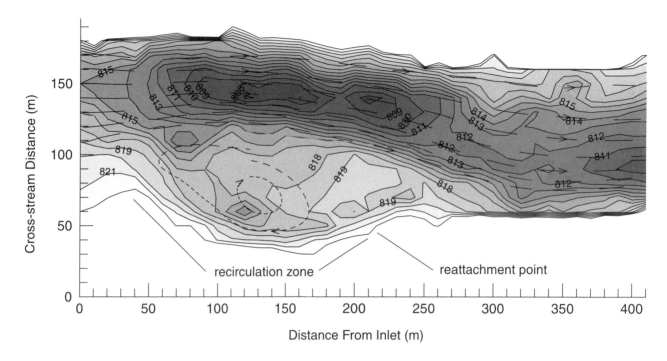

Figure 1. Bathymetric map of the of the Carbon reach at the start of the controlled flood. Contour interval is 1 m. Dashed lines show calculated stream lines at 1270 m³/s. Darker shaded areas are deeper.

located at river mile 63; the Carbon reach, named after the tributary that formed the constriction at the upstream boundary of the reach, is located at river mile 65.

The model was applied to the Carbon reach (Figure 1) using an upstream sand boundary condition derived from field measurements made during the 1996 controlled flood [*Konieczki et al.*, 1997], and the model results were compared with field bathymetric measurements. These results were compared to modeled predictions of bed topography that would have occurred in response to the highest sand concentrations measured prior to the closure of Glen Canyon Dam. The model was also used to predict the Carbon reach bed topography that would have occurred if sand concentrations had been as low as those characteristic of the sustained period of high releases from Glen Canyon during 1983 [*Topping*, 1997]. For the Salt reach (Figure 2), model results for the controlled flood were compared to the topography that resulted from large contributions of sand from a flood on the Little Colorado River [*Wiele et al.*, 1996]. Table 1 shows a summary of the modeled conditions presented in this paper.

2. STUDY REACHES

The channel morphology and flow in the 2 pools modeled in this study are characteristic of the Colorado

River in the 7 km below the confluence with the Little Colorado River. The Carbon and Salt sites are located 2.4 and 5.2 km, respectively, downstream from the LCR [*Webb et al.*, this volume]. *Schmidt and Rubin* [1995] reported that the average channel width of this 7 km reach is 105 m and that the average width of the valley that is bounded by bedrock is only 125 m. They estimated that 75 percent of all alluvial deposits in this 7 km reach are in recirculating zones. The reach is characterized by a series of pools typically less than 1 km long that are bounded upstream and downstream by debris flows that constrict the channel and create rapids or riffles [*Dolan et al.*, 1978; *Webb et al.*, 1989]. Downstream from these fans are recirculation zones that are effective traps of the suspended load. Pools in this reach characteristically have deep holes downstream from the constriction that forms the pool inlet. These deep holes have the potential to effectively, if temporarily, store sand at low water discharge or at high sand transport rates within the main channel, providing a ready source of sand at higher water discharges.

3. MODEL

The numerical methods used in the model are described by *Wiele et al.* [1996] and are based on the finite volume method of *Patankar* [1984] which features a staggered grid

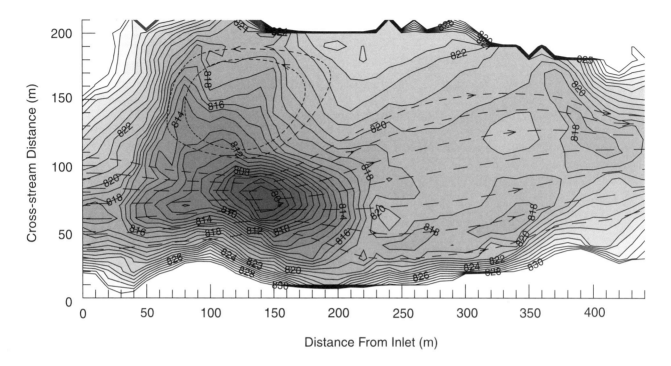

Figure 2. Bathymetric map of the Salt reach at the start of the controlled flood. Contour interval is 1 m. Dashed lines show calculated stream lines at 1270 m³/s. Darker shaded areas are deeper.

and upwind differencing to solve the 3 equations of motion for the vertically-averaged 2-dimensional flow field. A key attribute of the finite volume method is that it conserves mass, a crucial requirement for sediment transport applications. A diffusion-advection equation is used to calculate the 3-dimensional sand concentration field from which the local suspended sand transport can be determined. A turbulence closure is applied to recover the vertical structure of the turbulent mixing and the velocity profile. Local bedload is also calculated based on the local shear stress at the bed, local bed slope, and the grain size. The change in bed configuration over a small time step is calculated from the divergence of the total sand transport rate.

The flow field is calculated by numerically solving the momentum equations in the downstream direction,

$$\frac{\partial u}{\partial t} + u\frac{\partial u}{\partial x} + v\frac{\partial u}{\partial y} - \frac{\partial}{\partial x}\varepsilon\frac{\partial u}{\partial x} - \frac{\partial}{\partial y}\varepsilon\frac{\partial u}{\partial y} +$$

$$g\left(\frac{\partial h + \eta}{\partial x} - S\right) + \frac{\tau_x}{\rho h} = 0 \qquad (1)$$

and cross-stream direction,

$$\frac{\partial v}{\partial t} + v\frac{\partial v}{\partial y} + u\frac{\partial v}{\partial x} - \frac{\partial}{\partial y}\varepsilon\frac{\partial v}{\partial y} - \frac{\partial}{\partial x}\varepsilon\frac{\partial u}{\partial x} +$$

$$g\frac{h + \eta}{y} + \frac{\tau_y}{\rho h} = 0 \qquad (2)$$

and with the continuity equation,

$$\frac{\partial h}{\partial t} + \frac{\partial uh}{\partial x} + \frac{\partial vh}{\partial y} = 0 \qquad (3)$$

where x is the direction normal to the upstream boundary, y is the direction normal to x, u is the vertically-averaged velocity in the x direction, v is the vertically-averaged velocity in the y direction, h is the flow depth, η is the bed-surface elevation, S is the average reach slope, ε is the eddy viscosity, g is the gravity, ρ is the density of water, τ_x is the shear stress in the x direction, and τ_y is the shear stress in the y direction

The shear stress, τ, is related to the velocity by a friction coefficient, c_f:

$$\tau = \rho c_f U^2 \qquad (4)$$

where $U = (u^2 + v^2)^{1/2}$ is the magnitude of the resolved velocity. The x and y components are determined from the relations:

$$\tau_x = \rho c_f u U \qquad (5)$$

and

$$\tau_y = \rho c_f v U \qquad (6)$$

The friction coefficient is defined as

TABLE 1. Flow and sediment conditions for the model applications presented in this paper

	Reach	
	Carbon reach	Salt reach
	controlled flood discharge (1270 m3/s) and sand concentrations	controlled flood discharge (1270 m3/s) and sand concentrations
	historical high sand concentrations at 1270 m3/s	discharge (950 m3/s) and sand concentrations during Little Colorado River flood
	historical low sand concentrations at 1270 m3/s	

$$c_f = \left(\frac{\kappa}{\ln \frac{h}{z_0} - 1} \right)^2 \qquad (7)$$

where κ is von Karman's constant, z is the distance normal to the bed, and z_0 is the roughness parameter, discussed in more detail below.

The eddy viscosity, ε, is defined as:

$$\varepsilon(z) = u_* \kappa z \left(1 - \frac{z}{h} \right) \qquad (8)$$

where u_* is the shear velocity, $(\tau/\rho)^{1/2}$, and is vertically-averaged for use in (1) and (2):

$$\varepsilon = \frac{u_* \kappa h}{6} \qquad (9)$$

The transport of suspended sand is governed by an advection-diffusion equation:

$$\frac{\partial c}{\partial t} + \frac{\partial c u}{\partial x} + \frac{\partial c v}{\partial y} - \frac{\partial}{\partial x} \varepsilon \frac{\partial c}{\partial x} - \frac{\partial}{\partial y} \varepsilon \frac{\partial c}{\partial y} +$$
$$\frac{\partial}{\partial z} \varepsilon \frac{\partial c}{\partial z} + w_s \frac{\partial c}{\partial z} = 0 \qquad (10)$$

where c is the sand concentration and w_s represents the sediment settling velocity.

The sediment eddy viscosity, ε, in (10) is assumed to be equal to the momentum eddy viscosity represented by (8).

Equation (10) is solved for a given flow field with 11 points in the vertical that are concentrated near the water surface and near the bed to resolve the gradients near the boundaries. The sand transport is represented by the median grain size, d_{50}. Equation (8) gives the eddy viscosity as a function of z. The velocity as a function of z is computed from the logarithmic velocity profile [Keulegan, 1938] from which (7) is derived. The numerical method used to solve (10) is similar to the one used for the flow equations [Patankar, 1980] extended to 3 dimensions.

The lower sediment-concentration boundary condition used in the solution of (10) is calculated by first deter-

mining a reference concentration, c_a, at the top of the bedload layer, where $z = z_a$. The reference concentration, c_a, is determined from the relations of Smith and McLean [1977]:

$$c_a = \frac{c_b \gamma s}{1 + \gamma s} \qquad (11)$$

where c_b is the bed concentration and s is the normalized excess shear stress:

$$s = \frac{\tau_{sf} - \tau_c}{\tau_c} \qquad (12)$$

where the subscript sf indicates skin-friction shear stress and τ_c is the critical shear stress for the initiation of significant particle motion [Shields, 1936]. The value of the constant has been updated to 0.004 by Wiberg [reported by McLean, 1992]. The distance above the bed corresponding to c_a, namely z_a, is determined from the expression presented by Dietrich [1982] with terms a_1 and a_2 as updated by Wiberg and Rubin [1985].

$$z_a = d \frac{a_1 + T_*}{1 + a_2 T_*} \qquad (13)$$

where: d = grain diameter, $T_* = \tau_{sf}/\tau_c$; and a_1 constant = 0.68. The coefficient a_2 is a function of the grain size in cm:

$$a_2 = 0.02035 \left[\ln(d) \right]^2 + 0.(02203) \ln(d) + 0.0709(\qquad (14)$$

The lower sediment-concentration boundary condition at $z = z_0$ used in solving (10) is calculated using the relation Rouse [1937] derived by neglecting the horizontal advection and diffusion terms in (10):

$$\frac{c}{c_a} = \left(\frac{z_a(h-z)}{z(h-z_a)} \right)^\phi \qquad (15)$$

where ϕ is the Rouse number, defined as:

$$\phi = \frac{w_s}{k u_{*sf}} \qquad (16)$$

The boundary condition at the water surface is c = 0, consistent with (8).

The evolution of the bed over time is calculated from the sediment continuity equation:

$$\frac{\partial \eta}{\partial t} = -\frac{1}{c_b}\left(\frac{\partial q_s}{\partial x} + \frac{\partial q_s}{\partial y}\right) \qquad (17)$$

where η is the bed elevation. The sediment discharge, q_s, is the sum of the sand transported as bedload and in suspension. The suspended sand discharge is determined by vertically integrating the product of the flow velocity and the sand concentration.

The bedload is determined by applying the *Meyer-Peter and Mueller* [1948] formula modified with the critical shear stress of the given grain size in place of their constant of 0.047:

$$\Phi = 8\left(\tau_* - \tau_{*c}\right)^{3/2} \qquad (18)$$

where Φ is the nondimensional bedload transport

$$\Phi = \frac{q_s}{d\sqrt{\left(\frac{\rho_s}{\rho} - 1\right)gd}}$$

τ_* is the nondimensional boundary shear stress

$$\tau_* = \frac{\tau}{(\rho_s - \rho)\,gd}$$

τ_{*c} is the nondimensional critical shear stress

$$\tau_{*c} = \frac{\tau_c}{(\rho_s - \rho)\,gd}$$

The grain diameter is represented by d; the median grain diameter, d_{50}, was used in the model results presented below. The density of the sand is represented by ρ_s. The boundary shear stress used in (18) is the magnitude of the vector sum of the shear stress calculated from the flow equations and an apparent stress due to gravity. The apparent stress due to gravity was calculated with a method proposed by *Nelson and Smith* [1989] in which

$$\tau_g = \tau_c \frac{\sin\gamma\nabla\eta}{\sin\phi|\nabla\eta|} \qquad (19)$$

where τ_g is the apparent gravitational stress, γ is the local maximum bed slope, and ϕ is the grain angle of repose. The x and y components of the bedload are determined from the respective components of the flow velocity and components of the local bed slope.

4. HYDRAULIC PARAMETERIZATION AND BOUNDARY CONDITIONS

As described by *Wiele et al.* [1996], the local friction of the main channel for these calculations was related to variations of the bed surface taken from the measured bathymetry with respect to the planar surfaces between grid points. Rather than using a constant z_0, however, a constant h/z_0 was computed with z_0 taken, as in *Wiele et al.* [1996], as $\sigma/10$, where σ is two standard deviations of the magnitude of the variations of the bed about the mean. This analysis yields a h/z_0 of 64.

A different procedure than the one described above was used in the recirculating zone, which is insulated from the flow in the main channel, and where the flow and sand transport more closely resemble that of alluvial streams. In this region, the local channel resistance and skin friction were calculated as functions of local flow, depth, and sand size. This procedure used the methods described by *Bennett* [1995] to estimate bedform dimensions and form drag. Bennett drew on the work of *van Rijn* [1984] who used u_* to distinguish between ripples, dunes, and upper plane bed and to estimate the dimensions of the bed forms, if present. Given bed form height and wavelength, the local friction and skin friction can be determined using the relations of *Smith and McLean* [1977] and *Nelson and Smith* [1989]. Relating local flow resistance and skin friction to bed forms is an improvement over the use of values derived only from the local hydraulics, but errors may be induced by uncertainties in the relations used and in the assumption of equilibrium between the local flow and the bed forms.

A large fraction of the total shear stress at the bed is exerted as form drag on large roughness elements, such as the extreme irregularity of the bedrock channel, bedforms, and boulder-size talus and bed material. This form drag must be deducted from the total shear stress to arrive at the skin friction portion of the total shear stress that transports sediment. For the model calculations, a sand-rating curve computed for the Grand Canyon gaging station [*Topping*, 1997] was used to calculate a sand-transport rate that represents the long-term average of the river reach that includes both pools. Before the controlled flood, the discharge had been peaking at a discharge close to the upper end of discharges at which sand transport becomes significant [*Smith and Wiele*, unpublished manuscript] and there had been no major tributary inputs since the previous fall that would have increased the local transport above the long-term average. Under these conditions, the bed conforms to a near-equilibrium bed coverage such that the sand transport rates are in near-equilibrium with the water discharge. The percentage of the total shear stress exerted as skin friction was calculated by running the model at the discharges that occurred prior to the controlled flood and computing the fraction of the total shear stress that is required to support a sand discharge equal to that predicted by the sand rating curve of *Topping* [1997]. This calculation yielded a skin friction that was 15% of the total shear stress. This low value is consistent with the extremely large channel roughness and accompanying form drag. Calculating the skin friction in this manner provides an integral constraint on the sand transport in that the calculated sand transport matches measured values for known conditions. The model then uses physically-based algorithms to extrapolate from these known values, thereby minimizing errors that accumulate as a result of uncertainty in local boundary conditions and parameters used in the model.

Measurements of sand concentration made at the gaging station near Grand Canyon [*Konieczki et al.*, 1997] (Figure 3) during the controlled flood were used as the upstream boundary condition for modeling sand discharge in the 2 pools. The first measurements of suspended sand were not made at the gage near Grand Canyon until about 12 hrs after the high flow arrived because of darkness and the abundance of floating debris in the river. Given the low discharges, by historical standards, prior to the controlled flood, the initial bed coverage was likely similar to that prior to the dam, resulting in a high peak sand discharge at the start of the high flow that dropped off rapidly over the first few hours. This is also supported by running the model for four pools (the same pools modeled for different sand boundary conditions in *Wiele et al.* [1996]) in the reach downstream from the confluence with the Little Colorado River using the output from one reach as the input to the following reach. This produced a pattern of high initial sand concentrations that dropped off rapidly in succeeding pools. The initial sand discharge, therefore, was estimated from pre-dam records of suspended sand concentration measured at the controlled flood discharge during the early stages of spring floods.

5. MODEL RESULTS

Topographic maps of model-predicted bathymetry (Figure 4) and measured bathymetry (Figure 5) compare well for the Carbon reach; both show scour of the main channel and deposition in the recirculation zone. Other model runs show that the shape of the eddy bar in the recirculation zone is sensitive to the concentration of the sand in the river. The model calculations using a lower sand concentration show deposition occurred in the reattachment zone, but with less redistribution of sand elsewhere in the recirculation zone (Figure 6). At higher sand concentrations

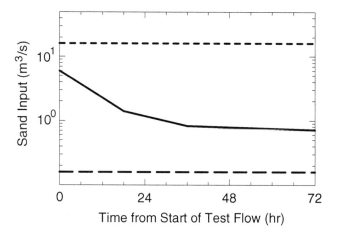

Figure 3. Sand discharges into the Carbon reach for model runs with high (top line), controlled flood (intermediate line), and low (bottom line) sand influxes.

than occurred during the controlled flood, the model predicts more complete infilling of the recirculation zone and more deposition along the channel sides. The main channel in the Carbon reach scoured (Figure 7) with these higher sand concentrations.

A comparison of the change in channel cross-sectional area as a function of distance from the inlet (Figure 8) shows that the model faithfully reproduces channel response during the controlled flood, with the exception of the region near the reach inlet. The main differences between the high and low sand concentration model runs represented by the change in cross-sectional area occur in the topography of the recirculation zone. The high-sand concentration run predicts a much higher deposit volume and bar elevation whereas the low sand concentration deposition just balances losses, mostly coming from the main channel.

Discrepancies between the measured and computed bathymetry during the controlled flood occurred in the upstream part of the channel and are mostly due to differences in the predictions of main channel bathymetry (Figure 9a). This discrepancy might be a result of inaccuracies in the parameterizations in the model which lead to overestimation of the sand transport rates in this region or may be due to underestimation of the initial sand discharge in the model input. Potential sources of error in the model that could lead to an underestimation of deposition in the main channel are most likely in the calculation of shear stress and in the fraction of the shear stress that is exerted as skin friction. The integral constraint applied to the sand transport in the determination of the skin friction suggests that, although there are uncertainties in the model parame-

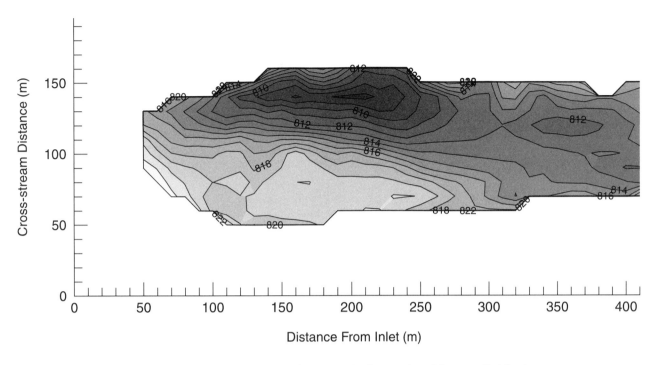

Figure 4. Bathymetric map of the Carbon reach after one day of the controlled flood.

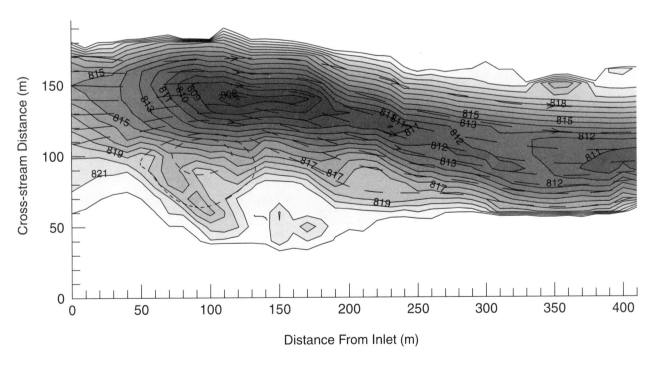

Figure 5. Map of the model-predicted topography of the Carbon reach after one day of the controlled flood using sand input derived from measurements. Contour interval is 1 m. Dashed lines show calculated stream lines at 1270 m³/s. Darker shaded areas are deeper.

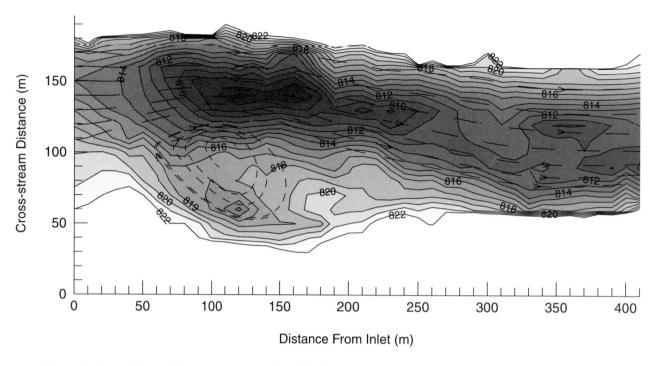

Figure 6. Map of the model-predicted topography of the Carbon reach bathymetry after one day using the low sand input. Contour interval is 1 m. Dashed lines show calculated stream lines at 1270 m³/s. Darker shaded areas are deeper.

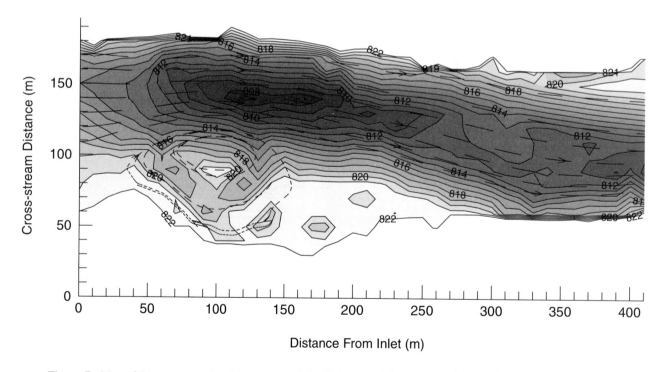

Figure 7. Map of the model-predicted topography of the Carbon reach bathymetry after one day using the high sand input. Contour interval is 1 m. Dashed lines show calculated stream lines at 1270 m³/s. Darker shaded areas are deeper.

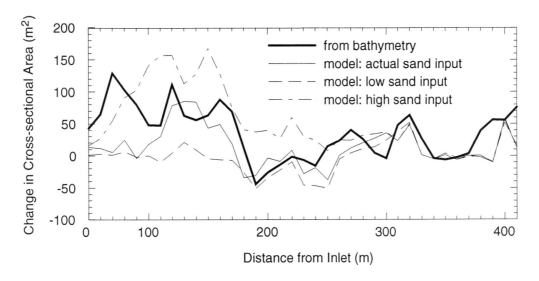

Figure 8. Change in cross-sectional area as a function of distance from the inlet for the Carbon reach after one day.

terizations, the significant difference between computed and measured bathymetry in this region is a result of another process. This deposit could be a result of a high influx of sand from upstream, such as would occur as a result of rapid removal of upstream bars [*Andrews et al.*, this volume]. Such an influx would likely occur as a pulse, however, rather than as a stream of sand supply persisting over days. Another and perhaps more likely possibility is that this discrepancy is a result of the transport of coarser material mobilized by the high shear stresses generated by the high flow. An upper limit of the size of this material can be estimated by calculating the largest grain sizes that could be transported by the shear stresses in this reach, that is, the grain sizes for which the shear stress would be at critical, called the critical grain size in the discussion below. Figure 9b shows the result of this calculation at the cross section located 120 m below the inlet. The shear stress is greater than critical for the sand d_{50} over most of the cross section. The maximum critical grain size is about 1 cm at a location slightly to the river-left of the thalweg. This result and the model's prediction of the evolution of bed topography suggest that sediment as coarse as fine gravel may have been transported through the reach during the controlled flood, and that changes in bathymetry may not solely be due to changes in sand storage.

Model predictions for the Salt reach [*Wiele*, 1997] compare well with bathymetric measurements made during the controlled flood, but these patterns differ from the bathymetry following the 1993 flood on the Little Colorado River (Figure 10). The 1993 flood supplied sufficient sand to the river to increase sand concentration to about twice the

concentration observed during the 1996 controlled flood. Water discharge in the main stem in 1993 reached a peak of about 950 m³/s [*Wiele et al.*, 1996], about 25 percent less than the controlled flood. In 1993, the combination of lower peak discharge, lower shear stresses, but higher sand concentrations led to a large amount of deposition in the main channel and in the recirculation zones as shown in the model results (Figure 11). The bathymetry for the 1993 flood was predicted by modeling and compares well with 5 cross sections measured before and after the LCR flood [*Wiele et al.*, 1996]. Patterns of deposition in 1993 and 1996 in the recirculation zone differ significantly. Sand was deposited in the main channel in 1993, whereas sand was eroded in the main channel in 1996. During the 1996 controlled flood, the main focus of deposition within the recirculation zone was in the reattachment zone (Figure 12), which is a pattern similar to that of the Carbon site. Sand was distributed from the main channel to the reattachment point, where it settled from suspension and was then redistributed within the recirculation zone primarily as bedload. The higher sand concentrations and lower shear stresses during the 1993 LCR flood led to deposition in the deep depression in the main channel, followed by deposition along the boundary between the main downstream flow and the recirculation zone where velocities and shear stresses are low. At the Salt site in 1993, the initial focus of the deposition was primarily near the separation point. Deposition subsequently extended downstream and in towards the left bank.

During the controlled flood, sand was redistributed in the Carbon reach but there was little net accumulation (Figure

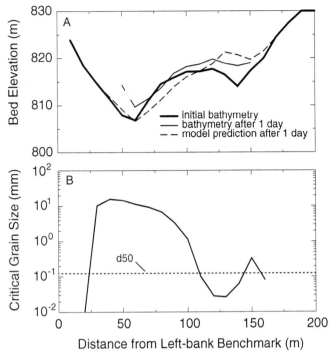

Figure 9. Cross section at x = 120 m prior to the controlled flood, from bathymetry after one day, and from the model after one day (a). Grain size for which the calculated shear stress is at critical at the cross section at x = 120 m (b).

13a). The volume eroded from the pool was only slightly less than the volume deposited in the pool. The erosion was primarily in the main channel and the deposition primarily in the recirculation zone. The model shows that the bulk of the sand was eroded and deposited within the first few hours of the controlled flood, which is consistent with the measurements of bathymetry after one day [*Andrews et al.*, this volume] and anecdotal evidence from researchers throughout the Grand Canyon. Rapid erosion of the main channel resulted from the sudden increase in discharge that scoured the main channel deposits that had formed in quasi-equilibrium with the lower discharges characteristic of the period before the controlled flood. Rapid deposition in the recirculation zone was a result of the high initial sand concentrations (initial measurements were about 10 times higher than predicted by rating curves) and the availability of depositional sites created by stages that were higher than any in the Canyon since high releases in 1983 through 1986. After about 1-1/2 days, sand erosion in the Carbon pool ceased, while the deposition volume of sand continued to increase at a slow rate. The rate at which sand was deposited in the recirculation zone was limited by the rate of redistribution within the recirculation zone once the initial filling near the reattachment point had taken place.

Computed volumes of deposition and erosion over time for the low sand concentration (Figure 13b) follow a similar

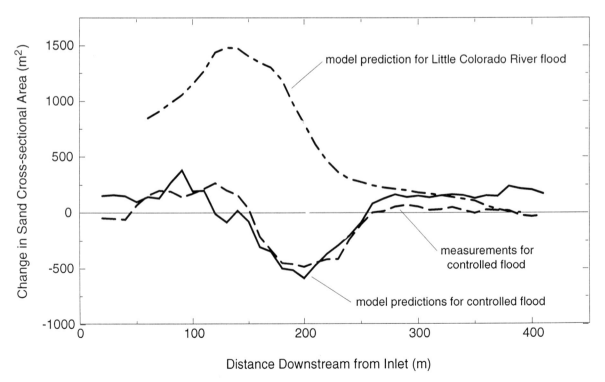

Figure 10. Change in cross-sectional area as a function of distance from the inlet for the Salt reach after three days.

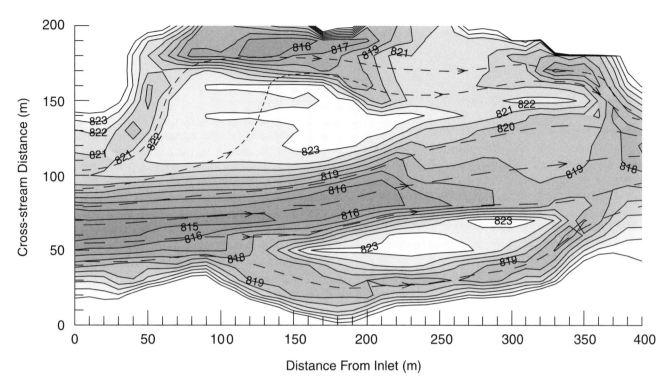

Figure 11. Map of the model-predicted topography of the Salt reach after three days during the LCR flood. Contour interval is 1 m. Dashed lines show calculated stream lines at 950 m³/s. Darker shaded areas are deeper.

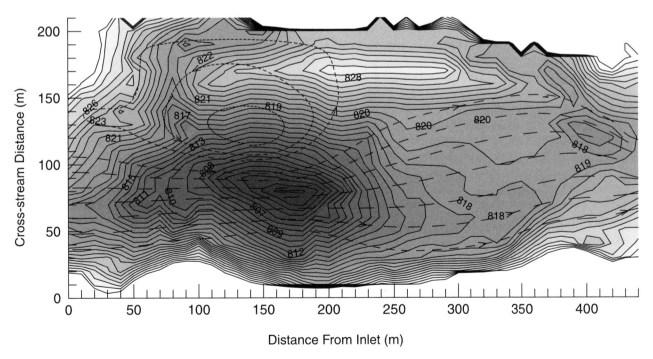

Figure 12. Map of the model-predicted topography of the Salt reach after three days of the controlled flood. Contour interval is 1 m. Dashed lines show calculated stream lines at 1270 m³/s. Darker shaded areas are deeper.

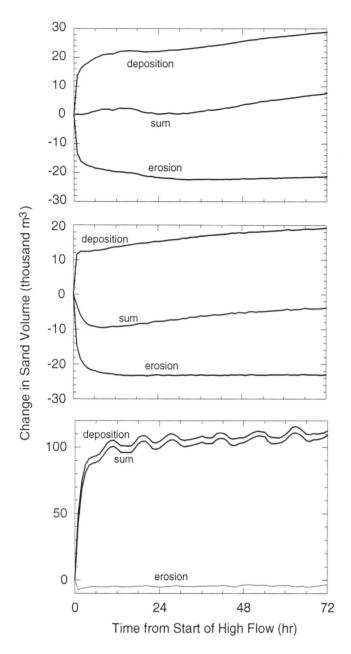

Figure 13. Changes in sand volume in the Carbon reach as a function of time for sand inputs derived from measurements during the controlled flood (a), for the low sand input (b), and for high sand input (c) calculated with the model.

pattern to those calculated for the controlled flood. The volume of erosion in the main channel shows the same initial rapid increase in magnitude, but the model predicted a higher erosion rate with the lowest sand concentration between about 6 and 24 hours than predicted with the sand concentration that occurred during the controlled flood. The

difference in erosion rate during this period is a result of the high initial sand concentrations that occurred during the controlled flood. A high influx of sand reduces the magnitude of the divergence of the sand transport rate which determines the rate at which sediment is deposited or scoured, as represented by (17).

Using a higher sand concentration (Figure 13c), the model predicted that the erosion is small compared to the volume of deposition in the Carbon pool. All erosion is confined to the main channel. The deposit within the recirculation zone was trimmed back along its boundary with the main flow with the low and controlled flood sand concentrations, but the recirculation deposit was slightly expanded towards the main flow at the highest sand concentrations (Figures 4, 6, and 7). The total volume of deposition at the higher sand concentrations was nearly 4 times greater than the deposition volume calculated for the controlled flood sand concentration.

The total predicted change in sand volumes within the recirculation zone at the Carbon pool for the 3 modeled cases and calculated from the measured bathymetry are shown in Figure 14. The model result using the controlled flood sand concentrations agrees well with bathymetry measured on the first day. Field measurements during the second and third, however, show that there was much less sand in the recirculation zone than predicted by the model. This discrepancy between the model predictions and field measurements is a result of rapid evacuations of sand from the recirculation zone not predicted by the model. *Andrews et al.* [this volume] suggest that slumping of newly deposited sand was a common occurrence during the controlled flood, a process not included in the model.

Model predictions of sand accumulation with a low sand supply show that the net change in sand volume would increase very slowly for the first 18 hours, but would be followed by a slight increase in sand accumulation rate (Figure 14). This initial slow increase is a result of the accumulation being offset by erosion of the outer part of the recirculation deposit. During the first 18 hours, the model predicted sand would be redistributed within the recirculation zone, with the higher stage and increased shear stresses moving sand up the reattachment bar closer to the bank.

The model predicted that with the highest sand concentrations the available deposition sites would fill more rapidly and more fully than with the controlled flood sand supply (Figure 14). After the first day, the calculated rate of further net accumulation of sand was only slightly higher with the high sand supply than it was with the controlled flood sand supply, suggesting that redistribution of sand within the recirculation zone away from the reattachment

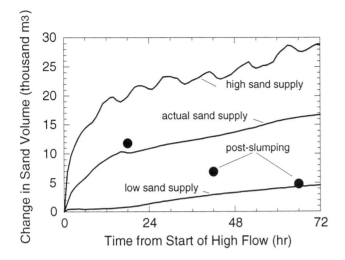

Figure 14. Change in sand volume in the recirculation zone in the Carbon reach. The solid lines are model predictions. The circles are taken from the bathymetric measurements. Andrews and others (this volume) propose that the decrease in sand on the second and third days was a result of slumping, a process not included in the model.

point controls the accumulation rate. The model predictions for deposition in the recirculation zone are consistent with the observations of *Hazel et al.* [1997, this volume], who found that the average response of 33 eddy bars distributed throughout the Grand Canyon that they surveyed before and after the controlled flood gained volume primarily by an increase in deposit elevation rather than area. The decreasing rate of deposition in the recirculation zone was observed by *Schmidt et al.* [1993] in their flume experiments. They attributed the decrease in deposition rate to the filling of the depositional sites and the accompanying reduction in redistribution rates.

Sand concentration also affected the total net change in sand volume (Figure 15). With the lowest sand influx, the net loss of sand immediately after the start of the controlled flood is twice the volume introduced at the upstream boundary as a result of the high shear stresses mobilizing sand in the main channel and along the outer margin of the recirculation zone without sufficient sand coming into the reach to replace it. With the highest sand supply, the reach retains more sand than it loses. The highest retention of sand occurs immediately after the start of the controlled flood when the depositional sites are most vacant and the accumulated sand input is low, and tapers over time to near zero after the depositional sites fill and the accumulated sand input has increased. The model results with the actual sand input are nearly balanced, with the reach showing little net accumulation.

Deposition along the channel margins predicted by the model indicates that sand can be readily deposited along the channel margins if the sand concentration is sufficiently high as a result of the high roughness. The deposition shown by the model along the left bank in the lower half of the Carbon reach also shows sensitivity to small curvature in the nearly straight reaches. Even slight bend curvature will shift flow to the outside bank, creating an effect similar to that of a channel expansion and promoting deposition along the inner bank.

The volume of deposition calculated by *Wiele et al.* [1996] and extrapolated to the reach of similar morphology indicated that the volume of sand contributed by the LCR was equivalent to the volume of sand deposited within 20 km downstream of the LCR confluence. Deposits along the channel sides observed to Lake Mead were attributed to the mining of sand from the channel bed. *Howard and Dolan* [1981] noted a similar occurrence as a result of a tributary flood in 1973 that left fresh deposits of sand from below the mouth of the LCR to Lake Mead, and suggested that most of the sand was derived from sand residing on the bed and stored in banks prior to the tributary flood.

6. SUMMARY AND CONCLUSIONS

The complicated flow patterns that are characteristic of reaches in the Colorado River in the Grand Canyon create depositional environments that have a wide range of responses to variations in flow and suspended sand. A wide range in sediment concentrations and local flow patterns leads to rates of erosion and deposition that exceed those typically encountered in alluvial rivers. In the part of the channel carrying the main downstream discharge, the sand on the bed is subject to the full brunt of the river's sand-carrying capacity. In these areas, sand scour or fill can be rapid, depending on the concentration of sand in suspension entering the reach. Consistent with measured bathymetry, model predictions show that deposition of suspended sand from the main channel within the recirculation zone was focused primarily near the reattachment point. The redistribution of sand throughout the recirculation zone occurred more slowly. Floods of short duration are efficient in that the time-averaged deposition rates are higher as a result of high initial deposition rates, but continued high flow may be necessary to distribute sand within recirculation zones, as pointed out by *Anima et al.* [1998].

The pattern of deposition observed during the 1996 controlled flood and predicted by the model in two reaches are similar to those described in field studies [*Schmidt and Graf,* 1990; *Schmidt,* 1990, and *Rubin et al.,* 1990]. With

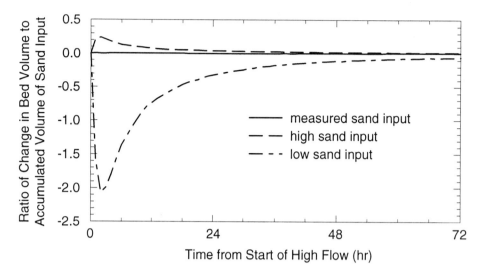

Figure 15. Net change in sand volume normalized by the accumulated sand input as a function of time for the Carbon reach.

this pattern, the recirculation zone is supplied by sand that is transported primarily in suspension to the reattachment point where it is then redistributed by the slower flow near the bank into the recirculation zone or downstream along the channel margin. Separation deposits formed from sand transported from the reattachment point through the recirculation zone. This pattern was predicted by the model in the Carbon reach for the high, controlled flood, and low sand concentrations used as sand boundary conditions at the controlled flood discharge of 1270 m^3/s and for the Salt reach under the conditions that occurred during the controlled flood. In contrast, the computed bed topography for the Salt reach during the LCR flood shows that the recirculation zone filled by the downstream and shoreward migration of a bar that initiated at the flow separation point just downstream form the inlet on river left. In addition, the model predicts, in agreement with cross sections measured after the LCR flood, that the main channel accumulated about 12 m of sand. This result is in contrast to the Salt reach during the controlled flood and model predictions for the Carbon reach which showed main channel scour. This difference in depositional pattern in the Salt reach during the LCR flood is a result of the lower water discharge during the LCR flood combined with higher sand concentrations that led to rapid deposition in the channel expansion downstream from the inlet. The decreasing rates of deposition predicted by the model for all cases are consistent with the observations of *Schmidt et al.* [1995] of their flume experiments of deposition in recirculation zones.

Differences between observed bathymetry and model predictions suggest the evacuations of sand observed by *Andrews et al.* [this volume] are not a result of the declining sand concentrations during the controlled flood, which is consistent with their proposal that sand was lost as a result of mass failure. Differences between observed bathymetry and model predictions of the bathymetry in the main channel of the Carbon reach during the controlled flood suggest that a significant volume of bed material coarser than the sand used in the calculations, in the very coarse sand to fine gravel range, was transported during the controlled flood.

Acknowledgments. The authors gratefully acknowledge Mark Gonzales, Kenton Grua, and Chris Johnston for their work on the collection and reduction of bathymetric measurements made during the 1996 controlled flood. Reviews by John C. Schmidt and three anonymous reviewers greatly improved the manuscript. This work was funded by the GCES II under the direction of Dave Wegner.

REFERENCES

Anima, R.J., Marlow, M.S., Rubin, D.M., and Hogg, D.J., *Comparison of sand distribution between April 1994 and June 1996 along six reaches of the Colorado River in Grand Canyon, Arizona*; U.S. Geol. Surv. Open-File Rept., 1998.

Bennett, J.P., Algorithm for resistance to flow and transport in sand-bed channels, *J. Hydr. Engin.*, ASCE, 121(8), 578-590, 1995.

Dietrich, W.E., *Flow, boundary shear stress, and sediment transport in a river meander*, Unpublished Ph.D. dissertation, Univ. Washington, Seattle, WA, 1982.

Dolan R., A.D. Howard, and D. Trimble, Structural control of rapids and pools of the Colorado River in the Grand Canyon, *Science*, 202, 629-631, 1978.

Hazel, J.E., M. Kaplinski, M.F. Manone, R.A. Parnell, A.R. Dale, J. Ellsworth, and L. Dexter, *The effects of the 1996 Glen Canyon Dam beach/habitat-building test flow on Colorado River sand bars in Grand Canyon*, Final Rept. to Glen Canyon Environ. Studies, N. Ariz. Univ., Flagstaff, AZ, 1997.

Howard A.D., and R. Dolan, Geomorphology of the Colorado River in the Grand Canyon, *J. Geol.*, 89(3), 269-298, 1981.

Keulegan, G.H., Laws of turbulent flow in open channels, Res. Paper RP1151, *National Bureau of Standards*, 21, 707-741, 1938.

Konieczki, S.D., Graf, J.B., and Carpenter, M.C., *Stage, streamflow, and sediment transport data collected to determine the effects of a controlled flood in March and April 1996 on the Colorado River between Glen Canyon Dam and Diamond Creek, Arizona*, U.S. Geol. Surv. Open-File Rept. 97-224, 1997.

McLean, S.R., On the calculation of suspended sediment load for noncohesive sediments, *J. Geophys. Res.* 97 (C4), 5759-5770, 1992.

Meyer-Peter, E., and R. Mueller, Formulas for bed-load transport, *Intern. Assoc. Hydr. Res.*, Second Meeting, Stockholm, 1948.

Nelson, J.M., and J.D. Smith, Evolution and stability of erodible channel beds, in *River Meandering*, ed. S. Ikeda and G. Parker, pp. 69-102, Amer. Geophys. Union, Mono. 12, Washington, D.C., 1989.

Nelson, J.M., and J.D. Smith, Mechanics of flow over ripples and dunes, *J. Geophys. Res.*, 89(C6), 8146-8162, 1989.

Patankar, S.V., *Numerical heat transfer and fluid flow*, Hemisphere, Washington, D.C., 1980.

Rouse, H., Modern conceptions of the mechanics of turbulence, *Trans. Am. Soc. Civil Engin.*, 102, 1937.

Rubin, D.M., J.C. Schmidt, and J.N. Moore, Origin, structure, and evolution of a reattachment bar, Colorado River, Grand Canyon, Arizona, *J. Sed. Pet.*, 60, 982-991, 1990.

Schmidt, J.C., Recirculating flow and sedimentation in the Colorado River in Grand Canyon, Arizona, *J. Geol.*, 98, 709-724, 1990.

Schmidt, J.C., and J.B. Graf, *Aggradation and degradation of alluvial sand deposits, 1965 to 1986, Colorado River, Grand Canyon National Park, Arizona*, U.S. Geol. Surv. Prof. Paper 1493, 74 pp., 1990.

Schmidt, J.C., D.M. Rubin, and H. Ikeda, Flume simulation of recirculating flow and sedimentation, *Water Res. Res.*, 29, 2925-2939, 1993.

Schmidt, J.C., and D.M. Rubin, Regulated streamflow, fine-grained deposits, and effective discharge in canyon with abundant debris fans, in *Natural and anthropogenic influences in fluvial geomorphology*, ed. J.E. Costa, A.J. Miller, K.W. Potter, and P.R. Wilcock, pp. 177-195, Amer. Geophys. Union, Geophys. Mono. Ser., 89, Washington, D.C., 1995.

Shields, A., *Application of similarity principles and turbulence research to bed-load movement*, (in German), Mitteil. Preuss. Versuchanst. Wasser, Erd, Schiffsbau, Berlin, 26, 1936.

Smith, J.D., and S.R. McLean, Spatially averaged flow over a wavy surface, *J. Geophys. Res.*, 84(12), 1735-1746, 1977.

Topping, D., Flow, sediment transport, and channel geomorphic adjustment in the Grand Canyon, Arizona gage reach of the Colorado River during the Grand Canyon flood experiment, *Abstracts and executive summaries, Symposium on the 1996 Glen Canyon Dam Beach/Habitat-Building Flow*, Flagstaff, AZ, 1997.

van Rijn, L.C., Sediment transport. Part III: suspended load transport, *J. Hydr. Engrg.*, ASCE, 110(11), 1613-1641, 1984.

Webb, R.H., P.T. Pringle, and G.R. Rink, *Debris flows from tributaries of the Colorado River, Grand Canyon National Park, Arizona*, U.S. Geol. Surv. Prof. Paper 1492, 39 pp., 1989.

Wiberg, P.L., and D.M. Rubin, Bed roughness produced by saltating sediment, *J. Geophys. Res.*, 94(C40), 5011-5016, 1989.

Wiele, S. M., Graf, J.B., and Smith, J.D., 1996, Sand deposition in the Colorado River in the Grand Canyon from flooding of the Little Colorado River, *Water Res. Res.*, 32(12), 3579-3596.

Wiele, S.M., *Modeling of flood-deposited sand distribution for a reach of the Colorado RIver below the Little Colorado River, Grand Canyon, Arizona*, U.S. Geol. Surv. Water Res. Invest. Rept. 97-4168, 1997.

S.M. Wiele, U.S. Geological Survey, Denver Federal Center, MS 413, Lakewood, CO 80225; email: smwiele@usgs.gov

E.D. Andrews and E.R. Griffin, U.S. Geological Survey, 3215 Marine St., Boulder, CO 80303

Changes in the Number and Size of Campsites as Determined by Inventories and Measurement

Lisa H. Kearsley

SWCA, Inc., Flagstaff, Arizona

Richard D. Quartaroli and Michael J.C. Kearsley

Northern Arizona University, Flagstaff, Arizona

We used three methods to evaluate the effects of the 1996 controlled flood on the number, size, and longevity of large riverside sand deposits used as campsites in Grand Canyon National Park. We performed an inventory of all campsites 2 weeks before, 2 weeks after, and 6 months after the flood. We also performed a reconnaissance analysis of system-wide flood-induced changes to 89% (194 of 218) of all established campsites. We also mapped the area of 50 established campsites 2 weeks before, 2 weeks after, and 6 months after the flood, and of 34 newly-created campsites 2 weeks after and 6 months after the flood. The controlled flood created 84 new campsites and destroyed three. Six months after the flood, 44% of the new deposits were no longer usable as campsites. Loss of new campsites primarily occurred by erosion of the newly-deposited sediment; in some cases, excessively steep riverside slopes made otherwise sufficiently large sand bars inaccessible. The remaining new campsites were on average half the size they had been 2 weeks after the flood. The system-wide reconnaissance analysis show that 2 weeks after the flood 50% of the campsites were substantially larger than they had been before the flood, 39% remained the same size, and 11% were smaller. The 50 established campsites we measured were, on average, 48% larger 2 weeks after the flood than 2 weeks before the flood. After 6 months, most of those 50 campsites had eroded; these sites were, on average, 17% larger than they had been before the flood.

1. INTRODUCTION

Sand bars along the Colorado River in Grand Canyon serve as campsites for river runners, as habitat for vegetation and wildlife, and as temporary storage sites for sand that is transported through the system. While river use has increased in the past 30 years to approximately 22,000 people per year, the number and size of campsites have markedly decreased [*Kearsley et al.,* 1994]. As a result, campsites in some stretches of the river are extremely limited, causing severe competition and excessive use of the few that are available. Campsites are an essential component of river trips because the rest of the shoreline consists of talus slopes, vertical to near-vertical bedrock, or sand bars which are too densely vegetated to use.

Campsites are also an important resource the National Park Service has been mandated to preserve [*U.S. Department of the Interior,* 1916].

1.1. Grand Canyon Campsite Characteristics

Only a portion of the sand bars in Grand Canyon are used as campsites. With the exception of 5 campsites on bedrock ledges, campsites are sand bars which are large enough and high enough to be used by boating parties. They represent the largest and most stable sand bars in Grand Canyon [*Schmidt and Graf,* 1990]. These sandbars need to be large in order to accommodate the following uses: an area for a kitchen, which consists of several tables, stoves, and boxes carried up from the boats; numerous sleeping areas with space in between the sleeping areas for privacy; and, an area that is out of sight of the rest of the campsite for a portable toilet. These campsite uses also necessitate that the sandbars have little or no vegetation on much of their area and that they are accessible to boating parties from the river. Many large sand bars are covered with dense vegetation, which precludes using the bar for camping. Other sand bars are inaccessible because the side of the bar is too steep to climb or because sand between the bar and the river has completely eroded away, exposing cobbles or boulders which, in combination with wave action below rapids or riffles, make docking of boats impossible. We have formulated these requirements into the following definition of a campsite: a sand bar above normal river fluctuations that is accessible from the river, has sufficient space for a kitchen and 10 or more people, and is not overgrown by vegetation. This definition is similar to that used in each previous campsite inventory in Grand Canyon. Figure 1 depicts a large campsite in Grand Canyon.

In addition to the above requirements of a campsite, other characteristics of a site make it more desirable. With summer temperatures regularly exceeding 40°C, access to shade is important. Many sites contain large vegetation such as tamarisk (*Tamarix chinensis*) or are located next to cliffs which provide shade. Vegetation and boulders protect boating parties from wind, which is often very strong. Vegetation and boulders also provide screening between sleeping areas and between the portable toilet area and the rest of the camp. While a completely unvegetated, exposed sand bar is usable as a campsite, it is much less desirable than a site that has some amount of vegetation, boulders, or cliffs that provide protection from the elements.

"Campsite area," as measured in this study, is the portion of the campsite that one could easily use as a kitchen or to sleep on. It is defined as a smooth substrate (mostly sand)

Figure 1. A large campsite ("Papago, " RM 75.8R) in the Grand Canyon, showing a standard kitchen in the foreground and sleeping areas in the background.

with no more than an 8° slope and with little to no vegetation. The top of the sand bar usually consists of a large area where the kitchen and several sleeping areas are located. More vegetation and rocks are found along the perimeter of the sand bar and on the slopes above, which break up campsite area into separate, smaller areas, that are used as sleeping areas. The smallest of these outlying areas consist of a small patch of sand just large enough for one or two people to lie down. Figure 2 depicts an aerial photograph of a large campsite and a map of the same site showing campsite area.

1.2. Geomorphology Associated with Grand Canyon Campsites

The Colorado River is distinctive in that it is a narrow river that is partially blocked by abundant debris fans that have formed at the mouths of tributaries [*Howard and Dolan,* 1981]. The association of these debris fans with the river creates fan-eddy complexes that consist of 1) a low-velocity backwater upstream from a debris fan, 2) a constriction caused by the debris fan which often forms rapids, 3) eddies and eddy bars downstream from the debris fan, where the channel widens, and 4) a gravel bar downstream from the eddies [*Schmidt and Rubin,* 1995]. Different types of alluvial deposits are formed in association with these complexes. Separation bars, which are formed where downstream flow separates from the shoreline, mantle the downstream end of the debris fan. Reattachment bars, which are formed where downstream

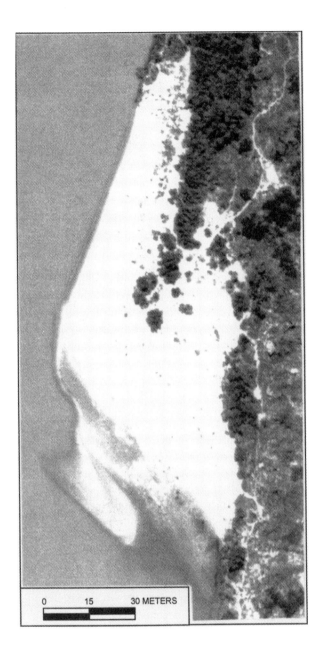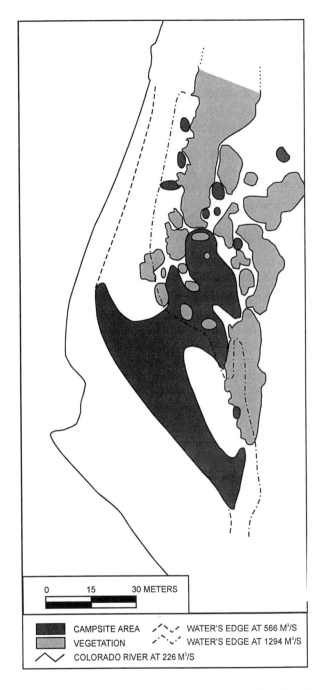

Figure 2. A 1:1200 scale aerial photograph of a large campsite, "lower Saddle," (RM 47.3R), and a map of the same site showing campsite area above 556 m³/s. Large vegetation associated with campsite area is also shown. Sand areas in the photograph that are not included as campsite area in the map are too steep to use (>8° slope).

flow reattaches to the shore, are located at the downstream end of eddies [*Schmidt and Graf,* 1990]. Undifferentiated eddy deposits lack the morphology of separation and reattachment bars [*Leschin and Schmidt,* 1995]. Outside of the fan-eddy complex, channel-margin deposits discontinu-

ously line the river channel [*Schmidt and Graf,* 1990]. When they are large enough and high enough, all these deposit types can be used as campsites.

Debris fans and their associated alluvial deposits remain stationary over a time scale of 10-100 years [*Melis et al.,*

1994, *Webb*, 1996]. The sizes and elevations of the alluvial deposits, however, vary widely over short time scales in response to sediment availability and flow magnitude [*Schmidt et al.*, 1995; *Schmidt and Rubin*, 1995].

The width and depth of the river channel and the distribution and characteristics of debris fans vary greatly in response to the bedrock lithology and structure [*Howard and Dolan*, 1981; *Melis*, 1997]. Resistant rocks, such as limestone and schist create a narrow river [*Howard and Dolan*, 1981] with relatively few tributary fans [*Dolan et al.*, 1978]. Erodible shales create a relatively wide river [*Howard and Dolan*, 1981]. *Schmidt and Graf* [1990] defined 11 reaches between river miles 0 and 225, based on the type of bedrock at river level, and they categorized reach width as "narrow" or "wide." Melis rearranged these 11 reaches into six reaches based primarily on average channel width but also on tributary fan area, fan shape, and channel constriction. He agreed with *Howard and Dolan* [1981] and *Schmidt and Graf* [1990] that bedrock lithology and structure are the main controlling factors in reach differentiation and fan-eddy complex characteristics. These geomorphological differences in reaches strongly affect the number, size, and characteristics of campsites. Wide reaches, particularly those with sloping-talus shorelines that co-occur with closely-spaced fan-eddy complexes, provide a foundation for many large campsites. Narrow reaches, particularly those with vertical bedrock shorelines, provide few opportunities for sand deposition large enough to be used as campsites. Also, higher rates of erosion of debris fans and alluvial deposits may occur in narrow reaches because of increased competence of the river [*Howard and Dolan*, 1981]. *Kearsley et al.* [1994] independently divided the river into critical and non-critical reaches based solely on recreational considerations. Critical reaches are river sections where campsites are infrequent and small in size, causing intense competition for sites. Non-critical reaches are river sections where campsites are numerous and usually large in size. Kearsley et al.'s critical and non-critical reach designations correspond closely to both *Schmidt and Graf* [1990] and *Melis'* [1997] narrow and wide reach designations (Figure 3).

1.3. Historical Changes to Grand Canyon Campsites

Decreased size of sand bars and consequently campsites in the past 30 years is well documented and results from the existence and operations of Glen Canyon Dam [*Weeden et al.*, 1975; *Howard and Dolan*, 1981; *Beus et al.*, 1985; *Schmidt and Graf*, 1990; *Kearsley et al.*, 1994; *Kaplinski et al.*, 1995]. Glen Canyon Dam, completed in 1963, traps sediment and reduces floods [*Howard and Dolan*, 1981;

Andrews, 1991]. The dam traps essentially all sediment that previously entered Grand Canyon from upstream, and the sediment resupply to Grand Canyon now comes from flash floods and debris flows in tributaries. The Paria and Little Colorado Rivers provide most of the post-dam sediment input to the Colorado River in Grand Canyon [*Andrews*, 1991]. These sources, however, contribute only a tenth of the pre-dam sediment load, as measured from 1941-1957 [*Andrews*, 1991]. Prior to Glen Canyon Dam, annual flooding averaged 2400 m^3/s [*Howard and Dolan*, 1981]. Post-dam maximum discharges rarely exceed 892 m^3/s; however, infrequent unplanned flooding has occurred. In 1983, a peak discharge of 2572 m^3/s occurred. Between 1984 and 1986, peak discharges ranged from 1355 m^3/s to 1747 m^3/s.

The inventory described in this paper was compared with the results from inventories and analyses of air photos dating to 1965. Aerial photograph analysis and campsite inventories show a 30-year trend of diminishing campsites punctuated by infrequent flood-induced increases. Between 1965 and 1973, air photograph comparisons show that nearly 33% of all campsites ceased to exist or decreased substantially in size due to erosion [*Kearsley et al.*, 1994]. The first campsite inventory was conducted in 1973 and provided a field-based reference against which to make subsequent comparisons; it documented the existence of 333 campsites [*Weeden et al.*, 1975]. The second inventory was conducted two months following recession from the June 1983 peak discharge; this inventory documented 438 campsites, a 34% increase from the number in 1973. The increased number of campsites from the number recorded in 1973 is assumed to be due to the depositional effects of the 1983 flood [*Brian and Thomas*, 1984]. Aerial-photograph analysis showed that these flood-induced increases were short-lived, however. One year after the inventory most of these new and larger campsites had substantially eroded [*Kearsley et al.*, 1994]. The most recent inventory was conducted in 1991 and documented 226 campsites, a 32% reduction in campsite number since 1973, and a 48% reduction since 1983 [*Kearsley et al.*, 1994]. The long-term trend between 1973 and 1991 was that all critical reaches decreased in carrying capacity, as defined as the total number of people accommodated in all of the camps in each reach [*Kearsley et al.*, 1994].

Monitoring of campsites since 1991 detected a low rate of decrease in campsite size. In 1991 and in 1994, 93 campsites were measured by on-site mapping and aerial photograph analysis. The measured campsites lost, on average, 16% of their campsite area above the water stage at 708 m^3/s during this time, primarily due to erosion but also due to invasion of riparian vegetation (L.H. Kearsley,

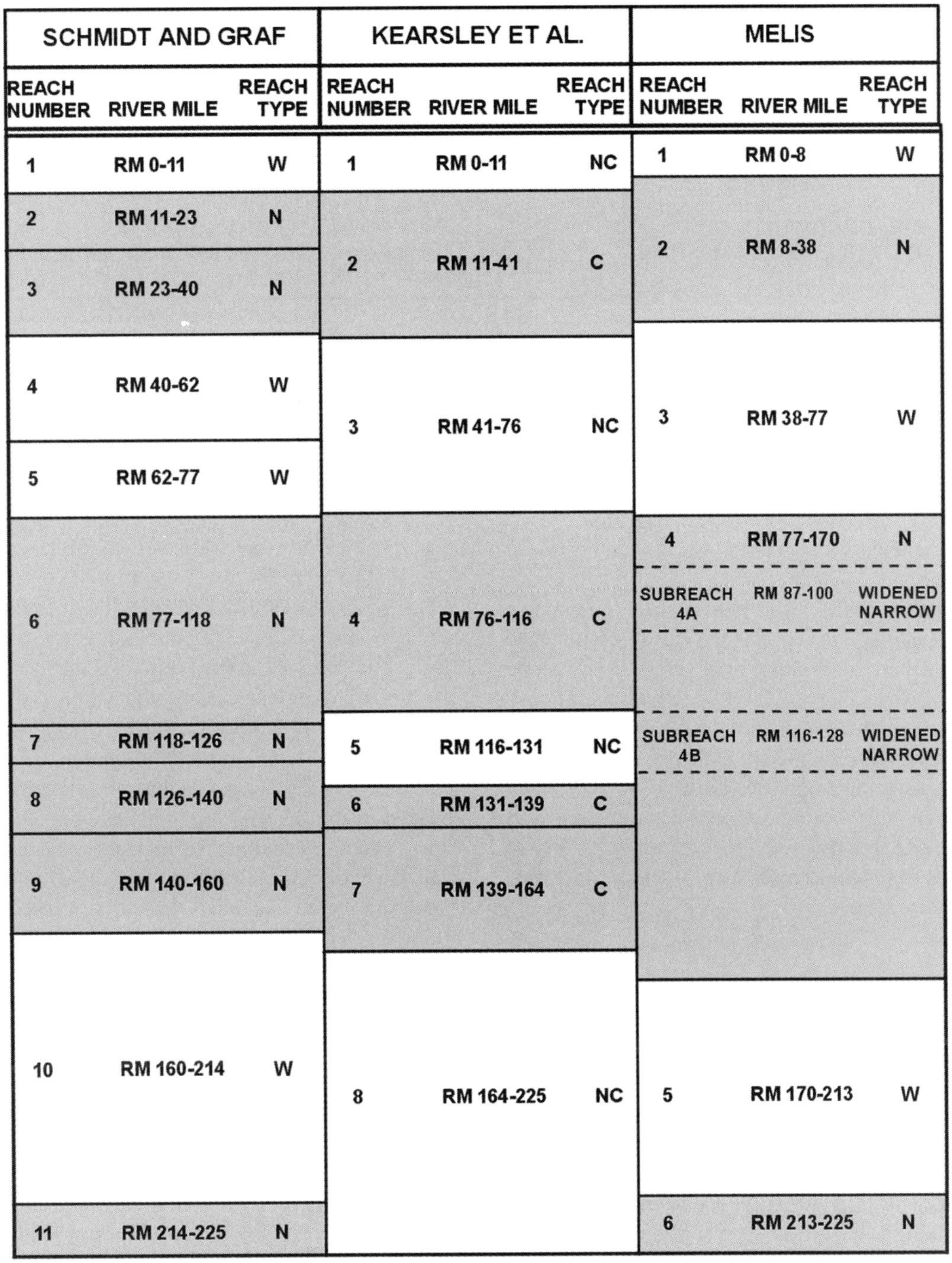

SCHMIDT AND GRAF			KEARSLEY ET AL.			MELIS		
REACH NUMBER	RIVER MILE	REACH TYPE	REACH NUMBER	RIVER MILE	REACH TYPE	REACH NUMBER	RIVER MILE	REACH TYPE
1	RM 0-11	W	1	RM 0-11	NC	1	RM 0-8	W
2	RM 11-23	N	2	RM 11-41	C	2	RM 8-38	N
3	RM 23-40	N						
4	RM 40-62	W	3	RM 41-76	NC	3	RM 38-77	W
5	RM 62-77	W						
6	RM 77-118	N	4	RM 76-116	C	4	RM 77-170	N
						SUBREACH 4A	RM 87-100	WIDENED NARROW
7	RM 118-126	N	5	RM 116-131	NC	SUBREACH 4B	RM 116-128	WIDENED NARROW
8	RM 126-140	N	6	RM 131-139	C			
9	RM 140-160	N	7	RM 139-164	C			
10	RM 160-214	W	8	RM 164-225	NC	5	RM 170-213	W
11	RM 214-225	N				6	RM 213-225	N

Figure 3. Reach designations along the Colorado River in Grand Canyon by *Schmidt and Graf* [1990], *Kearsley et al.* [1994], and *Melis* [1997]. Length of columns is scaled to reflect river length for each reach. Reach type is narrow and wide for *Schmidt and Graf* and *Melis*, and is critical and non-critical for *Kearsley et al.* Narrow and critical reaches are shaded. River-mile designations are rounded off to the nearest mile.

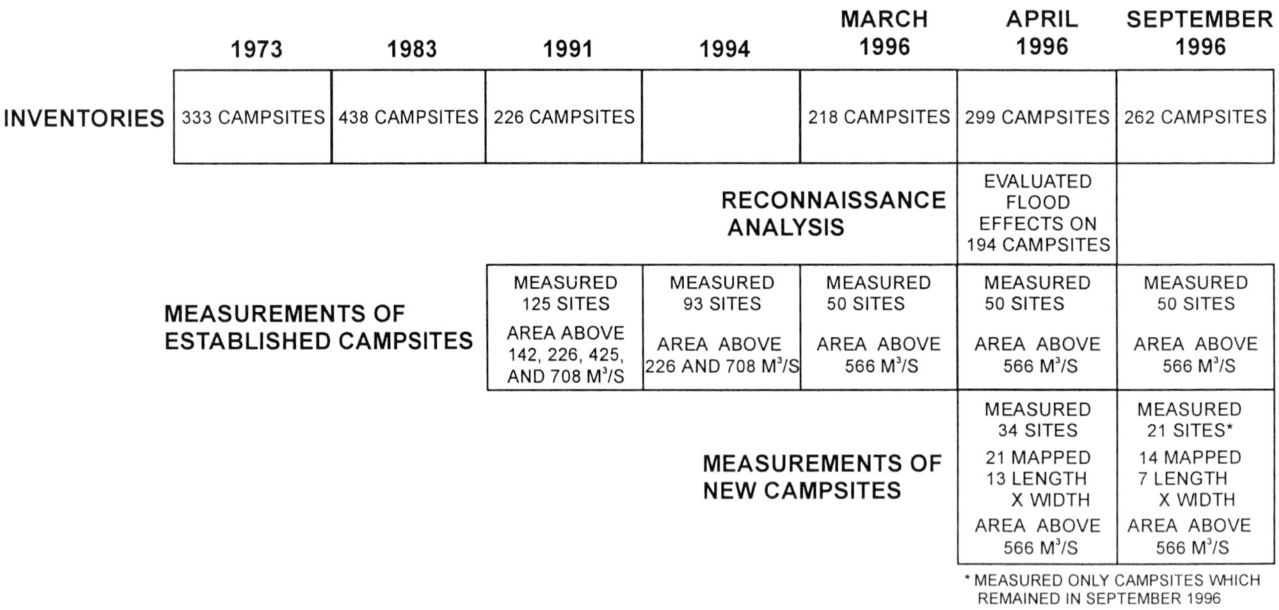

	1973	1983	1991	1994	MARCH 1996	APRIL 1996	SEPTEMBER 1996
INVENTORIES	333 CAMPSITES	438 CAMPSITES	226 CAMPSITES		218 CAMPSITES	299 CAMPSITES	262 CAMPSITES
					RECONNAISSANCE ANALYSIS	EVALUATED FLOOD EFFECTS ON 194 CAMPSITES	
MEASUREMENTS OF ESTABLISHED CAMPSITES			MEASURED 125 SITES AREA ABOVE 142, 226, 425, AND 708 M³/S	MEASURED 93 SITES AREA ABOVE 226 AND 708 M³/S	MEASURED 50 SITES AREA ABOVE 566 M³/S	MEASURED 50 SITES AREA ABOVE 566 M³/S	MEASURED 50 SITES AREA ABOVE 566 M³/S
MEASUREMENTS OF NEW CAMPSITES						MEASURED 34 SITES 21 MAPPED 13 LENGTH X WIDTH AREA ABOVE 566 M³/S	MEASURED 21 SITES* 14 MAPPED 7 LENGTH X WIDTH AREA ABOVE 566 M³/S

* MEASURED ONLY CAMPSITES WHICH REMAINED IN SEPTEMBER 1996

Figure 4. Summary of all of the different campsite evaluation techniques and when they occurred.

unpublished data, 1995). Through repeat topographic monitoring, *Kaplinski et al.* [1995] also documented erosion of sand bars during this time. In 1993, a natural flood from the Little Colorado River raised the mainstem's discharge downstream from the Little Colorado River to 930 m³/s [*Wiele et al.,* 1998]. Half of all measured campsites downstream from the Little Colorado River increased in area after this flood. A year later, most of this increased campsite area had been eroded; however, some of these campsites were still larger in 1994 than they had been in 1991 [*Kearsley,* 1995].

2. METHODS

2.1. Inventories of Campsites

We counted the number of campsites in Grand Canyon 2 weeks before, 2 weeks after, and 6 months after the controlled flood (Figure 4). Since 2 of the 3 authors conducting these inventories had also conducted the 1991 inventory, differences in personal perception of which sites fit the criteria for a campsite were eliminated. During our inventory 2 weeks after the flood, we also stopped at all large newly-created sand bars to evaluate whether they could be used as campsites. We estimated the capacity of each new campsite by walking on each site and counting and evaluating sleeping areas. During our inventory 6 months after the flood, we again stopped at these new

campsites and reevaluated whether they could still be used as campsites, and we again estimated their carrying capacity. Campsite capacity estimates are fairly subjective and are an approximation of the number of people the campsite can accommodate. Our inventory numbers differ slightly from those printed elsewhere [*Kearsley and Quartaroli,* 1997]. Our conclusions, however, have not changed qualitatively. The initial analyses included several "low water campsites," campsites which were only available at discharges at or lower than 425 m³/s, and which therefore were excluded from the present analysis. Likewise, "ledge campsites," campsites on bedrock ledges, had been excluded from the initial analysis. These were added in this analysis.

2.2 Reconnaissance Analysis of Changes in Campsite Area Immediately After Flood

We assessed the change in area caused by the flood of 89% (194 out of 218 that existed prior to the flood) of the existing campsites in Grand Canyon (Figure 4). Two weeks after the flood, we visually inspected each site while floating by and assessed whether substantial changes to campsite size had been caused by the flood. We estimated whether the sites appeared to have gained or lost at least 10% of their pre-flood campsite area, or whether they appeared to be the same. Our assessments were semiquantitative; however, what they lack in precision, they makeup for in large sample size.

2.3. Measurement of Established Campsites

We measured the area of 50 campsites 2 weeks before, 2 weeks after, and 6 months after the flood (Figure 4). These 50 sites were randomly selected from a set of 93 that had been measured in 1994; the original 93 had been randomly selected from the 226 campsites that had existed at that time. At each site, we mapped campable area above 556 m^3/s, the maximum allowable discharge during the period of the study. We mapped pre-flood campsite area on 1:1200 scale copies of 1:4800 scale aerial photographs that had been taken in May 1995. We mapped post-flood campsite area on the same photographs. We later transferred our data onto 1:1200 scale copies of aerial photographs that were taken one week after the flood. We estimated the elevation of flow at 556 m^3/s from cutbanks that were widespread 2 weeks before the flood; these cutbanks had formed in response to flows that had been approximately 556 m^3/s for several weeks. We took photographs of these cutbanks during this pre-flood investigation to ensure accurate relocation of the 556 m^3/s stage for post-flood measurements. The areas of the largest polygons were later calculated using a geographic information system (GIS). The length and width of smaller outlying sleeping areas, usually less than 15 m^2, were measured on-site.

We also measured the areas of 34 new flood-created campsites 2 weeks after and 6 months after the controlled flood. However, since we usually did not have prior knowledge of the new campsites' locations, we did not have a basemap from which to map campsite area, so we drew maps onto a blank sheet of paper. Whenever possible, we included trees, bushes, and boulders that could be used as a reference when drawing our maps. We also measured distances from these features to the perimeters of campsite area and measured length and width of campsite areas. Once the post-flood aerial photographs became available, we transferred our maps onto 1:12000 scale copies of the photographs. We transferred maps of 21 sites onto the post-flood photographs. We used length and width measurements to estimate campsite area for the other 13 sites.

Horizontal mapping accuracy in GIS is estimated at ± 3 m for the 50 established sites that were measured. Based on our judgements of how accurately we could draw campsite area perimeters from 1:1200 scale photographs, field mapping was estimated to have an error of ± 1 m while the Arizona state plane orthophotos have an error of ± 2 m. Mapping of new campsites is less accurate because areas were originally drawn without a basemap.

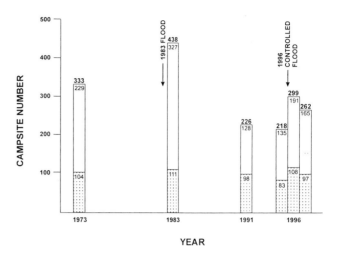

Figure 5. Numbers of campsites during different years as determined by inventories. Stippled bars show numbers of campsites in critical reaches; clear bars show numbers of campsites in non-critical reaches.

3. RESULTS

3.1. Inventories of Campsites

Based on the pre- and post-flood inventories, 84 campsites were created and 3 campsites were destroyed. There were 218 campsites in March 1996, just 8 fewer than counted in the 1991 inventory; there were 299 campsites in the April inventory (Figure 5). Thirty-four of these newly-created campsites had previously existed in 1973, 1983, and/or 1991 and had been larger at that earlier time. These sites had subsequently eroded and were recreated by the 1996 flood. These newly-created campsites accommodated on average 20 people per site. Many of the new campsites were created where deposition occurred at low-elevation bars that were not exposed above 556 m^3/s prior to the flood. Other new campsites were created where separation bars were deposited on debris fans, covering rocky substrate that had been unsuitable for camping (Figure 6). A few sites were created by deposition of sand on top of a bar that was previously overgrown with vegetation. The new sand buried most of the vegetation, creating ample camping space. Of the new campsites, 27% were separation bars, 54% were reattachment bars, 5% were channel-margin bars, and 14% were undifferentiated eddy deposits. Many of the reattachment bars were barren sand, offering no protection from sun or wind, which made them undesirable campsites. The 3 campsites destroyed by the flood, at river miles 61.7R ("below LCR island"), 164.8L ("below Tuckup") and 196.5L ("below Frogy Fault"), were

a

b

c

Figure 6. Photographs of a newly-created campsite (RM 21.5L) (a) 2 weeks before , (b) 2 weeks after , and (c) 6 months after the controlled flood showing sand deposition and subsequent erosion on an alluvial fan.

no longer usable because of extensive erosion that obliterated most of their campsite area.

These new campsites were not uniformly distributed longitudinally or by reach type. Twice as many campsites were created per river kilometer in non-critical reaches, which averaged one new campsite every 3.3 km, than in critical reaches, which averaged one every 6.8 km (Figures 5 and 7). Also, 42% of the new sites were created in a 40 km section between river miles 43-68, with an average of 1 new site/km (Figure 7).

While the number of new flood-created sites was substantial, these bars rapidly eroded after the flood receded. Six months after the flood, only 56% (47 of 84) of the new sites were still large enough to be suitable as campsites. There were 262 campsites inventoried in September 1996 (Figure 5). There was a trend for an increase (Figure 5) and subsequent decrease in campsite number during the controlled flood to occur primarily in non-critical reaches; however these differences were not significant (Chi-squared test, $X^2_{(1)}=0.184$, $p>0.05$).

Several factors caused the loss of new campsites. Of the 37 sites that were no longer large enough to be used as campsites in September 1996, 70% (26) had either eroded completely or to a size too small to be used as a campsite, 24% (9) were still of adequate size but had such steep slopes along the river that made access from the river infeasible, and 5% (2) were too densely vegetated to use. Vegetation that had been buried by sand at the new sites grew robustly during the summer, possibly benefitting from either the flood, the high summer flows or both [*Kearsley and Ayers,* this volume]. One of the new campsites (RM 35.2R) was eroded by a flash flood from a tributary.

3.2. Reconnaissance Analysis of Changes in Campsite Area Immediately After Flood

There was a pronounced system-wide increase in the area of campsites throughout Grand Canyon. Fifty percent (96 of 194) of the evaluated sites were larger by at least 10%, 39% (76 of 194) did not change, and 11% (22 of 194) were smaller than they had been prior to the flood. Thick deposition of sand did not necessarily cause an increase in campsite size. Forty-one of the 76 sites that did not change in area had extensive deposition; thus, the campsite was higher but not wider. This geomorphic pattern was also measured by the *Hazel et al.*[this volume] topographic surveys of sand bars. There was a substantial increase in sand at 72% (140 of 194) of the evaluated campsites, little to no change at 22% of the campsites, and a decrease at 6% of the campsites. We checked the accuracy of our reconnaissance analysis technique by comparing the assessments

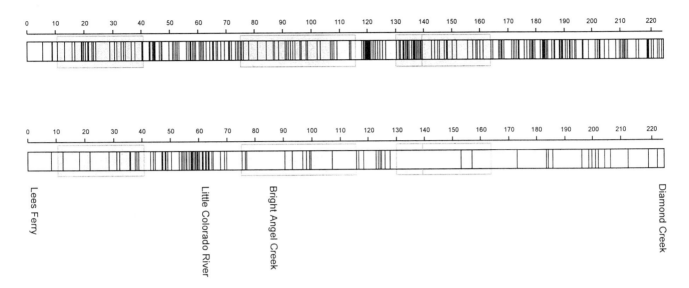

Figure 7. Schematic showing distribution of campsites 2 weeks before the 1996 controlled flood (top) and distribution of newly-created campsites (bottom) between Lees Ferry and Diamond Creek. Thin vertical bars represent locations of campsites. Shaded areas indicate critical reaches. Numbers are river mile locations.

of the 194 sites with the size changes measured at 50 sites and found the reconnaissance analysis to be less sensitive to change than the measured sites ($X^2_{(2)} = 7.17$, p<0.05), but not biased towards an increase or decrease in area. Of the measured sites, 30 (60%) increased in area by more than 10% of their pre-flood area, 10 (20%) were the same size, and 10 (20%) decreased in area.

3.3. Measurements of Established and New Campsites

The established campsites we measured were much larger 2 weeks after the flood than they had been 2 weeks before the flood. These bars eroded during the subsequent 6 months, but the extent of erosion was not as great as the deposition. Thus, on average, sites were larger 6 months after the flood than they had been immediately before the flood. Two weeks after the flood deposition of sand increased the campable area by a mean of 159 m^2, increasing on average by 48% (Figure 8). By September 1996 most sites had eroded so that when compared to their pre-flood size, they were larger, on average, by 17%. Loss of campsite area due to vegetation growth was negligible, with vegetation growth accounting for only 2% of all campsite area loss. In Figure 9, photographs illustrate these changes. A histogram showing the numbers of sites which increased by different percentages for the 3 time periods is shown in Figure 10.

The 34 new campsites we measured were approximately half the size of the established campsites, with a mean

campsite area of 470 m^2. By September 1996, campsite area of the measured sites decreased so that they were on average 41% of the size they had been 2 weeks after the flood (Figure 8). The new campsites eroded much faster than the established campsites (Wilcoxon rank-sum test statistic, W=829.5, p<0.0001). Also, in contrast to the small loss of campsite area in established sites due to vegetation growth, 27% of the new campsite area loss was attributed to vegetation growth.

4. DISCUSSION

The 1996 controlled flood increased the number and size of sand bars large enough to be used as campsites

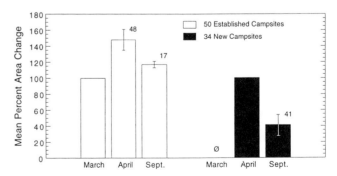

Figure 8. Mean percent area changes for the 50 established measured campsites 2 weeks before, 2 weeks after and 6 months after the controlled flood. Also, area changes for the 34 new campsites measured 2 weeks after and 6 months after the flood.

throughout Grand Canyon; the capacity of these bars to accommodate boating parties increased. The controlled flood created many new campsites, increased the area of more than four times as many campsites as the number that eroded, and increased the total campable area by nearly 50%. Thus, similar floods that resulted in similar geomorphic change would temporarily benefit recreational users in Grand Canyon.

Flooding also temporarily reversed the trend of slow decrease in campsite area that occurs during non-flood years. When compared to campsite area changes during non-flood years, changes to campsite area due to the controlled flood consisted of net increases rather than net decreases, and were more pronounced. Between 1991 and 1994, campsite area above the 708 m^3/s stage of the same sites as those measured during the test flood had decreased on average by 16% (L.H. Kearsley, unpublished data, 1995). As opposed to a 16% area loss over this 3-year time span, measured campsites had a 48% and 17% area gain during a 2-week and 6-month post-flood time span, respectively.

The 1996 flood also improved the campsites aesthetically. Particularly in a National Park, boating parties prefer to travel through a more natural, dynamic system that is periodically flooded (Jeri Ledbetter, Grand Canyon River Guides, pers. commun., 1996). While steep slopes and cutbanks formed at many of the sites, these higher, steeper sandbars are common along natural rivers in debris fan affected canyons. Also, in addition to sand deposition, redistribution of sand on campsites cleansed the sites, making them more appealing, particularly in heavily used portable toilet locations.

The trend for new flood-created campsites to predominate in non-critical reaches is not surprising, since the river in these reaches is wide and has many more depositional sites for large sand bars [Melis, 1997]. This has also been the pattern of change in campsite number for previous inventories (Figure 5) [Kearsley et al., 1994]. However, new campsites were created in critical reaches, even in the reach upstream from the Little Colorado River, which has less sediment replenishment [Andrews, 1991]. The new and larger campsites in critical reaches helped alleviate the campsite shortage in these areas.

While flood-induced benefits to campsites were substantial, erosion of new deposits occurred fairly quickly. Six months after the flood, nearly half of the new campsites were no longer usable, the remaining new campsites were half their initial size, and most of the increased area on the measured established campsites had eroded. The relatively high flows of 556 m^3/s that occurred on most days throughout the summer probably accelerated the erosion

Figure 9. Photographs of an established campsite (RM 134.6L) 2 weeks before, 2 weeks after, and 6 months after the controlled flood (9a) show sand deposition and subsequent erosion of the site. Maps of the same site during the same periods (9b) demonstrate how these depositional and erosional processes affected campsites.

Prior to Test Flow (March 1996)

Campable Area = 1135 m2

After Test Flow (April 1996)

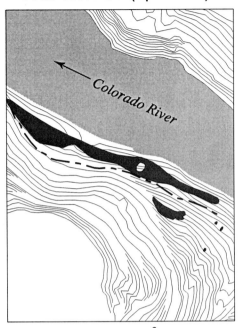

Campable Area = 1910 m^2

6 Months After Test Flow (Sept. 1996)

Campable Area = 948 m2

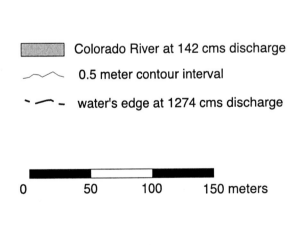

Colorado River at 142 cms discharge

0.5 meter contour interval

water's edge at 1274 cms discharge

0   50   100   150 meters

b

Figure 9 (continued)

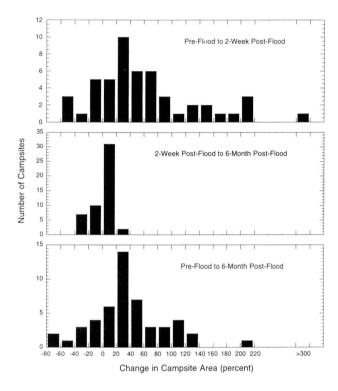

Figure 10. Distribution of campsite area changes of the 50 established measured campsites for three time periods during the study. The values represent the lower part of a range; e.g., a value of "0" represents a 0-20% increase in area, "20" represents a 20-40% increase.

process. But even without these high flows, rapid erosion of newly deposited sediment occurs [*Beus et al.,* 1985; *Schmidt and Graf,* 1990], and we must anticipate that most of the new campsites and the increased area on campsites will subsist for only a short time.

A comparison of these changes in campsite numbers with those in previous inventories following floods shows that while the increased number and area of campsites resulting from the controlled flood was substantial, the increases were not as large as the increase in campsite number due to the 1983 high flows (figure 5). Those flows, however, were twice as high and occurred after tributary sediment is estimated to have accumulated on the river channel for approximately 20 years [*Randle et al.,* 1993]. The new 1983 deposits were also short-lived, with a significant decrease in campsite size and number by 1984 [*Kearsley et al.,* 1994]. While the 1996 inventory number changes were not as dramatic as they have been in the past, considering the current conditions of more stable existing bars, moderate flood levels, and fewer years for sediment to accumulate on the river channel, they are nevertheless substantial.

It is important to evaluate the influence of the controlled flood on the pattern of campsite change over the entire post-dam period. *Kearsley et al.* [1994] documented an initial decrease in the number and size of many campsites between 1965 and 1973 followed by a decreasing rate of erosion as campsites stabilized. The documentation of slow erosion rates from 1991 to 1994 [*Kearsley,* 1995; *Kaplinski et al.,* 1995] along with a very small decrease in campsite number from 1991 to 1996 (Figure 3) provide further evidence of a stabilization of sand bars used as campsites. The increase in campsite number and size resulting from both the 1983 flood and the 1996 controlled flood have been of short duration. Therefore, the long-term trend from these data suggest that similar floods will temporarily increase campsite number and size, then the campsites will quickly return to their pre-flood condition and continue to erode slowly. While flood effects to campsites are temporary, they are the only feasible means of depositing sediment above normal river fluctuations, where sand is otherwise continually eroding.

Acknowledgments. The authors wish to express their thanks to Sue Markham, Diana Kimberling, Bill Dennis, and Keith Ryan for assistance in the field. We thank John C. Schmidt, Ted S. Melis, Robert H. Webb, and one anonymous reviewer for reviewing earlier versions of this document. We also wish to thank David L. Wegner, Project Manager for the U.S. Bureau of Reclamation's Glen Canyon Environmental Studies, for financial and logistical support.

REFERENCES

Andrews, E.D., Sediment transport in the Colorado River Basin, in *Colorado River ecology and dam management*, ed. G.R. Marzolf, Natl. Acad. Press, Washington, DC, 1991.

Beus, S.S., S.W. Carothers, and C.C. Avery, Topographic changes in fluvial terrace deposits used as campsite beaches along the Colorado River in Grand Canyon, *J. Ariz.-Nev. Acad. Sci.,* 20,111-120, 1985.

Brian, N.J., and J.R. Thomas, 1983 *Colorado River beach campsite inventory*, Natl. Park Serv., Div. Res. Manage., Grand Canyon National Park, Grand Canyon, AZ, 1984.

Dolan, R., A.D. Howard, and D. Trimble, Structural control of the rapids and pools of the Colorado River in the Grand Canyon, *Science,* 202, 629-631, 1978.

Howard, A.D., and R. Dolan, Geomorphology of the Colorado River, *J. Geol.,* 89, 269-298, 1981.

Kaplinski, M., J.E. Hazel Jr., and S.S. Beus, *Monitoring the effects of Interim Flows from Glen Canyon Dam on sand bars in the Colorado River Corridor, Grand Canyon National Park, Arizona,* Final Rep. to the Natl. Park Serv., N. Ariz. Univ., Flagstaff, AZ, 1995.

Kearsley, L.H, J.C. Schmidt, and K.D. Warren, Effects of Glen Canyon Dam on Colorado River sand deposits used as campsites in Grand Canyon National Park, USA, *Regul. Rivers: Res. Manage.*, 9, 137-149, 1994.

Kearsley, L.H., *Monitoring the effects of Glen Canyon Dam Interim flows on campsite size along the Colorado River in Grand Canyon National Park*, Natl. Park Serv., Div. Res. Manage., Grand Canyon National Park, AZ, 1995.

Kearsley, L.H. and R.D. Quartaroli, *Effects of a beach/habitat-building flow on campsites in the Grand Canyon*, Final Rept. to Natl. Park Serv., Grand Canyon National Park, AZ, 1997.

Leschin, M.F. and J.C. Schmidt, *Description of map units to accompany maps showing surficial geology and geomorphology of the point Hansborough and Little Colorado River Confluence reaches of the Colorado River, Grand Canyon National Park, Arizona*, Final Rept. to Bur. Reclam., Glen Canyon Environ. Studies, Flagstaff, AZ, 1995.

Melis, T.S., R.H. Webb, P.G. Griffiths, and T.J. Wise, *Magnitude and frequency data for historic debris flows in Grand Canyon National Park and vicinity, Arizona*, U.S. Geol. Surv. Water Res. Invest. Rept. 94-4214, 1994.

Melis, T.S., *Geomorphology of debris flows and alluvial fans in Grand Canyon National Park and their influence on the Colorado River below Glen Canyon Dam, Arizona*, Ph.D. thesis, Univ. Ariz., Tucson, AZ, 1997.

Randle, T.J., R.I. Strand, and A. Streifel, Engineering and environmental considerations of Grand Canyon sediment management, in *Engineering Solutions to Environmental Challenges: Thirteenth Annual USCOLD Lecture*, U.S. Committee on Large Dams, Chattanooga, TN, 1993.

Schmidt J.C., and J.B. Graf, *Aggradation and degradation of alluvial sand deposits, 1965-1986, Colorado River, Grand Canyon National Park, Arizona*, 74 p, U.S. Geol. Surv. Prof. Paper 1493, 1990.

Schmidt, J.C., and D.M. Rubin, Regulated streamflow, fine-grained deposits, and effective discharge in canyons with abundant debris fans, in *Natural and anthropogenic influences in fluvial geomorphology*, ed. J.E. Costa, A.J. Miller, K.W. Potter, and P.R. Wilcock, pp. 177-195, Amer. Geophys. Union Mono. 89, 1995.

Schmidt, J.C., P.E. Grams, and R.H. Webb, Comparison of the magnitude of erosion along two large regulated rivers, *Water Res. Bull.*, 31, 617-630, 1995.

U.S. Department of the Interior, National Park Service, Organic Act, 1916.

Webb, R.H., *Grand Canyon, a century of change*, 290 pp., Univ. Ariz. Press, Tucson, AZ, 1996.

Weeden, H., F. Borden, B. Turner, D. Thompson, C. Strauss, and R. Johnson, *Grand Canyon National Park campsite inventory*, Progress Report #3, Contract #CX 001-3-0061, Natl. Park Serv., Penn. State Univ., University Park, PA, 1975.

Lisa H. Kearsley, SWCA Inc., Environmental Consultants, 114 N. San Francisco Street, Flagstaff, AZ 86001; email: lkearsley@swca.com

Richard D. Quartaroli, Special Collections Cline Library, Northern Arizona University, Flagstaff, AZ 86011

Michael J.C. Kearsley, Department of Biological Sciences Northern Arizona University, Flagstaff, AZ 86011

Topographic and Bathymetric Changes at Thirty-Three Long-Term Study Sites

Joseph E. Hazel Jr., Matt Kaplinski, Roderic Parnell, Mark Manone, Alan Dale

Department of Geology, Northern Arizona University, Flagstaff, Arizona

This study documents the geomorphic response to the 1996 controlled flood of 33 study sites in fan-eddy complexes distributed throughout Grand Canyon. Repeated topographic mapping was used to quantify sediment redistribution at sites within three distinct parts of the fan-eddy complex: high-elevation sand bars, submerged areas of the recirculation zone (eddy), and the adjacent main channel. The topographic data show that the 1996 controlled flood rebuilt previously eroded high-elevation bars, regardless of location, bar type, or width of the canyon. The average thickness of new, high-elevation deposits was 0.64 m. Because there was an average thickness decrease in the main channel of 0.45 m, we conclude that large volumes of sediment were scoured from storage locations on the bed. Although there was variability between the behaviors of individual sites, two response styles were identified: (1) deposition filled or nearly filled all areas of small recirculation zones, and (2) extensive low-elevation scour caused net erosion in large recirculation zones. The average recirculation zone change at all sites was net aggradation. Aggradation was greatest in narrow reaches where stage change was greatest. The magnitude of deposition was greatest in the Marble Canyon reach between the Paria River and the Little Colorado River. Sites located downstream from the Little Colorado River still remained partly filled with sediment by the 1993 Little Colorado River floods so the potential for bar filling was limited compared to the sites located upstream. Our results suggest that the amount of high-elevation deposition is in part controlled by the space available for deposition in the recirculation zone before the flood, which we term "accommodation space." Erosion rates of newly-deposited sand were initially high following the 1996 controlled flood but decreased with time. Our results show that controlled flooding restores eroded bars downstream from Glen Canyon Dam, and produces sand volume increases in bar storage that are at least partially maintained for nearly a year afterward.

1. INTRODUCTION

The main objective of the 1996 controlled flood was to test the hypothesis that controlled floods in Grand Canyon can be used to entrain sand from the channel bed and deposit it higher along the channel margin. Of primary concern was the rebuilding of sand bars, which are funda-

The Controlled Flood in Grand Canyon
Geophysical Monograph 110
Copyright 1999 by the American Geophysical Union

mentally important environmental resources along the banks of the Colorado River. Bars provide substrate for some habitats of endangered and native fish species [*Valdez and Ryel*, 1995], riparian vegetation, marsh and wetlands [*Stevens et al.*, 1995], and are also used as recreational campsites [*Kearsley et al.*, 1994]. We test the hypothesis stated above by utilizing repeated topographic mapping of bar and channel topography. Surveys were conducted before and after the flood at study sites distributed throughout the river corridor. In this paper, we (1) describe spatial changes in bar and channel morphology at 33 long-term monitoring sites, (2) compare these results to temporal changes in sediment storage determined from 5 prior years of biannual surveys, and (3) assess the stability of the rebuilt deposits upon resumption of normal dam operations over the 10-month period that immediately followed the flood.

2. BACKGROUND

2.1. Post-Dam Patterns of Sand Bar Change

Over two decades of studies have examined geomorphic changes of sand bars in Grand Canyon. Terrestrial surveying began in 1974 when *Howard* [1975] used tape and transit to survey topographic profiles at 20 sites. *Howard and Dolan* [1981] repeated the profile surveys and suggested that the erosional trend evident following closure of Glen Canyon Dam had stabilized by the late 1970s [*Webb et al.*, this volume]. Continued monitoring of the profile lines in the 1890s by *Beus et al.* [1985] and *Schmidt and Graf* [1990] demonstrated that the largest system-wide geomorphic event of the post-dam era, up to that point, was the 1983 flood which created or enlarged many sand bars and aggraded many sites that had eroded after dam closure.

However, by 1986 many sites had suffered net erosion and some bars used as campsites were completely eroded. Following the 1983 flood, stratigraphic and sedimentologic studies indicated that high releases between 1984 and 1986 deposited little sediment on sand bars [*Rubin et al.*, 1990; 1994]. It was believed that the high discharges of the period 1983 to 1986 depleted main channel sediment that had accumulated in the canyon since 1965 [*Randle et al.*, 1993].

Several studies have shown that bars differ in susceptibility to erosion. Separation bars are more stable than reattachment bars, but erosion of either bar type is greatest in narrow reaches, especially those closest to Glen Canyon Dam [*Schmidt and Graf*, 1990; *Schmidt et al.*, 1995; *Webb*, 1996]. As a result, the number and size of campsites in the

post-dam era has decreased the most in narrow reaches [*Kearsley et al.*, 1994].

Since 1986, the only event, excluding the 1996 controlled flood, that resupplied sediment to bars at elevations greater than the stages that occur at normal power-plant capacity (approximately 891 m^3/s) were the January and February 1993 Little Colorado River floods. Deposition from these floods filled many recirculation zones immediately downstream from the confluence [*Hazel et al.*, 1993; *Kaplinski et al.*, 1994; *Schmidt and Leschin*, 1995; *Wiele et al.*, 1996]. In this reach, suspended-sand concentrations in the Colorado River were comparable to the highest measured pre-dam concentrations (D. Topping, U.S. Geological Survey, Denver, personal commun., 1998). Because the majority of the Little Colorado River-supplied sediment was probably deposited within the first 20 km downstream from the confluence, *Kaplinski et al.* [1995] and *Wiele et al.* [1996] reasoned that deposition on bars in the more downstream reaches of Grand Canyon resulted from redistribution of previously stored channel-bed sediment by the increased mainstem discharge, which peaked at 965 m^3/s. Many questions remained, however, and the effectiveness at which dam releases could be manipulated to transfer sediment from submerged main-channel storage to channel-margin sand bars was uncertain.

2.2. Recent Topographic Surveys

The importance of sand-supply limitations and examination of flow alternatives were the bases of monitoring and research conducted for the Glen Canyon Dam Environmental Impact Statement [*U.S. Department of Interior*, 1995]. With the advent of newer total station survey instruments and computer-based topographic modeling software, more detailed sand bar topographic data at 29 sites were collected during research test flows in 1990 and 1991 by *Beus et al.* [1992]. Surveys conducted twice-monthly allowed comparisons of sand bar volume changes between test flows. Although that study did not demonstrate which specific flow alternative best maintained sand bars, the study did correlate bar behavior with sediment transport capacity and antecedent conditions. Test flows characterized by large daily fluctuations resulted in aggradation when tributaries contributed sediment. Flows following aggradational periods were more likely to erode bars.

After the 1990-1991 test flows were completed, the sand bar monitoring program of *Kaplinski et al.* [1995] was initiated to examine the effects of the interim operating criteria for Glen Canyon Dam implemented in 1991 [*Schmidt et al.*, this volume]. These surveys were conducted twice-yearly with the same methods and study sites as *Beus et al.* [1992]. The sample size was expanded to 35 sand bars located at 33 fan-eddy complexes. *Kaplinski et al.* [1995] added a hydrographic survey system, developed by the Bureau of Reclamation's Glen Canyon Environmental Studies, to create detailed topographic maps of the submerged portions of sand bars and adjacent channel areas.

Results of the 1992-1995 monitoring indicated that, despite limits placed on dam operations, high-elevation parts of bars continued to erode, primarily by bank retreat. Loss of high-elevation sand also resulted from slope-failure mechanisms whereby large areas of sand bars eroded in less than one day [*Cluer*, 1995; *Dexter et al.*, 1995]. To observe these short-term events, the remote camera system utilized by *Dexter et al.* [1995] was incorporated into the Northern Arizona University sand bar monitoring project in 1995 [*Parnell et al.*, 1996].

Understanding the apparent role of antecedent conditions in determining the degree to which a site aggrades or degrades is important for predicting flood impacts and designing effective bar-building strategies. We examine this concept by using the potential space available for high-elevation deposition, termed "accommodation space," at each of our study sites to examine the relation between antecedent conditions and volume of deposition at each.

3. METHODS

3.1. Study Site Selection

In order to make long-term comparisons, we used the study sites of *Beus et al.* [1992] and *Kaplinski et al.* [1995] to study the effects of the 1996 controlled flood. *Beus et al.* [1992] listed the original criteria for selection of these measurement sites as (1) even distribution throughout 13 bedrock-defined, geomorphic reaches originally identified by *Schmidt and Graf* [1990], (2) even distribution of bar types [*Webb et al.*, this volume], (3) availability of historical topographic data, and (4) variation in recreation use intensity and vegetation cover.

At the time of site selection in 1990, little was known about the variability of site response to change in flow regime. *Schmidt and Leschin* [1995] studied reach-scale sand bar change using aerial photographs and determined that there may be wide variability in response from site to site. Within a reach, some sites aggrade at the same time others erode. We have not yet established that the response of our study sites accurately represents the system-wide condition. Nonetheless, both methodologies — *Schmidt*

and Leschin [1995] reach-scale analyses of every recirculation zone and *Kaplinski et al.* [1995] repeated topographic mapping of selected sites — led to similar conclusions concerning the effects of interim flow operations between 1991 and 1995 and the effects of the 1993 Little Colorado River floods. In addition, analysis of topographic change at a smaller subset of bars than used in our study detected large-scale trends caused by the high flows of 1983 [*Beus et al.*, 1985] and the system-wide erosion of bars following 1986 [*Schmidt and Graf*, 1990]. More importantly, *Schmidt et al.* [this volume] compare individual site and reach-scale behavior and show that our sites located within their comprehensively-mapped reaches responded in a manner similar to the average behavior of the adjacent reaches. Thus, we believe our 35-bar dataset provides the best representation of system response presently available.

Study site locations and geographic place names are shown in *Webb et al.* [this volume] and are provided in Table 1. Study site reference numbers use river mile location. One site is located in Glen Canyon, the reach between Glen Canyon Dam and Lees Ferry (river miles -15 to 0). Twelve sites are located in Marble Canyon, the reach between Lees Ferry and the confluence with the Little Colorado River (river miles 0 to 61.5). Sediment inputs to Marble Canyon are largely from the Paria River [*Andrews*, 1990]. Twenty sites are located in Grand Canyon, the reach between the confluence with the Little Colorado River and the Grand Wash Cliffs (river miles 61.5 to about 280). Grand Canyon receives sediment from the Paria River, Little Colorado River, and other tributaries. No sites were examined downstream from Diamond Creek (river mile 226).

Each study site is located at what *Schmidt and Rubin* [1995] termed a fan-eddy complex. Bar characteristics, flow patterns, and recirculation-zone response to changing discharge are described by *Webb et al.* [this volume]. In this study, we define recirculation zone as that area of the fan-eddy complex containing an eddy sand bar and recirculating flow. Our sample sites include separation, reattachment, and undifferentiated eddy bars. By February 1996, riparian and marsh vegetation had expanded to cover most of the remaining high-elevation portions of the study sites, except at the most heavily utilized campsites [*Kearsley and Ayers*, this volume]. However, lower elevations were typically bare sand and had a surface regularly inundated and reworked by dam releases.

3.2. Data Collection

Pre-flood topographic surveys of the study sites were conducted in February 1996. Before then, most of the sites

TABLE 1. Channel geometry and geomorphic characteristics for selected sites

Site Reference Number	River Mile[1]	Site Name	Deposit Type[2]	Expansion Ratio[3]	Stage Change (m)[4]	Reach/Relative Width[5]
-6	-6.5	Hidden Sloughs	U	1.17	3.64	0/W
3	2.6	Cathedral Wash	R	1.97	4.05	1/W
8	7.9	Lower Jackass	S	1.67	3.69	1/W
16	16.4	Hot Na Na	S	2.11	3.68	2/N
22	21.8		R	1.25	6.89	2/N
30	30.0	Fence Fault	R	1.57	5.86	3/N
32	31.6	South Canyon	U	2.24	3.52	3/N
43	43.1	Anasazi Bridge	R	3.00	4.82	4/W
45	44.6	Eminence Break	S,R	2.28	4.79	4/W
47	47.1	Lower Saddle	R	2.57	4.27	4/W
50	50.0	Dino	S,R	1.43	4.64	4/W
51	51.2		R	2.22	4.15	4/W
55	55.5	Kwagunt Marsh	R	1.88	2.95	4/W
62	62.4	Crash Canyon	R	2.47	4.41	5/W
68	68.2	Tanner	U	1.76	2.82	5/W
81	81.1	Grapevine	U	1.56	4.35	6/N
87	87.5	Cremation	U	1.36	4.88	6/N
91	91.1	Above Trinity	S	1.29	5.01	6/N
93	93.3	Upper Granite	U	1.87	3.46	6/N
104	103.9	Upper 104 Mile	R	1.27	4.57	6/N
119	119.1		R	1.79	5.45	7/N
122	122.2		R	2.84	4.96	7/N
123	122.7	Upper Forster	R	1.85	5.10	7/N
137	136.7	Middle Ponchos	R	1.53	4.90	8/N
139	139.0	Upper Fishtail	U	1.59	5.41	8/N
145	145.1	Above Olo	R	1.70	5.90	9/N
172	172.2	Below Mohawk	R	1.56	4.16	10/W
183	182.8		R	1.63	5.00	10/W
194	194.1		R	2.36	4.22	10/W
202	201.9	202 Mile	S	2.38	4.61	10/W
213	212.9	Pumpkin Spring	U	2.64	6.89	10/W
220	219.9	Middle Gorilla	U	1.75	3.47	11/N
225	225.3		R	2.00	3.65	11/N

[1]Distance downstream or upstream from Lees Ferry in river miles. From *Stevens* [1983].
[2]Deposit type: R - reattachment bar, S - separation bar, U - undifferentiated eddy bar.
[3]Average channel width in expansion divided by average channel width in constriction at ~594 m^3/s.
[4]Difference in water surface elevations between the 142 m^3/s and the 1274 m^3/s stage.
[5]Geomorphic reach (0-11) and channel width (W-wide, N-narrow) from *Schmidt and Graf* [1990].

had been monitored biannually since 1991. Following the flood, sites were resurveyed in April and September 1996, and in February 1997. Daily photographs were obtained at each site with the remote camera system. Topographic mapping was accomplished using electronic total stations equipped with digital data collectors. Bathymetric data were collected with a hydrographic system that consists of a total station modified to track a target mounted on a raft, a sonic depth sounder, a digital/analog receiving unit, and a computer that controlled the data collection process. Hydrographic surveys followed pre-set 7.5 to 10 m transects perpendicular and parallel to the bank to form a grid. Ground points were collected from 1 m below the water surface up to the elevation reached by a discharge of 1274 m^3/s or higher. Ground and bathymetric points were combined and a topographic surface model of recirculation zone and adjacent channel was created using the triangulated irregular network method of contouring with surface modeling software [*Datacom Software Research Limited*, 1997]. Site size and topographic complexity determined the point density needed to form proper topographic models.

Accuracy and precision of these techniques have been assessed by *Beus et al.* [1992] and *Kaplinski et al.* [1995], who also verified all benchmark and back-site relationships. Verification of horizontal position and elevation data found that ground points have a horizontal error of <0.1 m and a vertical error <0.05 m, and that hydrographic data have a horizontal error of <1.0 m and a vertical error < 0.25 m. Volume calculations derived from topographic surface models created from replicate daily surveys at one site were shown to vary less than 3% [*Beus et al.*, 1992].

3.3. Data Analysis

The topographic surface model of each site was used to generate profiles, comparison maps, and area and volume calculations. To quantify changes in sediment redistribution within the recirculation zone and the main channel, areas and volumes were calculated within boundaries that approximate the dimensions of the recirculation zone and adjacent channel (Figure 1a). A fixed boundary was established between the main channel and recirculation zone by estimating the position of the eddy fence, the streamline dividing downstream flow and the eddy, and by assuming this zone extends vertically to the bed (Figure 1b). Eddy fence location was determined by visual observation of oblique daily photographs, by aerial photographs, and by the positions of separation and reattachment points surveyed in the field at different discharges. This general approximation of eddy-fence location best represents the

dimensions of the recirculation zone at flows between 566 and 1274 m^3/s.

Areas of deposition and erosion in the recirculation zone were calculated above different topographic levels (Figure 1b). An empirically derived stage-discharge relation determined by *Kaplinski et al.* [1995] at each site was updated and used to define the elevation range of specific flows. The 142 m^3/s stage elevation is the minimum discharge at which dam releases occur and is hereafter referred to as baseflow. Baseflow stage was used to separate calculations of bar changes from changes in the deeper, continuously inundated portion of the recirculation zone, which we term the "eddy," and the main channel. We use the term "sand bar" for that part of the recirculation zone above baseflow. Volume and area changes were also calculated for the highest parts of sand bars above the 566 m^3/s stage elevation, and we refer to this as the high-elevation bar. This topographic level was chosen to define high-elevation sand bars, because this discharge was the highest operating limit for Glen Canyon Dam in the 1990s under the interim operating criteria. In addition, this was also the discharge above which bars were considered campsites by *Kearsley and Quataroli* [1997], *Kearsley et al.* [this volume], and *Thompson et al.* [1997].

This method permitted comparison of bed changes as sediment was transferred between different topographic levels within recirculation zones and with the main channel. As bar size increases or decreases our computational eddy size also changes but the total area of the recirculation zone remains the same.

The size of each study site differs. To compare sites of different size, changes in area and volume were normalized and expressed as a percentage of the pre-flood survey by

$$\Delta = \left[100 \frac{(V_j - V_i)}{V_i} \right] \qquad (1)$$

where V_i is the volume or area prior to the flood and V_j is the volume or area calculated from surveys at other times. Areas are only reported for sand bars. Because of boundary conditions imposed by the bedrock or talus-confined channel, eddy and channel computational areas do not change appreciably. Changes in elevation of the channel bed are assumed to result from the removal or addition of sand-size or smaller sediment.

We also converted changes in volume and area to net change in thickness. High-elevation sand bar thickness was determined by dividing the volumetric change between surveys by the maximum area of the sand bar measured at that site during the entire monitoring period. To examine recirculation zone thickness, the total volume change (bar

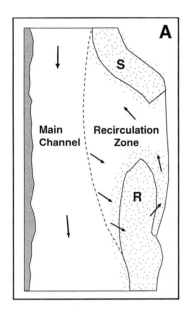
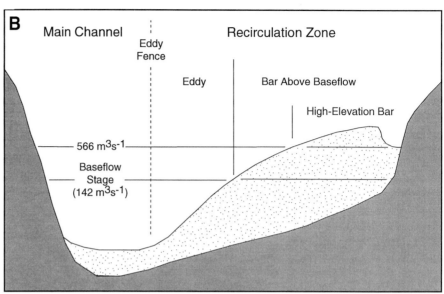

Figure 1. Definitions of boundaries at a typical fan-eddy complex for purposes of computations in planview (A), and in cross-section (B). Generalized flow patterns are depicted by arrows. The dashed line is the approximate location of the eddy-fence boundary that separates the main channel from the recirculation zone. Within the recirculation zone is the separation bar (S) and the reattachment bar (R). Sand-bar volume and area were calculated above different topographic levels that correspond to baseflow (142 m³/s) and high-elevation (566 m³/s). Main channel and eddy volumes were calculated below baseflow.

and eddy) was divided by the area of the recirculation zone. Likewise, main channel thickness was calculated by dividing volume change of the channel bed by the measured area of the channel. In calculating net changes, we assumed that no change occurred in the interval of time between the surveys and the flood. The maximum time interval between pre-flood surveys and flood onset was 42 days in upper Marble Canyon, and, after flood cessation, 30 days in lower Grand Canyon.

4. STYLES AND CHARACTERISTICS OF CHANGE

4.1. Overview of Changes, February-April 1996

The study sites were substantially modified by the 1996 controlled flood. Generally, sand bars were aggraded and adjacent main channel areas were degraded or remained unchanged (Figure 2a). The volume and area of sand at high-elevation (above 566 m³/s) in bars increased an average of 164% and 67%, respectively (Table 2). The average thickness of new deposits at high elevation was 0.64 m (Figure 2b). However, many sites that aggraded at high elevation also had large areas of low-elevation erosion, as indicated by decreases in area of sand above baseflow (above 142 m³/s). As a result, volume and area

changes of sand in bars above baseflow were smaller than high-elevation gains, and increased an average of 37% and 5%, respectively. Large volumes of sediment below baseflow were also scoured from eddies. The average recirculation zone thickness change, which combines the scour of sediment from below baseflow with high elevation aggradation, was 0.28 m. The volume of sediment in the main channel adjacent to the recirculation zone decreased an average of 14%. This decrease in volume corresponds to an average thickness change of the channel bed of -0.45 m.

There was substantial variability in response from site to site, but deposition consistently occurred at high-elevations (Figure 2b). Thirty-three out of 35 bars had a net high-elevation thickness change of 0.10 m or greater. Despite widespread high-elevation aggradation, overall thickness change in the recirculation zones ranged from -1.48 m to +2.23 m. Twenty-two out of 33 recirculation zones had a net thickness change of 0.10 m or greater. Because of the variability in site response there was no significant correlation between the size of the recirculation zone and amount of net change in sand thickness. However, all but one recirculation zone larger than 7500 m² had a net loss of sand (Figure 3a). All but two recirculation zones smaller than 7500 m² had a net gain or were relatively unchanged. Figure 3a shows that small recirculation zones (<7500 m²)

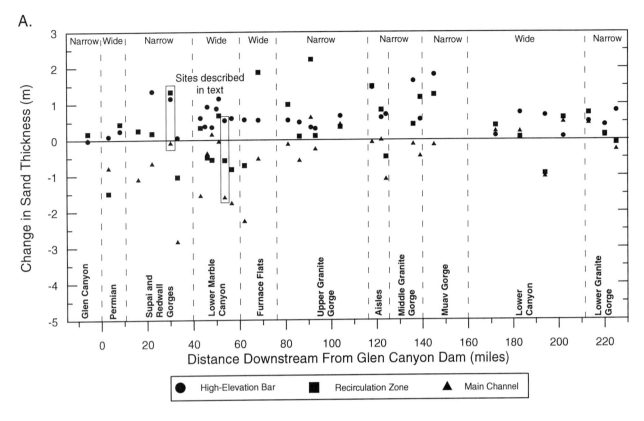

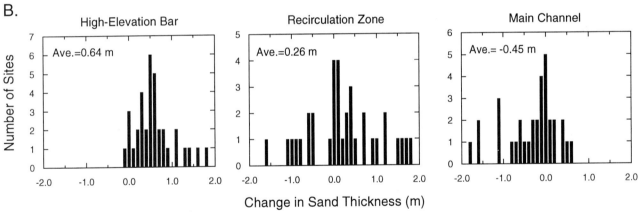

Figure 2. Graphs showing changes in sand thickness between February and April 1996. (A) The change at each site with distance downstream. Geomorphic reaches are from *Schmidt and Graf* [1990]. Sites described in text are highlighted with boxes. (B) Frequency distributions for sand thickness change at high-elevation sand bars, recirculation zones, and main channel. See Figure 1 for definitions of these areas.

had significant deposition during the flood (0.53 m ± 0.15 m). Large recirculation zones (>7500 m²) were significantly eroded during the flood (-0.55 m ± 0.17 m).

4.2. Responses at Specific Bars, February-April 1996

We attribute the variability in behavior of the study sites to two response styles. Deposition in recirculation zones either (1) aggraded the existing bar to such an extent that it filled or nearly filled all areas of the zone or (2) aggradation in the downstream part of large reattachment zones was accompanied by upstream scour of the reattachment bar platform. However, erosion in the upstream portion of recirculating flow in recirculation zones did not affect the high-elevation separation bars in this study.

TABLE 2. Area and volume changes from February to April 1996

Site Reference Number	RECIRCULATION ZONE										MAIN CHANNEL	
	High-elevation sand bar (Above 566 m³/s)				Sand bar above baseflow (Above 142 m³/s)				Eddy (Below 142 m³/s)		(Below 142 m³/s)	
	Area		Volume		Area		Volume		Volume		Volume	
	(m²)	(%)	(m³)	(%)	(m²)	(%)	(m³)	(%)	(m³)	(%)	(m³)	(%)
-6	-10	-2	-10	-2	260	7	670	19	---	---	200	1
3	40	11	40	16	-120	-9	140	7	-8300	70	-8400	-29
8	430	60	290	83	140	9	770	38	---	---	---	---
16	540	657	160	858	250	20	690	39	700	11	-9100	-32
22	400	85	1170	173	340	17	1710	36	-600	-9	-5300	-30
30	1280	179	2290	208	770	24	5190	79	2700	135	-500	-4
32	280	49	60	20	-810	-23	230	7	-8300	-62	-11,200	-56
43	190	19	730	107	-240	-11	710	16	---	---	-30,500	-57
45 S	600	49	700	56	30	1	1960	31	-14,600	-69	-7900	-13
45 R	320	19	1890	341	-120	-2	2410	30	---	---	---	---
47	1470	91	1120	68	250	3	4740	53	-17,300	-45	3600	6
50 S	230	75	470	507	390	41	930	57	200	14	-300	-2
50 R	750	187	1329	208	220	12	2577	91	---	---	---	---
51	240	4	3890	69	-1350	-17	340	3	-13,900	-58	-46,200	-46
55	1850	29	5000	272	-2240	-17	3830	23	-18,300	-74	-54,800	-54
62	-10	-4	140	80	-100	-10	50	4	-10,200	-37	-18,100	-47
68	1410	87	1690	347	400	8	9060	155	---	---	-7300	-32
81	170	19	1130	144	30	2	1390	46	---	---	-500	-7
87	0	1	160	51	-20	-4	60	4	---	---	-4900	-21
91	20	7	100	22	-40	-7	100	8	4300	105	3900	33
93	60	7	300	57	-10	-1	200	7	---	---	-2600	-14
104	80	30	230	80	-10	-1	250	17	---	---	1700	25
119	1710	218	3650	269	810	24	6890	101	2500	86	-300	-4
122	590	29	1740	106	1150	27	4120	38	1700	77	100	1
123	90	16	460	93	130	16	460	93	-3500	-39	-9100	-41
137	520	85	1860	357	-10	85	1860	357	-400	-22	-500	-8
139	410	188	360	274	480	188	360	274	---	---	-4700	-30
145	160	60	750	405	-20	60	750	405	---	---	-500	-23
172	310	27	190	21	1250	27	190	21	-200	-3	4000	14
183	-90	-8	810	41	-70	-8	810	41	-300	-9	2600	30
194	-260	-10	1670	105	-1150	-10	1670	105	-6100	-65	-17,900	-47
202	-200	-21	80	10	-460	-21	80	10	8300	53	4200	22
213	510	110	500	79	290	110	500	79	300	10	2800	13
220	10	1	290	45	-50	1	290	45	---	---	700	21
225	-50	-5	910	166	570	-5	910	166	-1500	-48	-4500	-19
mean	401	67	1033	164	27	5	1742	37	-3764	-5	-6915	-14
s.e.[1]	90	20	199	30	111	3	363	6	1562	13	2427	5

[1]Standard error

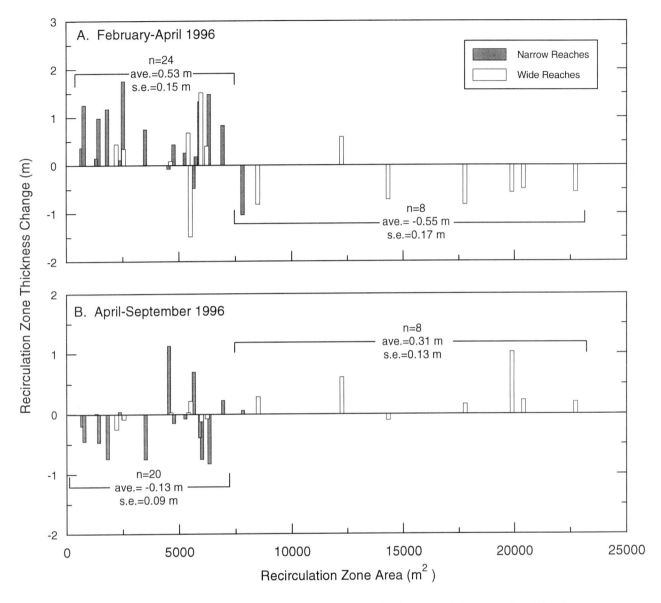

Figure 3. Graphs showing net recirculation zone thickness change related to recirculation zone size. (A) February to April 1996 comparison. (B) April to September 1996 comparison. Wide and narrow reach designations are those of *Schmidt and Graf* [1990]. The sample size, average, and standard error are shown for recirculation zones less than or greater than 7500 m in area.

The changes at site #30 are typical of the first style (Figure 4). This fan-eddy complex is located in the narrow Supai and Redwall Gorges geomorphic reach of Marble Canyon. The area of the recirculation zone within the computational boundaries is 5900 m^2. The stage change is 5.86 m between flows of 141 and 1274 m^3/s and the ratio of the width of the upstream constricting debris fan to the downstream channel expansion and recirculation zone (expansion ratio) is 1.57, which is comparable to other fan-

eddy complexes of similar size in this study (Table 1). In the pre-flood February 1996 survey, the area of reattachment bar above baseflow (141 m^3/s) was 3270 m^2 and the bar filled 55% of the recirculation zone above this stage. The return-current channel was poorly defined at high-elevation. At lower elevations, a deep, wide channel separated the bar crest from the bank. This bar was almost completely inundated by the flood (Figure 5).

Deposition was greatest in the upstream part of the eddy

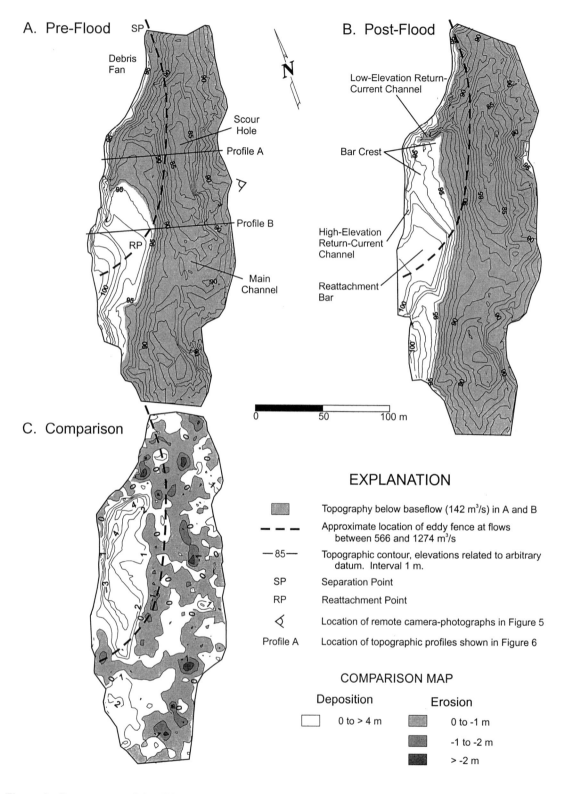

Figure 4. Contour maps of site #30. (A) Pre-flood survey, 16 February 1996. (B) Post-flood survey, 17 April 1996. (C) Comparison map showing areas of erosion and deposition between the surveys in A and B. Flow in main channel is from top to bottom.

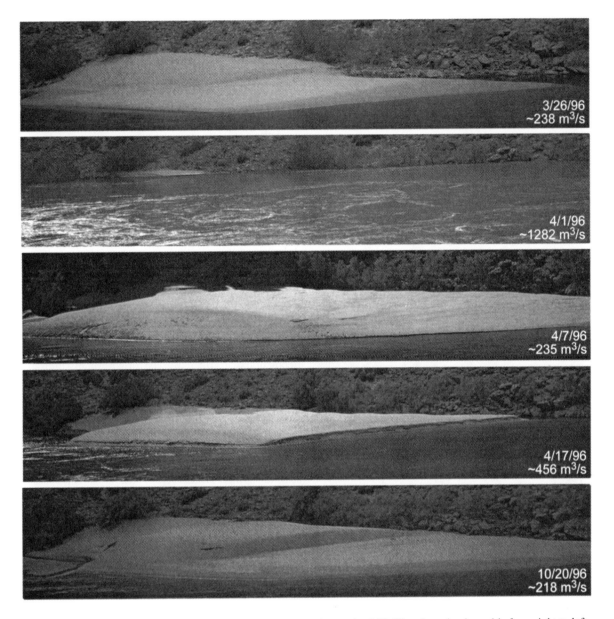

Figure 5. Selected daily photographs from the reattachment bar at site # 30. Flow in main channel is from right to left. Records from the stream-flow gaging station on the Colorado River at Lees Ferry and indirect calculations of travel time were used to estimate the indicated flow at the time of the photograph.

where as much as 4 m of fill occurred on the bar crest (Figure 4b and c). At this relatively low elevation part of the recirculation zone, the return-current channel was partially filled as the reattachment bar aggraded toward the bank (Figure 6a). In response to the high stage of the flood, a new return channel formed at high elevation (Figure 6b). The volume of sand above baseflow increased 79%, a gain of 5190 m³ (Table 2). The area of the bar above baseflow increased 24%, a gain of 770 m². At high-elevation, bar

volume and area increased 208% and 179%, respectively. As a result, the average high-elevation deposit thickness was 1.15 m. Deposition below baseflow in the recirculation zone increased the eddy-stored volume of sand by 2700 m³, an increase of 135%. The total area and volume increase resulted in a net change of 1.33 m in the thickness of sand in the recirculation zone. As a result, the percentage of sand filling the recirculation zone above baseflow increased from 55% to 68%. The bar aggraded to 0.10 to 1 m of the flood

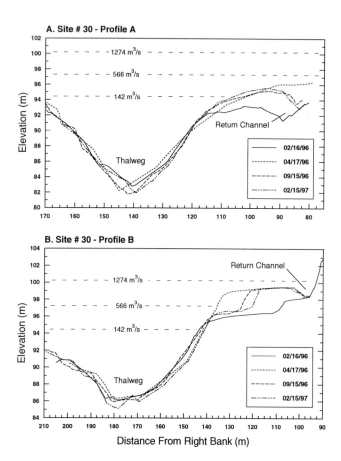

Figure 6. Topographic profiles from site #30. (A) Profile A. (B) Profile B. Elevation datum is arbitrary. Flow stage elevations for baseflow (142 m³/s), high elevation (566 m³/s), and the 1996 controlled flood (1274 m³/s) are shown. Location of profiles is shown on Figure 4a.

water surface over much of this area (Figures 5 and 6b). The main channel pool at this site was not substantially scoured. Net erosion of the bed was 400 m³, a decrease of 4%; this change is not readily discernible in Figure 4c.

The changes at site #51 are typical of the second style of change: high-elevation aggradation but net loss of sand in the recirculation zone (Figure 7). This large fan-eddy complex is located in the wide, lower Marble Canyon geomorphic reach. The computational area of the recirculation zone is approximately 20,160 m². At the time of the February 1996 survey, the area of reattachment bar exposed above baseflow was 11,800 m². Prior to the flood, the bar filled 62% of the recirculation zone above this stage. The stage change at the site is 4.2 m between flows of 141 and 1274 m³/s and the expansion ratio is 2.3, which is comparable to other large fan-eddy complexes in this study (Table 1). Before the flood, the elevation of the reattachment bar

was relatively uniform with a broad, gently sloping surface. The bar had a well-defined return current channel that separated much of the deposit from the bank (Figures 7a and 8). Low-elevation deposition on the upstream part of the bar during the low discharges typical of the early 1990s had effectively closed off the mouth of the channel, thereby limiting flow into the return channel except at high discharges. The high-elevation bar platform was completely inundated during the flood to a depth of approximately 1 to 1.5 m (Figure 8b). The size of this fan-eddy complex is roughly 3 times that of site #30 which was discussed above.

Flood deposition was greatest on the downstream part of the reattachment bar near the location of the 1274 m³/s reattachment point (Figure 7c). More than 3 m of fill occurred on this part of the bar where it slopes into the main channel. Sand was deposited over much of the bar platform upstream and downstream from the reattachment point. Deposition of 3,890 m³ of sediment resulted in a 69% increase in volume of the high-elevation portion of the bar. In contrast, high-elevation bar area increased by 240 m², an increase of 4%. The average thickness of high-elevation deposition was 0.55 m. The downstream, high-elevation part of the bar aggraded to within 0.10 to 1 m of the water surface. The return-current channel was partially filled in by migration of the bar crest towards the bank (Figure 8).

The flood removed more sediment than was deposited at this site because the lower elevations in the recirculation zone were eroded (Figures 7c and 8a). More than 5 m of sediment was eroded from the upstream part of the reattachment bar. Scour reduced the volume of eddy sediment below baseflow by 13,900 m³, a 58% volume decrease (Table 2). The area of the bar above baseflow decreased by 1350 m², a -17% area decrease. The extent to which the bar filled the eddy above baseflow decreased from 62% to 55%. The scour at low-elevation was greater than deposition at high-elevation, so this recirculation zone thickness changed -0.57 m. Nonetheless, the higher elevations of the bar aggraded vertically.

At site #51, more than 5 m of sediment was scoured downstream from the scour hole in the widest part of the main channel (Figure 9). Erosion of the slope between the recirculation zone and the thalweg resulted in deepening and widening of the thalweg (Figure 8). Net removal of channel bed sediment was 46,200 m³ at this site (Table 2). The general pattern of main channel change at site #51 during the flood was typical of that at other scoured sites.

4.3. Post-Flood Changes, April 1996-September 1997

Following the flood, the newly aggraded sand bars were eroded. Area and volume changes during this period are

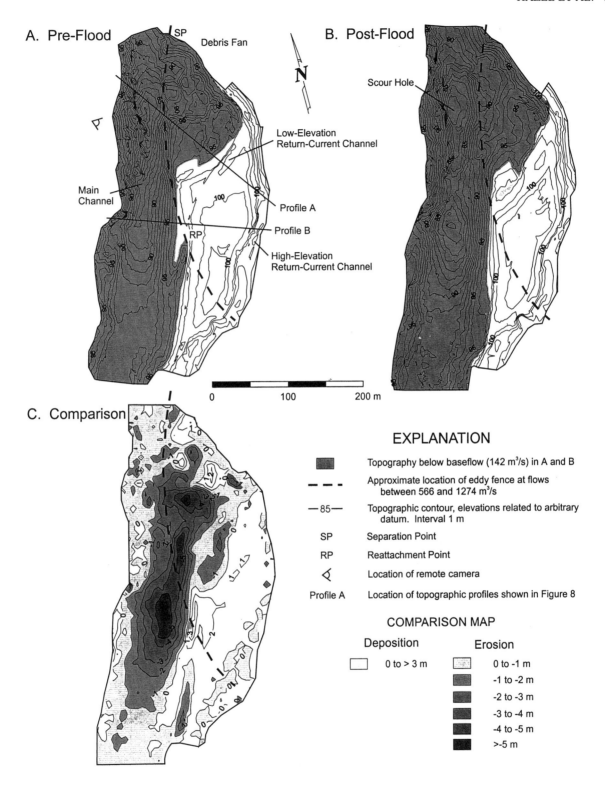

Figure 7. Contour maps of site #51. (A) Pre-flood survey, 19 February 1996. (B) Post-flood survey, 20 April 1996. (C) Comparison map showing areas of erosion and deposition between the surveys in A and B. Flow in channel is from top to bottom.

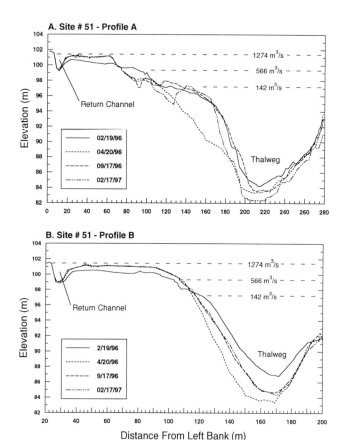

Figure 8. Topographic profiles from site #51. (A) Profile A. (B) Profile B. Elevation datum is arbitrary. Stage elevations for baseflow (142 m³/s), high elevation (566 m³/s), and the 1996 controlled flood (1274 m³/s) are shown. Location of profiles is shown on Figure 7a.

included in *Hazel et al.* [1997] and for brevity are not reported here. These data are summarized in Figure 10. In general, high-elevation sand eroded during this period was redistributed to low-elevation within recirculation zones and the main channel. Thickness changes illustrate this pattern (Figure 10a). By September 1996, the average change in thickness of the high-elevation deposits was -0.23 m (Figure 10b).

There was substantial variability in recirculation zone response during this interval (Figure 10b). Figure 3b shows that small recirculation zones (<7500 m²) which filled during the flood had the greatest loss of sand thickness following the flood (-0.13 m ± 0.09 m). In contrast, large recirculation zones (>7500 m²) that were scoured at low

elevation during the flood had the largest post-flood gains in sand thickness (0.31 m ± 0.13 m). As a result of this style of response, approximately one-third of the sites had a net loss and two-thirds had a net gain in recirculation zone thickness, an overall average thickness change of -0.02 m (Figure 10b).

This style of change in recirculation zones, erosion of flood-deposited sediment at high elevation and deposition at lower elevations, was well demonstrated at sites #30 and 51. The daily photographic record shows that immediately following the controlled flood cutbank erosion occurred at the stage elevation reached by normal powerplant operations (Figure 5). Between April and September 1996, the average daily fluctuation range was 421 to 523 m³/s. During this period, bar platforms were lowered throughout the system 1 to 2 m, as cutbanks migrated toward the bank (Figure 6b). The downstream parts of reattachment bars were especially vulnerable to erosion. After high flow events, these areas are abandoned by recirculation zones, which decrease in size during lower flows. This portion of the bar is then subjected to downstream-directed current and eroded material is transferred to the main channel.

Within large recirculation zones, however, high-elevation sand was redistributed to lower elevations. Deposition below baseflow resulted in rapid recovery of flood-degraded, low-elevation areas (Figure 8a). For example, the scour of 13,900 m³ of sediment from eddy storage below baseflow at site #51 during the flood was followed by deposition of 20,900 m³ of sediment within 5 months. Daily photographs of all study sites indicate gradual accumulation at low elevation in recirculation zones. In contrast to large recirculation zones, the erosion of high-elevation sand at smaller recirculation zones did not simply redistribute sediment to lower elevations. These smaller sites do not have low elevation areas of sufficient size to accommodate the sediment eroded from high-elevations; thus, a net loss of sand stored in small recirculation zones occurs (Figure 3b).

Changes in main channel bed thickness over the same 5 month interval were variable, with some sites showing net deposition and others net erosion (Figure 10). The average thickness change was -0.07 m. Individual sites that showed partial recovery of flood-eroded bed sediment in September 1996 were unchanged or scoured back to the post-flood bed elevation in February 1997 (Figure 8). Erosion of the bed typically occurred downstream from scour holes in and near the thalweg, similar to erosional patterns during the 1996 controlled flood (Figure 9).

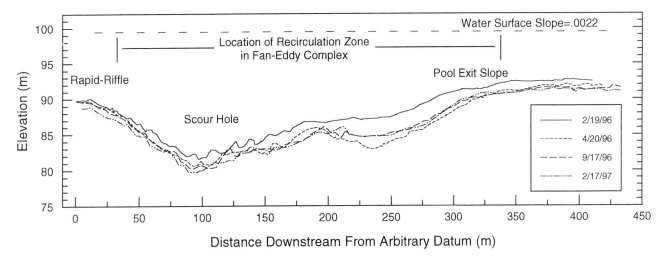

Figure 9. Bathymetric thalweg profile from site #51. Water-surface slope is at a discharge of ~603 m³/s. Approximate location of the recirculation zone in relation to geomorphic attributes of the channel bed are shown. Flow in main channel is from left to right.

5. SPATIAL AND TEMPORAL PATTERNS OF CHANGE

5.1. Differences in Deposition Between Geomorphic Reaches and Bar Types

The two response styles during the 1996 controlled flood, filling of small recirculation zones and low-elevation scour in large recirculation zones, occurred to different extents in narrow and wide reaches (Figure 3a). The largest net gains in small recirculation zones occurred in narrow reaches (for example, Supai and Redwall Gorges and Aisles; Figure 2a). Low-elevation scour of large recirculation zones primarily occurred in wide reaches, where most of the largest recirculation zones are located (for example, lower Marble Canyon). Regardless of reach width, however, high-elevation aggradation occurred throughout the river corridor, except where negligible change was measured at site #-6. We conclude that the amount of sand eroded from large recirculation zones in wide reaches did not have a significant impact on high-elevation bars in those same eddies. In addition, the rapid 5-month recovery of scoured areas indicates that while net loss may have occurred at individual sites, there was not long-term net loss of recirculation zone sand in our measured recirculation zones as a result of the flood. These observations demonstrate that the 1996 controlled flood objective of transferring sediment from low-elevation areas to high-elevation sand bars was achieved.

Net flood deposition occurred at all types of bars (Figure 11a). Vertical aggradation, as evidenced by greater volume increases than area increases, was similar for all bar types. Separation bars were not scoured at low-elevation, and had the greatest average increases in area and volume at high-elevation; however, they also showed the most site to site variability. The sample of separation bars studied is considerably smaller than the sample of reattachment bars studied and may not be as representative as the change measured for other bar types.

5.2. Depositional Trends and Accommodation Space

We integrated the results of this flood study with similar data collected since 1991 to produce a time series (Figure 12). We divided our sample sites into two populations: those in Marble Canyon, upstream from the Little Colorado River, and those in Grand Canyon, downstream from the Little Colorado River. One site, located in the sediment-depleted Glen Canyon reach upstream from the Paria River (site #-6) was not included in this analysis. Thus, the time series are averages of the 12 sites located in Marble Canyon and 20 sites in Grand Canyon. We used volume changes rather than the smaller area changes to evaluate spatial and temporal patterns of change. The time series data were normalized to the pre-flood condition measured in February 1996. As a result, the plotted points for the pre-flood survey are 0.

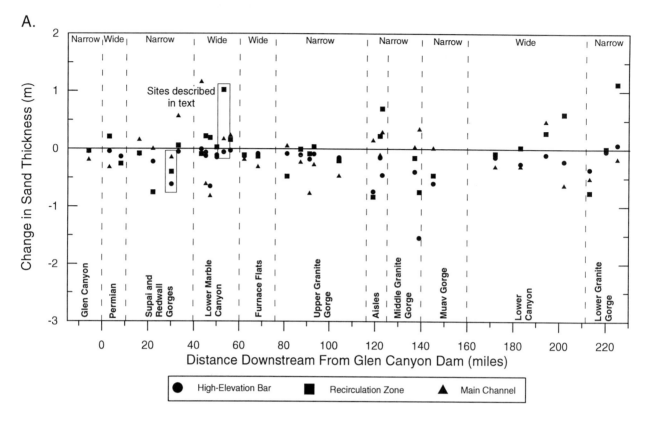

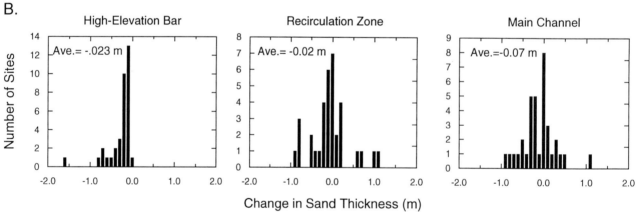

Figure 10. Graphs showing 5-month post-flood changes in sand thickness between April and September 1996. (A) The change at each site with distance downstream. Geomorphic reaches are from *Schmidt and Graf* [1990]. Sites described in text are highlighted with boxes. (B) Frequency distributions for sand thickness change at high-elevation sand bars, recirculation zones, and main channel. See Figure 1 for definitions of these areas.

The time series demonstrate that the 1996 controlled flood was the most significant aggradational geomorphic event to have occurred between 1991 and 1997. In Marble Canyon, the average high-elevation bar volume increased 213%, 75% greater than the 138% increase measured in Grand Canyon. Although sites in Marble Canyon increased more than sites downstream from the Little Colorado River, the site to site variability was substantially greater downstream. The high-elevation thickness change on bars in Marble Canyon was positively correlated to the magnitude of stage change ($r^2=0.59$, significant at the 95% confidence level) (Figure 13a). Thicker deposits were

A. February-April 1996

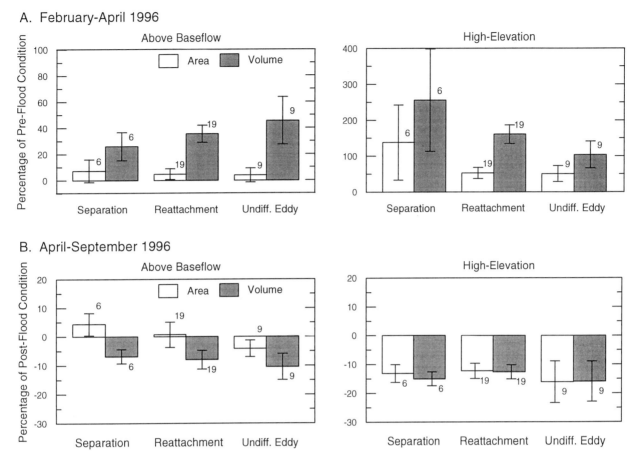

B. April-September 1996

Figure 11. Average sand bar area and volume changes for different bar types. (A) Changes between February and April 1996. (B) Changes between April and September 1996. Changes above baseflow (>142 m³/s) and at high-elevation (>566 m³/s) are shown. Sample population sizes are shown at the top or bottom of each bar graph. Error bars are standard error about the mean. Note that the graphs in A are a percentage of the pre-flood condition and in B are a percentage of the April post-flood condition.

formed in narrow reaches where stage change was greatest. There was no significant correlation for bars downstream from the Little Colorado River(Figure 13b).

We hypothesize that the lack of correlation of stage change and deposition for sites downstream from the Little Colorado River is because some of these sites were already partly filled with sediment from the 1993 Little Colorado River floods. The time series indicate that about one-fourth (24%) of the high-elevation volume increase downstream from the Little Colorado River in April 1993 was still present nearly 3 years later, in February 1996 (Figure 12). In addition, various investigators have shown that bars in upper Marble Canyon have eroded more than elsewhere, increasing available accommodation space there [*Schmidt and Graf*, 1990; *Schmidt et al.*, 1995; *Webb*, 1996]. Thus,

the sites in upper Marble Canyon could aggrade more than sites further downstream.

To test this hypothesis, we determined the accommodation space available for high-elevation deposition at each site (Table 3). We define accommodation space as the maximum volume of sand that can possibly be stored within high-elevation bars. Accommodation space for a particular bar is defined in terms of the storage potential, the largest measured area of high-elevation bar during the entire monitoring period (1991-1997) multiplied by the local stage change for flows between 566 m³/s and 1274 m³/s. The measured volume of sand deposited at high-elevation for each survey was divided by the storage potential and expressed as a percentage of the accommodation space filled at that site.

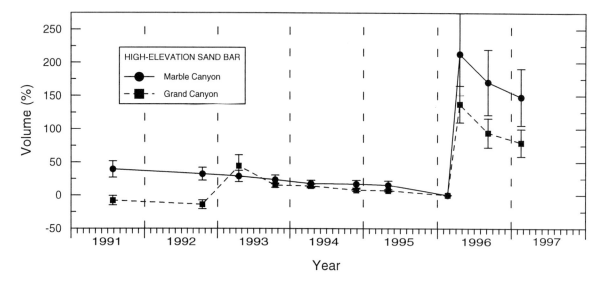

Figure 12. Temporal sequence of average high-elevation bar volume changes. Error bars are standard error about the mean.

The results in Table 3 show that because of the 1993 flood deposition, less accommodation space was available for high-elevation deposition at sites located downstream from the Little Colorado River. The percentage of accommodation space filling was greatest at sites that had (1) eroded the most in the preceding years, or (2) the greatest volume of accommodation space was available (i.e., they had already been eroded prior to 1991). In Marble Canyon, 13 of 14 bars were less than 30% filled in February 1996, whereas in Grand Canyon, 12 of 20 bars, were less than 30% full (i.e., had greater than 70% accommodation space available). More importantly, sites downstream from the Little Colorado River had substantially less accommodation space available for deposition in 1996 than in 1991. As a result, less accommodation space was available on bars downstream from the Little Colorado River. These results show that accommodation space is an important factor in determining flood deposition response.

5.3. Longevity of Rebuilt Bars

The post-flood style of sand bar change — erosion of flood-deposited sediment at high-elevation and deposition at lower elevations in recirculation zones — occurred throughout Grand Canyon. The rate of high-elevation erosion decreased with time, based on our measurements made 5 and 10 months after the flood. The decreases in high-elevation area and volume were similar for averages of different types of sand bars (Figure 11b). The transfer of sediment to lower elevation resulted in slight increases in area of separation and reattachment bars above baseflow.

To examine bar stability following the 1996 controlled flood, we used the time series to estimate average erosion rates of high-elevation sand bars both upstream and downstream from the Little Colorado River (Figure 12). In the 5 months following the 1996 controlled flood, bar erosion rates were -9%/mon in Marble Canyon and in Grand Canyon. From September 1996 to February 1997, erosion rates declined to -4%/mon in Marble Canyon and -2%/mon in Grand Canyon. By comparison, bars eroded at a rate of -1%/mon from 1991 to 1996 in Marble Canyon. After deposition of high-elevation sand by the 1993 Little Colorado River floods, bars downstream from the confluence eroded at a rate of -5%/mon between April and October 1994, then declined with time to -1%/mon between October 1994 and February 1996. The rates of post-flood adjustment are similar between the 1993 Little Colorado River floods and the 1996 controlled flood, although the rate of erosion in Grand Canyon following the 1996 controlled flood was higher than after the 1993 Little Colorado River floods (-9%/mon vs. -5%/mon). This suggests that the rate of erosion immediately following a flood is related to the magnitude of the high flow and amount of high-elevation deposition. Erosion rates decrease as the availability of erodible sediment decreases [*Schmidt et al.*, 1995] or as a stable slope is reached [*Budhu and Gobin*, 1994].

5.4. Channel-Stored Sediment

Time series of average channel-bed volume changes since 1992 show that main-channel storage at our

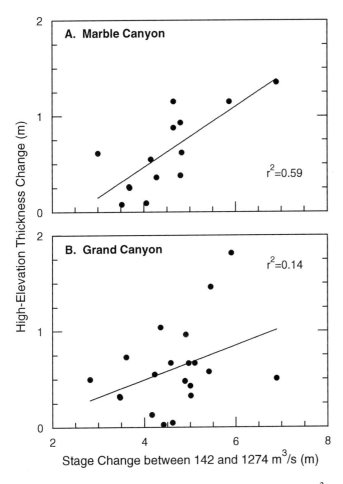

Figure 13. The relation between the stage change from 142 m³/s (baseflow) to 1274 m³/s (flood peak) and high-elevation thickness change between February and April 1996. (A) Sites in Marble Canyon. (B) Sites in Grand Canyon.

TABLE 3. Accomodation space available for high-elevation deposition of sand

Site Reference Number	Storage Potential (m^3)	Percentage of Filling (%)					
		July 1991	April 1993	Feb. 1996	April 1996	Sept. 1996	Feb. 1997
Glen Canyon							
-6	1054	33	28	27	26	24	24
Marble Canyon							
3	733	32	31	29	34	32	28
8	2448	21	19	14	26	21	18
16	1161	4	2	2	16	12	10
22	2857	27	25	24	65	59	57
30	5862	26	26	19	58	43	38
32	1273	30	26	22	26	24	22
43	2933	28	27	23	48	48	46
44S	4696	29	28	26	41	37	35
44R	5212	10	---	11	47	45	44
47	6855	30	26	24	40	23	22
50S	1288	12	11	7	43	38	32
50R	2797	44	41	23	70	66	63
51	14,716	41	40	38	65	62	60
55	12,331	---	---	15	55	54	54
Grand Canyon							
62	12,643	---	66	8	7	7	7
68	4764	10	10	10	46	41	42
81	2487	39	42	32	77	61	56
87	744	42	44	41	62	57	56
91	786	52	61	56	68	62	62
93	1604	24	37	33	52	47	45
104	773	39	37	37	66	60	50
119	7230	23	26	19	69	51	40
122	6473	17	29	25	52	46	41
123	2085	20	29	24	46	35	39
137	4865	3	38	11	49	40	36
139	1751	8	10	8	28	9	8
145	1322	12	23	14	71	54	50
172	2831	28	38	31	37	32	28
183	4847	30	44	41	58	53	57
194	6341	16	28	25	51	48	45
202	4003	30	34	19	21	17	16
213	3570	23	28	18	32	25	19
220	1630	40	57	39	56	55	59
225	2001	11	28	28	73	78	71

measurement sites responds to both floods and normal dam operations (Figure 14). The most significant changes as a result of the 1996 controlled flood occurred in Marble Canyon, where the average volume of main channel sediment at 11 of 12 sites decreased 30%. The magnitude of scour in Marble Canyon was substantially greater than downstream from the Little Colorado River. Five months after the 1996 controlled flood, sites located downstream from the Little Colorado River were still being eroded and by February 1997 had degraded by a similar percentage as Marble Canyon. Despite partial recovery at some sites during this period (Figures 8 and 9), the average volume changes indicate little main channel aggradation.

The time series indicate that channel storage systematically increases after tributary sediment inputs and declines gradually in the following months. Estimates of monthly

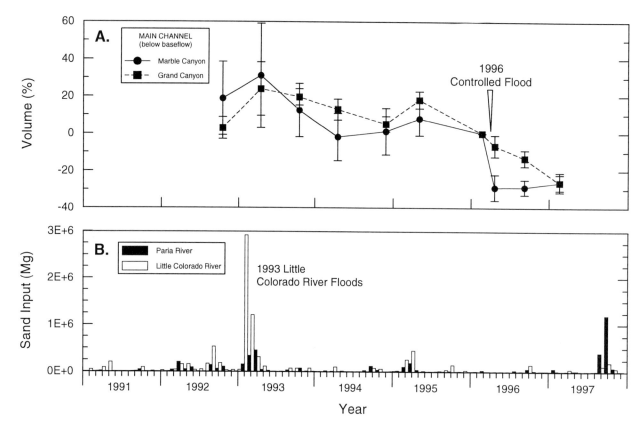

Figure 14. Temporal sequence of average main channel volume changes in A. Error bars are standard errors. B. Monthly estimates of sand inputs from the Paria River and Little Colorado River.

tributary sand inputs are shown in Figure 14b. The sand-transport relations for the Paria and Little Colorado Rivers were derived by *Randle and Pemberton* [1987]. Because of possible shifts in rating relations, these estimates are not intended to represent actual transport in the system but only to provide a measure of potential tributary sediment delivery to compare to channel sediment volumes. Our integrated time series of channel bed changes averaged from single sites may be too small a sample set to accurately represent reach-scale or canyon-wide patterns. Nonetheless, the trends between 1992 and 1996 we calculated (Figure 14) are similar to changes documented by other investigators. For example, measurements of continuous reaches downstream from the mouths of the Paria River and Little Colorado River show that sediment inputs from the Paria River in 1992 were quickly removed from the channel bed, but inputs from the Little Colorado River in 1993 moved more slowly downstream [*Graf et al.*, 1995; *Jansen et al.*, 1995]. A similar temporal trend is

indicated by our time-series; the volume of channel bed sediment decreased at a much slower rate downstream from the Little Colorado River than upstream between 1992 and 1994 (Figure 13a). In addition, pre- and post-flood bed elevation changes measured at the U.S. Geological Survey cross section network [*Konieczki et al.*, 1997] are in general agreement with the changes presented here. They document scour of sand from the deepest part of the channel but variable changes elsewhere. These changes indicate that the post-flood channel was relatively scoured and possibly depleted of sand-sized sediment at many locations. However, comparison maps of channel-bed sediment distributions in 6 main-channel pools downstream from the Little Colorado River indicate that rather than being depleted of sand, sand was more widely distributed on the bed [*Anima et al.*, 1998]. This survey was conducted 3 months after the controlled flood and post-flood changes in sand coverage may have resulted from the transfer of sand from eroding bars to the channel bed.

6. DISCUSSION

The widespread deposition of high-elevation sand bars during the 1996 controlled flood indicates that accommodation space for potential deposition is at least as important as sediment availability in determining the magnitude and persistence of flood impacts. The greatest amounts of deposition at our study sites occurred upstream from the Little Colorado River, despite the fact that measured suspended-sand concentrations were greatest downstream from the Little Colorado River [*Konieczki et al.*, 1997]. Based strictly on campsite considerations, *Kearsley et al.* [this volume] showed that the density of new campsites created by the 1996 controlled flood in Marble Canyon was more than twice the number per mile than below the Little Colorado River. Our calculations of accommodation space of eroded bars in this reach indicates that the difference in deposition may be the direct result of greater storage potential. Sites located in Marble Canyon, especially in the upstream, narrow reaches, were more degraded before the flood than sites located downstream from the Little Colorado River, where recirculation zones were already partly filled with sediment by the 1993 Little Colorado River floods.

An important unresolved question concerns the longevity of rebuilt bars and the frequency and magnitude of future controlled floods required to maintain them. Although erosion rates of high-elevation sand declined with time, nearly half of the aggradation by the 1996 controlled flood had been lost in the 10-month period following the flood (Figure 12). Furthermore, our results show that bars erode system-wide following a bar-building event, and erosion rates are not related to distance downstream or location of sediment supplying tributaries. These findings are similar to the conclusion of *Schmidt and Graf* [1990]: flows following bar-building flood events result in rapid erosion of deposits of all types throughout Grand Canyon.

Monitoring and research planning as part of adaptive management in Grand Canyon have identified the average size of sand bars in 1991 as a baseline condition upon which to evaluate sand bar changes. Based upon long-term erosion rates of high-elevation sand bars since 1991, floods need to be conducted, on a minimum, every decade for the desired average condition to be maintained. However, unavoidable hydrologic conditions may necessitate more frequent floods in high runoff years. Accurate prediction of the outcome of high-discharge, regulated releases is important if floods are to be used as a tool for resource management. While the results in this study show that the accommodation space available for deposition is an important factor in predicting the magnitude and persistence of flood impacts, site-specific benefits are also dependent on local sediment supply and channel geometry. Potential benefits for the Grand Canyon ecosystem from the limited sediment supply are achieved by maximizing the volume of sand stored in channel margin sand bars, but an available main channel sediment supply is required if future flow releases are designed to build bars. The time series for average main channel changes suggests that in February 1997 the system was more sediment-depleted than at any other time during the monitoring period (Figure 14).

7. CONCLUSIONS

High releases from Glen Canyon Dam may be used to rebuild eroded high-elevation bars along the Colorado River in Grand Canyon. The results from repeated topographic mapping presented here support the hypothesis that channel-stored sediment can be entrained by controlled flooding and redistributed to the channel margin for bar restoration purposes. Although there was variability in the behavior of individual sites, two styles of response to the flood were identified: (1) deposition filled or nearly filled all areas of small recirculation zones, and (2) low-elevation deposits in large recirculation zones were extensively scoured. High-elevation deposition exceeded that from any event since 1991. The thickness of new high-elevation deposits typically approached or exceeded 1 m in thickness, regardless of reach characteristic or bar type. Vertical aggradation resulted in sand bars that were higher, not wider than before the flood. Differences in deposition were directly correlated with storage potential at high elevation. Sites that were in a more degraded condition prior to the flood, especially in the narrow reaches of Marble Canyon, had greater increases in net deposit thickness. More importantly, the long-term erosional trend of high-elevation bars characteristic of the 1990s was mitigated by net deposition.

Following the 1996 controlled flood, readjustment of the newly aggraded bars to lower flows led to rapid but declining rates of erosion. High-elevation sand deposited by the flood still remained after nearly 1 yr of post-flood Glen Canyon Dam operations. However, the beneficial effects of the 1996 controlled flood will be of limited duration and future high flows will be required to maintain high-elevation sand bars downstream from Glen Canyon Dam. These results show that, in addition to sediment availability, the status of recirculation zone accommodation space prior to a flood may be an important factor in determining the magnitude and persistence of flood impacts.

Acknowledgments. This study was supported by the U.S. Bureau of Reclamation's Glen Canyon Environmental Studies. It is offered in tribute to Stanley S. Beus, whose original work initiated this study, and to Dave Wegner, for his support of our monitoring program. Discussions over the past several years with Larry Stevens, Tim Randle, Stan Beus, Jack Schmidt, Paul Grams, Ted Melis and others have contributed greatly in formulating the ideas presented here. Of the numerous individuals who contributed field assistance on river trips, the authors especially thank Greg Sponenburgh, Hilary Mayes, Eric Kellerup, Jeff Bennett, Grant Pierce, Paul Umhoefer, and the professional river guides without whose logistical support this study would not have been possible. We also wish to thank Mark Gonzales, Glen Canyon Environmental Studies surveyor, for his careful and dedicated hydrographic surveying. Constructive and insightful reviews of early versions of this manuscript by Jack Schmidt, Ted Melis, Paul Grams and two unidentified reviewers are gratefully acknowledged.

REFERENCES

Andrews, E.D., The Colorado River: a perspective from Lees Ferry, Arizona, in *Surface Water Hydrology*, ed. M.G. Wolman and H.C. Riggs, pp. 304-310, Geol. Soc. Amer., The Geology of North America, v. 0-1, 1990.

Anima, R.J., M.S. Marlow, D.M. Rubin, and D.J. Hogg, *Comparison of sand distribution between April 1994 and June 1996 along six reaches of the Colorado River in Grand Canyon, Arizona,* U.S. Geol. Surv. Open-File Report 98-141, 33 pp., 1998.

Beus, S.S., C.C. Avery, L.E. Stevens, M.A. Kaplinski, and B.L. Cluer, The influence of variable discharge regimes on Colorado River sand bars below Glen Canyon Dam, in *The Influence of Variable Discharge Regimes on Colorado River Sand Bars below Glen Canyon Dam,* ed. S.S. Beus and C.C. Avery, Final Rept. to Glen Canyon Environ. Studies, N. Ariz. Univ., Flagstaff, AZ, 1992.

Beus, S.S., S.W. Carothers, and C.C. Avery, Topographic changes in fluvial terrace deposits used as campsite beaches along the Colorado River in Grand Canyon, *J. Ariz.-Nev. Acade. Sci.,* 20, 111-120, 1985.

Budhu, M., and R. Gobin, Instability of sandbars in Grand Canyon, *J. Hydr. Eng.,* 120, 919-933, 1994.

Cluer, B.L., Cyclic fluvial processes and bias in environmental monitoring, Colorado River in Grand Canyon, *J. Geol.,* 103, 411-421, 1995.

Datacom Software Research Limited, *SDR map Software User Guide--SDR Mapping and Design system:* Sokkia Corp., Overland Park, Kansas, 1997.

Dexter, L.R., B.L. Cluer, and M.F. Manone, Using land-based photogrammetry to monitor sandbar stability in Grand Canyon on a daily time scale, in *Proceedings of the Second Biennial Conference on Research in Colorado Plateau National Parks,* ed. C. Van Riper III, pp. 67-86, Nat. Park Ser. Trans. Procs. Ser. NPS/ NRNAU/NRTP-95/11, 1995.

Graf, J.B., J.E. Marlow, G.G. Fisk, and S.M.D. Jansen, *Sand storage changes in the Colorado River downstream from the Paria and Little Colorado Rivers, June 1992 to February 1994,* U.S. Geol. Surv. Open-File Report 95-446, 61 pp., 1995.

Hazel, J.E. Jr., M. Kaplinski, S.S. Beus, and L.A. Tedrow, Sandbar stability and response to Interim Flows after a bar-building event on the Colorado River, Grand Canyon, Arizona: implications for sediment storage and sandbar maintenance (abstract), *Eos Trans. AGU,* 74, 320, 1993.

Hazel, J. E. Jr., M. Kaplinski, M.F. Manone, R.A. Parnell, A.R. Dale, J. Ellsworth, and L. Dexter, *The effects of the 1996 Glen Canyon Dam Beach/habitat-building flow on Colorado River sand bars in Grand Canyon,* Final Rept. to Glen Canyon Environ. Studies, 76 pp., N. Ariz. Univ., Flagstaff, AZ., 1997.

Howard, A.D., *Establishment of benchmark study sites along the Colorado River in Grand Canyon National Park for monitoring of beach erosion caused by natural forces and human impact,* Univ. Vir. Grand Canyon Study, Techn. Rept. no. 1, 182 pp., 1975.

Howard, A.D., and R. Dolan, Geomorphology of the Colorado River, *J. Geol.,* 89, 269-298, 1981.

Jansen, S.M.D., J.B. Graf, J.E. Marlow, and G.G. Fisk, *Monitoring channel sand storage in the Colorado River in Grand Canyon,* U.S. Geol. Surv. Fact Sheet FS-120-95, 1 pp., 1995.

Kaplinski, M., J.E. Hazel, Jr., S.S. Beus, C.J., Bjerrum, D.M. Rubin, and R.G. Stanley, Structure and evolution of the "Dead Chub Eddy" sand bar, Colorado River, Grand Canyon (abstract), *Geol. Soc. of Amer.,* 26, p. 21, 1994.

Kaplinski, M., J.E. Hazel, Jr., and S.S. Beus, *Monitoring the effects of Interim Flows from Glen Canyon Dam on sand bars in the Colorado River Corridor, Grand Canyon National Park, Arizona,* Final Rept. to Glen Canyon Environ. Studies, 62 pp., N. Ariz. Univ., Flagstaff, AZ., 1995.

Kearsley, L.H., J.C. Schmidt, and K.D. Warren, Effects of Glen Canyon Dam on Colorado River sand deposits used as campsites in Grand Canyon National Park, USA, *Regul. Rivers,* 9, 137-149, 1994.

Kearsley, L.H., and R. Quartaroli, *Effects of a beach/habitat-building flow on campsites in the Grand Canyon,* Final Rept. to Glen Canyon Environ. Studies, 18 pp., App. Tech. Assoc., Flagstaff, AZ., 1997.

Konieczki, A.D., J.B. Graf, and M.C. Carpenter, *Streamflow and sediment data collected to determine the effects of a controlled flood in March and April 1996 on the Colorado River between Lees Ferry and Diamond Creek, Arizona,* U.S. Geol. Surv. Open-File Rept. OF-97-224, 55 pp., 1997.

Parnell, R.A., L.R. Dexter, M. Kaplinski, J.E. Jr. Hazel, M.F. Manone, J.S. Ellsworth, and A.R. Dale, *Bridging the gap to long-term monitoring: transitional monitoring of sandbars along the Colorado River during Fiscal Year 1996,* Final Rept. to Glen Canyon Environ. Studies, N. Ariz. Univ., Flagstaff, AZ., 1996.

Randle, T.J., and E.L. Pemberton, *Results and analysis of the STARS modeling efforts of the Colorado River in Grand Canyon,* Bur. Reclam., Glen Canyon Envir. Studies Tech Rep. NTIS PB88-183421, 41 pp., Salt Lake City, UT, 1987.

Randle, T.J., R.I. Strand, and A. Streifel, Engineering and environmental considerations of Grand Canyon sediment management, in *Engineering Solutions to Environmental Challenges,* Thirteenth Annual US COLD Lecture, Chattanooga, TN, 1993.

Rubin, D.M., J.C. Schmidt, R.A. Anima, K.M. Brown, R.E. Hunter, H. Ikeda, B.E. Jaffe, R.R. McDonald, J.M. Nelson, T.E. Reiss, R. Sanders, and R.G. Stanley, *Internal structure of bars in Grand Canyon, Arizona, and evaluation of proposed flow alternatives for Glen Canyon Dam,* U.S. Geol. Surv. Open-File Rept. OF-94-594, 16 pp., 1994.

Rubin, D.M., J.C. Schmidt, and J.N. Moore, Origin, structure, and evolution of a reattachment bar, Colorado River, Grand Canyon, Arizona, *J. Sed. Pet.,* 60, 982-991, 1990.

Schmidt, J.C., and J.B. Graf, *Aggradation and degradation of alluvial sand deposits, 1965 to 1986, Colorado River, Grand Canyon National Park, Arizona,* U.S. Geol. Surv. Prof. Paper 1493, 74 pp., 1990.

Schmidt, J.C., P.E. Grams, and R.H. Webb, Comparison of the magnitude of erosion along two large regulated rivers, *Water Resour. Bull.,* 31, 617-630, 1995.

Schmidt, J.C., and M.F. Leschin, *Geomorphology of post-Glen Canyon Dam fine-grained alluvial deposits of the Colorado River in the Point Hansbrough and Little Colorado River confluence study reaches in Grand Canyon National Park, Arizona,* rept. to Glen Canyon Environ. Studies, Logan, UT., 1995.

Schmidt, J.C., and D.M. Rubin, Regulated streamflow, fine-grained deposits, and effective discharge in canyons with abundant debris fans, in *Natural and Anthropogenic Influences in Fluvial Geomorphology: the Wolman Volume, Geophys. Monogr. Ser.,* vol. 89, ed. J.E. Costa, A.J. Miller, K.W. Potter, and P.R. Wilcock, pp. 177-195, Amer. Geophys. Union, Washington, D.C., 1995.

Stevens, L.E., *The Colorado River in Grand Canyon,* 115 pp., Red Lake Books, Flagstaff, AZ., 1983.

Stevens, L.E., J.C. Schmidt, T.J. Ayers, and B.T. Brown, Flow regulation, geomorphology, and Colorado River marsh development in the Grand Canyon, Arizona, *Ecol. Applications,* 5, 1035-1039, 1995.

Thompson, K., K. Burke, and A. Potochnik, *Effects of the beach/habitat-building flow and subsequent interim flows from Glen Canyon Dam on Grand Canyon camping beaches, 1996: a repeat photography study by Grand Canyon river guides (Adopt-a-Beach Program),* Final Rept. to Glen Canyon Environ. Studies, 11 pp., Grand Canyon River Guides, Flagstaff, AZ., 1997.

U.S. Department of the Interior, *Operation of Glen Canyon Dam: Colorado River Storage Project, Arizona, Final environmental impact statement,* Bur. Recl., Salt Lake City, UT, 1995.

Valdez, R.A. and R.J. Ryel, *Life history and ecology of the humpback chub (Gila Cypha) in the Colorado River, Grand Canyon, Arizona,* Final report to Glen Canyon Environmental Studies, 1995.

Webb, R.H., *Grand Canyon, a century of change,* 290 pp., Univ. Ariz. Press, Tucson, AZ., 1996.

Wiele, S.M., J.B. Graf, and J.D. Smith, Sand deposition in the Colorado River in the Grand Canyon from flooding of the Little Colorado River, *Water Res. Res.,* 32, 3579-3596, 1996.

Joseph E. Hazel Jr., Matt Kaplinski, Roderic Parnell, Mark Manone, Alan Dale, Geology Department, Northern Arizona University, Box 4099, Flagstaff, AZ 86011; email: Joseph.Hazel@nau.edu

Variation in the Magnitude and Style of Deposition and Erosion in Three Long (8-12 km) Reaches as Determined by Photographic Analysis

John C. Schmidt, Paul E. Grams, and Michael F. Leschin

Department of Geography and Earth Resources, Utah State University, Logan, Utah

The 1996 controlled flood deposited sand along the edge of the Colorado River over an area of approximately 550,000 m^2 along 31 km in three study reaches. Deposition occurred in 218 eddies and as linear channel-margin levees. There was large variation in the response of individual eddies within each reach, and the average response among the reaches differed in some cases. These reaches were the 10.8-km long Point Hansbrough reach in lower Marble Canyon, and the 8.0-km long Tapeats Gorge and 12.1-km long Big Bend reaches near the Little Colorado River confluence. Eddies were the largest depositional environment in the two reaches where debris fans are most abundant; 72 and 80% of the area of new sand was deposited in eddies in the Point Hansbrough and Tapeats Gorge reaches. In the Big Bend, only 49% of new sand was deposited in eddies. New sand bars were emergent at low flow in less than 50% of the area of potential deposition in eddies. The wide variation in response of individual eddies makes it difficult to determine longitudinal trends in the magnitude of deposition, although the data suggest that the extent of new deposition was greater downstream from the Little Colorado River than upstream. Variation in eddy response also poses a challenge to river managers in assessing the "success" of the flood. Where specific sites are of great value, managers may have to decide between the goals of restoring average conditions along a reach and restoring the characteristics of specific sites.

1. INTRODUCTION

One of the objectives of the 1996 controlled flood in Grand Canyon was to increase the area of sand suitable for use as recreational campsites by increasing the area of sand that is exposed above the stage of normal powerplant operations. River managers and scientists hoped that the magnitude of this deposition would reverse the long-term decrease in size and number of sand bars that followed completion of Glen Canyon Dam (GCD) in 1963 [*Kearsley et al.*, 1994]. Increasing the area of bare, emergent sand bars would also return the river landscape to one more similar to that which existed prior to completion of the dam.

Several studies, each conducted at a different spatial scale, measured the magnitude of deposition caused by the controlled flood. *Kearsley et al.* [this volume] inventoried 200 out of 218 bars that were frequently used as camps before the flood and identified new camps created by the flood. *Hazel et al.* [this volume] measured the topography and bathymetry of 33 eddies and adjacent channels throughout the canyon before and after the flood; they compared their measurements with ones made since 1991.

TABLE 1. Selected characteristics of each of the study reaches

Characteristic	Point Hansbrough	Tapeats Gorge	Big Bend
Reach length (km)	10.8	8.0	12.1
Number of geomorphically significant tributaries[a]	16	26	20
Number of geomorphically significant tributaries pcr km	1.5	3.3	1.7
Number of persistent eddies	98	56	64
Number of persistent eddies larger than 1000 m^2	38	40	35
Number of persistent eddies larger than 1000 m^2 per km	3.5	5.0	2.9
Area of controlled flood deposits (m^2)			
all deposits	171,500	122,500	253,000
deposits within persistent eddies	124,000	97,750	124,500
deposits within persistent eddies larger than 1000 m^2	111,500	94,500	118,000
channel-margin deposits	47,500	25,000	128,500
Area of significant deposition and erosion in persistent eddies (m^2)			
significant deposition	76,500	68,500	90,000
significant erosion	64,500	62,000	37,000

[a] Geomorphically significant tributaries were determined by *Melis et al.* [1995].

Andrews et al. [this volume] measured the change in topography of five eddies every day during the flood. We mapped newly formed sand deposits and determined areas of significant erosion and deposition throughout three long (8-12 km) reaches of the river. In this paper, we estimate the proportion of newly-deposited sand that accumulated within eddies. We also describe the characteristics of newly-deposited eddy bars, and we measure the variability in the magnitude of flood-caused deposition and erosion among and within the study reaches.

2. FAN-EDDY COMPLEXES

Schmidt and Rubin [1995] argued that fan-eddy complexes are the fundamental geomorphic assemblage in canyons with abundant debris fans. *Schmidt et al.* [1995] and *Webb* [1996] showed that the eddies that occur downstream from debris fans, and their associated sand bars, persist for ten's to hundred's of years. Because the depositional locations are stationary, the measured size of individual bars can be compared with measurements made of the same bar at other times.

Fan-eddy complexes are composed of an area of ponded flow upstream from a debris fan, a constricted channel near the fan apex, a channel expansion where large eddies occur along the bank, and mid-channel or bank-attached gravel bars further downstream. Sand bars occur along the banks of the ponded flow and in the large eddies downstream from the fans. The topography of sand bars within eddies is very consistent from one fan-eddy complex to another. The highest elevation parts of separation bars [*Schmidt and Graf*, 1990] and reattachment bars [*Rubin et al.*, 1990] occur at the upstream and downstream ends, respectively, of eddies. At high flow, each reattachment bar may extend upstream to the primary eddy return-current channel [*Schmidt and Graf*, 1990]. At low flow, a stagnant embayment may partially inundate this channel and be used as habitat for nursery-age humpback chub [*Brouder et al.*, this volume]. The range of discharges over which return current channels are "backwater habitat" depends on the depth of the channel and the height of the reattachment bar that blocks flow into the channel. *Leschin and Schmidt* [1995] found that some separation and reattachment bars can not be distinguished from one another; therefore, they also mapped "undifferentiated eddy bars." Channel-margin deposits typically occur as channel-parallel levees along the margins of the ponded flow and do not form within eddies [*Schmidt and Rubin*, 1995].

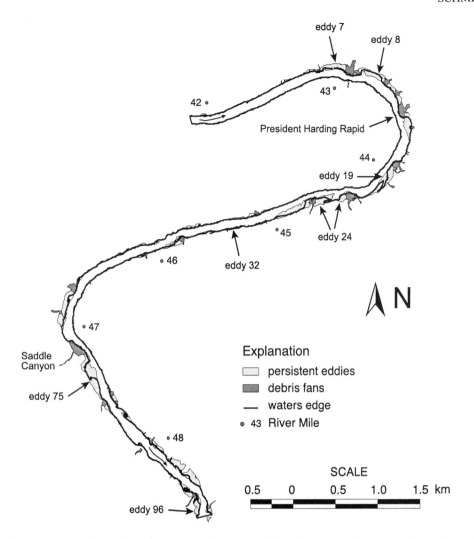

Figure 1. Map showing the Point Hansbrough reach. Persistent eddies where scour chains were located are labeled by eddy number. The Colorado River flows from top to bottom of the page. The water's edge is from the April 4, 1996, aerial photos taken at between 385 and 440 m³/s.

3. THE STUDY REACHES

We mapped sand bars and analyzed their changes in three study reaches where 1:2400 scale topographic (0.5-m contour interval) and orthophoto data are available (Table 1). The 10.8-km Point Hansbrough reach begins 92 km downstream from GCD and 68 km downstream from Lees Ferry, Arizona (Figure 1). The Tapeats Gorge (8.0 km) reach begins 124 km downstream from GCD and 100 km downstream from Lees Ferry (Figure 2). The Big Bend reach is immediately downstream from the Tapeats Gorge and is 12.1 km long (Figure 3). In some cases, we report the combined data from these two adjacent reaches as the Little Colorado River (LCR) confluence reach.

The Point Hansbrough reach is entirely within what *Schmidt and Graf* [1990] called lower Marble Canyon, which is one of the 11 geomorphic reaches that they identified. Lower Marble Canyon has the second-flattest reach-average channel gradient and second-largest channel width of these reaches. The width of the alluvial valley, measured as the distance between bedrock outcrops, is between 150 and 300 m, and bedrock at river level is the Cambrian Muav Limestone. The average channel width is about 100 m at a discharge of about 680 m³/s. As measured on the large-scale topographic maps used in this study, the average gradient of the Point Hansbrough reach is 0.0008. Debris fans formed by tributaries with a drainage basin area greater than 0.01 km² occur at a frequency of 1.5 fans/km

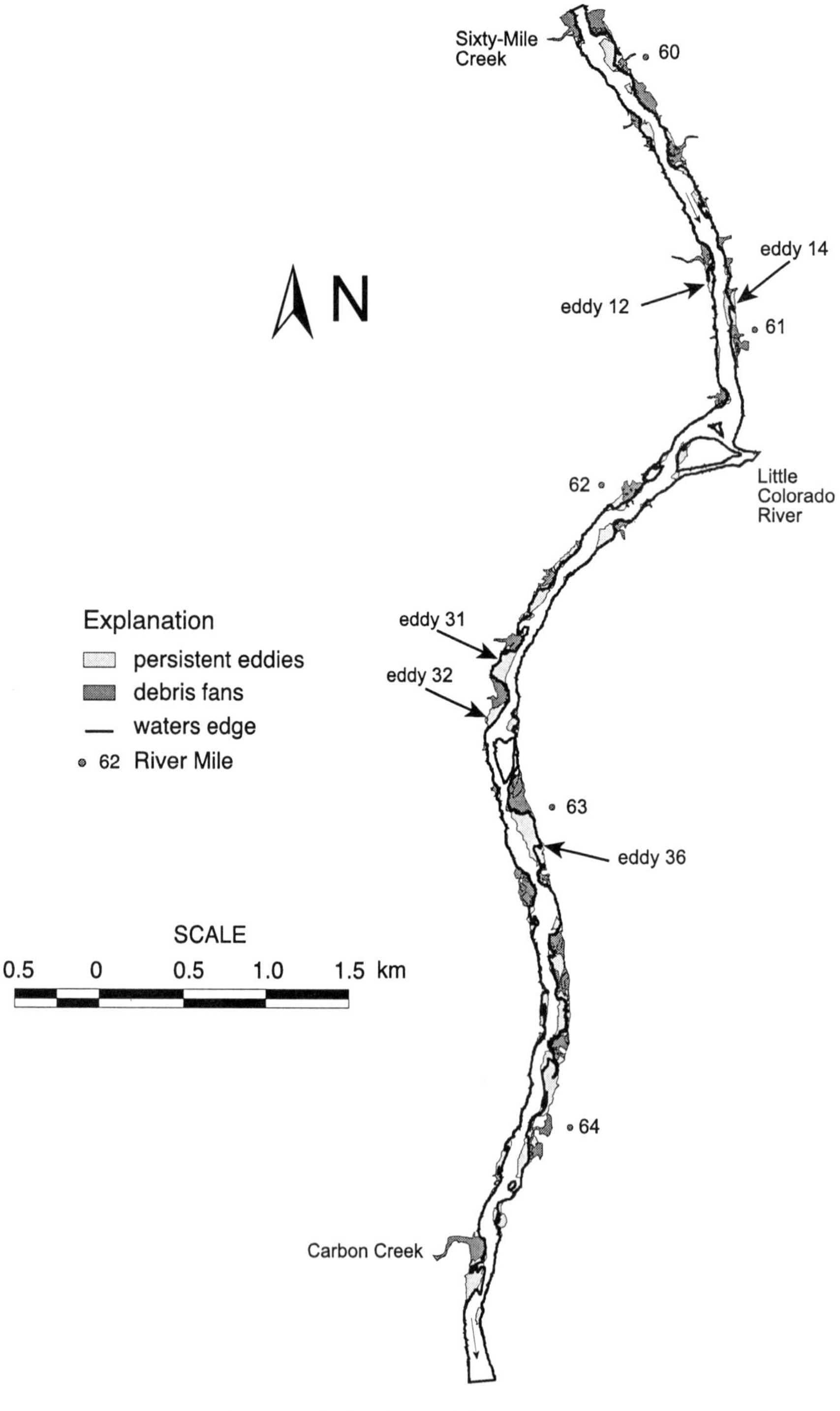

Figure 2. Map showing the Tapeats Gorge reach. Persistent eddies where scour chains were located are labeled by eddy number. The Colorado River flows from top to bottom of the page. For the upstream 4 km, the waters edge is from the April 4, 1996, aerial photos taken at between 385 and 440 m³/s. For the rest of the reach, the water's edge is from April 6, 1996, aerial photos taken at 245 m³/s.

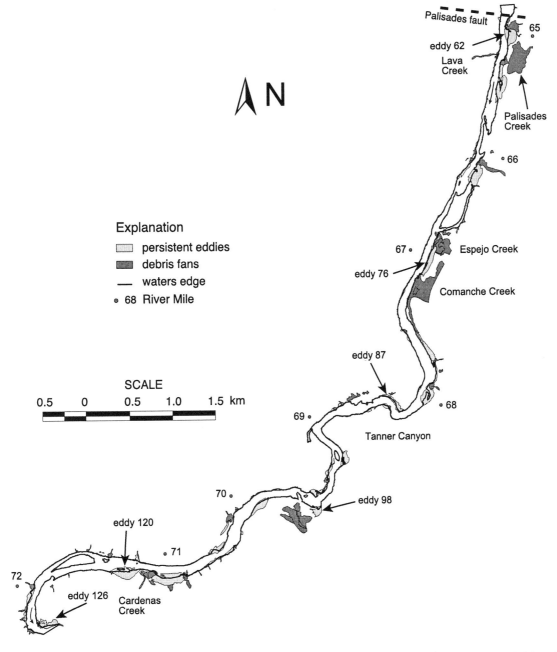

Figure 3. Map showing the Big Bend reach. Persistent eddies where scour chains were located are labeled by eddy number. The Colorado River flows from top to bottom of the page. The water's edge is from April 6, 1996, aerial photos taken at 245 m³/s.

[*Melis et al.,* 1995], and nearly all of the drop in channel gradient occurs near these fans.

Schmidt and Graf [1990] considered the LCR confluence to be the boundary between lower Marble Canyon and Furnace Flats. We determined, however, that significant geomorphic change of the Colorado River occurs near

Palisades Creek (Figure 3) where the Colorado River crosses the Palisades Fault and monocline [*Billingsley and Elston,* 1989]. Upstream from this fault in the Tapeats Gorge, bedrock at river level is the resistant Cambrian Tapeats Sandstone or the lower member of the Precambrian Dox Sandstone. Vertical cliffs and ledges dominate the near-river environment, and average alluvial valley width is between 120 and 180 m, which is narrower than the Point Hansbrough reach. Debris fans occur at a frequency of about 3.3 fans/km [*Melis et al.,* 1995], twice the frequency of the Point Hansbrough reach. The reach average gradient of the Tapeats Gorge is 0.0016 and is also twice that of the Point Hansbrough reach.

Downstream from the Palisades Fault and monocline is the Big Bend, which has more gently sloping riverside hillslopes than does the Tapeats Gorge. We have adopted the term Big Bend, used by *Billingsley and Elston* [1989], rather than the term Furnace Flats that is used by river runners. The alluvial valley is between 240 and 470 m wide in this reach. Bedrock at river level is the erodible upper part of the Dox Sandstone, the overlying Precambrian Cardenas Basalt, and cemented Quaternary gravels. Debris fans occur at a rate of about 1.7 fans/km [*Melis et al.,* 1995], which is less frequent than in the Tapeats Gorge. Individual debris fans are among the largest that occur anywhere in Grand Canyon [*Hereford et al.,* 1996]. *Graf et al.* [1995] mapped the bathymetry of the entire reach between the LCR and Tanner Canyon, and their data fully depict the large changes in channel width and depth that occur within fan-eddy complexes.

Eddies are not uniformly distributed in the study reaches; they occur more frequently where there are more debris fans (Table 1). In this paper, we focus on the characteristics and history of change of eddies larger than 1000 m^2. Smaller eddies tend to be formed by bank irregularities such as talus cones and rock outcrops, store proportionally little sediment, and often become washed-out by downstream flow at high discharges.

The sediment budgets of the reaches differ, because the number of unregulated tributaries that resupply sediment to Grand Canyon increases downstream. These tributaries contribute little streamflow, but some are large sources of sand and finer sediment. The Paria River is the primary contributor of sediment to the Point Hansbrough reach and the 2 km of the Tapeats Gorge that are upstream from the LCR confluence. Much higher sediment loads occur in the downstream part of the Tapeats Gorge and in the Big Bend, because more sediment is delivered to the Colorado River from the LCR than from any other tributary in Grand Canyon [*Andrews,* 1991].

4. METHODS

4.1. Mapping of Sand Deposits

Surficial geologic field mapping, aerial photograph interpretation, field sedimentologic description, installation and recovery of scour chains, and computer-assisted geographic analysis were conducted. Aerial photographs were taken on March 24, 1996, just before the controlled flood, when discharge of the Colorado River was 240 m^3/s. Post-flood aerial photographs of the Point Hansbrough reach and the upstream 4 km of the Tapeats Gorge were taken on April 4, 1996, when discharge was between 385 and 440 m^3/s. The remainder of the Tapeats Gorge and the Big Bend reach were photographed on April 6, 1997, when discharge was 245 m^3/s.

Map units were established on the basis of topographic level and type of deposit (Table 2; Plates 1-3). Topographic level was inferred from stereoscopic inspection and the color differences, on photos, among submerged, wet, and dry sand (Figure 4). Aerial photos show submerged deposits when water clarity is high. Sand bars are typically of darker color near the water's edge, because the sand is damp. High parts of bars are dry, and appear white in photographs. The width of wet sand depends on the slope of the bar, the river discharge immediately prior to the time of photography, and the height of capillary rise of alluvial ground water, which in turn depends on the grain size of the bar. The grain size of bars slightly coarsened during the flood as the proportion of transported silt and clay decreased [*Topping et al.,* this volume], but the error in our analysis introduced by this change was small, as demonstrated below. We used the same pre-flood topographic-level definitions as *Schmidt and Leschin* [1995] who mapped topographic levels on 1984, 1990, 1992, and 1993 photographs (Table 2).

On the post-flood aerial photographs, submerged, wet, and dry bare sand near the river were interpreted to have been deposited by the 1996 controlled flood. Field inspections were made in late March, early April, and June 1996 to confirm or revise photo interpretations. In June 1996, we also recovered scour chains that had been installed in February in 7 eddies in the Point Hansbrough reach (Figure 1), 5 eddies in the Tapeats Gorge (Figure 2), and 7 eddies in the Big Bend (Figure 3). Sedimentologic analysis of the excavations at these recovered chains greatly aided our ability to identify flood deposits on the aerial photos, and also allowed us to measure scour and fill. Mapping was done on overlays of aerial photos, and these data were entered into an ARC/INFO database by referencing permanent features on the photos to the same features on the orthophoto base maps.

We determined the persistent eddy area as the maximum extent of sand bars in all years of available historical photography. We used the surficial geologic maps of *Leschin and Schmidt* [1995] to determine the distribution of separation, reattachment, and undifferentiated eddy bars in 1935, 1965, 1973, 1984, 1990, 1992, and 1993, and we used

Plates 1-3 are in supplementary envelope to this volume.

TABLE 2. Description of units used in pre- and post-controlled flood geomorphic maps

Pre-1996 deposits

submerged sand at 226 m³/s

Coarse- to fine- grained sand, underwater, and visible on aerial photos. Extent of deposits is partially dependent on the quality of each aerial photo, the angle of the sun in the photo, the distribution of shadows in each photo, the electomagnetic wavelength used for photography, and the depth and turbidity of the river at the time of photography.

wet sand, inundated at between 226 and 550 m³/s

Coarse- to fine-grained sand with some silt and clay. These deposits appear darker on aerial photos than adjacent or nearby subaerial deposits of similar type. This level typically occurs adjacent to the river or to submerged deposits.

fluctuating-flow sand, inundated at between 550 and 890 m³/s

Very-fine- to fine-grained sand with widely ranging colors of light gray, brown, and reddish brown. The deposits are typically separated from the river by a single scarp and slope smoothly down into wet or submerged deposits or directly into the river. Well-defined bedforms are occasionally visible.

Little Colorado River (LCR) flood sand, inundated at less than 990 m³/s

Mainstem alluvial deposits of the winter 1993 LCR flood occurs only downstream from the LCR confluence. Deposits are higher in elevation than fluctuating-flow sand. In the 1993 photos, these deposits have no new vegetation growing on them but may extend into previously vegetated areas.

high flow sand, inundated at between 890 and 1400 m³/s

Medium- to very-fine grained sand, with some silty layers. Deposited by 1984-1986 Glen Canyon Dam bypass releases. High-flow deposits are typically separated from adjacent fluctuating-flow deposits by a cutbank. Dune bedforms are sometimes present and are distinct from the smaller and sharper bedforms that occur on fluctuating-flow deposits.

flood sand of 1983, inundated at between 1400 and 2700 m³/s

Medium- to very-fine-grained sand, very well-sorted to well-sorted, distinctive very light gray with some salt-and-pepper coloring. Deposited by the 1983 spillway flood. Internal structures include ripples, climbing ripples, cross-laminations, and planar bedding. Smooth, planar sand deposits present in the 1984 aerial photos and higher in elevation than high-flow deposits were mapped as flood sand. The 1983 peak stage is often indicated by a driftwood line.

1996 Controlled-flood deposits (interpreted from aerial photos taken immediately after flood recession)

submerged sand at between 226 and 385 m³/s

Coarse- to fine-grained sand, underwater, and visible on aerial photos. Extent of deposits is partially dependent on the quality of each aerial photo, the angle of the sun in the photo, the distribution of shadows in each photo, and the turbidity of the river at the time of photography.

wet sand, inundated at between 226 and 550 m³/s

Coarse- to fine-grained sand with some silt and clay. These deposits appear darker on aerial photos than adjacent or nearby subaerial deposits of similar type. This level typically occurs adjacent to the river or to submerged deposits.

perched wet sand, inundated at greater than 550 m³/s

Fine-grained sand that appears wet in photos but is located far from the river. In some cases, occurs at locations known to be more than a vertical meter from the water surface at the time of photography.

controlled-flood sand, inundated at between 550 and 1274 m³/s

Coarse- to fine-grained sand appearing clean and fresh in photos. Deposit forms are generally sharp and well-defined. Deposits are typically lighter colored than the nearby older fine-grained deposits. In some vegetated areas and in some low-velocity areas deposits may appear wet or darker due to higher silt content.

a

b

Figure 4. Photographs showing the Saddle Canyon fan-eddy complex. The Colorado River flows from left to right. Persistent eddy #75 (Figure 1; Plate 1) is downstream from the debris fan that constricts the channel. The deposit is a reattachment bar and its highest elevation parts are white in these photographs. A separation bar mantles the downstream part of the Saddle Canyon debris fan. (A) March 24, 1996, at a discharge of 226 m^3/s. (B) April 4, 1996, at a discharge of 385 m^3/s.

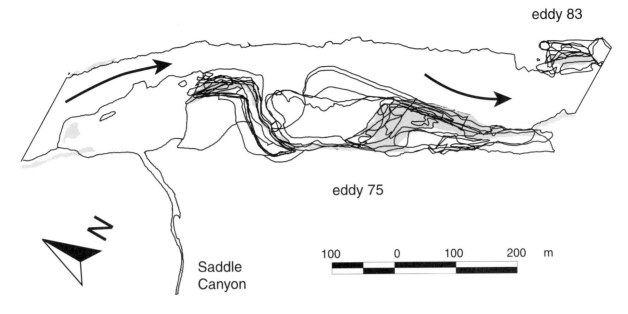

Figure 5. Map showing the maximum extent of sand deposits near Saddle Canyon and the method of determining the extent of the persistent eddy at this site. Location of this site is shown on Figure 1 and covers the same area as Figure 4. The areas of persistent eddies #75 and #83 are the light shaded area. The black lines within the shaded areas are the boundaries of eddy deposits in other years of aerial photography (1935, 1965, 1973, 1984, 1990, 1992, and pre-1996 controlled flood) that were used to define the total area of the persistent eddy. The area of the bars as mapped after the controlled flood (Figure 4a). is shown as the dark shaded area.

our own mapping of conditions before and after the flood. Using a geographic information system, we identified the largest contiguous area within which separation, reattachment, or undifferentiated eddy bars were mapped in any of the nine map series (Figure 5). We refer to these areas as persistent eddies, and we define a persistent eddy to be the largest area where there has ever been exposed sand in any year of historical aerial photography. The boundary of each persistent eddy is the maximum area within which there has been emergent separation, reattachment, or undifferentiated eddy bars in at least 1 year of historical photography. Each persistent eddy, regardless of size, was numbered so that we could account for the individual response of each eddy through time. Separation and reattachment bars that were not contiguous were assigned the same number if both bars form within the same eddy. For brevity, we use the term "eddies" to refer to these persistent eddies in the text below.

4.2. Measuring Topographic Change of Sand Deposits

Topographic change is typically measured by field survey or by photogrammetry. These strategies are not appropriate for the comprehensive evaluation of erosion and deposition in reaches that extend ten's of km, or which involve analysis of historical aerial photography that is often of poor quality. We used a method developed by *Schmidt and Leschin* [1995] to compare large-scale topographic change between pre- and post-flood conditions. This method does not require photogrammetric measurements of surface elevation, and it permits comparison among historical photos for which field data are unavailable.

Areas of significant erosion or deposition, and areas of no significant change, were determined by using a geographic information system to compare the topographic level and area of every map unit before and after the controlled flood (Plate 1). We used different algorithms to make this comparison, depending on how similar river discharge was in the pre- and post-flood photos (Figure 6). One algorithm was developed assuming that discharge was the same in both photo series; the other algorithm assumed that discharge in the post-flood photos was greater than in the pre-flood photos, as was the case in the Point Hansbrough reach and the upstream 4 km of the Tapeats Gorge.

We developed and calculated two metrics for each eddy. One metric was the ratio of actual deposition to potential deposition, termed the eddy filling ratio. We estimated the area of potential controlled flood deposition as the area of

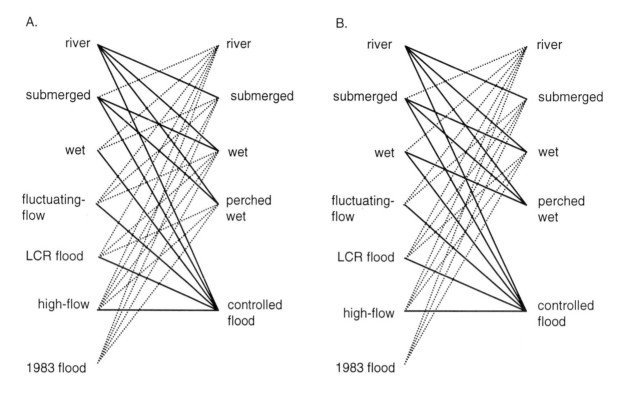

Figure 6. Diagrams that illustrate the algorithms used to determine areas of significant erosion and from the pre- and post-flood surficial geologic maps. (A) The algorithm used when the pre- and post-flood photographs were taken at approximately the same discharge. (B) The algorithm used when the post-flood photographs were taken at higher discharge. Solid lines indicate situations in which significant deposition was calculated and dashed lines indicate situations in which significant erosion was calculated. The map comparisons are illustrated on the accompanying plates. Map units are described in Table 2.

each persistent eddy lower in elevation than the upper margin of all 1984 high-flow deposits and 1996 controlled-flood deposits. The flood of 1984 was similar in magnitude to the controlled flood of 1996. The second metric was net-normalized aggradation (NNA), which was defined as:

$$NNA = (A_d - A_e) / A_{pe} , \qquad (1)$$

where A_d = area of significant deposition (m²), A_e = area of significant erosion (m²), and A_{pe} = area of the persistent eddy (m²).

We subdivided the eddies in the Tapeats Gorge, depending on their location upstream or downstream from the LCR sediment supply, in order to determine if the style of change was related to increased sediment supply from the LCR. Because the data were not normally distributed in all reaches, a Mann-Whitney U test was used to compare equivalency [*Davis*, 1986].

4.3. Analysis of Error in Determining Erosion and Deposition

We evaluated the accuracy of our map-based method of calculating areas of significant deposition and erosion by comparing our results with pre- and post-flood surveys measured by *Hazel et al.* [this volume] for 6 sand bars where both types of data are available. Our error analyses consisted of (1) visual comparison, (2) measurement of areas of agreement and disagreement, and (3) calculation of an error matrix.

Visual comparisons were made by overlaying our maps with ones we computed from the topographic data of *Hazel et al.* [this volume]. We compared mapped areas of significant deposition with the area where post-flood elevations surveyed by *Hazel et al.* [this volume] exceeded pre-flood elevations by more than 0.25 m; areas of significant erosion were compared with the area where post-flood elevations were at least 0.25 m less than pre-flood surveys. We defined areas of no significant change wherever the topographic data of *Hazel et al.* [this volume] indicated that pre- and post-flood elevations differed by less than 0.25 m. Visual inspection shows general agreement between our method and surveyed changes in topography for the site above Tanner Rapid (Figure 7), for example, and this agreement is typical of other sites. Errors primarily occur along the margins of areas of mapped deposition or erosion.

Error matrices were calculated for each site and summed (Table 3). This summed matrix demonstrates that our method of map-based analysis yields good results when compared to surveyed data. Our algorithms and map data correctly predict areas of significant change or areas of no significant change in 67% of the area for which comparisons were made. The largest error occurred where map analysis indicated no significant change and topographic surveys measured more than 0.25 m of deposition. Other large errors occurred where map analysis indicated significant deposition or significant erosion and actual elevation change was less than 0.25 m. Our algorithms incorrectly measured the style of bar change, i.e. we predicted significant erosion when significant deposition actually occurred, in only 6% of the evaluated area. The kappa coefficient, estimated by the k_{hat} statistic, is a measure of the actual agreement minus the agreement expected by chance [*Naesset*, 1996]. We calculated a k_{hat} value of 0.50 from Table 3 using the formulation of *Hudson and Ramm* [1987]. The possible values of k_{hat} range from $-\infty$ to 1, and values > 0.4 are considered to represent good agreement between the actual and predicted values.

5. RESULTS

5.1. Deposition in Persistent Eddies

There are 218 persistent eddies in the three study reaches (Figure 8). More than 50% of the eddies are smaller than 1000 m^2, and these small eddies account for a very small proportion of the total area of all eddies. Eddies larger than this size account for 95, 98, and 97% of the total area of persistent eddies in the Point Hansbrough, Tapeats Gorge, and Big Bend reaches, respectively. Five eddies larger than 1000 m^2 occur per km in the Tapeats Gorge, and the frequency of eddies of this size in the Point Hansbrough and Big Bend reaches is much less: 3.5 and 2.9 per km, respectively (Table 1). The largest individual eddies occur in the Point Hansbrough reach, but the median size of eddies is largest in the Tapeats Gorge (Figure 9).

When totaled for the three reaches, the controlled flood deposited more sand, by area, within eddies than elsewhere,

TABLE 3. Matrix comparing agreement between areas of significant erosion, deposition, and no change as measured by topographic survey and aerial photograph interpretation at 6 sites

		Area determined by topographic survey (m^2)		
		Deposition	No change	Erosion
Area determined by aerial photographs (m^2)	Deposition	19,611[a]	4699	1739[b]
	No change	7103	10,085[a]	3092
	Erosion	2418[b]	3965	16,388[a]

[a] Areas where the two methods are in agreement.
[b] Areas where the two methods substantially disagree.

but this proportion varied widely. Of the 171,500 m^2 of controlled-flood deposits mapped in the Point Hansbrough reach, 72% was within persistent eddies (Table 1). In the Tapeats Gorge, where there are more debris fans and more eddies, 80% of the 122,500 m^2 of controlled-flood deposits were in eddies. In contrast, eddy bars comprised only 49% of the 253,000 m^2 of new deposits in the Big Bend. Thus, channel-margin deposits comprised a much larger proportion of new deposits in the Big Bend than elsewhere.

In eddies, the area of significant deposition exceeded the area of significant erosion by 142% in the Big Bend but by only 19% and 11% in the Point Hansbrough reach and Tapeats Gorge, respectively (Table 1). The proportion of controlled flood deposits within eddies that were mapped as areas of significant deposition increased downstream. This proportion was 62% in the Point Hansbrough reach, 70% in the Tapeats Gorge, and 73% in the Big Bend.

5.2. Spatial Characteristics of Fine-Grain Flood Deposits

Controlled-flood deposits occurred as discontinuous patches along the river's edge throughout the study reaches (Plates 1-3). The largest deposits were typically reattachment bars, but there were also large mid-channel deposits in the Big Bend. Long, thin channel-margin deposits were mapped between River Miles (RM) 42 and 43 and between RM 46 and 47 in the Point Hansbrough reach and throughout the Big Bend. At the time the post-flood photographs were taken, no persistent eddies were entirely full of sand. However, the controlled flood could not possibly have deposited sand in the highest elevation parts of the persistent eddies, because some of these areas were not inundated. In a few cases, the low-elevation portion of persistent eddies was nearly filled with sand, such as #12 in the Point Hansbrough reach (Plate 1) and #32 in the Tapeats Gorge (Plate 2).

Both undifferentiated eddy bars and distinct separation and reattachment bars occurred within persistent eddies.

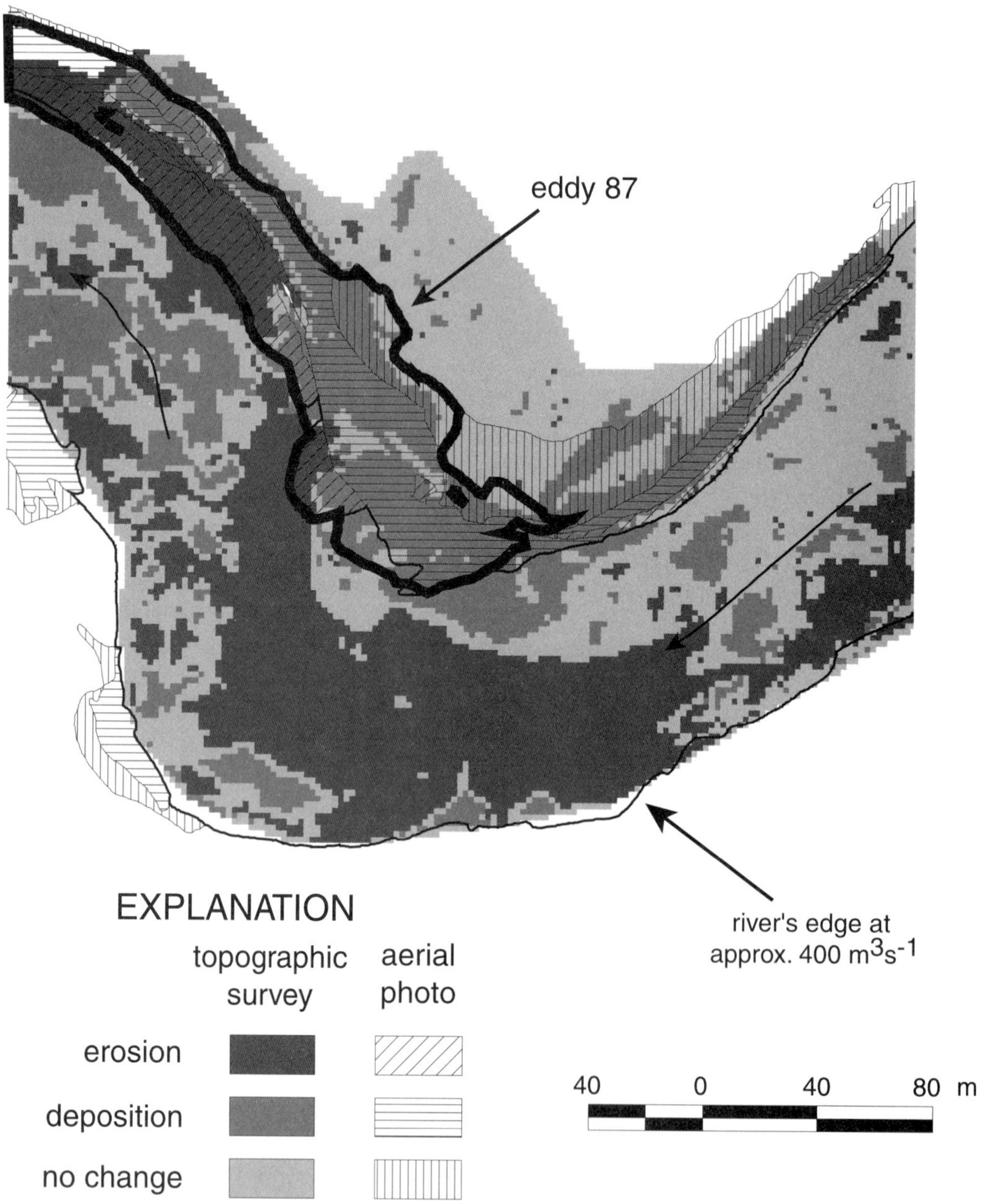

Figure 7. Map showing the distribution of areas of significant erosion and deposition caused by the 1996 controlled flood in eddy #87 upstream from Tanner Canyon (Plate 3), as determined by two methods. The persistent eddy is outlined by a heavy blue line. The areas shaded green, red, and blue show areas of deposition, erosion, and less than 0.25 m of change, respectively, as measured by topographic survey [*Hazel et al.*, this volume]. The horizontal, diagonal, and vertical lines show areas of significant deposition, erosion, and no change respectively, as measured by aerial-photo analysis. Only the areas where the methods overlap were compared. For example, areas that are shaded green and have horizontal lines are areas where both methods measured deposition.

The areas of significant deposition were typically located at the upstream or downstream ends of eddies, which *Schmidt and Graf* [1990] and *Rubin et al.* [1990] showed were the highest elevation parts of separation and reattachment bars, respectively. Typically, stagnant water in inundated return-current channels occurred between these large, well-defined separation and reattachment bars. Examples of this style of bar deposition include #19, #24, and #87 in the Point Hansbrough reach; #12 and #45 in Tapeats Gorge; and #64 in Big Bend. Some reattachment bars were mapped as wet

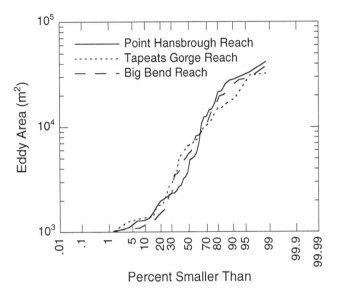

Figure 9. Graph showing the cumulative distribution of persistent eddies larger than 1000 m² in each reach.

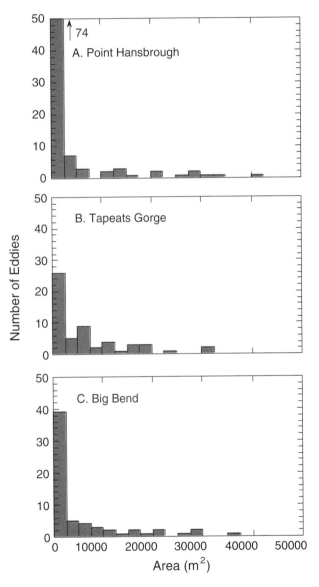

Figure 8. Graphs showing the distribution of sizes of all persistent eddies in each study reach. (A) Point Hansbrough. (B) Tapeats Gorge. (C) Big Bend.

sand and were of low elevation, such as #54 in Tapeats Gorge. Elsewhere, the exposed reattachment bar did not create a stagnant flow area, such as at #8 in the Point Hansbrough reach and #39 in the Tapeats Gorge. Sedimentary structures in these deposits were ripples and dunes whose bedform migration directions were consistent with those of reattachment and separation bars previously described by *Rubin et al.* [1990] and *Schmidt and Graf* [1990].

Although the total area of deposition exceeded the area of erosion in all reaches, some eddies had extensive areas of significant erosion. Sites where the area of erosion was greater than or similar to the area of deposition include #19, #24, and #75 in the Point Hansbrough reach and eddies #1, #30, and #36 in the Tapeats Gorge. In some cases, high-elevation reattachment bars did not form at all; examples include #14 and #44 in the Point Hansbrough reach and #26 and #47 in the Tapeats Gorge.

5.3. Variation in the Area of Flood Deposits and the Areas of Significant Erosion and Deposition

There was wide variation in the proportion of each persistent eddy filled by controlled flood deposits. Our data describe the entire population of eddy-deposited sand bars greater than 1000 m² in area in each reach and are not a statistical sample from an unknown larger population. Thus, these data reflect the actual variability on the amount of sand deposited in the eddies of the study reaches. The mean

TABLE 4. Summary of study reach characteristics and patterns of controlled flood deposition

Reach	Sediment sources	Major geomorphic characteristics	Eddy characteristics	Area of new deposits[a] (m²/km)	Eddy filling ratio[b]	Net normalized aggradation[c]
			average size (m²) number per km	all deposits large eddies	mean std. deviation % > 0.50	mean std. deviation % > 0.25
Point Hansbrough	Paria River, ungaged tributaries	wide valley; shallow slope; frequent, small fans	10,000 / 3.5	16,000 / 10,500	0.41 / 0.17 / 29	0.05 / 0.19 / 11
Tapeats Gorge, upstream from LCR	Paria River, ungaged tributaries	narrow valley; steep slope; very frequent, large fans	7250 / 4.6	10,500 / 7500	0.31 / 0.19 / 23	-0.05 / 0.25 / 15
Tapeats Gorge, downstream from LCR	Paria River, ungaged tributaries, LCR	narrow valley; steep slope; very frequent, large fans	9500 / 5.2	18,000 / 14,000	0.38 / 0.26 / 41	0.05 / 0.31 / 37
Big Bend	Paria River, ungaged tributaries, LCR	wide valley; steep slope; frequent, very large fans	9500 / 2.9	21,000 / 9500	0.50 / 0.25 / 49	0.15 / 0.22 / 29

[a] The area of all controlled flood deposits and controlled flood deposits in eddies larger than 1000 m², respectively, normalized by reach length.
[b] The ratio of mapped controlled flood deposits to the area of potential deposition (Figure 10).
[c] The area of significant deposition minus the area of significant erosion divided by the persistent eddy area (Figure 11).

eddy filling ratios ranged from 0.31 in the Tapeats Gorge upstream from the LCR confluence to 0.50 in the Big Bend (Table 4). The mean eddy filling ratio in the Big Bend reach was significantly larger than the mean ratios elsewhere (Mann-Whitney p = 0.01, 0.03, and 0.03). Differences in mean values among the other reaches were not statistically significant.

The variation of the eddy filling ratio was considerable within each reach and the distribution differed among the reaches (Figure 10). The modal eddy filling ratio was 0.35 in the Point Hansbrough reach, and the distribution was unimodal. Most eddies in this reach had between 30 and 60% of their area of potential deposition filled by bars. The range of variation was greater in the Big Bend where 10 eddies filled to greater than 60% of their potential. The largest variation in the proportion of eddies filled by new bars was in the Tapeats Gorge downstream from the LCR confluence. Many eddies had less than 20% of their potential depositional area filled by bars, yet the modal response was that between 50 and 60% of each eddy was filled by emergent bars.

There was wide variation in net normalized aggradation among the reaches (Figure 11). The mean NNA of the Big Bend reach was significantly greater than the means of the Point Hansbrough reach (Mann-Whitney p < 0.01) and the

Tapeats Gorge upstream from the LCR (Mann-Whitney p = 0.03).

There was also wide variation in NNA within each reach. In the Point Hansbrough and Big Bend reaches, the distribution of NNA values was unimodal. The range of NNA values in the Tapeats Gorge was much larger, especially downstream from the LCR (Figure 11). NNA values were not normally distributed in this reach and did not have a single mode; there were a greater number of eddies that were either extensively eroded or aggraded (e.g., Plate 2, #26 and #32). The area of significant erosion exceeded the area of significant deposition in at least 20% of the eddies in the Point Hansbrough and Big Bend reaches, but many of the eddies which were outliers of these distributions are less than 5000 m² in area. In these cases, small differences in the areas of erosion or deposition result in large proportional changes (e.g., Plate 1, #47). Other outliers of these distributions are eddies that are very long and narrow; these sites may not have actually been eddies during the controlled flood (e.g., Plate 1, #31). The number of eddies with negative NNA values in the Tapeats Gorge was higher for the sub-reach upstream from the LCR than in the subreach downstream from the LCR. Some of the outliers with high positive NNA values in the Tapeats Gorge were small (e.g., Plate 2, #50). However, some sites with high

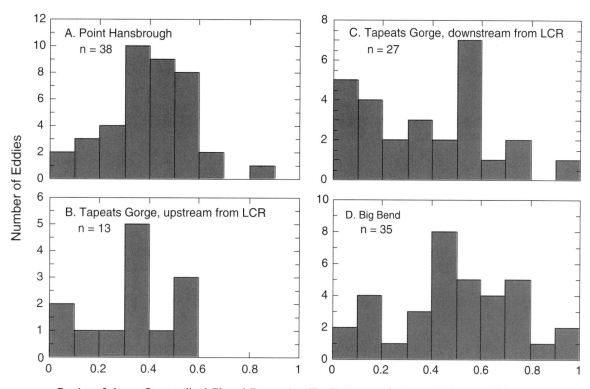

Figure 10. Graphs showing the eddy filling ratio, which is the ratio of 1996 controlled flood deposits to the area of potential deposition for eddies larger than 1000 m². The area of potential deposition was estimated as the area of each persistent eddy lower in elevation than the upper extent of 1984 high-flow deposits and 1996 controlled flood deposits. (A) Point Hansbrough. (B) Tapeats Gorge upstream from LCR confluence. (B) Tapeats Gorge downstream from LCR confluence (D) Big Bend.

negative NNA values were moderate to large size, and the area of erosion was large at these sites (e.g., Plate 2, #8 and #26; and Plate 3, #56).

5.4. Scour and Fill in Areas of Significant Deposition and Erosion

Areas where we mapped significant deposition were places where fill greatly exceeded scour during the flood, based on recovery of scour chains. At each bar where chains were recovered, near shore areas, especially at the downstream end of eddies, had significant deposition because there was almost no scour that preceded fill (Figure 12). Scour occurred at very few chain locations that were less than 20 m from the edge of water during the flood (Figure 13). Closer towards the center of the eddy, between 20 and 40 m from the edge of water, the thicknesses of scour and fill were approximately equal. Further offshore, in the center of eddies, the depth of scour was much greater, and fill was not measured at any chain located more than 60 m from the shoreline of the controlled flood.

6. DISCUSSION AND CONCLUSIONS

The 1996 controlled flood caused widespread deposition of new sand along the edges of the Colorado River. Most of this deposition occurred within eddies, except in the Big Bend. Nearly 550,000 m² of new sand were mapped along 31 km of the river, which is an average of about 18,200 m² along each km of the mapped reaches. Most new sand was deposited in discontinuous patches within eddies, and averages do not imply that a continuous band of sediment 9 m wide was deposited along each bank of the river. Approximately 60% of the new deposits within the three reaches were within persistent eddies larger than 1000 m²; the remainder occurred within small eddies and along channel-margins. Thus, eddies are an important depositional setting in Grand Canyon and the current focus of research on predicting the rates and styles of eddy deposition is appropriate.

On average, bars were exposed in less than half the area of potential deposition, as defined in this study. Flood deposits covered 41, 36, and 50% of the area of potential

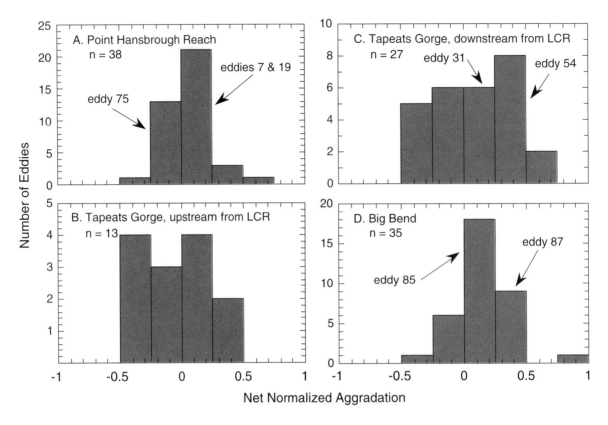

Figure 11. Graphs showing the net normalized aggradation values. (A) Point Hansbrough reach. (B) Tapeats Gorge upstream from LCR confluence. (C) Tapeats Gorge downstream from LCR confluence. (D) Big Bend.

deposition in eddies larger than 1000 m^2 in the Point Hansbrough, Tapeats Gorge, and Big Bend reaches, respectively. The failure of the 1996 flood deposits to create emergent bars within the entire area of potential deposition was probably due to four factors. First, we do not know if any individual flood is capable of depositing bars of sufficient area and volume such that they would be exposed at low flow in the entire area of each persistent eddy. Our definition of each persistent eddy is based on the cumulative history of the area of emergent sand as determined from nine photo series, and we have no evidence that sand has ever entirely filled an eddy at one time. Second, high deposition rates may not have been sustained for the full duration of the flood, even downstream from the LCR, and thus the rate of eddy deposition may have been too low to cause complete filling in some reaches during the 7-day flood. Third, erosion by mass failure may have redistributed sand from eddies to the main channel in those eddies that filled before recession of the flood [*Andrews et al.,* this volume]. Fourth, some parts of persistent eddies may not have had recirculating flow during the entire flood because

eddy circulation changes as eddies fill with sediment [*Schmidt et al.,* 1993; *Wiele et al.,* 1996].

Repeat measurements of bathymetry are the only way to determine which of these factors were most important in determining variation in eddy filling ratios, but the data collected in this study imply that the importance of these factors was not the same everywhere. Deposition rates in eddies must have declined during the flood, and rates may have been lower upstream from the LCR. *Smith* [this volume] showed that mainstem transport rates declined with time and that transport rates were lowest upstream from the LCR. *Schmidt et al.* [1993] showed that eddy deposition rates are proportional to mainstem transport rates.

Evacuation events are probably more likely to occur where there is a very wide range in eddy filling ratios, such as in the Tapeats Gorge downstream from the LCR, and are probably unimportant in reaches where there were no eddies with high filling ratios, such as in the Point Hansbrough reach and the Tapeats Gorge upstream from the LCR (Figure 10). High eddy filling ratios demonstrate

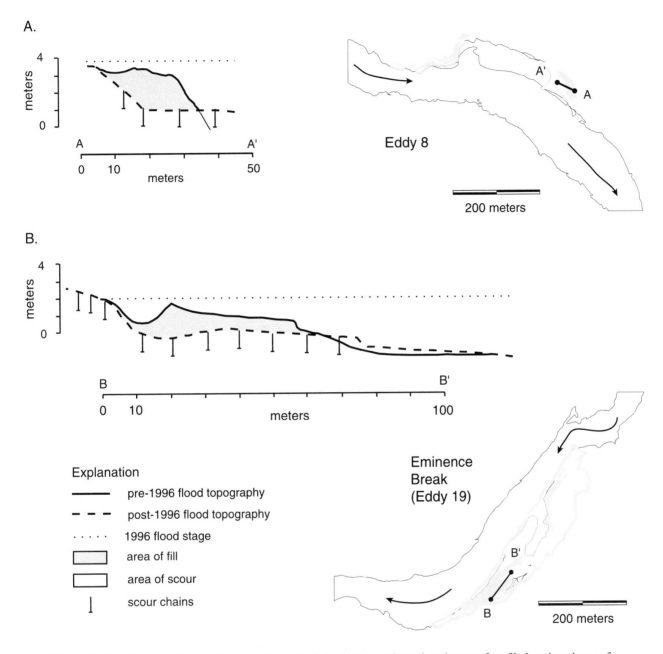

Figure 12. Stratigraphy at two persistent eddies in the Point Hansbrough reach and maps of profile location. Areas of scour and fill shown on profiles are as measured in the field by topographic survey at the time of chain excavation. Areas of scour and fill shown on location maps were determined from aerial photos and are the same as shown on Plate 1. (A) Persistent eddy #8. (B) Persistent eddy #19.

that some eddies in a reach have the potential to completely fill. Nearby eddies with very low eddy filling ratios may have been evacuated a short time before the post-flood photographs were taken. The effects of changes in eddy circulation on deposition rates are probably more important in smaller eddies, but every study reach has a very high proportion of small eddies.

The metrics developed in this study indicate that the magnitude of deposition downstream from the LCR was greater than upstream (Table 4). The total area of deposition per km, calculated as the area of all controlled flood deposits divided by reach length, was higher in the Big Bend and Tapeats Gorge downstream from the LCR than in reaches upstream from the LCR. More than 40% of the

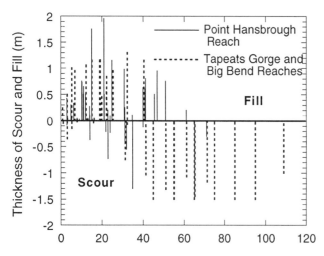

Figure 13. Graph showing measured scour and fill at every recovered scour chain in the study reaches. Scour chains were positioned along straight lines extending from the bank towards the eddy center at an oblique angle. The distances indicated are along these lines.

eddies downstream from the LCR mapped in this study had filling ratios greater than 50%, but less than 30% of the eddies upstream from the LCR had filling ratios that were as large. Twenty-nine percent or more of the eddies downstream from the LCR mapped in this study had NNA values greater than 0.25, but 15% or less of the eddies upstream from the LCR had NNA values as large.

Reach geomorphology affected the style of deposition in some reaches. The area per unit length of eddy and channel-margin deposits differed greatly between the Tapeats Gorge and the Big Bend, because eddy deposition dominated the Tapeats Gorge and channel-margin deposition was greatest in the Big Bend. Neither metric should be used to compare the overall characteristics of deposition in reaches whose geomorphology differs greatly.

The variability in eddy response was sufficiently great that the means of some metrics were not statistically different, even though the absolute values of the means differed greatly. The mean values of the eddy filling ratio and NNA of the Big Bend were significantly greater than that of the Point Hansbrough reach and the Tapeats Gorge upstream from the LCR. However, the mean values of the Tapeats Gorge downstream from the LCR was not significantly greater than the upstream reaches, because the variability of eddy response in this part of the Tapeats Gorge was so large. High variability may characterize reaches prone to evacuation events, as discussed above.

Variability in the response of individual eddies is sufficiently great that monitoring programs that measure detailed topography or bathymetry at a few sites risk use of a sample set that is not representative of average reach response. Fortunately in the case of the three reaches of this study, sites measured by *Hazel et al.* [this volume] and *Andrews et al.* [this volume] responded with a magnitude and style that was typical of these reaches. The representativeness of measurements elsewhere in Grand Canyon is not known.

The spatial variation in size of eddy sand bars affects how managers view the "success" of the 1996 controlled flood. In one sense, the widespread distribution of 1996 controlled flood deposits demonstrates success; extensive new sand deposits were exposed after the flood receded. However, the existence of outliers – eddies that had much greater amounts of erosion or deposition than the average reach response – means that river managers must establish clear objectives in determining flood success. "Success," as viewed by river users and river managers, must acknowledge the potential variation in individual site response. Managers and users will have to distinguish between their opinions about average reach response, such as the modal NNA value, and their opinions about changes at any specific eddy, because our results show that the 1996 controlled flood did not cause the same magnitude of change at every site. Although the average response of a reach may be towards deposition, individual sites within that reach may be extensively eroded. The significance of reach-average or individual data in determining "success" is the decision of the river manager.

REFERENCES

Andrews, E.D., Sediment transport in the Colorado River basin, in *Colorado River Ecology and Dam Management*, ed. G.R. Marzolf, pp. 54-74, Natl. Acad. Press, Washington, D.C., 1991.

Billingsley, G.H., and D.P. Elston, Geologic log of the Colorado River from Lees Ferry to Temple Bar, Lake Mead, Arizona, in *Geology of Grand Canyon, Northern Arizona [with Colorado River Guides] Lees Ferry to Pierce Ferry, Arizona*, ed. D.P. Elston, G.H. Billingsley and R.A. Young, pp. 1-36, Amer. Geophys. Union, 1989.

Davis, J.C., *Statistics and Data Analysis in Geology*, 646 p., John Wiley & Sons, New York, 1986.

Graf, J.B., S.M.D. Jansen, G.G. Fisk, and J.E. Marlow, *Topography and bathymetry of the Colorado River, Grand Canyon National Park, Little Colorado River Confluence to Tanner Rapids*, __p., U.S. Geol. Surv. Open-File Rept. 95-726, 1995.

Hereford, R., K.S. Thompson, K.J. Burke, and H.C. Fairley, Tributary debris fans and the late Holocene alluvial chronology of the Colorado River, eastern Grand Canyon, Arizona, *Geol. Soc. Amer. Bull.*, 108, 3-19, 1996.

Hudson, W.D., and C.W. Ramm, Correct formulation of the kappa coefficient of agreement, *Photogr. Engin. Remote Sensing, 53*, 421-422, 1987.

Kearsley, L.H., J.C. Schmidt, and K.D. Warren, Effects of Glen Canyon Dam on Colorado River sand deposits used as campsites in Grand Canyon National Park, USA, *Regul. Rivers, 9*, 137-149, 1994.

Leschin, M.F., and J.C. Schmidt, *Description of map units to accompany maps showing surficial geology and geomorphology of the Point Hansbrough and Little Colorado River confluence reaches of the Colorado River, Grand Canyon National Park, Arizona,* 6 p., Bur. Recl., Glen Canyon Environ. Studies, Flagstaff, AZ, 1995.

Melis, T.S., R.H. Webb, P.G. Griffiths, and T.W. Wise, *Magnitude and frequency data for historic debris flows in Grand Canyon National Park and vicinity, Arizona,* 285 pp., U. S. Geol. Surv. Water-Res. Invest. Rept. 94-4214, 1995.

Naesset, E., Use of the weighted kappa coefficient in classification error assessment of thematic maps, *Int. J. Geogr. Info. Syst., 10*, 591-604, 1996.

Rubin, D.M., J.C. Schmidt, and J.N. Moore, Origin, structure, and evolution of a reattachment bar, Colorado River, Grand Canyon, Arizona, *J. Sed. Petr., 60*, 982-991, 1990.

Schmidt, J.C., and J.B. Graf, *Aggradation and degradation of alluvial sand deposits, 1965 to 1986, Colorado River, Grand Canyon National Park, Arizona,* 74 p., U. S. Geol. Surv. Prof. Paper 1493, 1990.

Schmidt, J.C., P.E. Grams, and R.H. Webb, Comparison of the magnitude of erosion along two large regulated rivers, *Water Res. Bull., 31*, 617-631, 1995.

Schmidt, J.C., and M.F. Leschin, *Geomorphology of post-Glen Canyon dam fine-grained alluvial deposits of the Colorado River in the Point Hansbrough and Little Colorado River confluence study reaches in Grand Canyon National Park, Arizona,* 93 pp., Bur. Recl., Glen Canyon Environ. Studies, Flagstaff, AZ, 1995.

Schmidt, J.C., and D.M. Rubin, Regulated streamflow, fine-grained deposits, and effective discharge in canyons with abundant debris fans, in *Natural and anthropogenic influences in fluvial geomorphology*, ed. J.E. Costa, A.J. Miller, K.W. Potter, and P.R. Wilcock, pp. 177-195, Amer. Geophys. Union Mono. 89, 1995.

Schmidt, J.C., D.M. Rubin, and H. Ikeda, Flume simulation of recirculating flow and sedimentation, *Water Res. Res., 29 (8)*, 2925-2939, 1993.

Webb, R.H., *Grand Canyon, a century of change: Rephotography of the 1889-1890 Stanton expedition:*, 290 pp., Univ. Ariz. Press, Tucson, AZ, 1996.

Wiele, S.M., J.B. Graf, and J.D. Smith, Sand deposition in the Colorado River in the Grand Canyon from flooding of the Little Colorado River, *Water Res. Res., 32*, 3579-3596, 1996.

John C. Schmidt, Paul E. Grams, and Michael F. Leschin, Department of Geography and Earth Resources, Utah State University, Logan, Utah, 84322-5240; email: jschmidt@cc.usu.edu

Photosynthetic and Respiratory Processes: An Open Stream Approach

G. Richard Marzolf[1], Carl J. Bowser[2], Robert Hart[3], Doyle W. Stephens[4], and William S. Vernieu[5]

This investigation examined open-stream methods for detection of photosynthetically driven (light dependent) chemical change, and subsequently provided an estimate of flood effects on the photosynthetic community in the 25-km reach of the Colorado River below Glen Canyon Dam. Observations reported here confirm that the dynamics of oxygen concentration and pH are correlated because photosynthesis and respiration in the benthic community cause the diel patterns. Observations at the constant low flows immediately before and after the controlled flood permitted measurement of changes in stream chemistry caused by biological activity and provided a test of the hypothesis that plants were scoured from the channel by the flood. The diel amplitudes of oxygen concentration and pH change were decreased after the flood as the biomass was scoured. Patterns of oxygen production and carbon dioxide removal varied along the 25-km reach.

1. INTRODUCTION

One of the most dramatic and obvious effects of the construction of Glen Canyon Dam was the development of a popular trout sport fishery in the dam's tailwater, the 25 kilometers immediately downstream from the dam to the Colorado River confluence with the Paria River near Lees Ferry. Cold, clear water from Lake Powell passing through the dam established the conditions for this development. That water is clear because virtually all suspended sediment is deposited in Lake Powell and it is cold throughout the summer because it is drawn from Lake Powell's hypolimnion. This combination of conditions suited salmonid requirements and allowed for the establishment of benthic algae and submersed aquatic vegetation as primary producers. The dominant element of the trophic base is the benthic alga, *Cladophora*, its epiphytes, and the grazers [*Blinn and Cole,* 1991], but other plants include *Chara, Potomogeton,* and *Fontinalis* [for example, see *McKinney et al.,* this volume]. There is little information about the spatial or seasonal variation in biomass or productivity of this photosynthetic community or about what regulates the variation. The reach is large and spatial variation (depth, current velocities, and substrate types) makes statistical estimation of process rates from spot, or chamber, measurements virtually intractable [but see *Brock et al.,* this volume].

Measurement of the diel, light dependent, variation in O_2 and pH shows promise for estimation of reach-integrated estimates of river metabolism [*Odum,* 1956; *Beyers and Odum,* 1959, *Marzolf et al.,* 1995]. The development of computational techniques to generate useful estimates of river productivity for monitoring purposes is an ultimate goal of this investigation. Three objectives of this report are (1) to confirm that biological processes of photosynthesis and respiration proceed with sufficient magnitude to enable a measurable and interpretable diel chemical signal in the flowing river, (2) to use the change in that signal caused by

---

[1] U.S. Geological Survey, Reston, Virginia
[2] Dept. of Geology and Geophysics, Madison, Wisconsin
[3] U.S. Geological Survey, Flagstaff, Arizona
[4] U.S. Geological Survey, Salt Lake City, Utah
[5] Bureau of Reclamation, Flagstaff, Arizona

The Controlled Flood in Grand Canyon
Geophysical Monograph 110
Copyright 1999 by the American Geophysical Union

the controlled flood to document environmental change due to dam operation, and (3) to define further scientific issues associated with the use of dam operations to achieve management objectives.

Scientific issues center on (1) making appropriately accurate and precise estimates of diffusion of oxygen and CO_2 to the atmosphere and (2) developing information about flow velocity (travel time), channel geometry (depth and light penetration), canyon orientation (shadow effects), and seasonal photoperiod patterns so that reach-specific photosynthetic rates can be related to varying light regimes, and physiological condition of the benthic plant community that explain its adaptation to this set of conditions.

Preliminary observations [*G.R. Marzolf and D.W. Stephens,* unpublished data, 1994; and *G.R. Marzolf et al.,* unpublished data 1995] led us to extend *Odum*'s [1956] conceptual model of the photosynthetic control of diel oxygen changes to include effects of pH changes [*Beyers and Odum,* 1959] and suggested that we could make observations of pH and O_2 concentration at several locations during the low-flow periods before and after the controlled flood of 1996 as (1) a test of the idea that photosynthesis and community respiration could drive daily changes in O_2 concentration (Objective 1) and pH in a stream of this size and (2) a way to estimate the magnitude of scour effects of the flood on the benthic plant community.

The designed hydrograph [*Schmidt et al.,* this volume] of the controlled flood presented an extraordinary experimental opportunity to make measurements at the low flows (226 m^3/s) before and after the controlled flood through the reach. The low-flow periods provided conditions when benthic metabolic effects on gas concentrations would be most observable and when there would be virtually no effect from varying discharge. The high flow (1250 m^3/s) between these was a period when scouring of benthic plants was anticipated. If diel oscillations were caused by photosynthesis and respiration of benthic plants, then a consequence of the loss of biomass would be a reduction in the daily range of O_2 concentration and pH (CO_2 concentration).

2. METHODS

Purposes of this investigation required the measurement of small differences in gas concentrations in water (CO_2 and O_2) at short time intervals so that the time course of change could be interpreted as evidence of the processes (photosynthesis and respiration) causing the change.

Exchange of gases with the atmosphere is an unknown and uncontrollable process that often confounds the interpretation of such measurements, but in this case much confusion from this source was eliminated effectively because the flows before and after the flood were held constant at 226 m^3/s.

It was known from preliminary work that O_2 concentrations would be below the saturation concentrations and that carbon dioxide concentrations would be enriched relative to the atmosphere. Both of these conditions are related to the fact that water is drawn through Glen Canyon Dam from aphotic depths in Lake Powell reservoir below the mixed layer. We expected also that outgassing of CO_2 would be such a significant process in the river immediately downstream from the dam that it would be unlikely that photosynthetically caused change would be measurable for several kilometers downstream. Corollary to that was the expectation that light dependent (photosynthetically driven) changes in CO_2 or O_2 concentrations would become measurable only at some greater distance downstream as the water came closer to atmospheric equilibrium and had been exposed to daylight in the tailwater after its residence in Lake Powell at unlighted depths. Furthermore, we expected that the amplitude of the diel change, once detectable, would increase with increasing distance downstream at least as far as the first major riffle where the Paria River joins the Colorado River at Lees Ferry or at the first major set of rapids (Badger Creek Rapid at River Mile 8.0) where reaeration from the turbulent mixing would be great and gas concentrations might be driven to saturation.

Location of sampling sites was chosen, therefore, to provide for interpretation of results in view of these expectations. Convenient sites were available at Lees Ferry and at 4.8, 9.6, and 14.4 kms upstream, the site at Mile -9 being 9.6 km downstream from the dam. Locations on the Colorado River are identified conventionally as distance upstream from Mile 0 at Lees Ferry. Upstream from there to the dam the locations are in negative miles (designated explicitly) and downstream the miles are positive.

Sampling frequency through a solar day ranged from 5 min to 4 hrs. Intervals were chosen to be as frequent as possible with the most time consuming analytical method used. Where electronic data loggers were used the interval was short. Manual measurements made with analog meters were made at longer intervals. Field chemical measurements such as alkalinity were less frequently yet. The objective of these parallel and redundant methods was to achieve greater certainty and to guard against missing data in the event of mishaps.

2.1. Light

Photosynthetically active radiation was measured at 1-min intervals averaged for 5-min on a data logger. The purpose of this measurement was to document the comparability (or lack of it) of light intensity and duration on the two days that photosynthetic responses were measured. The two days were similar though integral solar energy was slightly lower on the sampling day before the flood because of early morning cloud cover. Any error due to this difference would, therefore, yield conservative results.

2.2. Dissolved Oxygen Concentration

Polarographic electrodes for sensing dissolved O_2 in water were used in two forms to provide redundant measurements. First, crews at each of the four stations deployed multiparameter, data-logging instruments configured to measure and log temperature, conductivity, pH, and O_2 concentration at fifteen minute intervals. These instruments were calibrated the night before the measurements were to be started and placed at a single location in the river to log these measurements from water at the same place for several hours. This was repeated at the end of the observation period. Second, crews at each of the four stations measured O_2 concentrations and temperature with hand-held probes at 0.5 hr intervals for 24 hrs. These instruments were recalibrated every 4 hrs. Additional redundancy was achieved at Mile -3.0 and Lees Ferry by collecting water samples for standard Winkler oxygen determinations at 0.5 hr intervals. Acidification and thiosulfate titrations were performed the day after the field measurements were completed.

2.3. pH

Hydrogen ion activity was determined using field electrometers (Orion Model 250A) and combination electrodes. The electrodes were new and purchased from the same production run. The electrodes were inserted into a syringe barrel of slightly larger diameter and contact with ambient air was restricted through use of a split tubing collar on the electrode just above the syringe tube top. Three-way valves and tygon tubing were used to connect the electrode assembly with the sample syringe to ensure minimal contact with air, and to flush the sample from the chamber after measurement. Samples were flushed through the pH electrode assembly and measured repeatedly until two successive sample injections agreed within 0.2 milli-

volts. Electrodes were calibrated using 3 buffers (pH 4.0, 7.0 and 10.0) to ensure linearity and nernstian slope calibration. The electrodes were stored in river water between measurements and held at or near the water temperature to minimize drift. Electrodes were recalibrated every 3 hrs.

2.4. Alkalinity

Alkalinities were measured both in the field and on samples brought back to the laboratory using the Gran titration technique. Field alkalinities were titrated immediately after collection with 0.1 molar HCl using a microburette (readable to 0.0001 ml) and 4.0 ml, stirred samples. Laboratory determinations were made using a Radiometer ABU 91 auto-burette system, with a Radiometer SAC 80 autosampler. Determinations were made using 0.1 N H_2SO_4 as titrant on 10-15 g samples. Major ion chemistry was measured on the same samples and provided a cross-check on these alkalinities. Agreement with ion balance criteria was within 5% and agreement between field and laboratory samples was excellent.

2.5. Dissolved Inorganic Carbon

The dissolved inorganic carbon was measured before the wet persulfate oxidation step during the process to measure dissolved organic carbon using a standard dissolved organic/inorganic carbon analyzer (OI Model 700).

2.6. pCO2 Calculations

Carbon dioxide partial pressures were calculated by two techniques as a check on the quality of the pH measurements; that is, measurements of pH, alkalinity, dissolved inorganic carbon, and major ion concentrations provided a redundant set of measurements. Ion balances from major ion and alkalinity data and the field and laboratory measurements of alkalinity indicate uncertainties of the alkalinity of less than 2.0%. Care to exclude air contact with samples during pH measurements and replicate measurements proved valuable, and further increase our confidence in the calculated pCO_2 values. Paired calculations of pCO_2 were made using pH-alkalinity and pH-dissolved inorganic carbon results. Excellent agreement was found between the two estimates of pCO_2 values. Figure 1 illustrates the results from both before and after the experimental flood. Results are shown on a comparable hour basis to illustrate the range of CO_2 concentrations before and after the flood.

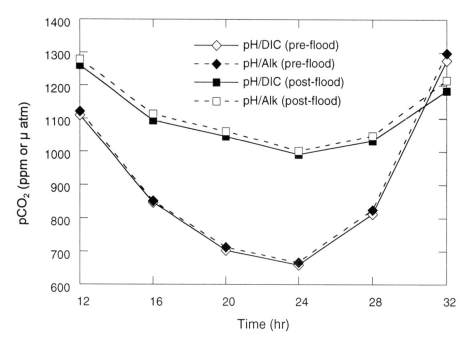

Figure 1. Changes in pCO_2 with time before the flood (lower curve) and after the flood (upper curve) as calculated from independent alkalinity and dissolved inorganic carbon data.

2.7. Total Carbon Calculations

The best estimate of the fraction of the total pool of inorganic carbon used by plants during photosynthesis in this study includes free CO_2, and half of the bicarbonate. Total CO_2 concentration (C_T) was calculated according to the following relation:

$$A = C_T \frac{K_{a1}[H^+] + 2K_{a1}K_{a2}}{[H^+]^2 + K_{a1}[H^+] + K_{a1}K_{a2}} + \frac{K_w}{[H^+]} - [H^+] \quad (1)$$

where A is the alkalinity [*Butler*, 1991]. Mean alkalinity of all measurements (n=24) was used in this calculation.

3. RESULTS

Preliminary lines of evidence indicate that diel changes in O_2 concentration and pH are caused by photosynthetic and respiratory activity in the river. These led us to conduct the reported observations before and after this experimental flood.

3.1. pH and O_2 Concentration

Changes in pH were observed along the 25 km reach from Glen Canyon Dam to Lees Ferry (Figure 2). Measurement of pH change along the reach documents that

water drawn from the depth of Lake Powell reservoir contained a high concentration of CO_2 and that, upon release into the tailwater river, the gas escapes to the atmosphere with a consequent rise in pH. These measurements were made at 1.6-km intervals on May 3, 1995, during a trip from the dam to Lees Ferry in the morning and returning to the dam in the afternoon. Measurements of CO_2 flux through the same reach documented that CO_2 gas was leaving the river in measurable amounts in the upper 8 km. The pH reached its maximum on the downstream leg of the trip at Mile -10, falling somewhat in the 8 km above Lees Ferry. The unexpected but definite increase in pH in the 16 km above Lees Ferry on the return trip later in the day indicated change with time of day and suggested that photosynthesis dominated CO_2 loss, while no change in river pH relative to the morning levels in the upper 8 km suggested that photosynthesis was not dominant and that CO_2 outgassing was of greater significance in the region near the dam.

The ranges of daily values of pH and O_2 concentration in the river are extraordinarily similar in phase and trend, suggesting a common cause. The data presented in Figure 3 was provided by the Bureau of Reclamation [*W.S. Vernieu*, unpublished data, 1994]. The long-term upward trend in the range — that is, the difference between the daily minimum and the daily maximum — is one of increasing range that, if due to photosynthesis, reflects increasing day length. The variation within the trend is thought to be associated with the interaction of varying discharge and/or cloud cover; the latter affects the light available for photosynthesis.

The daily minima in both pH and O_2 concentration in this same data set occurred in the morning (c. 0700) and the daily maxima occurred in the evening (c. 1800 - 2200); that is, the causal process is probably light dependent (Figure 4).

The reduction in the amplitudes of O_2 concentration and pH change at all four locations on the river where measurements were made was an expected result if photosynthetic biomass was scoured during the controlled flood (Figures 5 and 6).

Figure 2 is a graphic representation of pH, alkalinity, dissolved inorganic carbon concentrations, as redundant computations of pCO_2 concentrations. The close agreement of these calculations yields additional confidence in the pH measurements and provides additional dependent variable responses to the flood effects.

Figures 5 and 6 show the changing concentrations of O_2 and pH through a day-night period before and after the flood at four points in the Glen Canyon reach. The obvious result is that diel amplitude was reduced between March 23 and April 4, the period after the controlled flood [see also *McKinney et al.*, this volume].

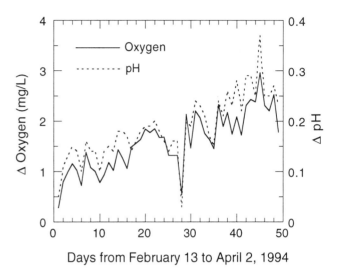

Figure 3. Daily ranges of O_2 concentration and pH at Lees Ferry from February 13 to April 2, 1994. The pH was scaled up by a factor of 10 to facilitate visual comparison of the congruence of these data. The increase in range through time is thought to be related to increasing day length and the variability is thought to be related to discharge and travel time.

There were two expectations: (1) that the amplitude of diel change would increase with distance downstream as far as Lees Ferry because water would have passed over more benthic plants with more carbon removed and more O_2 produced, and (2) the time at which maximum O_2 concentration and pH occurred would be essentially the same at all sampling points (the simplest case). Neither of these expectations was observed (see Table 1 and Discussion).

There are, thus, four components of these time traces that are notable for this preliminary interpretation: (1) the maximum value, (2) the minimum value, and (3) the times and (4) the places at which they occurred. Table 1 contains the values of these components during the period before the flood when the diel amplitude was greater. In the cases of both O_2 concentration and pH, the maxima occurred at Mile -6 at 1700 and 1730, respectively. This is 9.6 km upstream from where we expected the maximum at about the time we expected it. The peaks of the curves at Mile -3 and Mile 0 (Lees Ferry) were lower and later. The maximum values at Lees Ferry were delayed until near midnight, well after dark.

3.2. Flow Velocity

The velocity of flow through this reach has been measured on other occasions [*Graf,* 1995] and found to be 0.6 km/hr. This estimate was confirmed through an analysis

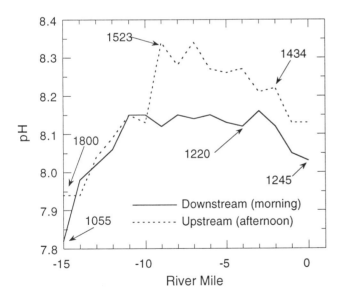

Figure 2. The distribution in space and through time of pH along the Glen Canyon reach from the dam (Mile -15) to Lees Ferry (Mile 0) on May 3, 1995. The times of day of some measurements are indicated by arrows. Note that pH was higher on the afternoon return trip in the lower nine miles than it was at the same places in the morning, but pH was virtually the same, morning and afternoon, in the upper 9.6 km.

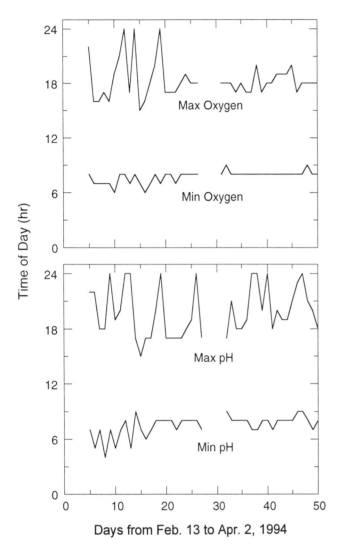

Figure 4. The times at which daily maxima and minima of O_2 concentration and pH occurred during the period shown in Figure 3.

of time-dependent conductivity changes that were conserved as the water mass flowed from the dam to Lees Ferry. The conductivity variation is associated with internal wave behavior in Lake Powell such that water from different layers of the halocline pass through the dam's penstocks and impose unique signatures on water masses moving through the reach. Records at the dam and at Lees Ferry were superimposed and adjusted by the time step until the regression of the Lees Ferry signal on the signal measured at the dam was maximized (C.J. Bowser et al., unpublished data, 1996. The time shift required to obtain the maximum regression coefficient is a powerful estimate of the travel time.

3.3. Expression of Metabolic Activity From These Data

We expected that the amplitude of the diel oscillation would increase progressively in a downstream direction. When we observed that it did not do so, we examined the variation in metabolism within the subreaches defined by the sampling design (in this case the subreaches are 4.8 km long). We compared metabolic processes involved in stream metabolism by reach with the methods of *Marzolf et al.* [1995].

3.4. Stream Metabolism

Calculation of stream metabolism from simple inspection of maximum and minimum values is, perhaps, too simplistic. Other methods exist that take into account the differences between two stations, the travel time of the water between the stations, and the exchange of gases in the river reach with the atmosphere [*Marzolf et al.,* 1995; *Young and Huryn,* 1997; and *Marzolf et al.,* 1998]. Preliminary estimates of gross primary production, community respiration, and their sum, an absolute value of total stream metabolism of the benthic community on March 22-23, 1996, are shown in the Table 2.

There are several issues to be resolved and missing data to be collected (beyond the scope of the present investigation) before accurate or precise calculation of carbon fixation from these data is possible, yet the preliminary differences in metabolism and its components among subreaches is striking.

3.5. Stoichiometry

The conversion of oxygen increase through time to the equivalent amount of carbon removed is straightforward. If the expressions

$$CO_2 + H_2O \rightarrow C_{organic} + O_2 \text{ (photosynthesis)}, \quad (2)$$

and

$$O_2 + C_{organic} \rightarrow CO_2 + H_2O \text{ (respiration)} \quad (3)$$

are stoichiometrically correct and O_2 and CO_2 are molar equivalents, then the gains and losses of O_2 can be expressed as gains and losses of carbon. The O_2 produced during photosynthetic carbon fixation plus the loss of O_2 associated with respiratory CO_2 gain at Lees Ferry is roughly quantifiable.

Figure 7 shows the time course of differences among pCO_2 values at the four stations in the reach. Maxima and

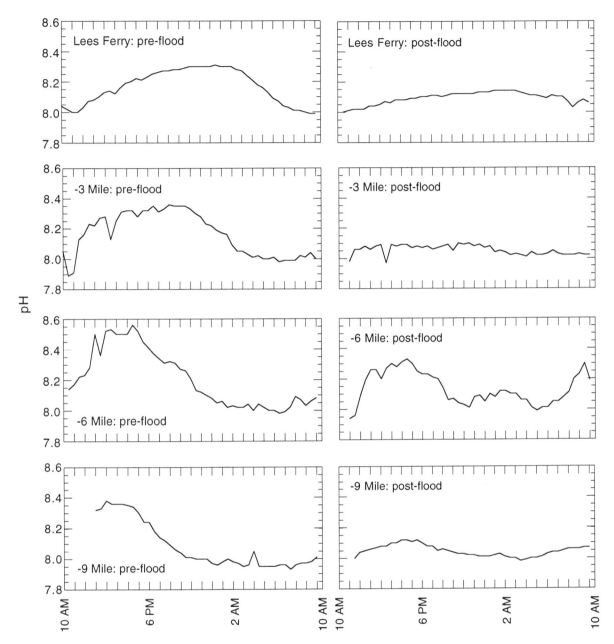

Figure 5. pH values measured at 0.5 hr. intervals at 4 sites for 24 hrs. before (left graph) and after (right graph) the 1996 controlled flood.

minima do not coincide in time until data sets were shifted for this figure (see below) to compensate for the travel time of the river. This observation is related to the fact that pH measurement at a point along a reach of a flowing river is not a measure of how pH has changed at that point since the last measurement but a measure of how pH has changed through the distance that the water has traveled from upstream since the last measurement; that is, it is a moving water mass that is changing. Note that all values are higher than atmospheric pCO_2 (c. 360 uatm) and that the slopes of increasing pCO_2 at night are similar at all sites (about 100 uatm/hr) and that the curves vary more during daylight. This pattern of variation suggests similarity related to respiration and that the difference among reaches resides in the photosynthetic process.

The time interval of the measurements establishes the distance over which the change is mediated depending on the velocity of flow; that is, water at Lees Ferry at dawn has

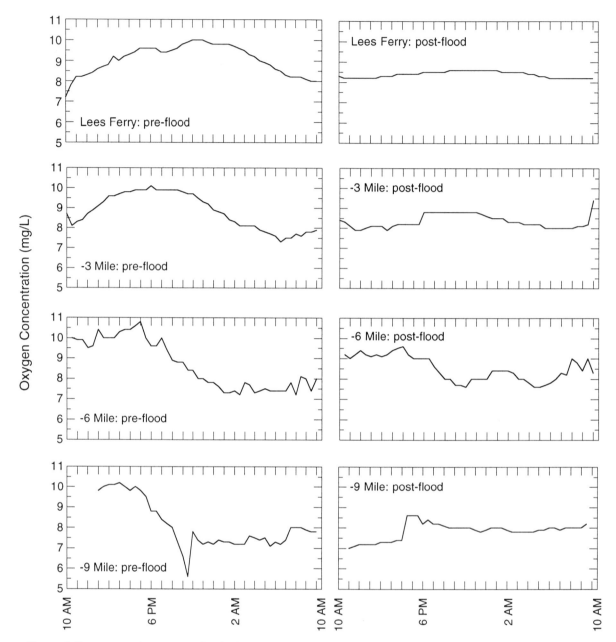

Figure 6. O$_2$ concentrations measured at 0.5 hr. intervals at 4 sites 24 hrs. before (left graph) and after (right graph) the 1996 controlled flood.

been flowing in the dark since it passed Mile -12 and at 0900 it has been flowing in the light from Mile -3 (travel time c. 3 hrs) after having flowed in the dark from Mile -15 for the previous 12 hrs, and so on.

The appropriate solution to this problem is to offset the diurnal curves measured three miles apart by the travel time of the water so that the plots can be compared. The curves from Lees Ferry were shifted forward in time by 9 hrs, the curves measured at Mile -3 were shifted forward 6 hrs, and

the curves at Mile -6 were shifted forward 3 hrs. This provides for the comparative plot of the shapes of the curves and of the maxima and minima in Figure 7.

4. DISCUSSION

At the time that this study was designed and planned, there was still some uncertainty that diurnal oscillation in

TABLE 1. Times and sites of O_2 and pH maxima and minima

Sample site	Maximum O_2 concentration (mg/L)	Time	Minimum O_2 concentration (mg/L)	Time	Maximum pH	Time	Minimum pH	Time
Mile 0	10.0	2200	7.80	1030	8.30	2230	8.02	1030
Mile -3	10.1	1800	8.10	1030	8.36	2005	7.91	1050
Mile -6	10.8	1700	7.20	0800	8.45	1730	8.03	0900
Mile -9	10.1	1400	7.10	0530	8.35	1600	7.93	0730

the chemical qualities of water, namely daily change in O_2 concentration and pH, were driven by photosynthesis and respiration in the benthic plant community. The inclusion of two low-flow periods, one before and one after, in the design of the controlled flood hydrograph presented the opportunity to perform a powerful test of that idea. If the expected scouring of benthic algae and macrophytes by the 1996 controlled flood diminished, the daily range of O_2 concentration and pH therefore would be greater certainty that the diel change was caused by photosynthesis and respiration. The daily range was diminished by about 80% (Figures 5 and 6) so we now conclude that net photosynthesis during the day is responsible for the rise in O_2 concentration and pH and the respiration causes the nocturnal decrease in O_2 concentration and pH.

Thus the first conclusion from this analysis, through the examination of several lines of evidence, is that the biological processes of photosynthesis and respiration cause the observed daily changes. That conclusion is preliminary but crucial. The second conclusion that the 80% decrease in the daily ranges of O_2 and CO_2 concentrations, measured at Lees Ferry provide evidence of biomass scouring. This confirms that the while the plant community suffered substantial loss, the photosynthetic signal imposed on the diurnal range of these gases was not obliterated. It is quite striking that the plant community could be reduced by 80% [see also *McKinney et al.*, this volume] but that the remainder would be so active. This means that the method is surprisingly sensitive and that there is a remnant of the plant community in place from which previous high rates of primary production will be recovered. *Brock et al.* [this volume] show by using chamber incubation techniques (more sensitive because exchange with the atmosphere is prevented) before and after the controlled flood that the rate of photosynthesis per unit area of *Cladophora* was higher after than before. They concluded that the flood scoured the physiologically less-active plant tissue, moribund plant tissues, and particulate organic matter associated with *Cladophora* thalli, thus leaving behind a *Cladophora*

population that was smaller but more efficient and a benthic community with reduced rates of microbial respiration.

The primary advantage of using this "open-stream" method is that the considerable variability within the reach is integrated into a single measurement, a measurement that can be made with tethered data loggers. The primary disadvantage is that the exchange of these gases with the atmosphere is not controllable, the rates are difficult to measure, the gradients of O_2 and CO_2 relative to the atmosphere are opposite in direction (CO_2 is lost and O_2 is gained). The differing interaction of CO_2 with water (as it enters into complex buffering reactions with carbonate solutions whereas O_2 does not) means that basic mechanisms of exchange may be importantly dissimilar.

Nevertheless, the limitation that this disadvantage imposes does not minimize the utility of the information for monitoring purposes. Consider that in the present example the flow and turbulence conditions during the periods when the measurements were made before and after the flood were designed to be the same, thus the results are fully comparable because the reaeration (diffusion) coefficients were the same. For comparative purposes the results are useful.

There were, however, unexpected results. First, the greatest daily ranges in O_2 concentration and pH were not measured at the downstream station at Lees Ferry (Figures 5 and 6). Second, the peaks of the daily ranges at the downstream station occurred long after sunset (Figures 5 and 6). These findings are further discussed below.

4.1. Greatest Daily Ranges Not Measured at the Downstream Station

The greatest range was expected at the most downstream site simply because the greater area of benthic photosynthetic community contributing to CO_2 removal or O_2 production should have been proportional to magnitude of the diurnal range. There are several issues involved. First, it is the photosynthetic process that drives the O_2 concen-

TABLE 2. Gross primary production, community respiration, and the absolute total metabolism at three sites in the Glen Canyon reach

Glen Canyon Reach	Gross Primary Production (kgO_2/d)	Community Respiration (kgO_2/d)	Total Metabolism, absolute (kgO_2/d)
Mile 0 to -3 Mile	5346	-2783	8129
-3 Mile to -6 Mile	17,260	-8073	25,332
-6 Mile to -9 Mile	88,563	-17,678	106,241

tration and pH up, but respiration (driving the curves downward) happens at the same time and includes not only respiration by the plants themselves but respiration by the microbial flora oxidizing dissolved and particulate organic substrates. Clearly an explanation will include the considerations that respiration increases with distance downstream and photosynthesis tends to decrease (perhaps, for example, in response to changing light conditions or the reduction of a limiting nutrient). This resulting effect was to dampen a large daily O_2 or pH peak imparted to the river upstream between Miles -9 to -6.

Water passing through Glen Canyon Dam has been at a depth of about 70 m in Lake Powell reservoir for an extended time. The time interval is not known precisely, but the scale is seasonal not diel. Light intensity at this depth is certainly low and would require sensitive instruments to measure. It is far lower than that required for photosynthesis, thus respiratory processes dominate and CO_2 concentration increases with time. The result is that pH is relatively low as the water enters the tailwater river. As the river degases with the flow downstream pH rises with little daily variation — the photosynthetic signal is masked (Figure 2). Additionally consider that a given water mass must move downstream for 16-km or so before the photosynthetic community can exchange gases with it for a full day.

The point is that, while the reach with the maximum daily rates of photosynthesis is expected to be upstream from Lees Ferry, it is not expected to be immediately downstream from the dam. In the present study the most productive reach is between Mile -6 and Mile -9 (Table 2). The recognition that water masses are being acted upon as they move downstream is important because it underlines that measurements made at a point are not the result of what

happened at that point but of what happened cumulatively upstream.

The light regime in this canyon river is clearly a critical variable. In addition to the seasonal variation, the length of the direct sunlit period in a day varies with the orientation of the canyon itself, with potentially shorter periods in N-S oriented reaches. The amount of light reaching the benthic plants on the bottom of the river is related further to channel geometry; wide, shallow reaches will have more wetted perimeter in brighter light than narrow, deep reaches for a given discharge.

4.2. The Peaks of the Daily Ranges at the Downstream Station Occurred Long After Sunset

The times of the peaks at the four sampling sites are presented in Table 1. The maxima occur earlier at the upstream stations and progressively later with distance downstream. This is fully consistent with the idea that respiration (that is not light dependent) becomes a more dominant process with distance downstream; thus, net photosynthesis is reduced. The peaks of pH and O_2 concentration do not grow in amplitude as the water mass moves downstream from the most productive reach. They are, quite simply, diminished and they arrive later.

The implications of this for monitoring are encouraging. The daily range at the downstream station will serve as a comparative measure of the production rate as it varies through time and the depression of the peak and the delay of its arrival may serve as an index of the relative contribution of respiration to stream metabolism.

Acknowledgments. The authors thank David Wegner and the Glen Canyon Environmental Studies of the Bureau of Reclamation for financial support for this work and the Glen Canyon National Recreation Area helped with logistic field support. The field work was possible only because of the willing 36-hr help of eleven volunteers: Alex Almario, Andy Ayers, Kevin Berghoff, Mike Dai, Susan Dodson, Gregg Fisk, Marilyn Flynn, Robert Forrest, Bob Gauger, Christine Miller, and Susan Thompson-McHugh. Norman Henderson and Allan "Blu" Pickard provided logistic help. Ron Antweiler and Howard Taylor performed alkalinity and major ion chemistry analyses. Jo Dyer and Lex Newcomb provided help with figures. Erich R. Marzolf provided the calculations on stream metabolism and made comments on a early draft of the manuscript. George W. Saunders also commented on that early draft. Colbert Cushing and Richard Valdez provided helpful reviews and suggestions.

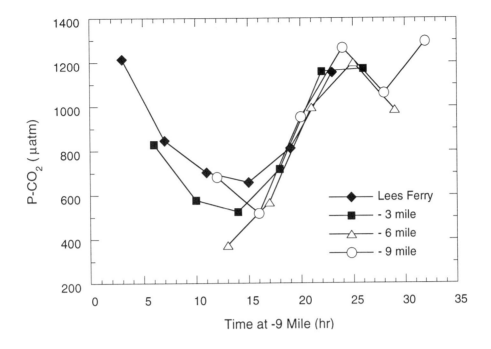

Figure 7. Daily variation in pCO$_2$ at the 4 locations. The time distribution of values at Lees Ferry, Mile -3, and Mile -6 is shifted by 9, 6, and 3 hrs, respectively (the travel time of water from Miles -9 , - 6 , and -3 to Lees Ferry). The result is that data are plotted as if the water mass at all locations is changing together. The change from one station to the next can be visualized by moving vertically along a time axis. Note that the changes through time during darkness are similar at all stations, while changes during daylight vary widely with location.

REFERENCES

Blinn, D.W., and G.A. Cole, Algae and invertebrate biota in the Colorado River: comparison of pre- and post-dam conditions, in *Colorado River ecology and dam management*, pp. 102-123. National Academy Press, Washington, DC, 1991.

Beyers, R.J., and H.T. Odum, The use of carbon dioxide to construct pH curves for the measurement of productivity, *Limnol. Ocean.*, 4, 499-502, 1959.

Butler, J.N, *Carbon dioxide equilibria and their applications*, Lewis Publishers, Chelsea, Michigan, 259 pp., 1991.

Graf, J.B., Measured and predicted velocity and longitudinal dispersion at steady and unsteady flow, Colorado River, Glen Canyon Dam to Lake Mead, *Water Res. Bull.*, 31, 265-281, 1995.

Marzolf, E.R., P.J. Mulholland, and A.D. Steinman, Improvements to the diurnal upstream-downstream dissolved oxygen change technique for determining whole-stream metabolism in small streams, *Can. J. Fish. Aqu. Sci.*, 51(7), 1591-1599, 1995.

Marzolf, E.R., P.J. Mulholland, and A.D. Steinman, Response to comment by Young and Heuryn, *Can. J. Fish. Aqu. Sci.*, 54, 1998.

Odum, H.T., Primary production in flowing waters, *Limno. Ocean.*, 1, 102- 117, 1956.

Young, R.G., and A.D. Huryn, Comment: Further improvements to the diurnal upstream-downstream dissolved oxygen change technique for determining whole-stream metabolism in small streams, *Can. J. Fish. Aquat. Sci.*, 1997.

G. Richard Marzolf, U.S. Geological Survey, 432 National Center, 12201 Sunrise Valley Dr., Reston, VA 20192; email: rmarzolf@usgs.gov

Carl J. Bowser, Dept. of Geology and Geophysics, 1215 Dayton Street, University of Wisconsin, Madison, WI 53706

Robert Hart, U.S. Geological Survey, 2255 N. Gemini Drive, Flagstaff, AZ 86001

Doyle W. Stephens, U.S. Geological Survey, 1745 West, 1700 South, Salt Lake City, UT 84104

William S. Vernieu, Bureau of Reclamation, 2255 N. Gemini Drive Flagstaff, AZ 86001

Periphyton Metabolism: A Chamber Approach

James T. Brock, Todd V. Royer, Eric B. Snyder, and Steven A. Thomas

Idaho State University, Pocatello, Idaho

In lotic ecosystems, the metabolism of periphyton is influenced strongly by natural and anthropogenic disturbances such as floods. Using recirculating metabolism chambers, we measured the metabolic activity of the *Cladophora glomerata*-dominated periphyton community in the Glen Canyon Dam tailwater, in relation to the 1996 controlled flood. Because scouring removes senescent plant material and detritus from periphyton, we hypothesized that productivity rates and the gross productivity/respiration (P/R) ratio of the periphyton community would be greater after the flood. Gross and net primary production (as chlorophyll-*a*) increased significantly after the flood and an approximately 2-fold increase was observed in net daily metabolism. Mean P/R ratio increased significantly from 1.3 in the pre-flood community to 2.6 in the post-flood community. Following the flood, periphyton on the rocks exhibited increased photosynthetic efficiency relative to measurements made before the flood. Given the importance of primary producers in desert rivers, such changes have implications for ecologically sound management of the Colorado and other rivers.

1. INTRODUCTION

Throughout the western United States, large rivers have been impounded for numerous reasons, including water storage, hydropower production, and flood control [*Benke*, 1990]. The total amount and rate of water discharged from the reservoirs are regulated by demands for electricity and irrigation, flood control, and the physical limitations of each dam. Glen Canyon Dam on the Colorado River was completed in 1963, and since that time, discharge from the dam has been regulated on a daily basis to meet demands for electricity during hours of maximum consumption [*Johnson and Carothers*, 1987]. A primary result of Glen Canyon Dam has been elimination of the annual flooding of the Colorado River which occurred each spring following snowmelt before the dam.

The structure and function of lotic periphyton communities are influenced strongly by natural and anthropogenic disturbances, including floods [*Peterson et al.*, 1994; *Fisher et al.*, 1982]. In desert streams, the response of periphyton to disturbance is particularly important because autotrophic production is often the primary source of energy for higher trophic levels [*Minshall*, 1978]. Additionally, *Fisher et al.* [1982] demonstrated that nutrient dynamics were associated with the recovery of periphyton metabolism following a flood in Sycamore Creek, Arizona. In autotrophic rivers, such as the Colorado River below Glen Canyon Dam, activities that influence the primary producers are likely to be more influential on ecosystem structure and function than in rivers that rely heavily on allochthonous inputs. Clearly, an elucidation of the effects of flood disturbance on periphyton metabolism is critical for integrating large-river ecology and resource management [*Johnson et al.*, 1995].

During the spring of 1996, an experimental flood was conducted to evaluate the effects of flow manipulation on channel morphology in the Grand Canyon [*Schmidt et al.*, this volume]. The 1996 controlled flood provided an opportunity to examine the metabolism of the periphyton community in the Glen Canyon Dam tailwater before and after a flood. Since completion of Glen Canyon Dam, *Cladophora glomerata* has become the dominant primary producer in the 25-km reach from the dam to Lees Ferry, Arizona [*Blinn and Cole,* 1991]. We measured the metabolism of the attached *Cladophora* communities to examine the effect of the flood on primary productivity in the Glen Canyon Dam tailwater. Specifically, we expected that net primary productivity and the gross productivity/respiration (P/R) ratio of the periphyton community would be greater after the flood due to scouring of the heterotrophic portion of the community (i.e., senescent filaments of *C. glomerata* and the associated decomposers).

2. METHODS

2.1. Description of Study Site

Glen Canyon extends 26 km downstream from the dam that forms Lake Powell. A description of the Colorado River in Glen Canyon along with reservoir release data collected during the study period (March - April 1996) are found in [*Webb et al.,* this volume; *Schmidt et al.,* this volume]. Unlike the high-gradient whitewater rapids of Marble and Grand Canyons, the Colorado River's channel in Glen Canyon consists of a series of sand-bottomed pools and runs with minimal surface wave disturbance. On the margins of the runs are armored cobble bars that cover about 16% of the total littoral habitat [*Angradi and Kubly,* 1993]. The attached filamentous green algae *Cladophora glomerata* forms abundant colonies on submerged portions of the cobble bars. The *Cladophora* community in Glen Canyon supports a diverse community of epiphytic diatoms [*Usher and Blinn,* 1990]. Rocks with attached periphyton were transported from two cobble bars to Lees Ferry for incubation; one bar was located at -4.1 Mile upstream from Lees Ferry and the second at -14.0 Mile, which is approximately 3 km downstream from the dam. The location of the cobble bars coincides with sampling stations used by Arizona Game and Fish Department [*McKinney et al.,* this volume]. Metabolism measurements were made at the boat launch at Lees Ferry, Arizona (river mile 0.0).

The physical characteristics of river water in Glen Canyon are dominated by the hypolimnetic release from the dam, which is cool and relatively clear [*Webb et al.,* this volume]. With respect to nutrients, the reach from the dam to Lees Ferry can be described as phosphorus poor (<10 µg/L) with ample inorganic nitrogen (>300 µg/L) [*Angradi and Kubly,* 1993].

2.2. Sample Collection and Metabolism Measurements

We used recirculating metabolism chambers to obtain estimates of periphyton metabolism [*Bott et al.,* 1997]. The chambers enabled us to directly measure net productivity and respiration of attached periphyton communities based on oxygen flux without the confounding effects of reaeration. The chamber measurements helped to evaluate functional attributes of specific components of the river ecosystem whereas concurrent measurements of open-channel dissolved oxygen (DO) and carbon dioxide flux measured whole ecosystem metabolism [*Marzolf et al.,* this volume]. Use of metabolism chambers before and after the flood made it possible for us to compare flood effects on attached *Cladophora* communities.

The metabolism chambers used in this study (Red Deer Chamber, Aliquot, Boise, Idaho) were constructed of clear plexiglass and held approximately 15 L of water. The chambers were water tight and unidirectional flow of water within the chambers was maintained by an in-line pump. The conditions within the chambers were maintained near ambient river conditions by: (1) selective submersion of chambers during incubation trials, (2) constant unidirectional flow of water, (3) the large volume of water relative to the amount of biological material, (4) periodic refreshing of the water inside the chamber with ambient river water, and (5) use of greenhouse shading placed over the chambers. In addition, colonized rocks were oriented to the direction of flow as closely as possible to that from which they had been collected.

Seven metabolism chambers were used, 6 of which contained rocks. The seventh chamber contained only river water and was the control to correct for metabolic activity in the water column. The metabolic values reported here represent the activity of the attached *Cladophora* community. All incubations were performed at Lees Ferry using ambient water from the Colorado River. All rocks were obtained from areas of the channel that are continuously inundated [*Angradi and Kubly,* 1993] and were stored in ambient river water during transportation to Lees Ferry. The pre-flood rocks were collected on 26 March 1996 and were held in plastic bags kept at river temperature until metabolic measurements were made over the ensuing several days. A set of post-flood rocks was collected from the river on 5 April 1996 and used for post-flood measurements.

Once the rocks had been placed inside, chambers were filled with river water, closed, purged of air bubbles, and lowered to a depth of approximately 10 cm below the surface of the water. Direct measurements of photosynthetically available solar radiation (PAR) indicated that about half of the incident solar irradiance was attenuated by a layer of greenhouse shading; this was similar to the light conditions at the depth from which the rocks were collected. Metabolism was measured as the change in the concentration of dissolved oxygen in the water inside the chambers using techniques described by *Bott et al.* [1978, 1997]. DO concentrations were measured continuously with either Royce (model 900) or Orion (model 840) DO probes, calibrated by means of a water-vapor saturated chamber. DO measurements were made every minute, averaged and recorded every 5 min with a Campbell Scientific data logger (model CR10). Incubations typically lasted >12 hrs and encompassed both light and naturally dark periods. Water temperature within the chambers and in the river adjacent to the chambers was monitored using thermistors. The rate of primary productivity is directly related to the amount of PAR. The data logger described above also continuously recorded PAR (LI-COR quantum sensors) at the surface of the river and at the depth of the chambers.

2.3. Sampling of Periphyton Biomass

Rocks used in the chambers were transported on wet ice to the Stream Ecology Center, Idaho State University, for determinations of periphyton chlorophyll-*a* and biomass after incubation. In the laboratory, three samples of periphyton were collected per rock (sample area = 3.14 cm^2) and filtered onto a Whatman GF/F filter [*Robinson and Minshall,* 1986]. For each rock, one sample was collected from an area of low, medium, and high periphyton density, as estimated visually. Concentration of chlorophyll-*a* was determined spectrophotometrically (Gilford model 2200). After extraction in 100% reagent-grade methanol for 24 hrs, a 3-ml subsample was taken for chlorophyll analysis which included correction for phaeophytin *a* [*Marker,* 1972; *Holm-Hansen and Riemann,* 1978; *Marker et al.,* 1980]. Periphyton biomass, as ash-free dry mass (AFDM), was determined gravimetrically using the filter and remaining 7 ml of extractant. The mass lost on combustion at 500°C for 2 hrs constituted AFDM [*Greeson et al.,* 1977].

2.4. Data Analysis

The change in DO concentration through time inside the chambers was used to estimate several metabolic variables, including gross primary productivity (GPP), community respiration (CR), net primary productivity (NPP), net daily metabolism (NDM), and gross productivity/respiration (P/R) ratio. Metabolic variables were estimated for the

periphyton community on each rock (chamber). Time constraints and a malfunctioning DO probe resulted in differing numbers of replicates for the pre- (n=4) and post-flood (n=8) trials. We observed no major differences in periphyton metabolism between rocks collected from -4.1 and -14.0 Mile, therefore we combined results from the two locations in order to provide a sample size sufficient for statistical analysis. A t-test was used to evaluate differences between pre- and post-flood measurements. Data were log(x+1) transformed for the t-test [*Zar,* 1984], except those used to express net daily metabolism, which contained both positive and negative values.

3. RESULTS AND DISCUSSION

For the rocks used in the chamber incubations, chlorophyll-*a* standing crop was significantly lower on post-flood rocks than that measured on pre-flood rocks, although AFDM was not different (Table 1; see *Blinn et al.* [this volume] for additional data on standing crop of primary producers). Mean gross primary productivity (GPP) was slightly greater in the post-flood community than in the pre-flood community (Table 1, Figure 1), although the difference was not significant (p=0.496). NPP increased from 12.4 to 16.6 g O$_2$ m^{-2} day^{-1} pre to post flood, respectively (Table 1, Figure 1). Community respiration (CR) decreased from 14 g O$_2$ m^{-2} day^{-1} prior to the flood to 10 g O$_2$ m^{-2} day^{-1} following the flood (p=0.07). The decrease in CR without a concomitant decrease in NPP suggests that a portion of the heterotrophic component of the periphyton community was removed by the flood.

Angradi and Kubly [1993] measured GPP in the Glen Canyon Dam tailwater using partial-day incubations and found values that ranged from 4 to 12 g O$_2$ m^{-2} hr^{-1} for permanently inundated rocks, such as those used in the present study. From the hourly rates, they predicted daily rates as high as approximately 20 g O$_2$ m^{-2} day^{-1}, under conditions of maximal PAR and periphyton biomass. We observed a mean GPP rate of 21.1 g O$_2$ m^{-2} day^{-1} following the flood, similar to the maximum value predicted by *Angradi and Kubly* [1993]. The Glen Canyon Dam tailwater displays GPP values that are extremely high, relative to values compiled from other North American streams [*Bott et al.,* 1985]. For example, in September of 1991 chamber measurements indicated GPP = 33.9 g O$_2$ m^{-2} day^{-1}, CR = 11.6 g O$_2$ m^{-2} day^{-1}, and the P/R ratio = 2.9 for periphyton at Lees Ferry (J.T. Brock, unpublished data, 1991). The GPP value of 33.9 g O$_2$ m^{-2} day^{-1} is the second largest reported from North America [*Bott et al.,* 1985]; the largest being 48.0 g O$_2$ m^{-2} day^{-1} in the Blue River, Oklahoma [*Duffer and Doris,* 1966]. Clearly, conditions in

TABLE 1. Summary table of algal abundance and the metabolic parameters measured from pre- and post-flood periphyton communities in the Glen Canyon Dam tailwater. Pre-flood measurements made 26-28 March, post-flood measurements made 4-9 April, 1996.

	Pre-flood (n=4)		Pre-flood (n=8)				
	Mean	SE	Mean	SE	t-statistic	df	p-value
Algal abundance							
Chlorophyll a (g m-2)[a]	0.45	0.05	0.37	0.05	2.524	11	0.03
AFDM (g m-2)[a]	105	32	95	19	0.409	11	0.69
Metabolism parameters							
NPP (g O2 m-2 d-1)[a]	12.4	5.5	16.6	4.2	-1.570	11	0.15
GPP (g O2 m-2 d-1)[a]	18.9	5.7	21.1	5.2	-0.704	11	0.49
CR24 (g O2 m-2 d-1)[a]	14.3	3.3	9.8	4.7	1.696	11	0.13
NDM (g O2 m-2 d-1)	4.6	5.8	11.2	4.4	-2.146	11	0.06
P/R	1.3	0.4	2.6	1.1	-2.517	11	0.03
NPP (g O2 g AFDM-1 d-1)[b]	0.14	0.09	0.18	0.04	-1.157	11	0.27
GPP (g O2 g AFDM-1 d-1)[b]	0.21	0.12	0.23	0.07	-0.507	11	0.62
CR (g O2 g AFDM-1 d-1)[b]	0.15	0.06	0.11	0.08	1.018	11	0.33
NPP (g O2 g Chl a-1 d-1)[b]	27.1	11.9	44.8	8.6	-2.835	11	0.02
GPP (g O2 g Chl a-1 d-1)[b]	41.5	11.6	57.1	11.5	-2.206	11	0.05

[a]Values ln(x+1) transformed prior to statistical analysis.
[b]Values square root(arc sin x) transformed prior to statistical analysis.

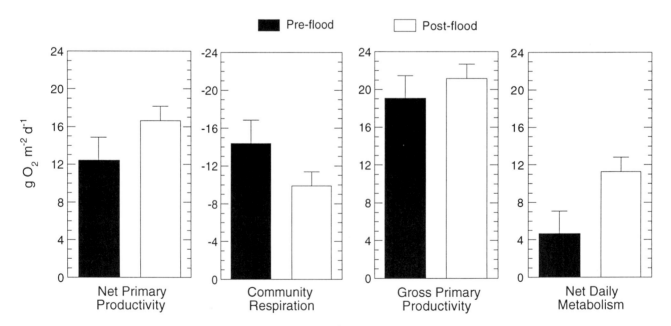

Figure 1. Metabolic variables measured from the periphyton community in the Glen Canyon Dam tailwater immediately before (26-28 March) and immediately after (6-8 April) the 1996 controlled flood. Error bars represent one standard error from the mean.

the Glen Canyon Dam tailwater support a highly productive autotrophic community.

Net daily metabolism (NDM) represents the amount of metabolic activity, in excess of respiratory requirements, that is available for conversion to biomass. We observed an approximately 2-fold increase (p=0.06) in the NDM of the post-flood periphyton community, relative to the pre-flood community (Table 1, Figure 1). These data suggest that following the flood there was a greater potential for conversion of primary productivity into algal biomass, as opposed to respiratory loss. Such an increase in available algal resource could result in increased secondary production, because algal resources are an important determinant in the distribution and abundance of benthic invertebrates in the Glen Canyon Dam tailwater [*Angradi,* 1994; *Shannon et al.,* 1994]. Although our data suggest the potential for increased secondary production associated with spates of a magnitude similar to the 1996 controlled flood, this hypothesis remains untested in the Glen Canyon tailwater.

Productivity to respiration ratios historically have been used to determine the relative autotrophic or heterotrophic state of a given stream or river and have sometimes been misinterpreted as representing the relative input of allochthonous vs. autochthonous energy sources [*Cummins,* 1974; *Rosenfeld and Mackay,* 1987; *Meyer,* 1989]. Due to the paucity of allochthonous organic matter in the Glen Canyon tailwater [*Angradi,* 1994], an increase in the P/R ratio likely indicates greater efficiency within the primary producers. Following the controlled flood, the P/R ratio of periphyton on rocks in the Glen Canyon Dam tailwater increased from 1.3 to 2.6 (p=0.03, Figure 2). In general, our results are in agreement with the patterns reported by *Fisher et al.* [1982], who described a rapid increase in P/R ratio following a flood in Sycamore Creek, Arizona, with the post-flood P/R ratio exceeding unity. However, we measured respiration of only the periphyton community, not of the entire ecosystem, as did *Fisher et al.* [1982]. Therefore, the P/R ratios presented here differ from the P/R ratio of the entire Glen Canyon tailwater ecosystem, which would include respiration associated with detrital material and the production and respiration of other primary producers such as aquatic macrophytes [*Marzolf et al.,* this volume].

The above measurements describe metabolism on an areal basis (i.e., per square meter of rock surface). The efficiency of the primary producers to convert solar radiation to chemical energy also can be expressed as productivity per unit mass of chlorophyll-*a*. The post-flood community displayed significantly (p=0.02) more productivity per gram of chlorophyll-*a* than did the pre-flood

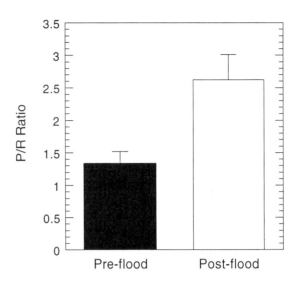

Figure 2. The ratio of gross productivity to respiration measured from the periphyton community in the Glen Canyon Dam tailwater immediately before (26-28 March) and after (6-8 April) the 1996 controlled flood. Error bars represent one standard error from the mean.

community, with an increase in NPP of nearly 20 g O_2/g chl-*a* following the flood (Figure 3). Net primary productivity, however, is dependent on the amount of respiration, and the results in Figure 3 may be due also to the decline in CR. We examined the photosynthetic efficiency of the periphyton communities before and after the flood by plotting biomass-specific GPP against PAR for one pre- and one post-flood rock (Figure 4). Although we cannot generalize for the river as a whole, the periphyton we examined after the flood was more efficient at producing energy with a given amount of solar radiation than was periphyton prior to the flood.

We postulate that the daily power peaking activity of Glen Canyon Dam maintained a productively vigorous periphyton community. Although the dam has been operated under constrained daily fluctuations since 1990, benthic communities measured in this study typically had been exposed to a daily variation in discharge of up to 40%. The flows recorded during winter 1996 varied between 300 and 500 m^3/s on a daily basis during the months prior to the controlled flood. Power peaking causes the daily inundation and flushing of a portion of the periphyton community, thereby removing senescent algae and potentially increasing rates of nutrient uptake, similar to the response observed in a Sonoran desert stream following a simulated spate [*Peterson et al.,* 1994]. High productivity in systems that are regularly and predictably disturbed are common (e.g., Florida Everglades, estuaries, and intertidal zones)

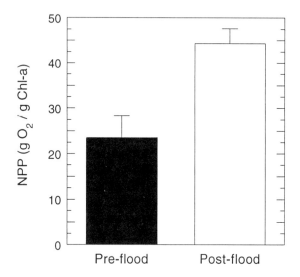

Figure 3. Net primary productivity per gram of chlorophyll-*a* measured from the periphyton community in the Glen Canyon Dam tailwater immediately before (26-28 March) and after (6-8 April) the 1996 controlled flood. Error bars represent one standard error from the mean.

due to the mechanism referred to by *Odum* [1971] as Pulse Stability. This is in contrast to the Kootenai River below Libby Dam (Montana, USA) which is a similar system (i.e., discharge range 142-566 m^3/s) but does not regularly experience daily power peaking [*Perry and Perry*, 1991; *Snyder and Minshall*, 1996]. Although productivity was not measured below Libby Dam, the periphyton mat was well developed and contained a large fraction of senescent material (E.B. Snyder, personal observation, 1996).

In conclusion, we found that the 1996 controlled flood did not deplete the algal resource due to scouring and bedload movement. This artificial spate did not shift ecosystem metabolism towards heterotrophy, as *Uehlinger and Naegeli* [1998] found for rivers subject to floods of sufficient magnitude to promote bed movement. *Marzolf et al.* [this volume] collected whole-stream measurements of daily ranges in dissolved O_2 and CO_2 pre- and post-flood and found that photosynthesis remained quite active following the flood. Our results from the chamber incubations indicate that productivity of periphyton was more efficient following the flood. Whether this increased productivity was transferred to higher trophic levels is unknown, but greater secondary production was possible, given the results of this study. The ecologically sound management of impounded rivers requires knowledge of biological responses to fluctuating flow regimes. In the Glen Canyon Dam tailwater, autochthonous production forms the base of the foodweb [*Angradi*, 1994] and

supports an economically important sport fishery. Based on our study of periphyton metabolism in relation to the 1996 controlled flood, future use of controlled floods in the Colorado River is unlikely to decrease productivity of the ecosystem. It is unknown if this result applies to other impounded rivers; further study is warranted before controlled flooding becomes a routine technique for the management of large rivers.

Acknowledgments. We thank Dick Marzolf (U.S. Geological Survey) for reviewing an earlier draft of the manuscript and for encouraging us to apply our interest in metabolism of the Glen Canyon Dam tailwater to the 1996 controlled flood. Rapid Creek Research, Inc. provided equipment and transportation for the study. The Stream Ecology Center at Idaho State University loaned metabolism chambers and related equipment. We thank G. Wayne Minshall for encouragement and suggestions. The helpful comments of two anonymous reviewers substantially improved this paper. Cost of travel and laboratory analyses was partly funded by the U.S. Geological Survey. Ted McKinney and Andrew Ayers (Arizona Game and Fish Department) kindly collected sample material, and Dianne Stanitski-Martin provided data on solar radiation.

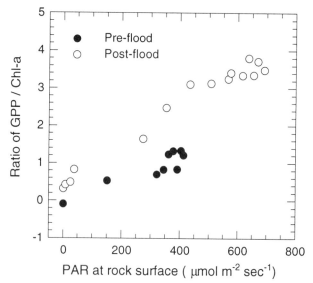

Figure 4. Photosynthetic efficiency at varying intensities of light (PAR) for a single rock (periphyton community) from immediately before (27 March) and a single rock immediately after (7 April) the 1996 controlled flood.

REFERENCES

Angradi, T.R, Trophic linkages in the lower Colorado River: multiple stable isotope evidence, *N. Amer. Benth. Soc.*, 13, 479-495, 1994.

Angradi, T.R., and D.M. Kubly, Effects of atmospheric exposure on chlorophyll-*a*, biomass and productivity of the epilithon of a tailwater river, *Regul. Rivers: Res. Manage.*, 8, 345-358, 1993.

Benke, A.C., A perspective on America's vanishing streams, *N. Amer. Benth. Soc.,* 9, 77-88, 1990.

Blinn, D.W., and G.A. Cole, Algae and invertebrate biota in the Colorado River: comparison of pre- and post-dam conditions, in *Colorado River and Dam Management*, ed. G.R. Marzolf, Natl. Acad. Press, Washington, D.C., 1991.

Bott, T.L., J.T. Brock, C.E. Cushing, S.V. Gregory, D. King, and R.C. Petersen, A comparison of methods for measuring primary productivity and community respiration in streams, *Hydrobiologia*, 60, 3-12, 1978.

Bott, T.L., J.T. Brock, C.S. Dunn, R.J. Naiman, R.W. Ovink, and R.C. Petersen, Benthic community metabolism in four temperate stream systems: An inter-biome comparison and evaluation of the river continuum concept, *Hydrobiologia*, 123, 3-45, 1985.

Bott, T.L., J.T. Brock, A. Baattrup-Pedersen, P.A. Chambers, W.K. Dodds, K.T. Himbeault, J.R. Lawrence, D. Planas, E. Snyder, and G.M. Wolfaardt, An evaluation of techniques for measuring periphyton metabolism in chambers, *Can. Fish. Aquatic Sci.*, 54, 715-725, 1997.

Cummins, K.W., Structure and function of stream ecosystems, *Bioscience*, 24, 631-641, 1974.

Duffer, W., and T.C. Dorris, Primary productivity in a southern Great Plains stream, *Limn. Ocean.*, 11, 143-151, 1966.

Greeson, P.E., T.A. Ehlke, G.A. Irwin, B.W. Irwin, B.W. Lium, and K.V. Slack, Methods for collection and analysis of aquatic biological and microbiological samples, *Techniques of Water-Resources Investigations of the United States Geological Survey*, Chapter A4, Book 5, U.S. Geological Survey, Washington D.C., 1977.

Fisher, S.G., L.J. Gray, N.B. Grimm, and D.E. Busch, Temporal succession in a desert stream ecosystem following flash flooding, *Ecol. Mono.*, 52, 93-110, 1982.

Holm-Hansen, O., and B. Riemann, Chlorophyll-*a* determination: Improvements in methodology, *Oikos*, 30, 438-447, 1978.

Johnson, B.L., W.B. Richardson, and T.J. Naimo, Past, present, and future concepts in large river ecology, *Bioscience*, 45, 134-141, 1995.

Johnson, R.R., and S.W. Carothers, External threats: the dilemma of resource management on the Colorado River in Grand Canyon National Park, USA, *Environ. Manage.*, 11, 99-107, 1987.

Marker, A.F.H., E.A. Nusch, H. Rai, and B. Riemann, The measurement of photosynthetic pigments in freshwaters and standardization of methods: Conclusions and recommendations, *Arch. Hydrobiologia*, 14, 91-106, 1980.

Marker, A.F.H., The use of acetone and methanol in the estimation of chlorophyll in the presence of phaeophytin, *Freshwater Bio.*, 2, 361-385, 1972.

Meyer, J.L., Can P/R be used to assess the food base of stream ecosystems? A comment on Rosenfeld and Mackay (1987), *Oikos*, 54, 119-121, 1989.

Minshall, G.W., Autotrophy in stream ecosystems, *Bioscience*, 28, 767-771, 1978.

Odum, E.P., *Fundamentals of ecology*, 3rd ed., Saunders, Philadelphia, PA 1971.

Perry, S.A., and W.B. Perry, Organic carbon dynamics in two regulated rivers in northwestern Montana, USA, *Hydrobiologia*, 218, 193-203, 1991.

Peterson, C.G., A.C. Weibel, N.C. Grimm, and S.G. Fisher, Mechanisms of benthic algal recovery following spates: comparison of simulated and natural events, *Oecologia*, 98, 280-290, 1994.

Robinson, C.T., and G.W. Minshall, Effect of disturbance frequency on stream benthic community structure in relation to canopy cover and season, *N. Amer. Benth. Soc.*, 5, 237-248, 1986.

Shannon, J.P., D.W. Blinn, L.E. Stevens, Trophic interactions and benthic animal community structure in the Colorado River, Arizona, U.S.A., *Freshwater Bio.*, 31, 213-220, 1994.

Snyder, E.B., and G.W. Minshall, *Ecosystem metabolism and nutrient dynamics in the Kootenai River in relation to impoundment and flow enhancement for fisheries management*, 102 pp., Dept. Biol. Sci., Idaho State Univ., Pocatello, ID 1996.

Rosenfeld, J.S., and R.J. Mackay, Assessing the food base of stream ecosystems: alternatives to the P/R ratio, *Oikos*, 50, 141-147, 1987.

Uehlinger, U., and M.W. Naegeli, Ecosystem metabolism, disturbance, and stability in a prealpine gravel bed river, *N. Amer. Benth. Soc.,* 17:165-178, 1998.

Usher, D.H. and D.W. Blinn, Influence of various exposure periods on the biomass and chlorophyll-*a* content of *Cladophora glomerata* (Chlorophyta), *Phyco.*, 26, 244-249, 1990.

Zar, J.H., *Biostatistical analysis*, 2nd ed., 718 pp., Prentice-Hall, Englewood Cliffs, NJ, 1984.

James T. Brock, Todd V. Royer, Eric B. Snyder, and Steven A. Thomas, Department of Biological Sciences, Idaho State University, Pocatello, ID 83209; email: jtbrock@micron.net

Mineralization of Riparian Vegetation Buried by the 1996 Controlled Flood

Roderic A. Parnell, Jr., Jeffrey B. Bennett

Geology Department, Northern Arizona University, Flagstaff, Arizona

Lawrence E. Stevens

Grand Canyon Monitoring and Research Center, Flagstaff, Arizona

Geochemical composition of beach groundwater, surface backwaters in return-current channels, and mainstem waters were measured before, during, and after the 1996 controlled flood in Grand Canyon. Similar, distinctive geochemical environments in return-channel sediments, back beach, and beach front positions within a reattachment-bar complex occurred at all study bars. The flood buried autochthonous vegetation at the study sites, resulting in significant increases in dissolved C, N, and P in beach groundwater and in surface backwaters. The trends observed in individual wells and site averages include measurable increases in dissolved ammonium, non-purgeable organic carbon (NPOC), and orthophosphate. Large standard errors resulted from large variation in concentrations between wells produced by their differing geomorphic settings. When comparing changes between pre- and post-flood samples from the same site, only DO, NPOC, and dissolved orthophosphate were significant. The largest pre- to post-flood changes occurred on a seasonal time scale. The magnitudes of seasonal changes following flooding were large compared to natural seasonal changes in nutrient concentrations previously observed. The nutrient composition changes occurring in groundwater and backwaters were not directly reflected in mainstem water compositions, although increases in mainstem primary productivity during high flows following the flood did correlate with periods of high nutrient concentrations in beach groundwater.

1. INTRODUCTION

The 1996 controlled flood in Grand Canyon provided a unique opportunity to observe changes in nutrient concentrations in bar groundwater and interconnected surface

waters resulting from a flood that buried autochthonous vegetation. Previous workers [*Bennett and Parnell*, 1995; *Parnell et al.*, 1996; *Bennett*, 1997] demonstrated that distinct geochemical differences exist between bar groundwater, backwater surface waters, and mainstem waters of the Colorado River in Grand Canyon. This study determined whether a controlled flood, which would bury large volumes of autochthonous organic matter below the average position of the water table, would initiate rapid rates of aerobic then anaerobic decay and release dissolved orthophosphate, ammonium, and organic carbon. The significance of this study was to determine if a flood that did not scour bar vegetation or backwaters, like the 1996 controlled flood, would produce a temporary increase in available N, P, and C, with the potential to positively affect terrestrial and aquatic productivity. We measured concentrations of dissolved oxygen, nitrate, nitrite, ammonium, orthophosphate, and organic carbon; pH; temperature; and conductivity. We used these parameters to monitor changing biogeochemical conditions within bars, under backwaters, within backwaters and mainstem surface waters. We hypothesized that this test flood would produce unique geochemical conditions because the size and duration of the event should maximize burial and minimize scour of autochthonous organic matter.

Because of the relative abundance of reattachment bars and the importance of return-current channels (RCCs) as backwater habitats [*Stevens et al.*, 1995, 1997], reattachment bar and eddy return-current channel complexes, as identified by *Schmidt and Graf* [1990], were selected for detailed study. Reattachment bars form along channel margins at the downstream terminus of eddies occurring below channel constrictions such as side canyon debris fans [*Schmidt and Graf*, 1990]. During high flows, eddy currents increase in size and migrate shoreward, reworking deposits through both erosion and redeposition [*Hazel et al.*, this volume]. Erosion along the canyon-wall edge of the bar forms a return-current channel. Volumetrically, reattachment bars are an important surficial deposit along the mainstem. Furthermore, during normal flows (213-708 m³/s), return-current channels provide slackwater environments with daytime surface water temperatures higher than mainstem temperatures [*Parnell et al.*, 1997a; *Stevens et al.*, 1997]. This study links to those of *Kearsley and Ayers* [this volume] and *Blinn et al.* [this volume] by providing a means of tracking nutrient concentrations released into terrestrial and aquatic ecosystems following the flood.

2. THE HYPORHEIC ZONE AND RIPARIAN GROUND-WATER NUTRIENT AVAILABILITY

Previous work by *White* [1993] suggests that in studies involving surface water/groundwater interactions, authors should define what they mean by hyporheic zone. In this study, the zone in which we studied nutrient concentrations was the zone of interaction between Colorado River surface water and groundwater in the sandy alluvial deposits occurring intermittently along channel margins, and to a lesser extent, the bottom of the Colorado River channel. The deposits studied were inundated by the controlled flood. These groundwater and surface-water interactions

were documented hydrologically by *Petroutson* [1997] and *Springer et al.* [1999]. They showed varying responses to river-stage fluctuations by wells 1-7 m deep sited throughout these deposits. Because we did not characterize associations of organisms in the subsurface, and because our depths of study below the water table (0.2-7 m) extended far below those of most hyporheic-zone nutrient studies [e.g., *Valett et al.*, 1990; *Hendricks and White*, 1991; *Campana et al.*, 1994; *Fraser et al.*, 1996], we used the term groundwater.

The nature of the subsurface water interacting with the bedrock-channeled Colorado River in Grand Canyon contrasts with water interactions in systems dominated by alluvial floodplains. Many hyporheic zone studies have been completed in the alluvial sediments underlying small streams [*Grimm and Fisher*, 1984; *Campana et al.*, 1994; *Hendricks and White*, 1995; *Holmes et al.*, 1994; *Wondzell and Swanson*, 1996; *Mulholland et al.*, 1997]. However, fundamental hydrologic, sedimentologic, and geochemical distinctions exist between the hyporheic environments underlying low-order streams and the systems described herein.

The Colorado River in Grand Canyon is dominated by bedrock reaches without continuous or abundant alluvial floodplains (i.e., *Stanford et al.'s* [1996] transition zones). Even in other studies that focused on these transition reaches, the emphasis was on downstream reaches with abundant alluvial deposits rather than on bedrock-channeled sections [e.g., *Valett et al.*, 1990; *Stanford and Hauer*, 1992, 1993].

Hydrologically, the Colorado River system differs from most previous studies in scale, in relative volumes of surface water and groundwater, and in the forces driving the mixing of waters. The absence of large tributaries or other means of resetting biophysical conditions in the aquatic system simplifies consideration of downstream transport processes in the Colorado. Scales of system patchiness in this constricted, bedrock-channeled system are defined by geomorphic reaches [*Schmidt and Graf*, 1990], which is consistent with the river continuum concept [*Sedell et al.*, 1989].

At the scale at which nutrient fluxes have been examined, the Colorado River differs from other systems examined in previous studies. The significance of hyporheic zone studies has been in the transfer of subsurface water to and from a different composition surface water. These transfers occur in upwelling and down-welling zones, which are, in the absence of floods, relatively fixed in time and space over intervals of weeks to months [*Grimm and Fisher*, 1984; *Hendricks and White*, 1991; *Valett et al.*, 1991; *Fraser et al.*, 1996]. These upwelling and downwelling zones are fixed by the under-lying sedimentology and hydraulic gradient of the stream bed. In contrast, the surface water/groundwater transfer zones along the Colorado River are driven by stage changes caused by river regulation. The same spot along a Colorado River alluvial deposit can be alternately inflowing or outflowing, depending upon rate and size of river stage change [*Springer et al.*, 1999]. *Wroblicky et al.* [1998] demonstrates that spatial and temporal changes in lateral hyporheic zones vary with deposit sedimentology and hydraulic conductivity. Furthermore, the sand bars we studied were primarily along the margins and not under-neath the channel, unlike the cases for most hyporheic zone studies [*Kaplinski et al.*, 1995]. The interstitial volume of Colorado River sand bars is small compared to the flow of water in the river [*Springer et al.*, 1999], in contrast to previous studies where the groundwater to surface water ratio exceeds one [e.g., *Valett et al.*, 1990; *Ritzenthaler and Edwards*, 1996]. Finally, the groundwater in these deposits was examined at depths reaching 7 m below water table, in contrast to hyporheic zones studies which extend 1-2 m below bed surface. The scales, patterns, and processes controlling the distribution of dissolved nutrients in ground-water, and in the transfer of these waters to surface waters, differ markedly from previous work.

In spite of these substantive differences, previous workers have demonstrated differences in water composi-tions, spatial heterogeneity of groundwater compositions, and responses to flooding, which we can apply to this study. Nitrate and soluble reactive phosphorus concentrations vary within groundwater as a function of mixing with surface waters [*Pinay et al.*, 1998], denitrification [*Holmes et al.*, 1996], organic matter availability in the subsurface, and N-uptake by riparian vegetation [*Ritzenthaler and Edwards*, 1996]. With some exceptions, groundwater acts as a net source of N and P to surface waters [e.g., *Hendricks and White*, 1995]. The form of dissolved N often changes with flow path, as nitrate concentrations decrease and ammonium concentrations increase with subsurface residence time [e.g., *Grim and Fisher*, 1984; *Fraser et al.*, 1996]. Previous workers have suggested the opposite relationship for dissolved organic carbon (DOC) [*Hendricks and White*, 1991], observing that groundwater does not seem to be a significant DOC source to surface water, although *Rutherford and Hynes* [1987] and *Schindler and Krabbenhoft* [1998] state that subsurface waters may be significant DOC sources to surface water.

Workers in smaller, alluvial Southwestern rivers have established the importance of flooding in impacting the availability of nutrients, especially nitrogen [e.g., *Grimm and Fisher*, 1986]. The transport, accumulation, and burial

of organic material caused by flooding may increase the availability of organic material for mineralization, increasing nutrient availability [*Elwood et al.*, 1983]. Previous work has established that flooding pushes signif-icant volumes of organic matter into subsurface environ-ments, generating a complex pattern of nutrient distribution after the flood [*Valett et al.*, 1990]. Floods bury allocth-onous and autochthonous organic matter, providing a source for nutrient release [*Crocker and Meyer*, 1987; *Valett et al.*, 1990; *Valett et al.*, 1993; *Stanley and Boulton*, 1995]. No attempt to quantify changes in dissolved-nutrient avail-ability in groundwater following flooding has been previ-ously attempted along the Colorado River in the Grand Canyon. However, work in Grand Canyon and the Rio Grande demonstrates that a loss of litter from terrestrial environments follows small-scale flooding events [*Molles et al.*, 1995; *Kearsley and Ayers*, 1997]. Flooding can affect whole-stream metabolism, generating increases in produc-tivity over less than 100 days [*Valett et al.*, 1993].

2.1. Nutrient Availability in the Colorado River in Grand Canyon

The serial-discontinuity concept of *Ward and Stanford* [1983] predicts a large-scale effect of Glen Canyon Dam on the Colorado River downstream by interrupting the transport of sediment and organic material. The entrapment of dissolved, suspended and bed load organic debris from the Colorado River in Lake Powell reservoir is well estab-lished [*U.S. Department of Interior*, 1995]. Allochthonous material supplied to the Colorado River below Glen Canyon Dam has been reduced, and the food base has been modified, shifting the system from heterotrophic toward more autotrophic energy sources [*Shannon et al.*, 1996; *Valdez and Ryel*, 1997]. With the loss of fine-grained sediment and particulate organic matter settling in the reservoir, nutrient transport and delivery to Grand Canyon under normal post-dam operating conditions has decreased [*U.S. Department of Interior*, 1995].

Delivery of organic matter to the Colorado River system, and its retention within Colorado River ecosystems, is expected to be determined by three factors: flow regime entering Lake Powell, reservoir processes, and the size and rates of change of releases from Glen Canyon Dam. Drift in the Colorado River is variable seasonally and correlates with dam operations [*Shannon et al.*, 1996]. Drift studies have stressed that cobble-riffle areas are the critical produc-tivity areas. However, these areas are not thermally advan-tageous for native fish, such as humpback chub (*Gila cypha*) [*Valdez and Ryel*, 1997]. More favored aquatic habitats for vertebrates include backwaters, that are physically, hydrologically, and geochemically linked to reattachment-bar/RCC complexes.

A 4-yr study of the hydrogeochemistry of reattachment-bar/RCC complexes at several sites (river miles -6R, 43.5L, 45L, 55.5R, 65R, 72L, and 194L) established that distinctly different geochemical patterns occur under the bar, under the RCC, in RCC surface water, and in mainstem surface water [*Bennett and Parnell*, 1995; *Parnell et al.*, 1996; *Bennett*, 1997]. groundwater beneath fined grained, organic-rich RCC sediments are less oxidized, are higher in ammonium and DOC concentrations, and are lower in nitrate and dissolved oxygen concentrations than groundwater in the coarser sands of the sand bar [back beach (BB) and beach front (BF), Figure 1]. Differences exist year-round, with similar relative differences between geochemical environments (RCC sediment versus beach back and front) regardless of season, although absolute chemical concentrations vary seasonally. Average concentrations were calculated for RCC surface water, RCC sediment groundwater, back beach (between the shore and the RCC) groundwater, beach front groundwater, and mainstem river water. Average concentrations for dissolved ammonium, nitrate-nitrogen, orthophosphate, and oxygen vary systematically across the bar (Figure 1). Concentrations represent average values of all 14 wells. Because of heterogeneity in the concentrations between individual wells, the aggregate average of all wells at a site sometimes produces large standard errors [*Bennett*, 1997]. Nonetheless, Figure 1 demonstrates that, under the RCC and bar, groundwater have low concentrations of dissolved oxygen and nitrate nitrogen and high concentrations of dissolved ammonium, especially in RCC sediments. As is the case in many previous studies, orthophosphate concentrations are highest in groundwater, especially in RCC sediments, and are at or near detection limits (5 μg/L) for surface waters. A similar trend occurs for dissolved organic carbon, with highest concentrations in groundwater, especially in RCC sediments, and lowest concentrations in surface waters [*Bennett*, 1997; *Parnell et al.*, 1997a, 1997b]. For all N, P, and organic and inorganic C species measured, RCC surface waters were intermediate in concentration between groundwater and mainstem waters [*Bennett*, 1997; *Parnell et al.*, 1997a, 1997b]. RCC surface waters are more like mainstem waters because they are replenished more rapidly by surface waters than by groundwater inflow under most river stages. Using discriminate-function analysis, physiochemical differences between RCC sediment, back-beach and beach-front groundwater produce significantly different discriminate categories. Thus, previous work has established the existence of three different groundwater bodies with much higher concentrations of dissolved N, P, and C

than exist in adjacent surface waters. Because mainstem waters have much lower concentrations than groundwater or backwaters, fluctuations in mainstem waters resulting from the flood [*Bowser et al.*, 1997] cannot increase the absolute concentrations or produce the concentration changes observed in groundwater and backwaters.

3. STUDY SITES AND SAMPLING METHODS

In order to compare flood-related changes to longer-term records, well-vegetated reattachment-bar/RCC complexes at river mile -6R (Hidden Slough) and 194L were used. New sites were established at river miles 45L, 55R (Kwagunt Marsh), and 65 to coordinate with other flood studies. At each site, 25.4-mm diameter, 46-cm long PVC well screens (0.001 slot screen size) or stainless steel screened drive points (0.006 slot) were installed individually at depths of 1.5, 3, and 7 m. These screened sections sat above 0.3-m sumps and were connected to PVC or steel pipes that rose above ground level. The tops of the pipes were sealed air-tight between samples. The individual wells were placed longitudinally along the axis of the RCC and cross-sectionally from the RCC toward the river, forming transects 40-150 m long. Intensive study sites at river miles -6R, 55R, and 194L had 14 to 19 wells distributed across the beach and RCC, while sites at 45L and 65R had 4 to 6 wells each. Wells were developed and allowed to stand for at least 24 hrs prior to purging and sampling. Wells were purged of at least 5 well volumes to allow constant temperature and conductivity readings immediately before sampling. Well-installation, development, and sampling procedures are given in *Bennett* [1997] and *Springer et al.* [1999]. At each site, wells were sampled within the same 6-hr period together with RCC and mainstem surface waters. Samples were taken 20 to 2 days before the flood and approximately 20, 80, 220, and 400 days after the flood.

3.1. Field and Laboratory Analyses

Because of the time required to deliver samples from the field to the laboratory (1-2 weeks), as many analyses as possible were completed immediately in the field. This approach increased sample processing time and allowed no more than 20% replication of individual well samples. The values presented are averages of multiple wells at a single site and time. Conductivity and temperature were measured in the field with an Orion model 122 meter, calibrated with 0.1N KCl. Dissolved oxygen and pH were measured in the field on unfiltered samples with an Orion model 290 meter and Orion dissolved oxygen and temperature-compensation pH triodes. The dissolved-oxygen electrode was calibrated

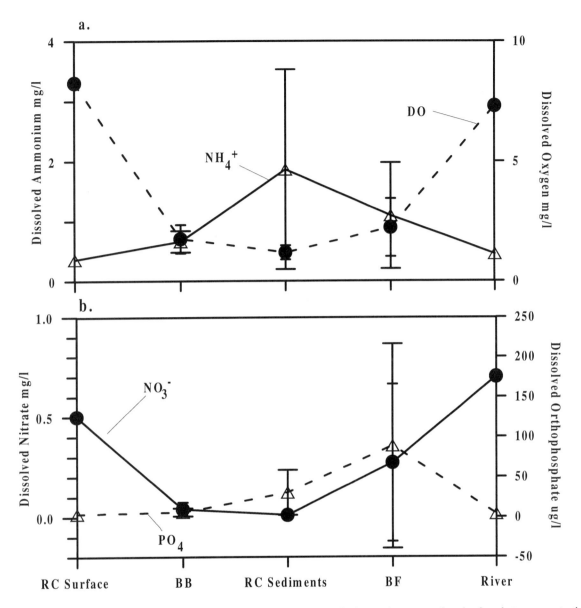

Figure 1. Ammonium and dissolved oxygen concentrations (top) and nitrate-nitrogen and orthophosphate concentrations (bottom) in a section of wells and surface waters across study site -6R in April 1995. "RC surface" is return-channel surface water, "BB" is the average of four back beach wells, "RC sediments" is the average of four wells in the return channel, "BF" is the average of 4 beach-front wells, and "River" is river surface water.

hourly using water-saturated air and was routinely checked against modified Winkler dissolved-oxygen titration [*Bennett*, 1997]. Measurements of pH were calibrated using 7 and 10 buffers.

Dissolved nitrate plus nitrate and nitrite concentrations were measured in the field on 0.2-μm filtered samples by UV-VIS spectrophotometry [*Hach*, 1992]. Dissolved nitrate was taken to be the difference between the two measurements. Dissolved ammonium concentrations in unfiltered samples were measured by Orion ion selective electrode [*Orion*, 1990]. Dissolved orthophosphate concentrations were measured on filtered samples with a LaChat ion chromatograph. Dissolved inorganic and organic carbon concentrations were measured on samples filtered with Whatman GFIC glass fiber filters using a Dohrmann DC 180 dissolved carbon analyzer.

The most sensitive and repeatable measure of dissolved organic carbon was non-purgeable organic carbon (NPOC).

TABLE 1. Comparison of pre-flood nutrient concentrations (n= 15) to long-term averages (n=47) for spring-season groundwater at river mile -6R. Averages and ±standard deviations are for all wells at -6R. NPOC is dissolved non-purgeable organic carbon, NH_4 is dissolved ammonium, DO is dissolved oxygen, NO_3^- is nitrate-nitrogen, and PO_4 is dissolved orthophosphate. Long-term NPOC data were not available.

	NPOC (mg/L)	pH (units)	NH_4 (mg/L)	DO (mg/L)	Conductivity (µS/cm)	NO_3^- (mg/L)	PO_4 (µg/L)
Pre-Flood	3.95±1.72	7.48± 0.80	0.7±0.8	1.8±0.3	1037±153	0.07±0.06	n/a
Long-Term	n/a	7.58±0.21	0.8±0.6	1.3±0.5	992±126	0.08±0.11	14.±10.

Degassing was used primarily to remove the dominant inorganic carbon fraction from samples prior to dissolved carbon analysis. Replicate analyses on preserved samples before and after degassing indicate that the purgeable organic carbon fraction was negligible.

Detection limits were 50µg/L for ammonium, 10 µg/L for DOC, 20 µg/L for nitrate, 5 µg/L dissolved orthophosphate, and 0.1 mg/L for dissolved oxygen [Bennett, 1997]. Detection limits come from manufacturers specifications but were confirmed by adding two times the standard error of multiple analyses of sample blanks to the average of the sample blanks as an approximation of the detection limit. Details of sample handling, preservation, analysis, and quality assurances/quality control appear in Bennett [1997].

Stratigraphy of bar and RCC sediments at the river miles -6R, 55R, and 194L were determined by excavation of soil pits down to the 213 m^3/s discharge water table (typically 0.5-1.5 m deep). One meter lengths of modified Livingston corer were taken sequentially to a depth of 4 m below the base of the pit at 55R to more completely determine the depth to and preservation of the vegetation buried by the flood.

3.2. Measurement of Pre-Flood Conditions

The morphology and geomorphic setting of each site studied was similar, allowing comparisons between sites and extrapolation of results to many reattachment-bar/RCC complexes [Kaplinski et al., 1995]. This similarity existed because of the similar geomorphic forms and geochemical and hydrological flow patterns which exist in most reattachment-bar/RCC complexes. Before and after the flood, silt- and clay-floored RCCs were separated from the mainstem by reattachment bars of fine to medium sand. At all sites, RCCs were typically drained of surface waters at lower discharges. However, the RCC had the lowest elevation relative to river stage at river mile -6R and the highest elevation relative to stage at 194L. The RCC at -6R was wet and well connected to the mainstem throughout the normal range of flows for the study period (1996-97). The RCC at 194L maintained surface water only at flows above

560 m^3/s. The RCCs at the other sites were intermediate in response to discharge fluctuations.

The pre-flood samples may be considered a valid, representative baseline for comparisons to samples during and after the controlled flood. A comparison of chemical concentrations in waters sampled between 20 and 8 days before the flood to the longer-term record of spring sampling periods revealed similar concentrations. Long-term average values for early spring season collections (n= 47) were compared with concentrations in samples taken immediately prior to the flood (n= 15; Table 1). Long-term averages were taken between 1993 and 1996. Direct comparisons between the pre-flood and long-term values were possible only at river mile -6 because new sites (at river miles 45, 55, and 65) were added in conjunction with other flood studies [e.g., Stevens et al., 1997], specifically for this study. Logistic constraints prevented pre-flood sampling of 194L. Comparison between the longer record and the pre-flood samples showed no significant differences for pH, nitrogen species, conductivity, or dissolved oxygen. The geochemical trends in concentrations of redox-sensitive species between individual wells across the bar were present during pre-flood sampling (Figure 1).

4. RESULTS AND DISCUSSION

4.1. Flood Deposition Patterns

The 1996 controlled flood in Grand Canyon resulted in burial of living and, to a small degree, detrital organic material under 0.2-1.95 m of sand at our study sites. The greatest depths of burial of vegetation (1.95 m) were recorded at river mile 194L and the shallowest depths (0.2m) were recorded at river mile -6R. The buried vegetation observed was always in place, with undisturbed roots, and so was assumed to be autochthonous. At 55R, the thickness of flood sand increased from the shoreward side of the RCC (25 cm) toward the post-flood beach face (94 cm) (Figure 2). Where the beach was moderately well-vegetated before the flood, pits exposed recognizable layers of freshly buried vegetation. Only in the 2 shoreward

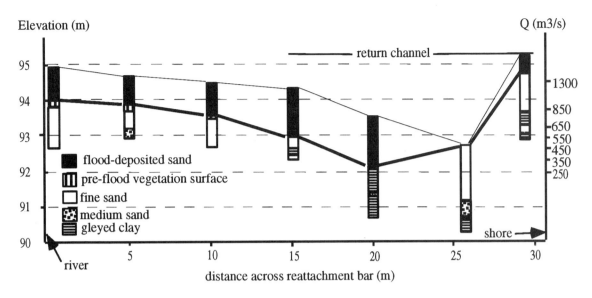

Figure 2. Cross-section through Kwagunt (55R mile) sand bar from the river (left) across the bar and return channel to shore (right). Flood deposits are fine sand and are shown in dark gray. The base of the flood deposits is shown by the heavy line connecting the individual profiles. Land surface on June 1996 is shown with the finer line connecting the tops of the individual profiles.

RCC pits were no observable vegetation mats preserved. However, pre-flood sampling at these sites (3/11/96) revealed no vegetation in the RCC; therefore, there was no vegetation to preserve. Measuring the depths to intact pre-flood clay layers in the 4 RCC sections allowed reconstruction of thickness of scour plus subsequent deposition. In the one section showing no flood-deposited sand, the depth through pre-1996 sand to gleyed, subsurface clay was the same before and after flooding, indicating no significant scour of pre-flood fine sands.

The results of these excavations contrast with those investigating previous floods of 1983, where no buried vegetation mats were observed in the beaches within a year after flooding [*Stevens et al.,* 1995]. The 1996 controlled flood appears to be a flood which did not scour RCC sediments or vegetation, in contrast to earlier flood events.

4.2. Short-Term Biogeochemical Responses to Flooding

Well data at all sites studied showed nutrient concentration changes after the flood. Groundwater samples were taken at river mile 55R within 24 hrs after flood recession and again within 72 hr. The responses of 4 wells to the flood recession showed these changes (Figure 3a-d). In each figure, the start of the flood at day 89 is shown by a vertical line. These 4 wells were the only wells at 55R which remained in place and in the same geomorphic setting (RCC for wells R3 and R4 and beach front for wells MS

and MD). The other 8 wells were either eroded or shifted geomorphic setting as the bar migrated shoreward during the flood. Because of large differences in chemical composition of groundwater across the beach [*Bennett,* 1997], temporal compositional data should only contrast wells in the same geomorphic setting. New wells were installed after the flood to reproduce the pre-flood well distribution. Three of the 4 wells (R4, MS, and MD) showed the same temporal trends, while R3 reacted differently during the first few days following recession. Well R3 had its screened interval in silty clay and responded differently than the other wells, hydraulically as well as chemically. Wells R4, MS, and MD all showed, within 24 hrs, increases in dissolved ammonium, NPOC, and orthophosphate (Figure 3). These wells show an initial, day-long increase in DO caused by groundwater recharge with oxygenated river water, followed by a rapid decrease within 72 hrs after recession. Over the summer growing season (days 120-220), DO concentrations increased with increasing terrestrial plant productivity and within increasing inflow into the beach of high stage summer river water.

Each individual well showed measurable changes over time. However, the concentration differences between individual wells at the same beach at any time were greater than the differences in composition that occurred within a single well over time. The degree of spatial heterogeneity across a beach was larger than the temporal changes occurring within any one well. For example, two of the 55R

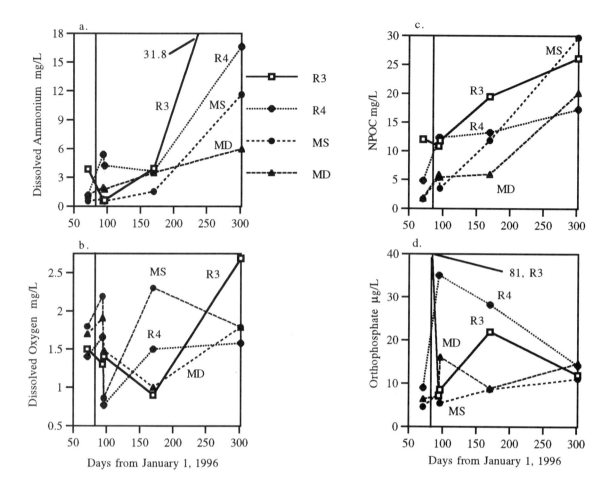

Figure 3. A. Dissolved ammonium concentrations in 4 individual wells at site 55R as a function of time. R3 and R4 are return-channel wells, and MS and MD are mid-beach wells. The vertical line represents the start of the flood. The dissolved ammonium concentration for R3 at 302 days is 31.8 mg/L. B. Dissolved-oxygen concentrations in 4 individual wells at site 55R as a function of time. C. Dissolved non-purgeable organic carbon (NPOC) concentrations in four individual wells at site 55R as a function of time. D. Dissolved orthophosphate concentrations in four individual wells at site 55R as a function of time. The pre-flood concentration for well R3 is 81 µg/L.

return channel wells and two of the 55R mid-beach wells showed wide ranges of concentrations of dissolved ammonium, NPOC, and orthophosphate at any one time (Figure 3). This pattern of site-specific compositional trends was repeated at all wells at all study sites. Thus, the spatial variability in groundwater composition was greater than the initial changes in composition following flooding.

The immediate post-flood compositional changes in nutrient concentrations were magnified with time; season-long changes in compositions of individual wells were much larger than the initial changes. These changes, through summer, fall, and winter, were the same trends as the immediate responses; increases in dissolved ammonium, NPOC, and orthophosphate for wells R4, MS, and MD (Figure 3). After the initial depression of DO

concentrations, these parameters for R4 and MS began to recover to pre-flood values during the summer (Figure 3b). With the exception of DO, the short-term compositional changes in groundwater observed within a week of the flood continued to occur throughout the next 10 months.

4.3. Post-Flood Seasonal Changes in Groundwater Compositions

The longer-term record of change in groundwater compositions at river mile 55R showed seasonal as well as short-term increases in NPOC, ammonium, and orthophosphate concentrations (Figure 4). These plots of concentrations over time are the average of all 14 wells at 55R. The trends observed in individual wells were reflected in the

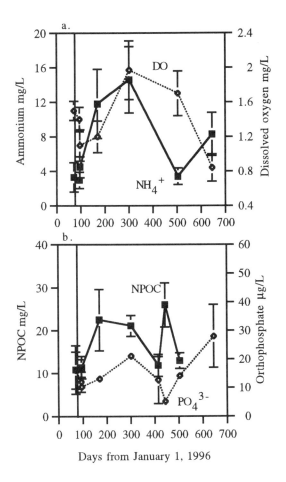

Figure 4. A. Averages in groundwater in 12 wells at site 55R for dissolved ammonium and dissolved oxygen. Error bars are standard errors. Vertical line indicates beginning of flood. B. Averages for dissolved non-purgeable organic carbon and orthophosphate in groundwater in 12 wells at site 55R.

trends of the averages; increased dissolved ammonium, NPOC, and orthophosphate (Figure 4). Large standard errors resulted from the large variation in concentrations among wells produced by their differing geomorphic settings. Ammonium and orthophosphate concentrations peaked during the fall following the flood and dropped to minima the following spring. After the depression in DO concentrations, immediately after the flood, DO increased to pre-flood concentrations.

The increasing concentrations of N in groundwater could be explained through increased nitrogen-fixation. However, direct measurements of nitrogen-fixation or nitrification were beyond the scope of this project. More importantly, the trends in N in groundwater were widely observed across all parts of the bar and RCC. Bacterial or cyanobacterial populations were limited to parts of the RCC. Even if they

were important sources of N in the RCC, they could not be used to explain increases elsewhere in the bar because of the extremely low hydraulic conductivity of the RCC sediments underlying the bacteria [*Petroutson*, 1997].

Similar trends were also observed for average groundwater nutrient compositions at the other study sites at river miles -6R (Figure 5) and 194L (Figure 6). Intervals between sampling were larger than at 55R (only one sample set within the first 100 days after the flood), yet the temporal trends were similar. Within 3 weeks of the flood, ammonium and NPOC concentrations increased well above the pre-flood averages, and DO concentrations dropped well below the pre-flood average (Figure 6). The -6R site was geomorphically similar to the 55R site. Thus, large standard errors were again present, due to the compositional heterogeneity between well waters of differing geomorphic setting.

Analysis of post-flood trends at river mile 194L was more problematic, due to the absence of a pre-flood sample shortly before the event. Samples from the previous year were compared with post-flood samples for this site (Figure 6a-b). Although we were unable to make a short-term, pre- to post-flood comparison, the same long-term, seasonal trend observed at 55R was apparent at 194L. Ammonium and NPOC concentrations increased in the months following the flood, peaking during the fall after the flood.

Over all the sites examined, the DO concentrations tended to recover to pre-flood levels more quickly than other parameters studied. The largest concentration gradient between groundwater and surface waters or groundwater and the atmosphere was for DO. Furthermore, higher river stages followed the flood, increasing the inwelling of river water into the beaches. Finally, as the growing season began, terrestrial productivity could have increased groundwater DO. As buried organic matter continued to decay, concentrations of NPOC and ammonium above pre-flood levels persisted through the seasons, for up to a year after the flood.

4.4. Statistical Analysis of Temporal Changes in Composition

Nutrient composition of samples from individual wells at the same site were compared over time. Samples were collected from the same well within weeks before the flood and again immediately and intermittently after the flood. The results were analyzed using a Wilcoxson pairwise comparison (Table 2). Only those comparisons resulting in a 95% confidence level or higher are shown. The number of individual wells included in each comparison varied from 6 to 13. These analyses use all the data for wells at a

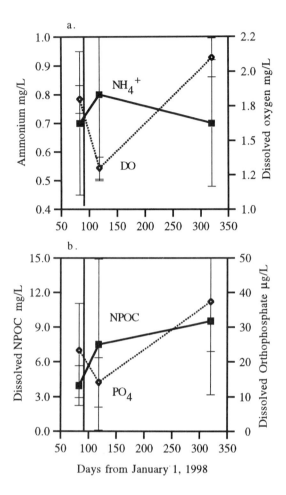

Figure 5. A. Averages in groundwater in 12 wells at site -6R (Hidden Slough) for dissolved ammonium and dissolved oxygen. Error bars are standard errors. Vertical line indicates beginning of flood. B. Averages for dissolved non-purgeable organic carbon and orthophosphate in groundwater in 12 wells at site -6R (Hidden Slough).

particular site, comparing pre-flood with post-flood 1996 samples. Figures 3-7 used averages of samples taken at a particular date and site, producing large standard errors. By using larger data sets in our statistical analyses, we observed significant differences not apparent in plots of single data sets for a site or time.

When using the entire spatially heterogeneous group of wells, few nutrient parameters showed significant short-term differences before versus after the flood. Because of an incomplete data base for ammonium and NPOC concentrations at river mile -6R, only DO concentrations were significantly different between pre- and post-flood samples. Only NPOC concentrations significantly increased across 55R. As previously noted, shoreward shift

of the reattachment bar at 55R during the flood changed the geomorphic position of the majority of wells, preventing a pre- to post-flood comparison of all wells. DO and orthophosphate concentrations showed significant differences following flooding at 194L, albeit with a much larger interval between pre- and post-flood samples.

Changes in concentrations of N, P, and C species immediately after the flood were small compared to the larger seasonal changes in concentrations following flooding. A Wilcoxson pairwise comparison for the 3 major sites (river miles -6R, 55R, and 194L) was also used to analyze sample sets gathered exclusively after the flood (Table 3). These data sets compared groundwater concentration differences resulting from of seasonal changes following flooding. Again at -6R, only DO concentrations showed significant changes. However, at 55R and 194L,

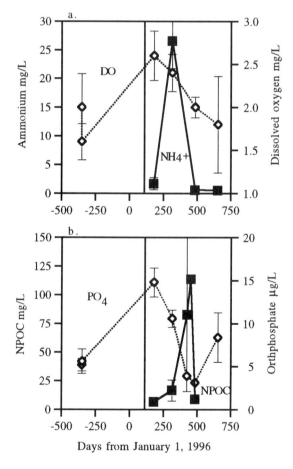

Figure 6. A. Averages in groundwater in 15 wells at site 194L for dissolved ammonium and dissolved oxygen. Error bars are standard errors. Vertical line indicates beginning of flood. B. Averages for dissolved non-purgeable organic carbon and orthophosphate in groundwater in 15 wells at site 194L.

TABLE 2. Wilcoxon pairwise comparisons showing significant P values for changes in chemical parameters between pre- and post-flood samples. Flooding ended on day 94 at site 55R. Comparison examines changes with time (before/after flooding) that occur within a site. NPOC is dissolved non-purgeable organic carbon, DO is dissolved oxygen, and PO$_4$ is dissolved orthophosphate.

Site	Parameter	Dates compared in Julian days after January 1, 1996	Z	P	Number of wells
-6R	DO	83-118	-2.803	0.0051	10
55R	NPOC	71-94	2.201	0.0277	6
55R	NPOC	71-96	2.366	0.0180	7
194L	DO	-350-178	2.253	0.0243	9
194L	PO$_4$	-350-178	3.181	0.0015	13

DO, NPOC, ammonium, and orthophosphate concentrations showed significant seasonal changes following flooding. NPOC, ammonium, and orthophosphate concentrations continued to increase following the flood until peaking in the fall of 1996. The magnitude of changes in compositions for these parameters was much greater on a seasonal scale than during the few weeks immediately around the flood (Figures 4, 5, and 6).

Although both seasonal and short-term changes in concentrations of nutrient species occurred, the cause of the seasonal changes in concentrations cannot be attributed solely to the impacts of the flood. Normal, seasonal fluctuations in groundwater compositions before the flood were a product of both biological activity and seasonal flow regulation at Glen Canyon Dam [Bennett, 1997]. However, these normal, seasonal fluctuations are not as large as those observed following flooding. Pre-flood seasonal differences at -6R were relatively small compared to the seasonal differences which followed flooding. Between April and November 1995, pre-flood well-nutrient concentrations increased by less than a factor of 2 for ammonium and nitrate and less than a factor of 4 for orthophosphate [Bennett, 1997]. Between April and November, 1996, over the time period immediately following the flood, groundwater nutrient concentrations increased by more than a factor of 6 for ammonium and 5 for orthophosphate; NPOC data prior to February 1996 does not exist. From these comparisons, and the statistical analysis, seasonal increases in ammonium, orthophosphate, and NPOC in groundwater following flooding can be attributed primarily to the impact of the flood and, to a lesser extent, seasonal variability. High seasonal composition changes could have been triggered by the short-term release of nutrients to groundwater immediately after the flood.

4.5. Changes Following Flooding in Surface-Water Compositions

The nature of changes following flooding in the nutrient concentrations in RCC and mainstem surface waters were predictable from previous work, although the concentration gradients produced were far different than other studies. In particular, ammonium and NPOC concentrations tended to be higher in groundwater and lower in surface waters than values observed in previous studies of other river systems [e.g., Hendricks and White, 1991, 1995; Pinay et al., 1998; Schindler and Krabbenhoft, 1998]. The reverse was true for nitrate; concentrations were low in groundwater and surface water compared to previous studies.

Surface-water samples from several sites along RCCs were taken when groundwater were sampled. Because these surface waters are well oxygenated (DO > 6 mg/L), no ammonium or nitrite was detected. All dissolved inorganic nitrogen was in the form of nitrate. Both dissolved inorganic nitrogen (nitrate-nitrogen) and NPOC increased in concentration after the flood, mirroring the changes observed in co-existing bar groundwater (Figure 7). However, when using the complete data base for RCC surface waters, only the increase in NPOC after the flood was statistically significant.

Only bar groundwater has sufficiently high concentrations of orthophosphate and NPOC to act as a source to produce the increases in backwater-nutrient concentrations. Evapotranspirative concentration through the summer can be discounted to explain these increases because no significant increase in electrical conductivity occurred in concert with these concentration increases. Furthermore, Petroutson [1997] demonstrated that groundwater discharges from the beaches into the RCCs. The increases in

TABLE 3. Wilcoxon pairwise comparisons for seasonal changes between post-flood visits. Comparison examines seasonal changes with time after flooding that occur within a site. NPOC is dissolved non-purgeable organic carbon, DO is dissolved oxygen, NH_4 is dissolved ammonium, and PO_4 is dissolved orthophosphate.

Site	Parameter	Dates compared in Julian days after January 1, 1996	Z	P	Number of wells compared
-6R	DO	118-321	-2.67	0.0080	11
55.5R	DO	170-302	1.96	0.0499	8
55.5R	DO	504-645	-2.599	0.0093	11
55.5R	NH_4	96-170	2.701	0.0069	9
55.5R	NH_4	170-302	2.521	0.0117	8
55.5R	NH_4	302-504	-2.197	0.0280	8
55.5R	NPOC	96-170	2.803	0.0051	10
55.5R	PO_4	302-504	-2.521	0.0012	7
55.5R	PO_4	504-645	2.49	0.0128	11
194L	DO	313-484	-2.103	0.0355	15
194L	NH_4	178-313	3.18	0.0015	14
194L	NH_4	313-484	-3.296	0.0010	15
194L	PO_4	313-484	-3.18	0.0015	13
194L	PO_4	484-653	2.272	0.0231	15

backwater nutrient concentrations were consistent with a flux of dissolved carbon and nitrogen from beach groundwater into the RCCs.

Mainstem surface waters collected at the same time as backwater and groundwater samples did not show the same significant changes in composition observed in the other water types. The absence of a significant change in mainstem composition in spite of high rates of mineralization and dissolved-nutrient release in interconnected groundwater may be readily explained. Shannon et al. [1997] demonstrated a significant increase in drift and benthic biomass following the flood. Increased productivity could buffer the concentration of critical nutrients in mainstem surface waters. Other authors demonstrate that increasing aquatic productivity allows large fluxes of nutrient-hyporheic zone water to flow into surface waters without a significant resultant change in surface water compositions [Schindler and Krabbenhoft, 1998]. Thus, a flux of nutrients from groundwater into surface water could occur without a change in concentration being detected.

Furthermore, the flux of groundwater into surface waters can be demonstrated to be relatively small compared to the flow of the mainstem. Darcy's equation can be used to estimate the groundwater flux from a stretch of shoreline into the Colorado River:

$$Q = (1.16 \cdot 10^{-5}) \cdot K \cdot A \cdot (dh/dl), \qquad (1)$$

where Q = discharge (m^3/s); K = hydraulic conductivity (m/day); A = flow area (m^2); and dh/dl = change in average groundwater head across the bar expressed as change in vertical elevation (m) per change in horizontal distance (m).

The distance from Glen Canyon Dam to Lees Ferry is 25.7 km, average river depth is 7.6 m at typical flows, and K was measured by Petroutson [1997] as an average of 30.5 m/day. The groundwater/surface water gradient into the river, measured at river mile 55R during a large drop in stage elevation, was 0.004 [Springer et al., 1998]. The resulting groundwater discharge along this 26-km stretch was 0.3 m^3/s, less than 1% of the Colorado River flow.

This groundwater inflow is small and would produce little change in mainstem compositions, in spite of the large nutrient-concentration differences between groundwater and surface waters. If all nutrients flowing into the river from groundwater were diluted by the entire volume of the mainstem river, it would take a concentration difference of at least 1000 times between groundwater and surface water, based on the calculation made above, to produce a measurable increase in mainstem concentration. However, inflow does not occur everywhere along the channel. Groundwater inflow to the river is restricted in location, producing "hot spots" of high nutrient-flux zones [Petroutson et al., 1997; Springer et al., 1999], analogous to upwelling zones identified in many hyporheic zone

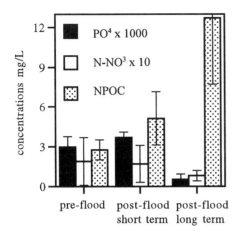

Figure 7. Average backwater concentrations of dissolved non-purgeable organic carbon (NPOC), 10 times the nitrate-nitrogen (N-NO$_3^-$), and orthophosphate (O-PO$_4$) for samples collected from return-current channels at river miles -6R, 45L, 55R, 65R and 194L. Collection periods were 2-20 days before, 0.5-83 days after, and 210-228 days after the controlled flood. Error bars are standard errors of the averages of multiple samples collected from the same backwater at the same date (see text). Only the NPOC shows a statistically significant difference at the 99% confidence level before vs. after flooding.

nutrient studies [e.g., *Grimm and Fisher*, 1984; *Campana et al., 1994*; *Hendricks and White*, 1995; *Holmes et al.*, 1994; *Wondzell and Swanson*, 1996; *Mulholland et al.*, 1997]. Thus, even though total mainstem concentrations are not significantly raised, the discharge areas from reattachment bars create restricted zones of higher dissolved NPOC, ammonium, and orthophosphate.

5. SUMMARY AND CONCLUSIONS

Increases in ammonium and NPOC concentrations and decrease in DO concentrations in bar groundwater occurred immediately following the flood. We observed freshly-buried vegetation below the water table in the same locations as the groundwater sampled. These observations are consistent with the hypothesis that increasing rates of microbial respiration within the beaches followed flooding. Mineralization of the buried organic material could begin immediately after burial, releasing more labile forms of N, C, and P into groundwater. The fact that maximum increases in ammonium occur months instead of days after flooding suggests that mineralization of buried organic material, and not de-nitrification, is the dominant inorganic nitrogen transformation [*Holmes et al.*, 1994, 1996]. In concert with these nutrient releases, DO concentrations are

decreased through microbial respiration. Natural seasonal fluctuations in microbial activity would act upon a larger mass of available organic carbon, magnifying post-flood changes in groundwater composition. The increased microbial activity following flooding releases more nutrients, which become available for uptake by terrestrial or aquatic vegetation.

The 1996 controlled flood buried standing, autochthonous organic material under 0.2-1.95 m of fine sand. The uncommonly high flows of the months following the flood caused water tables to rise in sand bars, placing the vegetation mat below the water table. Significant changes in dissolved C, N, and P occurred in bar groundwater and in surface backwaters following the flood. The changes observed were consistent with the hypothesis that the flood buried readily mineralizable organic material, increasing the potential microbial mineralization rate in beaches, beginning immediately after flood recession. The largest pre- to post-flood changes occurred on a seasonal time scale. The size of seasonal changes following flooding was large compared to natural seasonal changes in nutrient concentrations observed by *Bennett* [1997]. That seasonal changes in nutrient concentrations were amplified by the flood is supported by seasonally adjusted increases in terrestrial productivity (M.J.C. Kearsley, N. Ariz. Univ., pers. commun., 1997). Increased nutrient availability after the flood was demonstrated through terrestrial productivity increases (fertilization), as well as increases in groundwater nutrient concentrations. The nutrient-composition changes in groundwater and backwaters were not directly reflected in mainstem water compositions, although increases in mainstem primary productivity did correlate with periods of high nutrient concentrations in beach groundwater [*Shannon et al.*, 1997]. Increases in nutrient concentrations in groundwater were mirrored by increases in backwater concentrations. Thus, critical aquatic habitats received an increased flux of N, P, and C following the flood.

Because the 1996 controlled flood was ineffective in scouring vegetation from reattachment-bar/RCC complexes, the nutrients stored in vegetation in these complexes was not remobilized as particulate material. Instead, vegetation was buried and mineralized, releasing a pulse of N, P, and C to groundwater, which in turn flowed into RCC and mainstem surface waters. *Petroutson* [1997] has identified areas of groundwater discharge from bars into the river. These zones become "hot spots" where groundwater enriched in nutrients relative to surface waters enter surface waters. Groundwater inflow to surface waters, and therefore nutrient flux into surface waters, is highest during periods of gently falling river stages, which maintain a high hydraulic gradient. Because of the gradient between river and groundwater stages, groundwater inflows to the mainstem (and nutrient fluxes) are also high for constant low-stage flows and low for high-stage flows [*Petroutson*, 1997].

From the perspective of nutrient availability, the 1996 controlled flood was a unique event in the observed historical record. It was large enough to bury vegetation but too small to scour standing terrestrial vegetation. The controlled flood followed an extended period of regulated flows during which greater than pre-dam volumes of vegetation were established on reattachment-bar/RCC complexes [*Stevens et al.*, 1995]. Thus, more autochthonous organic matter was available for burial and mineralization. The results presented here should therefore represent an unusually high increase in groundwater NPOC, ammonium, and orthophosphate concentrations.

Acknowledgments. This work was supported by contracts to the authors from the Bureau of Reclamation, Glen Canyon Environmental Studies, and Grand Canyon Monitoring and Research Center. Discussions with Matt Kaplinski, Dick Marzolf, Joe Shannon, Dave Wegner, and Mike Yard helped to focus the study and improve our interpretations. The authors appreciate the cooperation of personnel from Grand Canyon National Park and Glen Canyon National Recreation Area in permitting this research. Valuable field assistance came from many including Chris Black, Betsy Gilbert, John Malussa, Sarah Rogers, Kirsten Rowell, Dana Strength and Amy Welty. The manuscript profited from the thoughtful comments contributed by two anonymous reviewers.

REFERENCES

Bennett, J.B., *A biogeochemical characterization of reattachment bars of the Colorado River, Grand Canyon National Park, Arizona*, M.S. thesis, N. Ariz. Univ., Flagstaff, AZ, 1997.

Bennett, J.B., and R.A Parnell, Biogeochemical cycling in riparian and aquatic environments in the flow-regulated Colorado River, Grand Canyon National Park, Arizona, USA, *Bull. Ecol. Soc. Amer.*, 76, 34, 1995.

Bowser, C.J., G.R. Marzolf, D.W. Stephens, R.J. Hart, and W.S. Vernieu, *Carbon and oxygen dynamics in the Glen Canyon Dam's tailwater: processes and observations during the 1996 experimental flood*, exec. summary, pp. 26, Glen Canyon Dam Beach/Habitat-Building Flow Symposium, Grand Canyon Monitor. Res. Center, Bur. Reclam., Flagstaff, AZ, 1997.

Campana, M.E., G.J. Wroblicky J.A. Morrice, C.N. Dahm, H.M. Valett, and M.A. Baker, Hyporheic zone mixing and residence time distributions, *Geol. Soc. Am. Absts. Prog.*, 26, 286, 1994.

Crocker, M.T., and J.L. Meyer, Interstitial DOC in sediments of a southern Appalachian headwater stream, *J. N. Amer. Benthol. Soc.*, 6, 159-167, 1987.

Elwood, J. W., J.D. Newbold, R.V. O'Neill, and W. Van Winkle, Resource spiraling: an operational paradigm for analyzing lotic systems, in *Dynamics of lotic ecosystems*, ed. T.D. Fontaine and

S.M. Bartell, pp. 3-28, Ann Arbor Sci. Publ., Ann Arbor, MI, 1983.

Fraser, B.G., D.D. Williams, and K.W.F. Howard, Monitoring biotic and abiotic processes across the hyporheic /groundwater interface, *Hydrogeology J.*, 4, 36-50, 1996.

Grimm, N.B. and S.G. Fisher, Exchange between interstitial and surface water: implications for stream metabolism and nutrient cycling, *Hydrobiologia*, 111, 219-228, 1984.

Grimm, N. B. and S. G. Fisher, Nitrogen limitation in a Sonoran Desert stream, *J. N. Amer. Benthol. Soc.*, 5, 1-12, 1986.

Hach, *Hach Water Analysis Handbook*, 2nd Edition, 831 pp., Hach Company, Loveland, CO, 1992.

Hendricks, S. P. and D.S. White, Physicochemical patterns within a hyporheic zone of a Northern Michigan river, with comments on surface water patterns, *Can. J. Fish. Aquat. Sci.*, 48, 1645-1654, 1991.

Hendricks, S.P., and D.S. White, Seasonal biogeochemical patterns in surface water, subsurface hyporheic, and riparian ground water in a temperate stream ecosystem, *Arch. Hydrobiol.*, 134, 459-490, 1995.

Holmes, R.M., S.G. Fisher, and N. B. Grimm, Parafluvial nitrogen dynamics in a desert stream ecosystem, *J. N. Amer. Benthol. Soc.*, 13, 468-478, 1994.

Holmes, R.M., J.B. Jones, S.G. Fisher, and N.B. Grimm, Denitrification in a nitrogen-limited stream ecosystem, *Biogeochem.*, 33, 125-146, 1996.

Kaplinski, M.A., J.E. Hazel, and S.S. Beus, *Monitoring the effects of interim flows from Glen Canyon Dam on sand bars in the Colorado River corridor, Grand Canyon National Park, Arizona*, GCES annual rept., Bur. Reclam., Flagstaff, AZ, 1995.

Kearsley, M.J.C., and T. Ayers, *Effects of the 1996 beach/habitat-building flow on vegetation, seed banks, organic debris, and germination sites*, exec. summary, pp. 64, Glen Canyon Dam Beach/Habitat-Building Flow Symposium, Grand Canyon Monitor. Res. Center, Bur. Reclam., Flagstaff, AZ, 1997.

Molles, M.C., C.S. Crawford, and L.M. Ellis, Effects of an experimental flood on litter dynamics in the middle Rio Grande riparian ecosystem, *Reg. Rivers*, 11, 275-281, 1995.

Mulholland, P.J., E.R. Marzolf, J.R. Webster, D.R. Hart, and S.P. Hendricks, Evidence that hyporheic zones increase heterotrophic metabolism and phosphorus uptake in forest streams, *J. N. Limnol.-Oceanogr.*, 42, 443-451, 1997.

Orion Instruments, Boston, 36 pp., 1990.

Parnell, R.A., Jr., J.B. Bennett, and D.A. Strength, Effects of the 1996 Glen Canyon Dam controlled release on nutrient spiraling along the Colorado River corridor, Grand Canyon, *Eos, Trans. Amer. Geophys. Union*, 47, 46, 1996.

Parnell, R.A., A. Springer, L. Stevens, J. Bennett, T. Hoffnagle, T. Melis, and D. Staniski-Martin, *Flood-induced backwater rejuvenation along the Colorado River in Grand Canyon, Arizona*, exec. summary, p. 41-51, Glen Canyon Dam Beach/Habitat - Building Flow Symposium, Grand Canyon Monitor. Res. Center, Bur. Reclam., Flagstaff, AZ, 1997a.

Parnell, R.A., A. Springer, J. Bennett, and L. Stevens, *Effects of the Glen Canyon dam controlled flood on nutrient spiraling along the Colorado River corridor, Grand Canyon, Arizona*, exec. summary, pp. 52-55, Glen Canyon Dam Beach/ Habitat-Building Flow Symposium, Grand Canyon Monitor. Res. Center, Bur. Reclam., Flagstaff, AZ, 1997b.

Petroutson, W.D., *Interpretive simulations of advective flowpaths across a reattachment bar during different Colorado River flow alternatives*, M.S. thesis, N. Ariz. Univ., Flagstaff, AZ, 1997.

Pinay, G., C. Ruffinoni, S. Wondzell, and F. Gazelle, Change in groundwater nitrate concentration in a large river floodplain: denitrification, uptake, or mixing, *J. N. Amer. Benthol. Soc.*, 17, 179-189, 1998.

Ritzenthaler, E.A., and R. T. Edwards, Hyporheic flows within a floodplain backchannel in the Queets River, WA, *Bull. N. Amer. Benthol. Soc.*, 13, 127, 1996.

Rutherford, J.E., and H.B.N. Hynes, Dissolved organic carbon in streams and groundwater, *Hydrobiologia*, 154, 33- 48, 1987.

Schindler, J.E., and D.P. Krabbenhoft, The hyporheic zone as a source of dissolved organic carbon and carbon gases to a temperate forested stream, *Biogeochem.*, 43, 157-174, 1998.

Schmidt, J.C., and J.B. Graf, *Aggradation and degradation of alluvial sand deposits, 1965 to 1986, Colorado River, Grand Canyon National Park, Arizona*, U.S. Geol. Surv. Prof. Paper 1493, 1990.

Sedell, J.R., J.E. Richey, and F.J. Swanson, The river continuum concept: a basis for the expected ecosystem behavior of very large rivers? in Proceedings of the international large river symposium, ed. D.P. Dodge, *Can. Spec. Publ. Fish. Aquat., Sci.*, 106, 49-55, 1989.

Shannon, J.P., D.W. Blinn, P.L. Benenati, and K.P. Wilson, *O*rganic drift in a regulated desert river, *Can. J. Fish. Aquat. Sci.*, 53, 1360-1369, 1996.

Shannon, J.P., D.W. Blinn, K.P. Wilson, P.L. Benenati, J. Hagan, and C. O'Brien, *Impacts of the spring, 1996 spike flow from Glen Canyon Dam on the aquatic food base in the Colorado River through Grand Canyon, Arizona*, exec. summary, pp. 29-40, Glen Canyon Dam Beach/Habitat-Building Flow Symposium, Grand Canyon Monitor. Res. Center, Bur. Reclam., Flagstaff, AZ.

Springer, A.E., W.D. Petroutson, and B.A. Gilbert, Spatial and temporal variability of hydraulic conductivity in an active sedimentary environment, *Ground Water*, in press, 1999.

Stanford, J.A., and F.R. Hauer, Mitigating the impacts of stream and lake regulation in the Flathead River catchment, Montana, USA: an ecosystem perspective, *Aquat. Conserv.*, 2, 35-63, 1992.

Stanford, J.A., and F.R. Hauer, Fifth International Symposium on Regulated Streams, *Regul. Riv.*, 8, 1-2, 1993.

Stanford, J.A., J.V. Ward, W.J. Liss, C.A. Frissell, R.N. Williams, J.A. Lichatowich, and C.C. Coutant, A general protocol for restoration of regulated rivers, *Regul. Riv.*, 12: 391-413, 1996.

Stanley, E.H. and A.J. Boulton, Hyporheic processes during flooding and drying in a Sonoran Desert stream, *Arch. Hydrobiol.*, 134, 1-26, 1995.

Stevens, L.E., J.C. Schmidt, T.J. Ayers, and B.T. Brown, Flow regulation, geomorphology, and Colorado River marsh development in the Grand Canyon, Arizona, *Ecol. Applic.*, 6, 1025-1039, 1995.

Stevens, L.E., J.P. Shannon, and D.W. Blinn, Colorado River benthic ecology in Grand Canyon, Arizona, USA: dam, tributary, and geomorphological influences, *Reg. Rivers*, 13, 129-149, 1997.

Stevens, L.E., T.J. Ayers, K. Christiansen, M.J.C. Kearsley, V.J. Meretsky, R.A. Parnell, A.M. Phillips, J. Spence, M.K. Sogge, A.E. Springer, and D. L. Wegner, Planned flooding and riparian trade-offs: the 1996 Colorado River Planned Flood, *Ecol. Applic.*, in press, 1999.

U.S. Department of the Interior, *Operation of Glen Canyon Dam - Final Environmental Impact Statement*, Colorado River Storage Project, Coconino County, AZ, 337 pp. + appendices, Salt Lake City, UT, 1995.

Valdez, R.A., and R.J. Ryel, Life history and ecology of the humpback chub in the Colorado River in Grand Canyon, Arizona, in *Proceedings of the 3rd Biennial Conference of Research on the Colorado Plateau*, ed. C. van Riper and E.T. Deshler, Natl. Park Service Trans. Procs. Series NPS/NRNAU/NRTP-97/12, 3-31, 1997.

Valett, H.M., S.G. Fisher, and E.H. Stanley, Physical and chemical characteristics of the hyporheic zone of a Sonoran Desert stream, *J. N. Amer. Benthol. Soc.*, 9, 201-215, 1990.

Valett, H.M., C.C. Hakenkamp, and A. Boulton, Perspectives on the hyporheic zone: integrating hydrology and biology, *J. N. Amer. Benthol. Soc.*, 12, 40-43, 1993.

Valett, H.M., S.G. Fisher, N.B. Grimm, and P. Camill, Vertical hydrologic exchange and ecological stability of a desert stream ecosystem, *Ecol.*, 75, 548-560, 1994.

Ward, J.V., and J.A. Stanford, *T*he serial discontinuity concept of lotic ecosystems, in *Dynamics of lotic ecosystems*, ed. T. D. Fontaine and S. M. Bartell, pp. 29-42, Ann Arbor Sci. Publ., Ann Arbor, MI, 1983.

White, D.S., Perspectives on defining and delineating hyporheic zones, *J. N. Amer. Benthol. Soc.*, 12, 61-69, 1993.

Wondzell, S.M., and F.J. Swanson, Seasonal and storm dynamics of the hyporheic zone of a 4th-order mountain stream, *J. N. Amer. Benthol. Soc.*, 115, 20-34, 1996.

Wroblicky, G.J., M.E. Campana, H.M. Valett, and C.N. Dahm, Seasonal variation in surface-subsurface water exchange and lateral hyporheic area of two stream-aquifer systems, *Water Res. Res.*, 34, 317-328, 1998.

Roderic Parnell and Jeffrey B. Bennett, Geology Department, Northern Arizona University, Box 4099, Flagstaff, AZ 86011, 520-523-3329, 520-523-9220 fax; email: Roderic.Parnell@nau.edu

Lawrence E. Stevens, Grand Canyon Monitoring and Research Center, P.O. Box 22459, Flagstaff, AZ, 86002-2459

Changes in Number, Sediment Composition, and Benthic Invertebrates of Backwaters

Mark J. Brouder, David W. Speas, and Timothy L. Hoffnagle

Research Branch, Arizona Game and Fish Department, Phoenix, Arizona

The 1996 flood in the Colorado River, Grand Canyon, Arizona, induced changes in the number, sediment composition and benthic invertebrates of backwaters. A before and after comparison at similar discharge (227 m³/s) indicated that more backwaters were present immediately following the flood. Following the flood, sediment composition changed from approximately equal proportions of sand and silt to predominantly sand, and mean percentage of coarse particulate organic matter decreased. No changes in backwater sediments were observed from April through September. Mean densities of mollusks and simuliid and ceratopogonid dipterans decreased following the flood, as did mean biomass of dipterans (including chironomids). Benthic invertebrates were negatively impacted by the flood but recovered to pre-flood levels by September. Due to their short longevity, newly created backwaters provided little shelter for larval and juvenile native fishes during the summer growing period. Modifying the hydrograph of future floods may improve the utility of this tool for benefitting native fishes and their habitats.

1. INTRODUCTION

Backwaters have become increasingly important as rearing areas for larval and juvenile native fishes in the Colorado River due to habitat alterations associated with river regulation [*Holden*, 1978; *Valdez and Clemmer*, 1982; *Minckley*, 1991]. Backwaters of the Colorado River in Grand Canyon are pockets of low-velocity water connected to the main channel and are usually formed by recirculation currents in eddy complexes [*Schmidt and Graf*, 1990]. These areas are calm, warm and productive habitats for larval and juvenile native fishes [*Arizona Game and Fish*, 1996].

The creation and longevity of backwaters are largely dependent on river discharge and sediment input and movement throughout the system. Since closure of Glen Canyon Dam in 1963, the Little Colorado and Paria rivers have become the primary sources of sediment in the Colorado River [*Andrews*, 1991]. Over time, backwaters may fill with fine sand and silt and ultimately become marshes [*Stevens et al.*, 1995]. Alternatively, fluctuating river discharge and high winds may degrade reattachment bars, eventually resulting in inundation by the main channel. Large magnitude floods rejuvenate backwaters through scouring of marsh deposits and formation of sand bars. Mobilization of sediments can cause short-term negative impacts on benthic invertebrates and, consequently, fish. *Giller et al.* [1991] reported that sediment instability was responsible for reductions in diversity, density, and biomass of benthic invertebrates.

The Controlled Flood in Grand Canyon
Geophysical Monograph 110

Sediment composition of backwaters affects species composition and abundance of benthic invertebrates. Species richness is low in sandy substrates and greater in silt-sand substrates, while muddy substrates support greater biomass but not necessarily more species [*Hynes*, 1970]. *Ward and Stanford* [1979] and *Petts* [1984] reported that substrate particle size and stability may be the dominant factors controlling macroinvertebrate distribution. *Arizona Game and Fish* [1996] reported higher total benthic invertebrate densities in backwaters than in main channel beachface habitats, where substrates are relatively unstable. Because benthic invertebrates are in close contact with sediments and are relatively long-lived, they display greater sensitivity to environmental disturbances than many other organisms [*Reice and Wohlenberg*, 1993].

From 22 March through 7 April 1996, the U.S. Bureau of Reclamation conducted an controlled flood in the Colorado River, Grand Canyon, Arizona [*Schmidt et al.*, this volume]. This experiment was expected to benefit native fishes by rejuvenating or creating backwaters [*Department of the Interior*, 1996]. This chapter examines changes in backwater number, sediment composition and benthic invertebrates immediately following the flood and throughout the summer and early fall of 1996. Additionally, the potential effects of changes in sediment composition on benthic invertebrates and ultimately native fishes are discussed.

2. METHODS

The study area included nearly 226 river miles of the Colorado River in Grand Canyon between Lee's Ferry (mile 0) and Diamond Creek (mile 225.6), Arizona. Six backwaters were sampled both before and after the 1996 controlled flood from mile 44.27 to 201.06 [see map in *Webb et al.*, this volume]. These sites were selected on the basis of two criteria: 1) most had been repeatedly sampled for fish, sediments and benthic invertebrates by Arizona Game and Fish over the past 5 years and 2) they would remain backwaters following the experiment.

In this paper, "the flood" refers collectively to the periods of steady low flows (227 m^3/s) that preceded and followed the flood peak (1274 m^3/s) as well as the flood release itself [*Schmidt et al.*, this volume]. Backwaters were sampled before (28 February - 14 March 1996) and after (18 April - 3 May 1996) the flood (pre- and post-flood, respectively) and during June (18 June - 4 July) and 8-23 September. To determine changes in backwater number, backwaters were identified using aerial videography and counted during pre- and post-flood periods of steady low flows (227 m^3/s).

During pre- and post-flood sampling periods, sediment samples (50 mL) were collected from the foot, center, and mouth of each backwater using a modified 60 mL syringe and preserved in 40% isopropanol. Sediment components were separated into one of 4 categories: coarse (>65 μm) inorganic matter (sand), coarse particulate organic matter (CPOM), fine (<65 μm) inorganic matter (silt), or fine particulate organic matter (FPOM) [*Birkeland* 1984]. Temporal changes in the percentages of sand, silt, CPOM and FPOM were analyzed using Mann-Whitney U-tests with significance at α=0.05 [*Sokal and Rohlf*, 1981].

Additionally, 3 replicate benthic invertebrate samples (foot, center and mouth) were collected from each backwater with a petite Ponar dredge (0.0232 m^2) and filtered through a 600 μm mesh sieve bucket. Organisms were enumerated by the following categories: Crustacea (Ostrocoda and *Gammarus lacustris*), Chironomidae, other Diptera (Simuliidae and Ceratopogonidae), Mollusca (Gastropoda and Bivalvia), Nematoda, Oligochaeta and miscellaneous invertebrates (unidentified Annelida, Cladocera, Trichoptera, Hemiptera, Hymenoptera, Ephemeroptera and Odonata). Samples were ashed at 500°C for 2 hours to determine ash-free dry weight (AFDW) biomass. Benthic invertebrate density and biomass were analyzed using Kruskal-Wallis tests to detect differences by sampling trip and location (river mile). Pearson product/moment correlation coefficients were calculated to quantify relationships between sediment composition (percent sand, silt, CPOM and FPOM) and benthic invertebrate density and biomass. For all tests, statistical significance was determined at α=0.05.

3. RESULTS

3.1. Backwater Number

Using aerial videography, 31 backwaters were identified during the pre-flood (steady 227 m^3/s) period and 39 during the post-flood (steady 227 m^3/s) period.

3.2. Backwater Sediment Composition

Sediment composition of backwaters differed among sampling trips, with 3 of the 4 components changing between pre- and post-flood sampling periods (P≤0.0196; Figure 1). Percent sand was significantly higher (P=0.0166) after the flood (76.9%) than before (48.5%). Percent silt was significantly higher (P=0.0183) before the experiment (49.8%) than after (22.0%). Coarse particulate organic matter was also significantly higher (P=0.0016) before (0.39%) than after (0.18%) the experiment. Following the

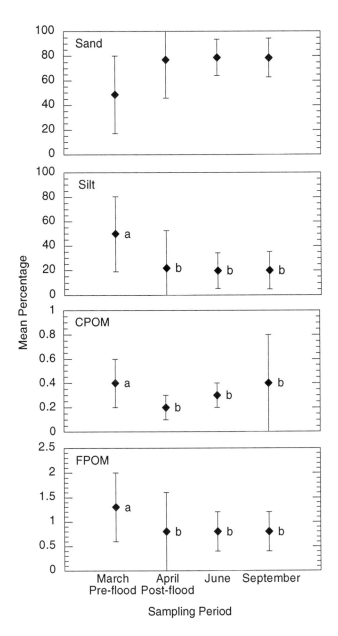

Figure 1. Mean (±1 SD) percentage of sand, silt, CPOM and FPOM in backwater sediments during pre-flood (March), post-flood (April), June and September sampling trips in 1996. Means with different letters are significantly different at a=0.05.

experiment, percent sand increased by 58.6% while percent silt decreased by 55.8% and percent CPOM decreased by 53.8%. There was no significant change (P=0.1336) in percent FPOM among sampling periods. No significant differences (P>0.0500) in percent sand, silt, CPOM or FPOM were detected among the post-flood, June and September samples.

3.3. Benthic Invertebrates

Mean densities of other dipterans (simuliids and certatopogonids) and mollusks decreased significantly after the flood (P≤0.0194; Figure 2). Densities of chironomids and other dipterans recovered to pre-flood densities by June, but mollusks did not recover until September. Densities of miscellaneous invertebrates were greater than pre-flood densities (P=0.0394) during June due to the presence of unidentified annelid worms.

Mean AFDW biomass of chironomids and other dipterans decreased after the flood (P≤0.0004) but attained pre-flood levels by June (Figure 3). Mean AFDW biomass of mollusks and total invertebrates were below pre-flood levels in June (P≤0.0055) but recovered to pre-flood densities in September.

Prior to the flood, total benthic invertebrate density varied by sampling location (P=0.0102), but not afterwards (Figure 4). Mean total AFDW biomass of invertebrates varied with sampling location prior to the experiment (P=0.0084) but not afterwards (P=0.1669). Pre-flood benthic invertebrate density and biomass were greatest from mile 44.27-60.85.

3.4. Sediment/Benthic Invertebrate Correlations

Correlations between pre- and post-flood sediment composition and changes in benthic invertebrate density and biomass were significant but weak. Mean densities of crustaceans and mollusks were inversely correlated with percentages of sand (R≤-0.4544; P≤0.0257) and positively correlated with percentages of silt in sediments (R≥0.4506; P≤0.0271). Also, mean densities of crustaceans were positively correlated with percentages of CPOM (R=0.4318; P=0.0351) and FPOM (R=0.4993; P=0.0130), mean densities of chironomids were positively correlated with percentages of CPOM (R=0.4636; P=0.0225) and mean densities of mollusks were positively correlated with percentages of FPOM (R=0.5388; P=0.0066). Mean AFDW biomass of crustaceans, miscellaneous dipterans, mollusks and total invertebrates were negatively correlated with percentages of sand (R≤-0.4689, P≤0.0208) and positively correlated with percentages of silt (R≥0.4653, P≤0.0219). Mean AFDW biomass of chironomids was positively correlated with percentages of CPOM (R=0.4944; P=0.0140), while mean AFDW biomass of crustaceans, mollusks and total invertebrates were positively correlated with percentages of FPOM (R≥0.5333, P≤0.0073).

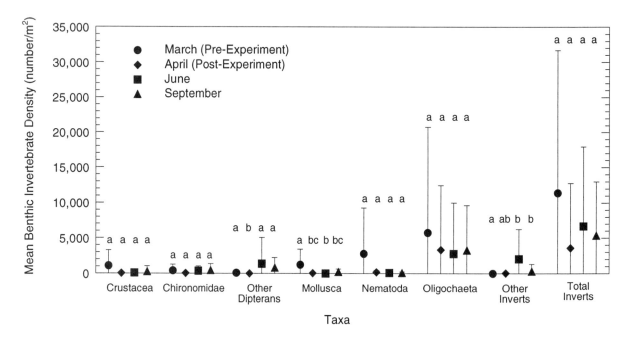

Figure 2. Mean total benthic invertebrate density during pre-flood (March), post-flood (April), June and September sampling trips in 1996. Means (within taxa) with different letters are significantly different at a=0.05.

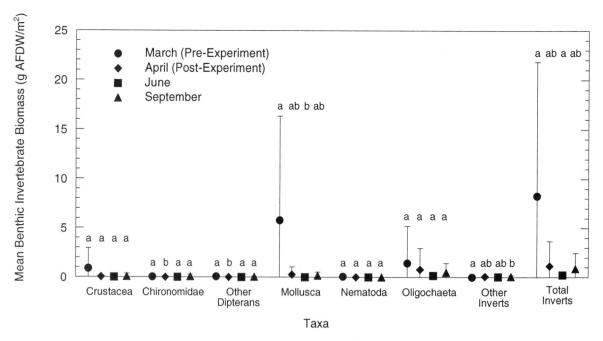

Figure 3. Mean total benthic invertebrate biomass during pre-flood (March), post-flood (April), June and September sampling trips in 1996. Means (within taxa) with different letters are significantly different at a=0.05.

4. DISCUSSION

Changes in backwater sediments and benthic invertebrates following the 1996 controlled flood were mostly as expected. Decreases in fine and organic sediments and benthic invertebrates were documented. The flood also created new backwaters; however, their longevity was very short.

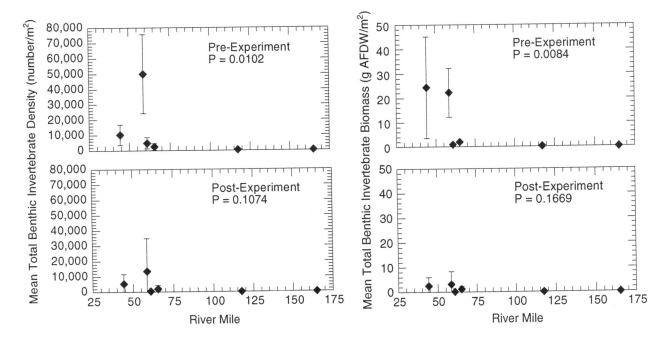

Figure 4. Mean total benthic invertebrate density and mean total benthic invertebrate biomass in backwaters at each river mile sampled during pre- and post-flood sampling trips in 1996.

Reattachment bars created by the flood were rapidly eroded due to dramatic differences in flow pattern between flood (1274 m³/s) and post-flood steady 227 m³/s flows and subsequent operating flows (227-552 m³/s). Most newly created backwaters were temporary and, by 2 weeks after the flood, were no longer available as rearing habitat for larval and juvenile native fishes. Floods with more gradual ramp rates, lower magnitude floods and/or floods of a different duration may lessen degradation of new reattachment bars and thus create more persistent backwaters [*Department of the Interior*, 1996].

Observed changes in sediment composition can be explained by the flood and the process of backwater formation. High percentages of silt in backwaters before the flood agree with data reported by *Stevens and Ayers* [1993], who found that in the absence of flooding, backwaters may fill with silt and detritus and ultimately become marshes. High water velocities during the flood scoured silt out of backwaters, while sand from the river channel bottom was suspended in the water column and deposited in low-velocity eddy recirculation zones, forming reattachment bars. Redeposition of FPOM must have occurred as the flood waters receded, as no change in percentages of FPOM were noted following the experiment. As water levels dropped, reattachment bars were exposed, partially isolating the eddy return-current channel, creating

or reforming backwaters [*Rubin et al.*, 1990; *Schmidt*, 1990; *Schmidt and Graf*, 1990].

Changes in density and/or biomass of benthic invertebrates are almost certainly related to the experiment, as sudden reductions in benthic invertebrate density and biomass during the month of April have not been observed in previous years (1991-1994; *Arizona Game and Fish*, 1996). However, it is unclear whether benthic standing stock losses were the direct result of sediment disturbance, desiccation during the periods of steady low flows (227 m³/s) that preceded and followed the flood, or a combination of both factors. Weak correlations between changes in sediment composition and changes in benthic invertebrate standing stocks suggest that alternative sources of variation (e.g., desiccation preceding and following flood releases) were responsible for much of the observed reductions in invertebrate density and biomass. *Blinn et al.* [1995] suggested that atmospheric exposure of benthos in the Colorado River in Grand Canyon is a more severe disturbance than flooding because benthic organisms are directly eliminated rather than displaced or buried. The latter processes allow mechanisms for rapid recolonization [e.g., from the hyporheic zone and/or downstream drift; *Williams and Hynes*, 1976; *Mackay*, 1992; *Palmer*, 1992]. Furthermore, invertebrate behavioral responses to drawdown (downstream drift, burrowing) were probably ineffective due to the unnaturally rapid rate of drawdown in the present experiment [*Perry and Perry*, 1986; *Yount and Niemi*, 1990]. Thus, it is likely that substantial portions of benthic standing stocks in backwaters were eliminated before the high release portion of the flood began.

Nevertheless, disturbance and redistribution of sediment during the flood certainly contributed in part to reduction of benthic standing stocks and, possibly, in their recolonization rates. Substrate type and stability are central to benthic invertebrate community organization and productivity [*Ward*, 1992]. Fine substrates, such as silt- and CPOM-enriched sand found in backwaters are typically productive but also unstable during floods. Thus, changes in sediment composition in backwaters likely affected the benthic community in two ways. First, the relative instability of backwater substrates during the high-release portion of the flood probably eliminated a substantial portion of standing stocks through stimulation of drift response, or passive displacement with fine sediments [*Yount and Niemi*, 1990]. Second, recolonization was probably hindered due to losses of finer, more organically enriched sediments.

Benthic-invertebrate standing stocks in backwater and main channel habitats are greatest at Lee's Ferry and decrease with distance downstream [*Shannon*, 1993; *Shannon et al.*, 1996; *Arizona Game and Fish*, 1996; *Stevens et al.*, 1997]. Our data from pre-flood sampling were consistent with this pattern; however, this productivity gradient was disrupted by the flood and was not observed in backwaters for the remainder of 1996. Pre-flood benthic-invertebrate densities were greatest from mile 44.27-60.85, but were reduced to levels which did not differ from downstream locations following the flood. Impacts of dam discharge on river ecosystems generally diminish with distance downstream [*Ward and Stanford*, 1983; *Blinn et al.*, 1995] and evidence from the present study indicates that this generality may apply also to planned flood releases.

Dipterans showed low resistance but high resilience to the 1996 controlled flood. *Ward* [1992] reports that dipterans suffer population losses of 77-78% during spates, but are capable of recovering to pre-flood levels within a median of 0.42 yr [*Niemi et al.*, 1990]. Recovery of dipterans in the Colorado River following the flood was rapid and well within the range of 0.01-20.0 years reported by *Niemi et al.* [1990]. *Shannon et al.* [1996] reported similar results for benthos collected from cobble riffles during the flood. Recovery of mollusks (primarily gastropods) following the flood took place well within their median recovery time of 0.79 yr [*Niemi et al.*, 1990]. Very little is known of molluscan ecology in the mainstem

Colorado River in Grand Canyon, although physid snails in its tributaries are thought to be particularly vulnerable to eradication by flash flooding [*Spamer and Bogan*, 1993]. Oligochaetes were largely unaffected by the flood, probably due to rapid recolonization from the drift [*Palmer et al.*, 1992]. Oligochaetes comprised a major proportion of the drift during the flood [*Shannon et al.*, 1996], and their proliferation probably accounted for the increase in miscellaneous invertebrates observed in June. Losses of dipterans following the flood may have impacted nutrition of larval and juvenile native fishes, particularly bluehead sucker (*Catostomus discobolus*), flannelmouth sucker (*C. latipinnis*), and the endangered humpback chub (*Gila cypha*), all of which utilize backwaters and consume benthic invertebrates found within [*Valdez and Ryel*, 1995; *Arizona Game and Fish*, 1996]. Chironomid and simuliid larvae are usually numerically dominant in gut contents of native fishes [*Kaeding and Zimmerman*, 1983; *Valdez and Ryel*, 1995; *Arizona Game and Fish*, 1996]. Based on our data, timing of the flood was successful in that impacts on the food base were probably minimized. Young-of-year native fish were found in the mainstem Colorado River during the latter half of June 1996 [*Arizona Game and Fish*, 1997], by which time critical food items in backwaters, such as dipterans, had recovered to pre-flood levels of density and biomass.

5. CONCLUSIONS

The 1996 controlled flood was responsible for considerable physical and biological modifications of the Colorado River ecosystem in Grand Canyon. Numerous sandbars and accompanying backwaters were created or enlarged, but they rapidly eroded under operating flows and thus were unavailable as native fish rearing habitats by late spring, 1996. Fortunately, negative impacts on the benthic invertebrate community were short-lived and, despite alterations to the sediment composition of backwaters, recovery of benthic invertebrates following the flood was rapid. The reduction in benthos densities did not overlap with dispersal of young-of-the-year fish from tributaries to mainstem backwaters. However, the reduction in available backwater habitat may have resulted in decreased growth and/or survival of juvenile native fishes in the mainstem Colorado River. Modifications of future flood hydrographs are necessary to create reattachment sandbars which are more resistant to erosion. For example, floods with more gradual ramp rates, lower magnitude floods and/or floods of a different duration may lessen degradation of new reattachment bars and thus create more persistent backwaters.

Acknowledgments. The authors would like to thank Dave Wegner, Mike Yard, Ted Melis and LeAnn Skrzynski of Glen Canyon Environmental Studies and Bill Persons of Arizona Game and Fish Department for their logistical support. Tom Dresser and a myriad of AGFD volunteers helped collect and analyze samples. Ric Bradford, Tony Robinson and Jody Walters (AGFD) and Rich Valdez (SWCA) and 3 anonymous reviewers reviewed and improved earlier drafts of this manuscript.

REFERENCES

Andrews, E.D., Sediment transport in the Colorado River Basin, in *Colorado River Ecology and Dam Management*, pp. 54-74, Natl. Acad. Press, Washington, D.C., 1991.

Arizona Game and Fish Department, *Ecology of Grand Canyon backwaters: Glen Canyon Environmental Studies Final Report*, Bur. Recl., Salt Lake City, UT, Coop. Agree. No. 9-FC-4007940, 155 pp., Ariz. Game Fish Depart., Phoenix, AZ, 1996.

Arizona Game and Fish Department, *Glen Canyon Environmental Studies 1996 Annual Report*, Glen Canyon Environ. Studies, Bur. Recl., Flagstaff, AZ, Ariz. Game Fish Dept., Phoenix, AZ, 1997.

Birkeland, P.W., *Soils and geomorphology*, 372 pp., Oxford Univ. Press, New York, 1984.

Blinn, D.W., J.P. Shannon, L.E. Stevens, and J.P. Carder, Consequences of fluctuating discharge for lotic communities, *J. N. Am. Benthol. Soc.*, 14, 2, 233-248, 1995.

Department of the Interior, *Glen Canyon beach/habitat-building test flows: Final environmental assessment and finding of no significant impact*, 67 pp., Bur. Recl., Upper Colo. Reg., Salt Lake City, UT, 1996.

Giller, P.S., N. Sangpraduh, and H. Twomey, Catastrophic flooding and macroinvertebrate community structure, *Verh. Internat. Verein. Limnol.*, 24, 1724-1729, 1991.

Holden, P.B., A study of the habitat use and movement of rare fishes in the Green River, Utah, *Trans. Bonn. Chap. Am. Fish. Soc.*, 107, 64-89, 1978.

Hynes, H.B., *The Ecology of Running Waters*, 555 pp., Liverpool University Press, Liverpool, UK, 1970.

Kaeding, L.R., and M.A. Zimmerman, Life history and ecology of the humpback chub in the Little Colorado and Colorado River of Grand Canyon, *Trans. Am. Fish. Soc.*, 112, 577-594, 1983.

Mackay, R.J., Colonization by lotic macroinvertebrates: a review of processes and patterns, *Can. J. Fish. Aquat. Sci.*, 49, 617-628, 1992.

Minckley, W.L., Native fishes of the Grand Canyon region: an obituary? in *Colorado River Ecology and Dam Management*, pp. 124-177, Natl. Acad. Press, Washington, D.C., 1991

Niemi, G.J., P. DeVore, N. Detenbeck, D. Taylor, A. Lima, J. Pastor, J.D. Yount, and R. J. Naiman, Overview of case studies on recovery of aquatic systems from disturbance, *Environ. Manage.*, 14, 5, 571-587, 1990.

Palmer, M.A., A.E. Bely, and K.E. Berg, Response of invertebrates to lotic disturbance: a test of the hyporheic refuge hypothesis, *Oecologia*, 89, 182-194, 1992.

Perry, S.A., and W.B. Perry, Effects of experimental flow regulation on invertebrate drift and stranding in the Flathead and Kootenai Rivers, Montana, USA, *Hydrobiol.*, 134, 171-182, 1986.

Petts, G.E., *Impounded Rivers: Perspectives for Ecological Management*, 326 pp., John Wiley Sons, New York, 1984.

Reice, S.R., and M. Wohlenberg, Monitoring freshwater benthic invertebrates and benthic processes: measures for assessment of ecosystem health, in *Freshwater Biomonitoring and Benthic Macroinvertebrates*, edited by D.M. Rosenberg and V.H. Resh, pp. 287-305, Routledge, Chapman Hall, Inc., New York, 1993.

Rubin, D.M., J.C. Schmidt, and J.N. Moore, Origin, structure, and evolution of a reattachment bar, Colorado River, Grand Canyon, Arizona, *J. Sed. Petrol.*, 60, 982-991, 1990.

Schmidt, J.C., Recirculating flow and sedimentation in the Colorado River in Grand Canyon, Arizona, *J. Geol.*, 98, 709-724, 1990.

Schmidt, J.C., and J.B. Graf, *Aggradation and degradation of alluvial sand deposits, 1965 to 1986, Colorado River, Grand Canyon National Park, Arizona*, U.S. Geol. Surv. Prof. Paper 1493, 74 pp., 1990.

Shannon, J.P., *Aquatic ecology of the Colorado River through Grand Canyon National Park, Arizona*, M.S. thesis, N. Ariz. Univ., Flagstaff, AZ, 1993.

Shannon, J. P., D.W. Blinn, K.P. Wilson, P.L. Benanati, and G.E. Oberlin, *Interim flow and beach building spike flow effects from Glen Canyon Dam on the aquatic food base in the Colorado River in Grand Canyon National Park, Arizona*, Annual Report, 156 pp., N. Ariz. Univ., Flagstaff, AZ 1996.

Spamer, E.E., and A.E. Bogan, *Mollusks of the Colorado River Corridor, Grand Canyon, AZ*, Rept. Glen Canyon Environ. Studies, 40 pp., Acad. Natur. Sci. Philadelphia, Philadelphia, PA, 1991.

Sokal, R.R., and F.J. Rohlf, *Biometry*, 2nd. ed., 859 pp., W.H. Freeman Company, New York, 1981.

Stevens, L.E., and T.A. Ayers, *The impacts of Glen Canyon Dam on riparian vegetation on soil stability in the Colorado River corridor, Grand Canyon, Arizona*, 1991 Final Admin. Rept., Natl/ Park Serv. Coop. Studies Unit, N. Ariz. Univ., Flagstaff, 1993.

Stevens, L.E., J.C. Schmidt, T.J. Ayers, and B.T. Brown, Flow regulation, geomorphology and Colorado River marsh development in the Grand Canyon, Arizona, *Ecol. Appl.*, 4, 1025-1039, 1995.

Stevens, L.E., J.P. Shannon, and D.W. Blinn, Colorado River benthic ecology in Grand Canyon, Arizona, USA: dam, tributary and geomorphological influences, *Reg. Riv. Res. Mgmt.*, 13, 129-149, 1997.

Valdez, R.A., and G.H. Clemmér, Life history and prospects for recovery of the humpback and bonytail chub, in *Fishes of the Upper Colorado River System: Present and Future*, ed. W.H.

Miller and H.M Tyus, pp. 109-119, Amer. Fish. Soc., Bethesda, MD, 1982.

Valdez, R.A., and R.J. Ryel, Life history and ecology of the humpback chub (*Gila cypha*) in the Colorado River, Grand Canyon, Arizona, BIO/WEST Rept. TR-250-08, 286 pp., BIO/WEST, Inc., Logan, UT, 1995.

Ward, J.V., *Aquatic Insect Ecology: Biology and Habitat*, 438 pp., J. Wiley Sons, New York, 1992.

Ward, J.V., and J.A. Stanford, Ecological factors controlling stream zoobenthos, in *The Ecology of Regulated Streams*, edited by T.D. Fontaine and S.M. Bartell, pp. 35-56, Ann Arbor Sci., Ann Arbor, MI, 1979.

Ward, J.V., and J.A. Stanford, The serial discontinuity concept of lotic ecosystems, in *Dynamics of Lotic Ecosystems*, edited by T.D. Fontaine and S.M. Bartell, pp. 29-42, Ann Arbor Sci., Ann Arbor, MI, 1983.

Williams, D.D., and H.B.N. Hynes, Recolonization mechanisms of stream benthos, *Oikos*, 27, 265-272, 1976.

Yount, J.D., and G.J. Niemi, Recovery of lotic communities and ecosystems from disturbance: a narrative review of case studies, *Environ. Manage.*, 14, 547-569, 1990.

Mark J. Brouder, David W. Speas, and Timothy L. Hoffnagle, Arizona Game and Fish Department, Research Branch, 2221 West Greenway Road, Phoenix, Arizona 85023; email: mbrouder@gf.state.az.us

Lotic Community Responses in the Lees Ferry Reach

T. McKinney, R.S. Rogers,

Arizona Game and Fish Department, Page, Arizona

A. D. Ayers, and W. R. Persons

Arizona Game and Fish Department, Phoenix, Arizona

Responses of periphyton, aquatic macrophytes, benthic macroinvertebrates, and rainbow trout to the 1996 controlled flood were investigated in the Lees Ferry tailwater reach below Glen Canyon Dam on the Colorado River. Lotic biota differed spatially and temporally in abundance and distribution following recession of flood waters, and there was no evidence that the flood benefitted trout or lower trophic levels. The flood was associated with short-term changes in lower trophic levels, but benthic vegetation and macrofauna with low resistance were resilient. Adverse impacts of the flood on lower trophic levels were greater and more prolonged in depositional areas than on cobble bar habitat, but recovery occurred in both habitat types 4-8 months after the flood. The flood likely resulted in some downstream displacement of smaller fish but had no effects on catch rate or condition indices of trout. Percentage of young-of-the-year trout 8 months after the event indicates that the flood did not prevent successful spawning. The flood had little direct influence on diets of trout, but relative gut volume increased in the week after the event, remained high in summer, and composition changed seasonally. Amphipods (*Gammarus lacustris*), chironomids, and snails were predominant food items, and *Gammarus* generally were eaten more often and comprised greater relative volume in the diet than other macroinvertebrate taxa.

1. INTRODUCTION

Floods are a common type of disturbance in unregulated streams and may dramatically influence the structure and dynamics of lotic communities [*Minshall*, 1988; *Steinman and McIntire*, 1990; *Wallace*, 1990; *Yount and Niemi*, 1990]. The response of lotic communities to floods differs widely, depending on frequency, severity and timing, but

The Controlled Flood in Grand Canyon
Geophysical Monograph 110

floods and spates generally reduce standing stocks of primary and secondary producers. Recovery of biotic assemblages following floods may require a few weeks to several years [*Wallace*, 1990; *Yount and Niemi*, 1990; *Pearsons et al.*, 1992], and frequency and timing of events influence recovery and resistance [*Tett et al.*, 1978; *Peterson*, 1996; *Peterson and Stevenson*, 1992; *Barrat-Segretain and Amoros*, 1995]. The best indices of ecosystem recovery remain uncertain, and no theoretical model is available to predict recovery by lotic communities following floods [*Yount and Niemi*, 1990].

Fish communities tend to be persistent in unregulated streams exposed to frequent flooding [*Matthews*, 1986; *Meffe and Minckley*, 1987; *Meffe and Berra*, 1988]. Floods can alter assemblage structure and abundance of fishes [*Pearsons et al.*, 1992] and may impact smaller fish more than adults [*Seegrist and Gard*, 1972; *Hanson and Waters*, 1974; *Harvey*, 1987]. Few investigation have addressed effects of floods on lotic biota in large regulated rivers [*Niemi et al.*, 1990; *Steinman and McIntire*, 1990; *Yount and Niemi*, 1990]. Natural floods are inherently unpredictable in duration, magnitude and timing, but floods incorporating discharge manipulations from hydroelectric power facilities allow timely comparison of the same sites prior to and following disturbance. Lack of pre-disturbance data commonly hinders flood-related investigations [*Lamberti et al.*, 1991].

Controlled floods are an element common to potential operating regimes for Glen Canyon Dam on the Colorado River, Arizona [*U.S. Department of Interior*, 1995]. The first controlled flood was implemented March 22 to April 7, 1996, to test effects of high discharges on sediment and lotic biota below the dam. This paper describes effects of this controlled flood on lotic communities in the Lees Ferry reach. The Lees Ferry reach extends between Glen Canyon Dam (-15.8 Mile) and Lees Ferry (0 Mile), and differs from the downstream system in that hypolimnetic releases from Lake Powell result in cold stenothermic, clear, and physicochemically stable tailwaters [*Stanford and Ward*, 1991; *Stevens et al.*, 1997]. We tested null hypotheses that the 1996 flood had no effects on periphyton, aquatic macrophytes, epiphytic diatoms, benthic macrofauna, and rainbow trout (*Oncorhychus mykiss*) in the Lees Ferry Reach.

2. METHODS

2.1. Periphyton and Aquatic Macrophytes

We collected samples of benthic vegetation prior to (March 22; pre-flood steady 227 m^3/s discharge) and following the controlled flood (April 4; post-flood steady 227 m^3/s discharge), July 28 and November 24. We collected cobbles (10-20 cm diameter) in a random manner along transects (-4.1 and -14.0 Mile) parallel to river flow (142 m^3/s flow stage). One to three 4.15 cm^2 areas (n = 15-16/site) of periphyton within a template were scraped from the upper surface of each cobble [*Angradi and Kubly*, 1993]. Pheophytin-corrected chlorophyll *a* content of periphyton (n = 6 March and April; n = 5, July and November) was determined spectrophotometrically [*Tett et al.*, 1975], and mass (ash free dry weight, AFDW) was determined by loss on ignition (550^0 C, 2 hrs).

We collected submerged macrophytes from depositional substrate (-3.5 Mile) using a Hess sampler (0.087 m^2, 0.25 mm mesh) at randomly-located points along a transect parallel to river flow (142 m^3/s flow stage). Subsamples (20 g wet weight) of macrophytes were added to 200 ml of deionized water and homogenized for 2 min in a blender. Aliquots (10 ml) of the homogenate were filtered onto a glass-fiber filter and analyzed for chlorophyll *a* [*Tett et al.*, 1975]. AFDW was determined for the unhomogenized portion of the sample by loss on ignition (550^0 C, 2 h). Chlorophyll *a* content and mass were expressed in units of mg/m^2.

We collected additional periphyton and macrophyte samples (142 m^3/s flow stage) for analyses of diatom epiphytes. We collected cobbles along the transects described above and pooled (N=1/site) 4.15 cm^2 quadrats of periphyton scraped from the surface of each cobble. We collected samples of (N=1/site) of *Chara contraria* from three locations along the transect at -3.5 Mile. We clipped stalks (N = 2-3 at each location) of the macroalga between the second and fifth nodes below the growing tip. We pooled within-site samples of periphyton and *Chara* and preserved them in Transeau's Solution.

Diatom epiphytes were identified initially at 1000x magnification under oil immersion. Subsequent identification and enumeration were done at 20x magnification using a Sedgewick-Rafter counting cell. Two or more 0.09 mm^3 fields were observed from each of three counting cells to identify and count a minimum of 500 algal units per sample, although counts often exceeded 1500 algal units. Diatom densities on cobbles were calculated on the basis of cells per square millimeter of the area sampled on the stones; densities on *C. contraria* were calculated on the basis of number of cells per milligram AFDW.

We surveyed submerged macrophytes along each shoreline between Lees Ferry and GCD in March 16-17, April 15-16, July 15-16, and November 13-14, 1996. Distribution and relative abundance were estimated visually, plotted on topographic maps and ranked ordinally: 1 = low vertical growth, patchy and sparse distribution; 2 = moderate vertical growth, moderate and occasionally patchy distribution; and 3 = higher vertical growth, extensive, generally continuous distribution.

2.2. Macroinvertebrates

We collected benthos samples (Hess sampler, 0.087 m^2, 0.25-mm mesh) during pre- and post-flood steady flows (N=5/site) and during July and November (N=3/site) at the same transects, flow stage, and on the same dates as for the periphyton and macrophyte samples. Samples were preserved in 10% formalin, sieved (0.25-mm mesh), and macroinvertebrates were identified and sorted.

TABLE 1. Mean±1 standard error of mass (AFDW, g/m^2) and chlorophyll a for periphyton at -14.0 and -4.1 Mile and macrophytes at -3.5 Mile before (March) and following (April to November) the 1996 controlled flood

Month	PERIPHYTON -14.0 Mile		PERIPHYTON -4.1 Mile		MACROPHYTES -3.5 Mile	
	AFDW (g/m^2)	Chlorophyll a (mg/m^2)	AFDW (g/m^2)	Chlorophyll a (mg/m^2)	AFDW (g/m^2)	Chlorophyll a (mg/m^2)
March	143.5±19.9	1787.6±182.1	135.8±15.2	1104.0±284.5	96.8±15.5	352.7±30.7
April	160.7±18.0	1106.2±144.1	99.6±19.0	669.1±53.1	n.d.	n.d.
July	161.7±17.4	1846.1±336.3	139.7±27.2	1322.1±438.3	n.d.	n.d.
November	97.2±18.0	858.8±126.6	108.4±19.3	672.9±172.2	44.4±10.1	244.8±41.6

n.d., no data.

2.3. Rainbow trout

We electrofished 14-15 transects (ca. 33 min/transect) between dusk and dawn and between -15.0 Mile and -3.0 Mile during pre- (March 23-25) and post-flood (April 5-7) steady discharges, then again during August 28-30 (discharge ca. 425 m^3/s) and November 18-20 (discharge ca. 227 m^3/s). Rainbow trout were measured as total length (TL), weighed, and released unless collected for analysis of diet. Stomachs N = 30-60/trip) were preserved in 10% formalin, and contents were identified to the lowest possible taxonomic category and measured (±0.1 ml) using volumetric displacement.

2.4. Data Analysis

Analysis of variance (ANOVA) was performed on means for macroinvertebrates and periphyton standing stock (AFDW, chlorophyll a), total lengths, weights, condition factors (K = weight x 10^5/TL^3), and relative gut volumes (RGV = volume of stomach contents in milliliters/fish length in meters; *Filbert and Hawkins*, 1995) and proportional composition of ingested items. Planned comparisons were conducted on data from the pre- and post-flood steady discharges. *Post hoc* analyses (Duncan's Multiple Range test) were used to explore patterns of differences between pairs of months. Analyses were performed on transformed data for benthic amphipods, oligochaetes and chironomid larvae (log[x+1]) and chironomid pupae and turbellarians (sqrt[x+1]+sqrt[x]).

Mean diatom densities and ordinal rankings of macrophyte abundance and distribution were compared between pre- and post-flood discharge periods using the Mann-Whitney U-test, and mean ordinal rankings were compared over all periods using the Kruskal-Wallis ANOVA. Chi-square tests were used to compare frequencies of occurrence of empty stomachs and of predominant taxonomic groups in the diet. Power of ANOVAs were computed for tests that failed to reject null hypotheses [*Peterman*, 1990; *Sokal and Rohlf*, 1995].

3. RESULTS

3.1. Periphyton and Aquatic Macrophytes

Mean mass of periphyton (predominantly *Cladophora glomerata*) did not differ between March and April or among sampling periods (Table 1), but power of the ANOVA was low (0.979). Mean chlorophyll a content differed among all sampling periods (P<0.01), declined from March to April (P<0.01), and was less (P<0.05) at one cobble bar (-14.0 Mile) in November than March (Table 1). Total diatom densities and densities of large/upright species declined in the reach (P<0.05), but densities categorized as small/adnate taxa were similar (P>0.05) between these months (Table 2).

Submerged macrophytes at -3.5 Mile were removed by the flood but colonized at the site by November (Table 1). Abundance and distribution of macrophytes throughout the reach (ordinal rankings based on river surveys) differed among sampling periods (P<0.02). Mean ±1 standard error of the ordinal rankings were: March, 1.5±0.2, N=19; April, 0.6±0.1, N=29; July, 2.1±0.1, N= 19; and November 1.7±0.1, N=34. *Chara contraria* was the most abundant macrophyte taxon in the tailwater prior to the flood. *Potamogeton pectinatus* occurred sparsely prior to the flood but colonized extensively and was dominant by July. Abundance and distribution of macrophytes declined between March and April (P<0.02) and increased in July. *Chara* and *Potamogeton* colonized extensively in November, when relative abundance and distribution of macrophytes did not differ significantly (P>0.05) from that

TABLE 2. Mean±1 standard error of densities of total and dominant small adnate and large upright diatoms on cobbles and *Chara contraria* before (March) and following (April to November) the 1996 controlled flood in the Lees Ferry reach. Densities for *C. contaria* in April and July reflect total loss of macrophytes from the sampling site.

	Diatoms	March	April	July	November
Cobbles	Total	32,310.5±6107.4	10,501.0±80.0	19,984.0±235.0	2725.0±863.1
(number/mm^2)	Small adnate	6106.5±565.5	3407.0±297.0	8281.0±250.3	775.5±171.8
	Large upright	21,330.0±7857	5432.5±1161.8	9402.0±1279.9	1479.0±1031.0
C. contaria	Total	68,328	0	0	377,722
(number/mg AFDW)	Small adnate	25,696	0	0	173,978
	Large upright	26,280	0	0	163,744

prior to the flood, and *Chara* hosted abundant epiphytic diatoms (Table 2). *Egeria densa* (Brazilian elodea) was observed infrequently prior to November, when abundance and distribution became extensive. Exposed sand substrata (no observable macrophytes) were virtually absent in March, but occurred extensively in April.

3.2. Macroinvertebrates

Macroinvertebrate densities (Table 3) differed among sampling periods (P<0.001), but interaction between transect locations and sampling periods was significant (P<0.02). Total benthic macroinvertebrate densities did not differ at any site between March and April (P>0.05). Densities of individual taxa (Table 3) except oligochaetes (Power=0.586) differed among sampling periods (P<0.01), and interaction was significant (P<0.01) between transects and sampling periods for all taxa except *Gammarus lacustris*. Densities of the amphipod declined (P<0.01) from March to April in depositional and cobble areas (-3.5 and -14.0 Mile), but were similar between months (P>0.05) at -4.1 Mile. Amphipod densities at all sites in July and November exceeded (P<0.01; -14.0 Mile) or were similar (P>0.05) from the pre-flood concentrations.

Although densities of chironomid larvae (Table 3) differed (P<0.001) among sampling periods, densities were similar (P>0.05) between March and April (Power=0.787). Densities in July exceeded (P<0.01) those in March at -4.1 Mile and -3.5 Mile, but were lower (P<0.001) at -14.0 Mile. However, densities of larvae in November were less than (P<0.001) pre-flood concentrations at -14.0 Mile and -3.5 Mile. Densities of chironomid pupae (Table 3) differed among sampling periods (P<0.001), but were not significantly different (P>0.05) between March and April at -14.0 Mile and -3.5 Mile (Power=0.787).

Gastropod densities differed among sampling periods (P<0.001) and declined (P<0.05) between March and April

at -4.1 Mile (Table 3). Densities in November exceeded (P<0.05) those in March on cobble bars (-4.1 and -14.0 Mile). Turbellarian densities in July and November exceeded (P<0.001) those in March at -14.0 Mile and -4.1 Mile, but were similar to pre-flood concentrations at -3.5 Mile.

3.3. Seasonal Patterns of Change

Comparisons of seasonal patterns of change between present data (Tables 1-3) and 1991-97 (Table 4) suggest that trends in standing stocks of periphyton, macrophytes, *Gammarus*, chironomids, gastropods, and oligochaetes (no seasonal pattern of change for oligochaetes) following the controlled flood exhibit seasonal trends comparable to those prior to the event. However, densities of diatoms (on cobbles: July 1994 = 45,300 /mm^2, November 1993 = 95,500 /mm^2; on *Chara*: November 1993 = 23,175 /mg; T. McKinney, unpublished data, 1997), gastropods, and macrophytes were lower than seasonal maxima during 1991-1997 (Tables 2, 3, and 4). All samples collected before March 1996 conformed with current procedures.

3.4. Rainbow trout

Catch per unit effort (CPUE, Table 4) of trout did not differ between March and April but was lower in August and November. The proportional catch of trout <152 mm TL [assumed to be mainly young-of-the-year; Arizona Game and Fish Department, unpublished data] increased more than 33% in November, compared to previous months (Table 5), consistent with a normal pattern of seasonal change. Prior to the controlled flood, trout <152 mm TL in late fall to early winter typically comprise 30-55% of fish captured by electrofishing (proportion of trout <152 mm TL in electrofishing samples: October 1993, 55%; December

TABLE 3. Mean±1 standard error of densities of benthic macroinvertebrates on cobble (-14.0 and -4.1 Mile) and depositional (-3.5 Mile) habitats in March (pre-flood) and April to November (post-flood) of 1996

		DENSITIES OF BENTHIC MACROINVERTEBRATES (number/m^2)					
Month	Site (Mile)	*Gammarus lacustris*	Oligochaetes	Gastropods	Chironomid Larvae	Chironomid Pupae	Turbellarians
March	-14.0	404.6±147.0	41.4±13.4	27.6±3.9	3931.0±717.9	480.5±92.4	0
	-4.1	1331.0±276.8	1354.0±27.2	154.0±(20.8)	114.9±34.1	0	1802.3±273.1
	-3.5	1448.3±389.1	4223.0±2460.8	59.8±(9.9)	310.3±88.4	39.1±16.9	1285.1±249.1
April	-14.0	121.8±51.3	48.3±8.5	32.2±(7.6)	2763.2±841.0	703.4±174.4	0
	-4.1	820.7±245.2	3634.5±2045.6	48.3±(15.2)	416.1±146.7	62.1±39.8	579.3±122.3
	-3.5	512.6±390.4	1519.5±699.3	87.4±22.0	223.0±53.8	80.5±31.1	669.0±256.7
July	-14.0	2046.0±278.1	931.0±496.0	95.8±23.3	160.9±74.8	57.5±57.4	888.9±534.4
	-4.1	1379.3±40.4	977.0±309.1	137.9±40.4	2444.4±1049.8	636.0±257.0	1578.5±400.2
	-3.5	444.4±93.9	30.7±20.3	7.7±7.7	2796.9±855.7	398.5±63.7	0
Nov	-14.0	5153.3±953.6	92.0±23.9	118.8±20.3	95.8±61.7	0	15.3±7.7
	-4.1	7835.3±2005.2	241.4±28.9	245.2±20.3	30.7±7.7	0	3632.2±344.9
	-3.5	2532.6±981.7	494.3±154.3	88.1±16.7	19.2±13.8	0	1103.5±481.2

1994, 30%; December 1995, 55%; and November 1996, 49%; T. McKinney, unpublished data, 1997).

We captured trout 46-593 mm TL, and mean lengths and weights differed (P<0.05) among sampling periods (Table 5). Trout caught in April were longer (P<0.001) and heavier (P<0.05) than those captured in March. Mean length was less in November than in March (P<0.05), but mean weights (Power=2.551) and condition factors (Power=0.819) were similar (P>0.05) between months (Table 5). Condition indices in all months and CPUE in August and November were similar to those observed in the tailwater during 1994-1995 [T. McKinney, unpublished data, 1997: condition factor, April 1994=0.97, December 1994=0.98, May 1995=0.93, December 1995=0.95; CPUE: July 1994=2.62, December 1994=1.90, May 1995=2.75, December 1995=2.87].

Stomachs were collected from trout 121-538 mm TL. Proportional composition of trout diets differed significantly among sampling periods (P<0.001), and individual components differed in patterns of change (Table 6). *Cladophora* tended to be the predominant ingested item except in November. Amphipods, chironomids and snails dominated macroinvertebrates in the diet, and other taxa (Diptera, oligochaetes, terrestrial invertebrates) each generally comprised less than 2% of stomach content volume. Proportional composition of individual taxa in the diet did not differ (P>0.05) between pre- and post-flood steady flows, but composition of *G. lacustris* (P<0.001) and

snails (P<0.02) increased above pre-flood levels in November, while proportional composition of chironomids and *Cladophora* declined (P<0.001).

Relative gut volume also differed (P<0.001) among sampling periods (Table 6). The RGV increased between pre- and post-flood steady discharges (P<0.01) and remained greater than prior to the flood in August (P<0.01), but declined from pre-flood levels in November (P<0.002). Frequency of occurrence of empty stomachs did not differ (P>0.05) among sampling periods (Table 6). Overall, more trout ate *Gammarus* than chironomids or snails (P<0.001), and more fish consumed amphipods and snails (P<0.05), but fewer ate chironomids and *Cladophora* (P<0.001) in November than prior to the flood.

4. DISCUSSION

Our results indicate that a controlled flood of brief duration and moderate magnitude in the spring influenced dynamics of the lotic system and produced short-term changes primarily in lower trophic levels. Some benthic vegetation and macrofauna were reduced by the event, but most were resistent, except that lower trophic levels were affected more on depositional than on cobble bar habitats. The flood also had little measurable impact on rainbow trout in the tailwater reach. No clear benefits of floods were apparent for lower trophic levels or trout. Conclusions for this study must be viewed with some caution due to limita-

TABLE 4. Mean±1 standard error of seasonal densities of macrophytes, periphyton collected in the Lees Ferry reach, 1991-97. Sample sizes: macroinvertebrates–spring = 132, summer = 132, fall = 58, winter =87; periphyton–spring = 175, summer = 257, fall = 249, winter = 175; macrophytes–spring = 15, summer = 34, fall = 40, winter = 15

	SEASONAL DENSITIES OF MACROPHYTES (g/m^2 AFDW) AND BENTHIC MACROFAUNA (number/m^2)			
Biota	Spring	Summer	Fall	Winter
Gammarus	1707.0±177.4	2030.1±139.3	5661.7±703.5	2175.5±161.2
Chironomids	1826.3±252.0	1070.3±150.5	434.3±76.3	709.1±141.0
Gastropods	111.1±11.2	244.2±34.7	497.7±93.5	326.7±58.6
Oligochaetes	928.1±256.3	822.6±137.2	729.1±108.6	952.4±173.4
Periphyton	153.9±6.4	171.8±7.7	114.0±5.5	123.4±5.6
Macrophytes	121.8±23.3	30.4±7.0	68.6±10.9	126.0±21.9

tions of data, including lack of true replicates and temporal control and generally low statistical power.

We found that changes in abundance and distribution of biota in the Lees Ferry reach differed spatially and temporally following the controlled flood, as evidenced by many interactions between month and site. The flood reduced diatom densities and standing stock of aquatic macrophytes and *G. lacustris*, but recovery occurred in 8 months or less. Other biota generally were resistant to the flood. All species examined followed expected seasonal patterns of change following the flood. Although it was not possible to establish an experimental control (not exposed to flooding) system, changes in benthic flora and macrofauna suggest that flooding can initiate a complex series of adjustments in regulated riverine environments [*Petts*, 1984].

4.1. Periphyton and Aquatic Macrophytes

Contrary to what has been reported for unregulated systems [*Whitton*, 1970; *Fisher et al.*, 1982; *Biggs and Close*, 1989; *Scrimgeour and Winterbourn*, 1989; *Steinman and McIntire*, 1990; *Lamberti et al.*, 1991; *Dodds and Gudder*, 1992], the controlled flood failed to reduce periphyton mass in the Lees Ferry reach. Lack of an effect likely was related to low magnitude of the flood and low entrainment of sediment particles that act to dislodge algae [*Steinman and McIntire*, 1990; *Stanford and Ward*, 1991; *Peterson*, 1996] and possibly reduce light penetration during the flood [*Fisher and Grimm*, 1988; *Shannon et al.*, in review]. Lower chlorophyll-*a* content of periphyton following the flood likely reflected loss of diatom epiphytes [*Haines et al.*, 1987; *Dodds*, 1991] due to water shear [*Biggs*, 1995]. *Tett et al.* [1978] also reported that phytopigment decreased abruptly after floods. Periphyton mass exhibited seasonal trends in July and November,

consistent with the natural seasonal cycle in streams [*Steinman and McIntire*, 1990].

Densities of total epiphytic diatoms and of large upright taxa were reduced by the flood, and densities in July and November were lower on cobbles, but higher in November on *Chara*, than in previous years. Diatom epiphytes often are reduced by floods [*Grimm and Fisher*, 1989; *Peterson et al.*, 1994], and large upright taxa are impacted more than are small adnate species [*Robinson and Rushforth*, 1987]. Small adnate taxa typically dominate diatom epiphyte assemblages following severe scour events, but colonization by both physiognomic groups may be rapid [*Power and Stewart*, 1987; *Peterson and Stevenson*, 1992; *Peterson*, 1996]. Diatoms also may have colonized denuded substrata [*Grimm and Fisher*, 1989; *Peterson et al.*, 1994], providing food base for benthic macrofauna. Diatoms with small adnate physiognomies are less available than large upright taxa to benthic grazers [*Gregory*, 1983; *Steinman*, 1996]. Thus, the food base for benthic macroinvertebrates that consume diatoms — for example, amphipods [*Blinn et al.*, 1992, 1994, 1995; *Angradi*, 1994; *Stevens et al.*, 1997] — likely was reduced in the Lees Ferry reach by the flood.

The flood reduced distribution and abundance of aquatic macrophytes in the tailwater reach and *Chara* did not recolonize until the fall. Colonization of flood-denuded areas by *Chara* spp. also required several months in an Oklahoma stream [*Power and Stewart*, 1987]. *Potamogeton* colonized extensively in the Lees Ferry tailwater soon after the flood, likely increasing habitat and food base above those associated with exposed sand substrate [*Menon*, 1969; *Krull*, 1970; *Pip and Stewart*, 1976; *Pip*, 1978; *Sand-Jensen et al.*, 1989; *Newman*, 1991; *Angradi*, 1994; *Wollheim and Lovvorn*, 1996]. Macrophyte communities often are decimated by floods, influencing trophic webs [*Bilby*, 1977; *Power and Stewart*, 1987; *Barrat-Segretain and Amoros*,

TABLE 5. Total catch, mean±1 standard error of total length, weight, and condition factor, and catch per unit effort (CPUE) and percentage of catch <152 mm for rainbow trout caught by electrofishing prior to (March) and following (April to November) the 1996 controlled flood in the Lees Ferry reach

Month	Total Catch (number)	Total Length (mm)	Weight (g)	Condition Factor (K)	CPUE (number/ min)	Percent of Catch <152 mm
March	1513	230.8±2.8	198.5±5.7	0.961±0.006	3.52	35.9%
April	1685	239.9±2.6	211.2±5.1	0.954±0.005	3.58	28.3%
August	1306	228.4±3.2	232.0±6.7	0.979±0.010	2.61	36.5%
November	1335	214.7±3.2	208.2±6.5	0.986±0.010	2.58	49.1

1995]. Loss of macrophytes can reduce habitat and the food base for fish and macroinvertebrates [*Menon*, 1969; *Pip and Stewart*, 1976; *Pip*, 1978; *Sand-Jensen et al.*, 1989; *Hanson*, 1990; *Newman*, 1991; *Blinn et al.*, 1992; *Angradi*, 1994; *Stevens et al.*, 1997].

4.2. Macroinvertebrates

Macroinvertebrate populations often are negatively impacted by flooding [*Meffe and Minckley*, 1987; *Scrimgeour and Winterbourn*, 1989; *Giller et al.*, 1991; *Lamberti et al.*, 1991; *Cobb et al.*, 1992; *Palmer et al.*, 1996], but we found only moderate evidence of this following the 1996 controlled flood. Resistance and recovery of benthic fauna differed spatially and temporally in the Lees Ferry reach following the flood. *Gammarus*, more than other taxa, lacked resistance to the flood, but the amphipod was resilient, and all taxa examined generally exhibited seasonal patterns of change in densities comparable to non-flood conditions. Recovery by *Gammarus* occurred more slowly in depositional than on cobble habitat and was coincident in the soft-sediment area with colonization by *Chara*, suggesting that an algal resource is important in colonization dynamics of macroinvertebrates [*Robinson et al.*, 1990]. Macroinvertebrate populations reduced by floods often recover rapidly [*Fisher et al.*, 1982; *Meffe and Minckley*, 1987; *McElravy et al.*, 1989; *Lamberti et al.*, 1991; *Mackay*, 1992].

4.3. Rainbow trout

The controlled flood had no direct influence on CPUE or condition indices of rainbow trout in the Lees Ferry reach. Condition indices in all months and CPUE in August and November were similar to those observed in the tailwater during 1994-95. However, mean size of trout captured increased, and the proportion of fish <152 mm TL declined in the week after the flood, suggesting some downstream

displacement of smaller-sized trout [*Seegrist and Gard*, 1972; *Harvey*, 1987; *Lamberti et al.*, 1991].

Rainbow trout in the Lees Ferry reach spawn primarily during late fall and winter, and larvae may remain in the gravel for a week to a month after hatching before emerging as free-swimming fry [*Kondolf et al.*, 1989]. Flood impacts likely would be greatest when eggs are in the gravel and when fry are emerging [*Seegrist and Gard*, 1972; *Hanson and Waters*, 1974; *Pearsons et al.*, 1992]. Thus, it is likely that most trout fry had emerged prior to the controlled flood. The high proportion of wild-spawned (stocked trout were distinguished by presence of coded wire tags implanted at the hatchery) young-of-the-year fish in our electrofishing samples during fall and proportional similarity in previous years indicate that the flood did not prevent successful spawning.

We electrofished only at night and assume that diel changes in feeding behavior [*Bisson*, 1978; *Angradi and Griffith*, 1990] likely had little influence on trout diets in our study. Increased food intake in the week following the flood likely reflected opportunistic feeding associated with greater drift of macroinvertebrates [*Elliott*, 1973; *Scullion and Sinton*, 1983; *Bres*, 1986; *Filbert and Hawkins*, 1995]. Food intake failed to correspond with declines in benthic macroinvertebrate densities following the flood, suggesting that reductions in the benthos had little impact on food availability to fish. Composition of stomach contents changed seasonally during the study, and the pattern was similar to that previously observed in the tailwater [*Angradi et al.*, 1992], suggesting long-term temporal stability and continued low diversity of available food in the reach. Among predominant food items, *Gammarus* generally were eaten more than other macroinvertebrates, and chironomids were consumed more than snails. *Cladophora* also was consumed frequently and comprised a large percentage of stomach contents, but the alga likely provides little direct nutritional benefit [*Angradi*, 1994], although epiphytic diatoms provide high lipid content that apparently is utilized by rainbow trout [*Leibfried*, 1988].

TABLE 6. Percent frequencies of occurrence and mean±1 standard error of percent composition by volume and relative gut volume (RGV) of predominant items in stomachs of rainbow trout prior to (March) and following (April to November) the 1996 controlled flood

	March (n=36)		April (n=30)		August (n=60)		November (n=54)	
Ingested Item	Frequency (%)	Volume (RGV)	Frequency (%)	Volume (RGV)	Frequency (%)	Volume (RGV)	Frequency (%)	Volume (RGV)
Gammarus	62.5	25.2±6.1	74.1	38.1±6.7	75.9	31.6±5.1	82.4	71.6±5.2
Chironomids	71.5	23.8±6.7	54.6	8.0±3.6	59.6	14.3±3.5	15.8	8.3±3.6
Gastropods	9.7	2.1±1.3	7.8	0.1±0.1	24.8	9.1±2.9	35.1	6.0±2.2
Cladophora	58.5	46.2±7.8	62.4	50.1±9.1	58.3	43.0±5.6	12.3	7.1±3.1
RGV		4.8±1.0		11.8±2.6		11.7±1.4		3.9±0.7
Empty Stomachs	16.7	0	16.7	0	9.0	0	22.9	0

In conclusion, the controlled flood of moderate magnitude and short duration had limited impact on, and no apparent benefits to, lotic biota. Future research and management decisions should address questions of spatial, temporal, and species variability in resistance and resiliency and the potential effects of frequency, severity and seasonal timing if we hope to predict influences of controlled floods on biota in the Lees Ferry reach.

Acknowledgments. Funding for this study was provided in part by the U.S. Bureau of Reclamation, Glen Canyon Environmental Studies, Cooperative Agreement No. 9-FC-40-07940.

REFERENCES

Angradi, T.R., Trophic linkages in the lower Colorado River: multiple stable isotope evidence, *J.N. Am. Benthol. Soc.*, 13, 479-495, 1994.

Angradi, T.R., and J.S. Griffith, Diel feeding chronology and diet selection of rainbow trout (*Oncorhynchus mykiss*) in the Henry's Fork of the Snake River, Idaho, *Can. J. Fish. Aquat. Sci.*, 47, 199-209, 1990.

Angradi, T.R., and D.M. Kubly, Effects of atmospheric exposure on chlorophyll *a*, biomass and productivity of the epilithon of a tailwater river, *Regul. Rivers*, 8, 345-358, 1993.

Angradi, T.R., R.W. Clarkson, D.A. Kinsolving, D.M. Kubly, and S.A. Morgensen, *Glen Canyon Dam and the Colorado River: responses of the aquatic biota to dam operations*, Research Rept., 155 pp., Ariz. Game Fish Dept., Phoenix, AZ, 1992.

Barrat-Segretain, M.H., and C. Amoros, Influence of flood timing on the recovery of macrophytes in a former river channel, *Hydrobiologia*, 316, 91-101, 1995.

Biggs, B.J.F., The contribution of flood disturbance, catchment geology and land use to the habitat template of periphyton in stream ecosystems, *Freshwater Biol.*, 33, 419-438, 1995.

Biggs, B.J.F., and M.E. Close, Periphyton dynamics in gravel bed rivers: the relative effects of flows and nutrients. *Freshwater Biol.*, 22, 209-231, 1989.

Bilby, R., Effects of a spate on the macrophyte vegetation of a stream pool, *Hydrobiologia*, 56, 109-112, 1977.

Bisson, P.A., Diel food selection by two sizes of rainbow trout (*Salmo gairdneri*) in an experimental stream, *J. Fish. Res. Board Can.*, 35, 971-975, 1978.

Blinn, D.W., L.E. Stevens, and J.P. Shannon, *The effects of Glen Canyon Dam on the aquatic food base in the Colorado River corridor in Grand Canyon, Arizona*, Rept. CA-8009-8-0002, 98 pp., Glen Canyon Environ. Studies, Flagstaff, AZ, 1992.

Blinn, D.W., L.E. Stevens, and J.P. Shannon, *Interim flow effects from Glen Canyon Dam on the aquatic food base in the Colorado River in Grand Canyon National Park, Arizona*, Final Rept., 136 pp., Glen Canyon Environ. Studies, Flagstaff, AZ, 1994.

Blinn, D.W., J.P. Shannon, L.E. Stevens, and J.P. Carder, Consequences of fluctuating discharge for lotic communities, *J.N. Am. Benthol. Soc.*, 14, 233-248, 1995.

Bres, M., A new look at optimal foraging behaviour: rule of thumb in the rainbow trout, *J. Fish Biol.*, 29, 25-36, 1986.

Cobb, D.G., T.D. Galloway, and J.F. Flannagan, Effects of discharge and substrate stability on density and species composition of stream insects, *Can. J. Fish. Aquat. Sci.*, 49, 1788-1795, 1992.

Dodds, W.K., Micro-environmental characteristics of filamentous algal communities in flowing freshwaters, *Freshwater Biol.*, 25, 199-209, 1991.

Dodds, W.K., and D.A. Gudder, The ecology of *Cladophora*, *J. Phycol.*, 28, 415-427, 1992.

Elliott, J.M., The food of brown and rainbow trout (*Salmo trutta* and *S. gairdneri*) in relation to drifting invertebrates in a mountain stream, *Oecologia*, 12, 329-347, 1973.

Filbert, R.B., and C.P. Hawkins, Variation in condition of rainbow trout in relation to food, temperature, and individual length in the Green River, Utah, *Trans. Am. Fish. Soc.*, 124, 824-835, 1995.

Fisher, S.G., and N.B. Grimm, Disturbance as a determinant of structure in a Sonoran Desert stream ecosystem, *Verh. Internat. Verein. Limnol.*, 23, 1183-1189, 1988.

Fisher, S.G., L.J. Gray, N.B. Grimm, and D.E. Busch, Temporal succession in a desert stream ecosystem following flash flooding, *Ecol. Mono.*, 52, 93-110, 1982.

Giller, P.S., N. Sangpradub, and H. Twomey, Catastrophic flooding and macroinvertebrate community structure, *Verh. Internat. Verein. Limnol.*, 24, 1724-1729, 1991.

Gregory, S.V., Plant-herbivore interactions in stream ecosystems, in *Stream Ecology: Application and Testing of General Ecological Theory*, ed. J.R. Barnes and G.W. Minshall, pp. 17-190, Plenum Press, NY, 1983.

Grimm, N.B., and S.G. Fisher., Stability of periphyton and macro-invertebrates to disturbance by flash floods in a desert stream, *J.N. Am. Benthol. Soc.*, 8, 293-307, 1989.

Haines, D.W., K.H. Rogers, and F.E.J. Rogers, Loose and firmly attached epiphyton: their relative contributions to algal and bacterial carbon productivity in a *Phragmites* marsh, *Aquat. Bot.*, 29, 169-176, 1987.

Hanson, D.L., and T.F. Waters, Recovery of standing crop and production rate of a brook trout population in a flood-damaged stream, *Trans. Am. Fish. Soc.*, 103, 431-439, 1974.

Hanson, J.M., Macroinvertebrate size-distribution of two contrasting feshwater macrophyte communities, *Freshwater Biol.*, 24, 481-491, 1990.

Harvey, B.C., Susceptibility of young-of-the-year fishes to downstream displacement by flooding, *Trans. Am. Fish. Soc.* 116, 851-855, 1987.

Kondolf, G.M., S.S. Cook, H.R. Maddux, and W.R. Persons, Spawning gravels of rainbow trout in Glen and Grand Canyons, Arizona, *J. Arizona-Nevada Acad. Sci.*, 23, 19-28, 1989.

Krull, J.N., Aquatic plant-macroinvertebrate associations and waterfowl, *J. Wildl. Manage.*, 34, 707-718, 1970.

Lamberti, G.A., S.V. Gregory, L.R. Ashkenas, R.C. Wildman, and K.M.S. Moore, Stream ecosystem recovery following a catastrophic debris flow, *Can. J. Fish. Aquat. Sci.*, 48, 196-208, 1991.

Leibfried, W.C., *The utilization of* Cladophora glomerata *and epiphytic diatoms as a food resource by rainbow trout in the Colorado River below Glen Canyon Dam, Arizona*, M.S. thesis, N. Ariz, Univ., Flagstaff, AZ, 1988.

Mackay, R.J., Colonization by lotic macroinvertebrates: a review of processes and patterns, *Can. J. Fish. Aquat. Sci.*, 49, 617-628, 1992.

Matthews, W.J., Fish faunal structure in an Ozark stream: stability, persistence and a catastrophic flood, *Copeia*, 1986, 388-397, 1986.

McElravy, E.P., G.A. Lamberti, and V.H. Resh, Year-to-year variation in the aquatic macroinvertebrate fauna of a northern California stream, *J.N. Am. Benthol. Soc.*, 8, 51-63, 1989.

Meffe, G.K., and T.M. Berra, Temporal characteristics of fish assemblage structure in an Ohio stream, *Copeia*, 1988, 684-690, 1988.

Meffe, G.K., and W.L. Minckley, Persistence and stability of fish and invertebrate assemblages in a repeatedly disturbed Sonoran Desert stream, *Am. Midl. Natur.*, 117, 177-191, 1987.

Menon, P.S., Population ecology of *Gammarus lacustris* Sars in Big Island Lake. I. Habitat preference and relative abundance, *Hydrobiologia*, 33, 14-32, 1969.

Minshall, G.W., Stream ecosystem theory: a global perspective, *J.N. Am. Benthol. Soc.*, 7, 263-288, 1988.

Newman, R.M., Herbivory and detritivory on freshwater macro-phytes by invertebrates: a review, *J.N. Am. Benthol. Soc.*, 10, 89-114, 1991.

Niemi, G.J., P. deVore, N. Detenbeck, D. Taylor, A. Lima, J. Pastor, J.D. Yount, and R.J. Naiman, Overview of case studies on recovery of aquatic systems from disturbance, *Environ. Manage.*, 14, 571-587, 1990.

Palmer, M.A., P. Arensburger, A.P. Martin, and D.W. Denman, Disturbance and patch-specific responses: the interactive effects of woody debris and floods on lotic invertebrates, *Oecologia*, 105, 247-257, 1996.

Pearsons, T.N., H.W. Li, and G.A. Lamberti, Influence of habitat complexity on resistance to flooding and resilience of stream fish assemblages, *Trans. Am. Fish. Soc.*, 121, 427-436, 1992.

Peterman, R.M., Statistical power analysis can improve fisheries research and management, *Can. J. Fish. Aquat. Sci.*, 47, 2-15, 1990.

Peterson, C.G., Response of benthic communities to natural physical disturbance, in *Algal Ecology: Freshwater Benthic Ecosystems*, ed. R.J. Stevenson, M.L. Bothwell, and R.L. Lowe, pp. 375-402, Academic Press, NY, 1996.

Peterson, C.G., and R.J. Stevenson, Resistance and resilience of lotic algal communities: importance of disturbance timing and current, *Ecol.*, 73, 1445-1461, 1992.

Peterson, C.G., A.C. Weibel, N.B. Grimm, and S.G. Fisher, Mechanisms of benthic algal recovery following spates: comparison of simulated and natural events, *Oecologia*, 98, 280-290, 1994.

Petts, G.E., *Impounded Rivers*, 326 pp., John Wiley Sons, NY, 1984.

Pip, E., A survey of the ecology and composition of submerged aquatic snail-plant communities, *Can. J. Zool.*, 56, 2263-2279, 1978.

Pip, E., and J.M. Stewart, The dynamics of two aquatic plant-snail associations, *Can. J. Zool.*, 54, 1192-1205, 1976.

Power, M.E., and A.J. Stewart, Disturbance and recovery of an algal assemblage following flooding in an Oklahoma stream, *Am. Midl. Natur.*, 117, 333-345, 1987.

Robinson, C.T., and S.R. Rushforth, Effects of physical distur-bance and canopy cover on attached diatom community structure in an Idaho stream, *Hydrobiologia*, 154, 49-59, 1987.

Robinson, C.T., G.W. Minshall, and S.R. Rushforth, Seasonal colonization of macroinvertebrates in an Idaho stream, *J.N. Am. Benthol. Soc.*, 9, 240-248, 1990.

Sand-Jensen, K., E. Jeppesen, K. Nielsen, L. Van Der Bijl, L. Hjermind, L.W. Nielsen, and T.M. Iversen, Growth of macro-phytes and ecosystem consequences in a lowland Danish stream, *Freshwater Biol.*, 22, 15-32, 1989.

Scrimgeour, G.J., and M.J. Winterbourn, Effects of floods on epilithon and benthic macroinvertebrate populations in an unstable New Zealand river, *Hydrobiologia*, 171, 33-44, 1989.

Scullion, J., and A. Sinton, Effects of artificial freshets on substratum composition, benthic invertebrate fauna and invertebrate drift in two impounded rivers in mid-Wales, *Hydrobiologia*, 107, 261-269, 1983.

Seegrist, D.W., and R. Gard, Effects of floods on trout in Sagehen Creek, California, *Trans. Am. Fish. Soc.*, 101, 478-492, 1972.

Shannon, J.P., D.W. Blinn, T. McKinney, P.L. Benenati, K.P. Wilson, and C. O'Brien, Aquatic food base response to the 1996 test flood below Glen Canyon Dam: Colorado River, Arizona, *Ecological Applications*, In Review.

Sokal, R.R., and F.J. Rohlf, *Biometry*, third edition, 887 pp, W.H. Freeman Co., New York, 1995.

Stanford, J.A., and J.V. Ward, Limnology of Lake Powell and chemistry of the Colorado River, in *Colorado River Ecology and Dam Management*, ed. G.R. Marzolf, pp. 75-101, Natl. Acad. Press, Washington, D.C., 1991.

Steinman, A.D., Effects of grazers on freshwater benthic algae, in *Algal Ecology: Freshwater Benthic Ecosystems*, ed. R.J. Stevenson, M.L. Bothwell, and R.L. Lowe, pp. 341-373, Academic Press, New York, 1996.

Steinman, A.D., and C.D. McIntire, Recovery of lotic periphyton communities after disturbance, *Environ. Manage.*, 14, 589-604, 1990.

Stevens, L.E., J.P. Shannon, and D.W. Blinn, Colorado River benthic ecology in Grand Canyon, Arizona, USA: dam, tributary and geomorphological influences, *Regul. Rivers*, 13, 129-149, 1997.

Tett, P., M.G. Kelly, and G.M. Hornberger, A method for the spectrophotometric measurement of chlorophyll *a* and pheophytin *a* in benthic microalgae, *Limnol. Oceanogr.*, 20, 887-896, 1975.

Tett, P., C. Gallegos, M.G. Kelly, G.M. Hornberger, and B.J. Cosby, Relationships among substrate, flow, and benthic microalgal pigment density in the Mechums River, Virginia, *Limnol. Oceanogr.*, 23, 785-797, 1978.

U.S. Department of Interior, *Operation of Glen Canyon Dam: Final Environmental Impact Statement*, 337 pp., Bur. Reclam., Salt Lake City, UT, 1995.

Wallace, J.B., Recovery of lotic macroinvertebrate communities from disturbance, *Environ. Manage.*, 14, 605-620, 1990.

Whitton, B.A., Biology of *Cladophora* in freshwaters, *Water Res.*, 4, 457-476, 1970.

Wollheim, W.M., and J.R. Lovvorn, Effects of macrophyte growth forms on invertebrate communities in saline lakes of the Wyoming high plains, *Hydrobiologia*, 323, 83-96, 1996.

Yount, J.D., and G.J. Niemi, Recovery of lotic communities and ecosystems from disturbance — a narrative review of case studies, *Environ. Manage.*, 14, 547-569, 1990.

T. McKinney, Arizona Game and Fish Department, P.O. Box 1651, Page, AZ 86040; email: tmckin@dcaccess.com

R.S. Rogers, Arizona Game and Fish Department, P.O. Box 1651, Page, AZ 86040

A.D. Ayers, Arizona Game and Fish Department, 2221 W. Greenway Road, Phoenix, AZ 85023

W.R. Persons, Arizona Game and Fish Department, 2221 W. Greenway Road, Phoenix, AZ 85023

Response of Benthos and Organic Drift to a Controlled Flood

Dean W. Blinn, Joseph P. Shannon, Kevin P. Wilson, Chris O'Brien, and Peggy L. Benenati

Department of Biological Sciences, Northern Arizona University, Flagstaff, Arizona

The controlled flood in the Colorado River below Glen Canyon Dam, Arizona, provided valuable information on short-term responses for both the riverine system and the biotic community, but the long-term effects of the flood on the aquatic food base were more difficult to assess. The 1274 m³/s discharge flushed the silt/clay fraction from the channel bottom throughout the river corridor. There were no significant differences in dissolved oxygen, specific conductance, and pH before and after the flood compared to during the flood. However, water clarity was dramatically reduced during the first 2 days of the flood event, but cleared after 7 days. Over 90% of the phytobenthos and ≥50% of the benthic invertebrates were scoured from the Lees Ferry reach, with biota associated with unstable fine sediment most vulnerable. Most of the dissolved organic carbon (DOC) and particulate organic carbon (POC) that passed through the river corridor was entrained in the initial hydrostatic wave; values for DOC and POC were significantly lower throughout the remainder of the flood. Stable isotope analyses indicated that riparian and upland vegetation made up most of the stream drift during the experimental flood, whereas phytobenthos was the dominant drift constituent during normal dam operations. Recovery rates to pre-flood levels were fast for phytobenthos (1 mon) and invertebrates (2 mon). We propose that the rapid recover rates and current high standing stock of aquatic benthos in the river corridor is more a function of higher water clarity, due to higher relatively constant dam releases, rather than solely related to the controlled flood. Our data indicate that consistent high discharges (≥400 m³/s) from Glen Canyon Dam mitigate the influence of suspended sediments delivered from tributaries on water clarity. Therefore, optimum conditions for management of the present exotic food base below Glen Canyon Dam may be achieved by steady discharges (~450 m³/s) with minimal fluctuation cycles (~50 m³/s).

1. INTRODUCTION

Natural floods play critical roles in the level of standing crop, community structure, and energy flow in lotic ecosystems [*Fisher et al.*, 1982; *Resh et al.*, 1988]. Extreme discharges can rearrange substrata [*Allen,* 1951], selectively scour benthic organisms from substrata [*Statzner and Higler*, 1986; *Power and Stewart*, 1987; *Robinson and Rushforth*, 1987; *Duncan and Blinn*, 1989], re-suspend nutrients and sediments [*Fisher and Minckley*, 1978; *Newcombe and MacDonald*, 1991; *Allan*, 1995], and modify geomorphic and riparian features of stream channels [*Schmidt and Rubin*, 1995; *Toner and Keddy*, 1997]. In some instances, high discharges may selectively benefit certain biota by increasing nutrient exchange within dense periphyton mats and filamentous algal tufts as well as reduce interspecific competition and predation from larger biota that are removed by hydraulic scour [*Statzner and Higler*, 1986; *Duncan and Blinn*, 1989; *Peterson*, 1996].

Prior to the closure of Glen Canyon Dam in 1963, flows in the Colorado River through Grand Canyon National Park fluctuated dramatically but predictably over an annual cycle. Typically, average annual maximum flows were ~2550 m³/s in the spring and early summer, while average minimum flows in the winter were ~114 m³/s [*Stanford and Ward*, 1986; *Carothers and Brown*, 1991; *Collier et al.*, 1996]. Every decade or so, a flood of 3400 m³/s raged through the canyon, with the largest gauged discharge at ~5664 m³/s during the summer of 1921.

After the closure of Glen Canyon Dam, standing crops of benthic algae and macroinvertebrates were high in the clear tailwaters and provided a trophy trout fishery in the Glen Canyon reach below the dam [*Leibfried and Blinn*, 1986; *Usher and Blinn*, 1990; *Blinn and Cole*, 1991]. However, native fish populations were being threatened due to alterations in habitat, cold water, and perhaps the reduced food base throughout the canyon [*Minckley*, 1991]. Due to declining resources in the Colorado River ecosystem, Congress passed the Grand Canyon Protection Act of 1992. Later, it was recommended that a "flood" discharge be released from Glen Canyon Dam to help restore the sediment and biological resources of the Colorado River ecosystem. The controlled flood event was conducted during 22 March to 7 April of 1996 and consisted of a 1274 m³/s discharge with a 4 day pre- and post-drawdown of 227 m³/s [*Schmidt et al.*, this volume].

This study measured selected physico-chemical features, stream benthos, and drift at selected sites along the 227 mile river corridor in Glen, Marble and Grand Canyons prior to, during, and after the controlled flood. Changes in food base composition below Glen Canyon Dam were also measured with stable isotopes during the controlled flood.

2. STUDY AREA

The Colorado River flows 460 km through northern Arizona, between Glen Canyon Dam and Lake Mead. The general geomorphology and ecology of the Grand Canyon river corridor as well as the discharge records for those gaging stations below the dam in operation during the 1996 controlled flood are described by *Webb et al.* [this volume] and *Schmidt et al.* [this volume], respectively.

Glen Canyon Dam impounds Lake Powell, a reservoir that receives inflow largely from the catchments of the

Colorado and San Juan Rivers [*Stanford and Ward*, 1991]. The constancy in light, nutrients and thermal properties of the water released from the upper hypolimnion of Lake Powell provide an environment for high standing crops of the filamentous green alga, *Cladophora glomerata* and associated epiphytic diatoms as well as chironomid larvae, *Gammarus lacustris*, and gastropods (namely *Physella*) in the upper 25 km tailwaters of Glen Canyon Dam [*Blinn and Cole*, 1991; *Stanford and Ward*, 1991; *Blinn et al.*, 1992; *Blinn et al.*, 1995; *Shannon et al.*, 1996a]. These biotic communities are greatly modified and reduced below the seasonally sediment-laden confluences of the Paria and Little Colorado rivers [*Shannon et al.*, 1994; *Shaver, et al.*, 1997; *Stevens et al.*, 1997).

3. METHODS

3.1. Physico-Chemical Measurements

Sediment (ca. 500 g) was collected at Lees Ferry (River mile, RM 0.0), Nankoweap (RM 52.7), and Spring Canyon (RM 203) pools with either a Peterson dredge or a Petite Ponar prior to and post (2 and 6 mon) the controlled flood to evaluate channel erosion during the 1274 m³/s discharge. Sediments were oven-dried at 60°C and mechanically sieved for percent particle size using the Wentworth Scale according to *Cummins* [1962] and *Minshall* [1984]: silt/clay (<0.063 mm), sand (0.063-0.5 mm), coarse sand (0.5-1.0 mm), and gravel (>1.0 mm).

Water temperature (°C), specific conductance (mS), dissolved oxygen (mg/L), and pH were measured with a Hydro-Lab Scout II® prior to, daily during the controlled flood and post flood at Lees Ferry. We also monitored light intensity (log lumens/m²) at Lees Ferry (RM 0.0) and below the Little Colorado River (LCR) at Tanner Canyon (RM 68) during the experimental flood with submersible On-Set® data loggers. These instruments were maintained 50 cm below the water surface during the flood to measure light attenuation. Secchi depth (water transparency) was also measured at selected intervals prior to and post the controlled flood at Lees Ferry, below the Paria River at Two-Mile Wash (RM 1.9), and at Tanner Canyon.

Samples for dissolved organic carbon (DOC) were collected at selected intervals immediately below Glen Canyon Dam (RM -16.0), Lees Ferry, Tanner, and Hells Hollow (RM 172.7) during the controlled flood. Three 50 ml samples were drawn directly from the top 0.5 m of the river away from shore in at least 1 m of water with a Millipore Swinex syringe and filtered through a Whatman GF/F glass fiber filter into a sterile glass bottle. The sample was then acidified (pH <2) with concentrated sulfuric acid. Two hundred fifty µml aliquots were injected into a

Rosemount/Dohrmann DC-180 Total Organic Carbon Analyzer at least three times or until the standard deviation was <10% of the mean and DOC (mg/L) estimates were calculated.

3.2. Benthic Collections

Collection sites for the benthos were selected in reaches with relatively high numbers of exotic and/or native fishes, as well as, above and below tributary confluences that seasonally deliver large loads of suspended sediments to the mainstem. General locations for these collections included: 1) Lees Ferry, 2) Two-Mile Wash, and 3) Tanner.

Six Hess substrate samples were collected from cobble bars during March 1995, October 1995, February 1996, 1 day prior to the controlled flood, and post flood (1, 2, 4, and 6 mon) at Lees Ferry, Two-Mile Wash and 2 and 6 mon at Tanner. All samples were placed on ice and processed within 24 hrs. At the time of each collection the following data were recorded: general habitat conditions, depth, current velocity, time of day, and discharge estimated on site and later verified from U.S.G.S. gauging station data.

Biotic samples were sorted into the following 11 categories: *C. glomerata*, cyanobacteria algal crust (*Oscillatoria* spp.), miscellaneous algae and macrophytes, detritus, chironomid larvae, *Gammarus lacustris*, gastropods, lumbriculids, tubificids, simuliids, and miscellaneous macroinvertebrates. The latter category included trichopterans, terrestrial insects, and unidentifiable animals. Each category was oven-dried at 60°C to a constant mass. Ash-free dry mass (AFDM) conversions were estimated from dry weight to AFDM regression equations [*Shannon et al.*, 1996a].

3.3. Organic Drift Collections

Drift samples (n=6) were collected at Lees Ferry, Two-Mile Wash, Tanner Canyon, and Hells Hollow at pre- (October 1995, March 1996, 1 and 4 days before) and post-controlled flood (4 days, June, and October 1996), once during the hydrostatic wave (river water in front of Lake Powell reservoir water), and three periods during the steady 1274 m^3/s discharge [*Schmidt et al.*, this volume]. Personnel were stationed at each site in order to collect organic drift from the same parcel of water as it passed through the river corridor. Collection times during the controlled flood were based on the speed of water travel according to equations established by *Wiele and Smith* [1996].

Near-shore drift samples (n=6) of fine particulate organic matter (FPOM) were collected in the river column between 1000 and 1500 hrs with a plankton net (30 cm diameter opening with 153 μm mesh) at each site. Samples were preserved in 70% ETOH and sorted in the lab with a dissecting scope into the following categories: detritus, Copepoda (Calanoida, Cyclopoida, Harpacticoida), Cladocera, Ostracoda, and miscellaneous zooplankton which included early instars of chironomid larvae and *Gammarus lacustris*, planaria, and hydra. Large samples were split with either 1 ml, 5 ml or 10 ml subsamples sorted from a 100 ml dilution. FPOM was sieved to remove >1 mm material, then sorted for organisms (<1.0 mm), with the remaining material filtered onto glass fiber filters (Whatman® GF/A) with a Millipore Swinex® system. These filters were oven-dried at 60°C and combusted for 1 hr at 500°C. Zooplankton was sorted and dried for dry mass estimates and converted to AFDM using a regression equation [*Shannon et al.*, 1996a, 1996b].

Near-shore surface drift samples (n = 6) of CPOM (0-0.5 m deep) were taken at the above sites with a conical tow net (48 cm diameter opening, 500 μm mesh) held in place behind a moored pontoon raft or secured to the river bank. Samples were placed on ice and processed live within 48 hrs and sorted into seven categories including: *Gammarus lacustris*, chironomid larvae, simuliid larvae, miscellaneous invertebrates, *C. glomerata*, miscellaneous algae/macrophytes, and detritus. Detritus was composed of both autochthonous (algal/bryophyte/macrophyte fragments) and allochthonous (tributary upland and riparian vegetation) flotsum. Drift samples were analyzed for size fractions after dry mass was obtained. Material from each collection interval and site was dry sieved into <1 mm, 1-9 mm and ≥10 mm size fractions. Each sample was manually shaken for 30 s which allowed for the separation of size fractions without particulate degradation. Samples of each size fraction were enumerated, oven dried at 60°C, weighed, ashed (500°C, 1 hr), and reweighed.

Current velocity and duration of each set were recorded so volumetric calculations could be made as g AFDM m^3/s. River discharge was determined from United States Geological Survey gauge data. The duration of all drift collections (n = 411) averaged 1.4 ± 0.06 min with an average of 9.2 ± 0.5 m^3 of water sampled through nets. The seemingly low duration and volume of water filtered resulted from the large amount of organic material drifting during the controlled flood. Nets were set for only a few seconds before the drift mass started to interfere with effective collection. Also, sample size would have been too great to process in a timely manner. The standard sampling error was within ±10% of the mean total drifting mass (0.218 ± 0.024 g m^3/s) which implies that collections were consistent and representative of the study site [*Culp et al.*, 1994].

We tested the hypothesis that CPOM organic drift is somewhat uniformly distributed across the river channel with simultaneous collections at two locations at the surface

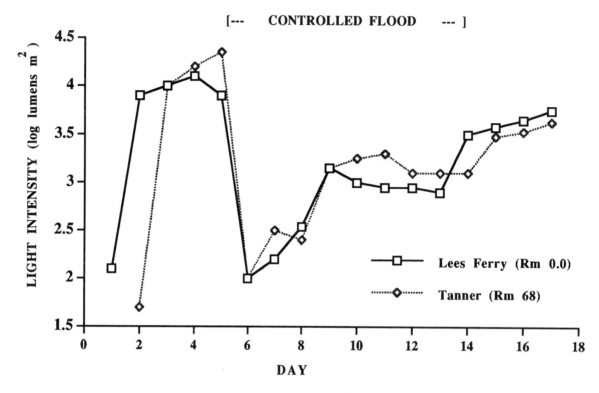

Figure 1. Light intensity (50 cm below the surface; log lumens/m²) in the Colorado River at Lees Ferry and Tanner Canyon prior to, during and 1 week after the controlled flood below Glen Canyon Dam, Arizona. Figure includes 4-day drawdown (227 m³/s) prior to and after the controlled flood.

and 3 m below the water surface at each site. Using the same drift nets as described above, two crews simultaneously made 25 collections at each location and depth (n = 100). Estimates of total organic drift were made by oven-drying each sample at 60°C, combusting the sample for 1 hr at 500°C, and determining ash-free dry mass (AFDM). Independent sample t-tests indicated no significant differences in organic drift between sites at either the surface (p = 0.07) or at a 3 m depth (p = 0.9). Nor was there a significant difference at either depth between sites (p = 0.1). Therefore, these data suggest that single location collections are representative of the channel. This is probably a result of a restricted channel, which is common in the Colorado River below Glen Canyon Dam. Surface collections had the most variability, possibly due to wind and erratic surface currents.

Our CPOM collections did not include large flotsam (>100 mm; riparian and upland vegetation) that occurred during the up-ramping and the first 2 days of the 1274 m³/s discharge. In an effort to quantify this portion of the organic budget, we examined interval photographs for large flotsum from the Grand Canyon beach survey program taken during the controlled flood [Hazel et al., this volume]. Cameras

that showed a 250 m area of river channel with a clear view not obstructed by rapids or canyon shadows were selected at approximately every 35 km, including RM 8.0L, 54.7R, 103.4R, 121.6R, 144.7L, 171.2L, and 200.8R. While viewing each frame on CD-ROM, which was enlarged to the best resolution possible, every noticeable bundle of flotsam was scored. The location and time of the picture taken was compared to the controlled flood hydrograph so that scores could be placed in relation to other drift collections.

3.4. Source of Organic Drift During the Experimental Flood

Drift samples of organic matter were collected during the controlled flood with similar nets as described above to help evaluate the source of stream drift with stable isotopes. We also collected triplicate plankton samples (≥2 g dry weight each) from Lake Powell with a 63 μm mesh net prior to the controlled flood. These plankton samples were taken to update ^{13}C and ^{15}N ratios of *Angradi* [1994] because of previous high in-flow years. All samples were air-dried immediately after collection in the field and ground to <0.5

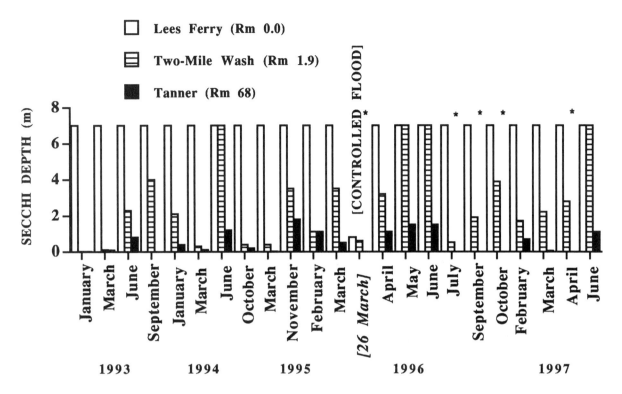

Figure 2. Secchi depth (m) measurements in the Colorado River at Lees Ferry, Two-Mile Wash, and Tanner Canyon at selected dates prior to, during and after the controlled flood. * indicates readings were not taken.

mm in particle size in the laboratory. Samples were analyzed by Robert Michner (Boston University) with a mass-spectrometer. PeeDee belemite and atmospheric nitrogen were used as carbon and nitrogen standards, respectively. The low nitrogen values for organic drift collected at Hells Hollow required using a cryo-foucs technique that extracts nanomoles of nitrogen [*Fry et al.*, 1996]. Ratios of ^{13}C and ^{15}N collected in drift during the controlled flood were compared to results presented by *Angradi* (1994) for Lake Powell plankton, river benthos, and riparian or upland vegetation from tributaries. *Angradi's* [1994] samples were also analyzed by Robert Michner at Boston University.

3.5. Statistical Analyses

Influence of the controlled flood, including pre- and post-flood collections, on benthic and organic drift estimates within collection sites were analyzed with a Kruskal-Wallis one-way analysis of variance. All calculations were performed with SYSTAT computer software on ln+1 transformed data [Version 5.1, *Wilkinson*, 1989].

4. RESULTS

4.1. Physico-Chemical Analyses

The controlled flood removed silt from the channel bottom of the Colorado River at all sites analyzed. The composition of sediment at Lees Ferry prior to the flood was approximately 5% very coarse sand and gravel, 75% sand and 20% silt/clay. However, sediment fractions were nearly all sand at Lees Ferry 2 mon after the experimental flood. Similar scouring patterns occurred at the two lower sites (RM 52.7 and 202.8) as well. Six months after the controlled flood, the silt/clay fraction returned at Lees Ferry (5%) and Spring Canyon (11%), but not at the Nankoweap site.

Underwater light intensity (log lumens/m^2; maintained at 50 cm below the surface) at Lees Ferry and at Tanner Canyon followed a predictable pattern during the controlled flood (Figure 1). Underwater light intensity increased during the drawdown, but decreased dramatically during the first day of the flood from >4 to 2 log lumens/m^2 at both sites. Within 1 day, underwater light intensity started to

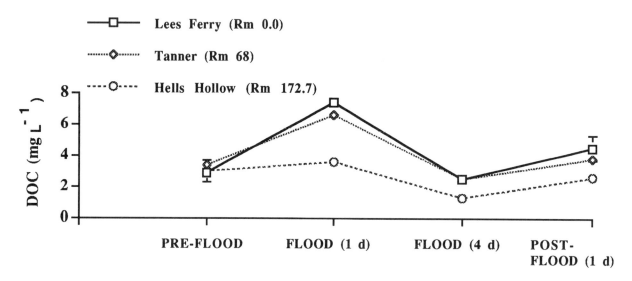

Figure 3. Average dissolved organic carbon (DOC, mg/L[1] ±SE) values in the Colorado River at Lees Ferry, Tanner Canyon and Hells Hollow taken 1 day prior to, during the hydrostatic wave (1 day), 4 days into the 1274 m^3/s discharge, and 1 day after the controlled flood (n = 3 for each collection).

increase and continued to increase until it nearly approximated pre-flood levels 4 days after the controlled flood. The overall pattern of light intensity during the flood was similar at both sites even though they were approximately 100 km apart. The reduced light intensity that occurred at both sites 3-4 days prior to the experimental flood resulted from a spate and fluctuating flows that influenced the entire river corridor (Figure 1).

Secchi depths (water transparency) in the Glen Canyon reach at Lees Ferry were ≥7 m (to the channel bottom) prior to and after the controlled flood (Figure 2). In fact, starting in November 1995 and continuing through mid-1997 after the controlled flood, water transparency was considerably higher throughout the canyon corridor, especially at Two-Mile Wash, compared to previous records [*Stevens et al.*, 1997; Figure 2]. Secchi depths below the confluence of the Paria River during this period were nearly one-half of those measured above the confluence in Glen Canyon (Figure 2). During the first 2 days of the controlled flood, Secchi depths were <1 m, but returned to near-normal levels (≥7 m) by day 4 of the flood.

There were no significant differences in dissolved oxygen, pH, and specific conductance in the Glen Canyon reach before and after the controlled flood event. Hydrogen-ion concentrations averaged 7.8 (SE ±0.1) prior to the flood compared to 7.9 (±0.1) after the flood. Dissolved oxygen was 11.7 mg/L (±0.7) prior to the flood and 11.5 mg/L (±0.8) after the flood, while specific conductance was 0.81 mS (±0.01) and 0.84 mS (±0.003) prior to

and after the flood, respectively. Although there was a significant difference (p <0.01) in water temperature before and after the controlled flood, the differences were quite small. Water temperatures were slightly lower (9.3 °C ± 0.2) after the flood compared to pre-flood water temperatures (10.0°C ±0.5).

Pre-flood concentrations of dissolved organic carbon (DOC) ranged from 2.9 to 3.5 mg/L and were not significantly different at the three collection sites throughout the river corridor (Figure 3). However, DOC concentrations increased during the first day of the controlled flood with highest concentrations at Glen Canyon Dam (7.4 mg/L ±0.4), but progressively decreased downstream. By day 4 of the controlled flood, DOC concentrations were slightly lower than pre-flood concentrations at all sites, but returned to near pre-flood levels (2.6 to 4.5 mg/L) 1 day after the controlled flood (Figure 3).

4.2. Patterns in Benthos

The standing crop of all biotic components in the stream benthos of the Colorado River was significantly (p <0.04) reduced at all sites as a result of the 1274 m^3/s controlled discharge (Figure 4). The standing crop of primary producers was >80 g AFDM m^{-2} at Lees Ferry prior to the controlled flood, 90% of which was removed during the event. Most of the aquatic plants lost during the flood were rooted macrophytes and filamentous green algae. Standing crops of benthic primary consumers were also reduced. We

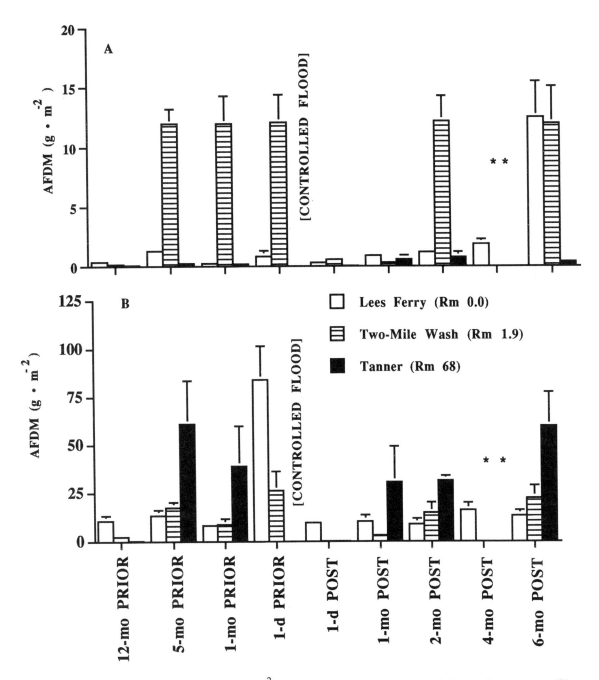

Figure 4. Average benthic biomass (AFDM, g/m^2 ±SE) of macroinvertebrates (A) and algae and macrophytes (B) on selected dates prior to, during, and after the controlled flood in the Colorado River below Glen Canyon Dam, Arizona (n = 6 for each collection). * indicates no collection was made. 12-mon prior = March 1995, 5-mon prior = October 1995, 1-mon prior = March 1996, 1-day prior = pre-drawdown, 1-day post = post-drawdown, 1-mon post = May 1996, 2-mon post = June 1996, 4-mon post = August 1996, 6-mon post = October 1996.

estimated that ≥50% of the invertebrate mass was removed at Lees Ferry and over 90% was removed at Two-Mile Wash below the confluence of the Paria River (Figure 4). The invertebrates most susceptible to the controlled flood were lubriculids and tubificids associated with fine sediments. Although losses occurred, the standing crop of the amphipod *Gammarus lacustris* was not significantly different prior to and after the controlled flood at Lees Ferry, but was at Two-Mile Wash (Figure 4).

Recovery rates to pre-flood levels after the controlled flood were relatively fast: 1 mon for primary producers and 2 mon for benthic invertebrates (Figure 4). In fact, standing masses for chironomids, *Gammarus lacustris*, gastropods, simuliids, and miscellaneous invertebrates all exceeded pre-flood levels. Gastropods showed the fastest response at Lees Ferry, while chironomid and simuliid larvae were quick to recover at the downstream sites.

4.3. Patterns in Organic Drift

The greatest mass of fine particulate organic matter (FPOM) was collected during the initial hydrostatic wave of the controlled flood at all three collection sites. Values below Glen Canyon Dam and at Lees Ferry during the hydrostatic wave were 0.15 (SE ±0.02) and 1.77 (± 0.18) g AFDM m^3/s, respectively, and 0.77 g AFDM m^3/s (± 0.45) at Hells Hollow 172.2 RM downstream. The dominant zooplankters collected at Lees Ferry during the hydrostatic wave of the controlled flood included early instars of chironomid larvae and *Gammarus lacustris*, planaria and snail eggs (~3200 animals m^3/s) as well as crustaceans (~2600 animals m^3/s) from the Lake Powell reservoir. Total numbers of zooplankton in the drift prior to and after the flood were typically <400 animals m^3/s with a mass of <0.0001 g AFDM m^3/s.

Size fractions of particulate organic matter changed with site and collection interval with a decrease in the largest size fraction (≥10 mm) at both Lees Ferry and Tanner during the controlled flood, but not at the most downstream site (RM 172.7; Figure 5). The ≥10 mm-sized fraction was primarily composed of filamentous algae, bryophytes, and rooted macrophytes, which reflected the high standing crops of each in the upstream benthos (see Section 4.2). Typically there was a strong positive correlation (r = 0.87; p <0.01) between mass of benthos and organic drift in the Colorado River system [*Shannon et al.*, 1994]. Proportions of large CPOM (≥10 mm) equalled that of pre-flood conditions within 1 week after the flood at Lees Ferry and exceeded pre-flood levels at the two downstream sites within 3 mon after the controlled flood; values were not measured at the two downstream sites at 1 week (Figure 5).

Total mass of organic drift during the controlled flood was several orders of magnitude higher in the hydrostatic wave (>2.5 g AFDM m^3/s) compared to values before and after the controlled flood (≤0.05 g AFDM m^3/s), most of which was detritus (Figure 6). An order of magnitude higher total drift mass was measured in the Lees Ferry reach compared to downstream stations due to higher benthic mass in the upper station (see Section 4.1). Also, detrital drift mass was present, if not dominant, during all collections periods at downstream sites, but most prevalent during the controlled flood at Lees Ferry instead of the usual phytobenthic mass (Figure 6).

No large CPOM bundles were observed during the pre-flood flows. Although large amounts of CPOM flotsam (>100 mm) were transported downstream during the controlled flood, drift fractions ≤10 mm in size contributed an order of magnitude more mass. For example, based on a rate of 22.2 bundles/hr, conservatively estimating each bundle to weigh ~4 kg, we estimated that 23,000 kg AFDM of large flotsum was transported downstream during the 7-day controlled flood compared to 970,000 kg AFDM of FPOM and 1,060,000 kg AFDM of CPOM.

4.4. Source of Organic Drift at Pre-Flood, During the Flood, and Post-Flood.

Under constrained hydrographs below Glen Canyon Dam, autothchonous algal matter was the primary source of organic drift in the upper tailwaters, whereas allochthonous energy from riparian and upland communities was more important during the flood event. The strongest $\delta^{13}C/\delta^{15}N$ signals in the upper tailwaters (1.9 RM) of the Colorado River prior to and after the controlled flood were for lake plankton and tailwater phytobenthos, compared to riparian and upland vegetation at downstream sites (≥61.3 RM, Table 1). Whereas, during the test flood, $\delta^{13}C/\delta^{15}N$ ratios were strongest for riparian and upland vegetation at all sites throughout the river corridor (Table 1). The composition of organic drift visually examined prior to, during, and after the controlled flood concurred with stable isotope analyses (see Section 4.3).

5. DISCUSSION

The 1274 m^3/s controlled flood below Glen Canyon Dam provided valuable information on short-term responses for both the riverine system and the biotic community, but the long-term effects on the aquatic food base were more difficult to assess. The experimental discharge selectively

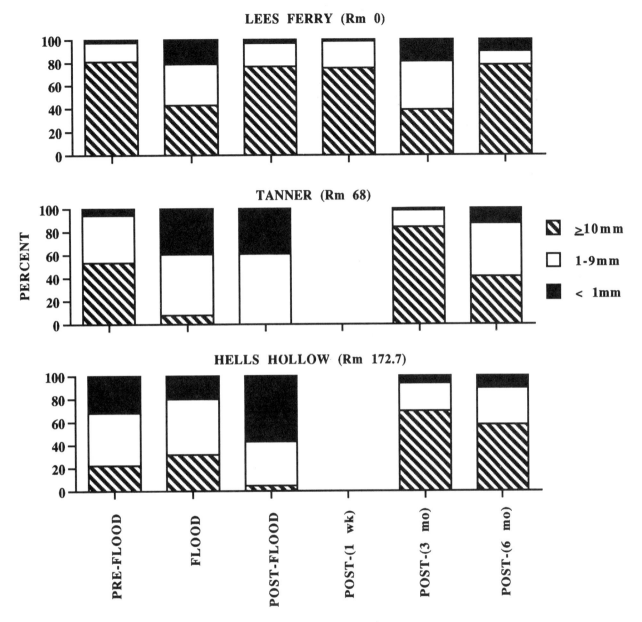

Figure 5. Relative size fractions (%) of coarse particulate organic matter at selected times prior to, during, and after the controlled flood at Lees Ferry, Tanner Canyon, and Hells Hollow in the Colorado River below Glen Canyon Dam, Arizona. Pre-flood = drawdown, post-flood = drawdown, post-1 week = 15 April 1996, post-3 mon = June 1996, post-6 mon = October 1996.

removed fine silt/clay from the channel bottom and along the varial zones of the Colorado River. This initially modified both water clarity and available habitat for biota. Most of the scour occurred within the first 2 days of the controlled flood as measured by underwater light intensity. During the initial hydrostatic wave and up to 48 hrs following the wave, light intensity was dramatically reduced (Figure 1). However, water clarity increased substantially after this period and has remained relatively high since the controlled flood due to elevated flows (Figure 2). The initial days of the controlled flood re-suspended and scoured much of the smaller-sized sediment, including FPOM and CPOM, from the river channel and upper shorelines. The re-suspended fine sediment fraction is particularly important in determining water clarity in lotic systems [*Morisawa*, 1968; *Newcombe and MacDonald*, 1991].

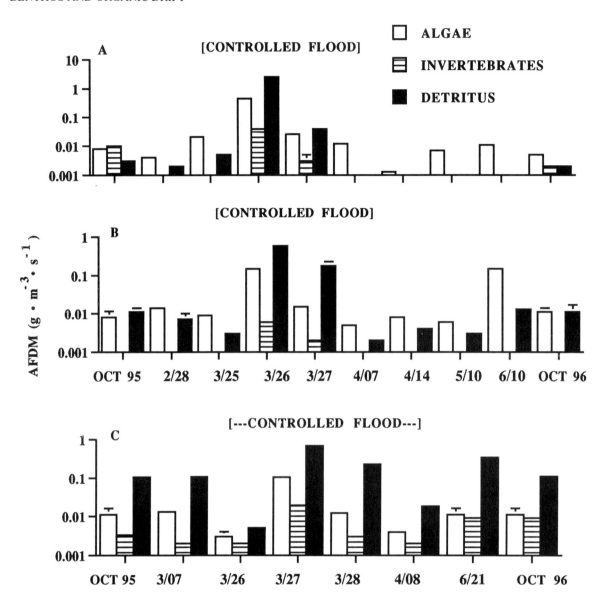

Figure 6. Average drift biomass (AFDM, g/m³s ±SE) of algae, invertebrates, and detritus on selected dates prior to, during and post the controlled flood in the Colorado River below Glen Canyon Dam, Arizona, at Lees Ferry (A), Two-Mile Wash (B) and Tanner Canyon (C) (n = 6). Collection dates for A and B are the same.

The high, relatively steady flows since the controlled flood have helped to maintain the higher than normal water clarity conditions throughout the river corridor. Although there have been comparable discharges from the Paria and LCR Rivers during the past year compared to previous years, the high flows in the Colorado River have provided a "dilution effect" on levels of suspended sediments delivered by tributaries which has resulted in higher water clarity. Water releases ≥ 450 m³/s from Glen Canyon Dam tend to reduce tributary influence on water clarity.

The "flashy" nature of major tributaries that deliver suspended sediments into the Colorado River throughout the canyon system and selectively influence water clarity along the river corridor was evident from the episodic event that occurred several days prior to the controlled flood. At this time, a large portion of the watershed in the Colorado River Basin received heavy precipitation which in turn was delivered, along with suspended sediments, to the Colorado River via tributaries. During this event, water clarity was dramatically reduced throughout much of the river corridor,

TABLE 1. Dual isotope [δ^{13}C / δ^{15}N (‰)] analyses for drifting organic material at selected stations along the Colorado River through Grand Canyon prior to, during, and after the 7-day controlled flood (1274 m³/s) during March 1996. (POM = particulate organic matter from Lake Powell; Phytobenthos = *Cladophora glomerata* and associated filamentous green algae; Riparian = tamarisk and cottonwood; Upland = floodplain litter from tributaries). The hydrostatic wave pushed river water ahead of the released Lake Powell reservoir flood water. Range of isotope ratios for organic sources are from *Angradi* [1994]: Lentic POM (-23.8 to 28.5 δ^{13}C / 6.8 to 9.6 δ^{15}N); Phytobenthos (-21.2 to -33.5 δ^{13}C /4.6 to 9.5 δ^{15}N); Riparian vegetation (-24.9 to -29.0 δ^{13}C / 5.0 to 7.4 δ^{15}N); Upland (-23.8 to -29.0 δ^{13}C / -0.1 to 5.0 δ^{15}N).

	DISCHARGE (m³/s)				
	227	1274	1274	1274	227
SITE (Rm)	BEFORE	WAVE	DAY 1	DAY 5	AFTER
1.9	-25.3/9.0	-26.5/6.8	-25.6/7.3	-27.1/6.2	-28.5/9.9
	(Lentic POM, Phytobenthos)	(Riparian)	(Riparian)	(Riparian)	(Lentic POM, Phytobenthos)
61.3	-25.2/2.9	-25.3/3.9	-24.9/2.0	-24.8/1.3	-27.1/7.4
	(Upland)	(Upland)	(Riparian, Upland)	(Upland)	(Riparian)
172.9	-15.8/6.0	-23.3/1.4	-24.5/1.4	-19.3/1.7	-24.1/2.0
	(Unknown)	(Upland)	(Upland)	(Unknown)	(Riparian, Upland)

but returned to pre-flood conditions after only several days (Figure 1). Both this natural spate event, as well as, the experimental controlled flood indicate the relatively short-term influence flood disturbances have on water clarity in the mainstem. This may imply that pre-dam conditions in the Colorado River were highly dynamic in which floods would scour benthos from the channel bottom and periodically clear to allow for re-colonization of photosynthetic algae. These dynamic interactions between suspended sediment, water transparency and algal primary production in the Colorado River have been recently modeled by *Yard et al.* [1995].

The hydraulic force of the experimental discharge was sufficient to scour a large proportion of the benthic community below Glen Canyon Dam. As much as 90% of the benthos was removed during the controlled flood in the Glen Canyon reach, however, removal rates were largely a function of pre-flood biomass levels, and perhaps the 4-day drawdown (227 m³/s) prior to the flood. Even short-term exposures of <12 hrs have been reported to be detrimental to the aquatic food base in the Colorado River system [*Usher and Blinn*, 1990; *Angradi and Kubly*, 1993; *Blinn et al.*, 1995]. Other workers have found similar relationships between mass and dislodgment [*Statzner and Higler*, 1986; *Duncan and Blinn*, 1989; *Allan*, 1995]. The biotic components most susceptible to the scour were those associated with the unstable fine sediments such as rooted macrophytes (i.e., *Elodea* and *Chara)*, as well as tubificid and lumbriculid worms. Whereas, biota associated with larger,

more stable cobble such as *Cladophora glomerata* and *Gammarus lacustris* were not as severely affected.

Levels of DOC, FPOM and CPOM all reflected the initial scour of benthos within 48 hrs of the controlled flood. There was a 2.5-fold increase in DOC and a two order of magnitude increase in drifting particulate material during the hydrostatic wave compared to pre- and post-flood periods as well as the remainder of the controlled flood. These high levels of exported dissolved and particulate organic carbon provided a short-term energy source to downstream aquatic communities during the controlled flood. Based on energy equivalents reported by *Blinn et al.* [1995], we estimated that >20,000 joules m³/s of energy were exported downstream as plant and animal mass during the hydrostatic wave. *Fisher and Minckley* [1978] found similar pulses in nutrients during the initial wave of a flood in Sycamore Creek—an ephemeral stream in central Arizona.

Stable isotope analyses lead to the inference that much of the drifting organic matter transported downstream during the controlled flood was derived from riparian and upland vegetation (allochthonous), whereas during normal dam operations, phytobenthos (autochthonous) is the dominant source of organic matter (Table 1). *Angradi* [1994] found similar patterns during dam operation for the upper tailwaters of Glen Canyon Dam. Under typical dam operations, dense tufts of *C. glomerata* become dislodged and are exported through rapids where they quickly become pulverized into FPOM as they drift along the river corridor

[*Shannon et al.*, 1996b]. This drifting autochthonous energy supports a downstream community of collectors including chironomid and simuliid larvae. During the controlled flood, allochthonous materials (namely tamarisk, willow and juniper) stranded along the upper shorelines and the confluences of tributaries were entrained into stream drift during early stages of the experiment.

Recovery rates for the aquatic benthos after the controlled flood disturbance were relatively fast. Both primary producers and consumers regained pre-flood standing crop levels within 2 mon after the flood and have since exceeded previously reported biomass estimates for algae (120 g AFDM m^{-2}) and macroinvertebrates (>14 g AFDM m^{-2}) below Glen Canyon Dam [*Usher and Blinn*, 1990; *Blinn et al.*, 1992; *Shannon et al.*, 1996a; *Stevens et al.*, 1997]. Comparable recovery rates under optimum conditions have been reported for algae [*Fisher et al.*, 1982; *Power and Stewart*, 1987; *Steinman and McIntire*, 1990; *Blinn et al.*, 1995; *Peterson*, 1996] and macroinvertebrates [*Wallace*, 1990; *Blinn et al.*, 1995], but recovery rates in the Colorado River following the controlled flood were somewhat faster than the 3-mon period following major disturbances in aquatic ecosystems as reviewed by *Yount and Niemi* [1990].

We propose that the rapid recovery rates and current high standing stock of aquatic benthos in the river corridor is more a function of higher water clarity, due to higher relatively constant discharges from Glen Canyon Dam, rather than solely related to the controlled flood. Although considerable energy was transported downstream on a short-term basis (≤1 week) during the controlled flood, the greatest result of the flood may have been to flush fine sediments from the system, which along with consistent high flows have sustained high water clarity for photosynthetic processes.

5.1. Summary and Management Considerations

The present aquatic foodweb below Glen Canyon Dam is highly modified from that of more natural riverine conditions in the arid Southwestern United States [*Haden* 1997]. Presently, filamentous green algae and associated epiphytic diatoms make up the primary producer trophic level rather than allochthonous energy, and exotic grazers such as the crustacean, *Gammarus lacustris*, dominate the primary consumers rather than caddisfly and/or mayfly assemblages [*Blinn and Cole*, 1991; *Shannon et al.*, 1994; *Haden*, 1997; *Stevens et al.*, 1997]. These modifications in the aquatic foodweb have largely resulted from changes in physicochemical conditions and sources of carbon energy below

the dam [*Blinn and Cole*, 1991; *Stanford and Ward*, 1991; *Haden*, 1997].

Optimum management of the present modified food base below Glen Canyon Dam can likely be achieved by steady discharges (~450 m^3/s) with minimal fluctuation cycles (~50 m^3/s). Previous studies on the Colorado River have found that both water clarity, as a function of sediment input from major tributaries, and dam management play a major role in determining levels of standing crop and community structure of the aquatic food base below Glen Canyon Dam [*Blinn et al.*, 1995; *Shannon et al.*, 1996; *Stevens et al.*, 1997, *Shaver et al.*, 1997]. Discharges >400 m^3/s tend to mitigate the influence of suspended sediment on water clarity and the higher discharges provide more wetted perimeter for colonization by benthos. Furthermore, minimum levels of fluctuation in discharge reduce the extent of the varial zone which is a zone that is lethal to the aquatic benthic community [*Angradi and Kubly*, 1993; *Blinn et al.*, 1995, *Shaver et al.*, 1998]. We recommend that future controlled floods eliminate the pre- and post drawdowns that expose large areas of benthos to desiccation and ultraviolet light. These conditions weaken holdfast systems of the phytobenthos and induce macroinvertebrates to abandon their habitats [*Usher and Blinn*, 1990; *Angradi and Kubly*, 1993; *Blinn et al.*, 1995].

Acknowledgments. This work was partly supported by the U.S. Bureau of Reclamation Glen Canyon Environmental Studies Program (GCES) in cooperation with Grand Canyon National Park and Glen Canyon National Recreational Area. We thank Dave Wegner and Mike Yard (GCES) and Allen Haden (Northern Arizona University, NAU) for logistical and field support, Jeff Bennett and Rod Parnell (NAU) for DOC analyses, Robert Michner (Boston University) for stable isotope analyses, and Mark Malone (NAU) for assistance with the Interval Photograph CD-ROM analyses of course flotsum material. We also appreciate the work of 24 volunteers that assisted in field collections on this project.

REFERENCES

Allan, J.D., *Stream Ecology: Structure and Function of Running Waters*, Chapman & Hill, London, UK, 1995.

Angradi, T.R., Trophic linkages in the lower Colorado River: multiple stable isotope evidence, *J. N. Am. Benthol. Soc.*, 13, 479-495, 1994.

Angradi, T.R., and D.M. Kubly, Effects of atmospheric exposure on chlorophyll a, biomass and productivity of the epilithon of a tailwater river, *Regul. Rivers*, 8, 345-358, 1993.

Blinn, D.W., and G.A. Cole, Algae and invertebrate biota in the Colorado River: comparison of pre- and post-dam conditions, in *Colorado River Ecology and Management*, ed. G.R.

Marzolf, pp. 102-123, Natl. Acad. Press, Washington, D.C., 1991.

Blinn, D.W., J.P. Shannon, L.E. Stevens, and J.P. Carder, Consequences of fluctuating discharge for lotic communities, *J. N. Amer. Benthol. Soc.*, 14, 233-248, 1995.

Blinn, D.W., L.E. Stevens, and J.P. Shannon, *The effects of Glen Canyon Dam on the aquatic food base in the Colorado River Corridor in Grand Canyon, Arizona*, Glen Canyon Environ. Studies Report CA- 8009-8-0002. Flagstaff, AZ, 1992.

Carothers, S.W., and B.T. Brown, *The Colorado River through Grand Canyon: Natural history and human change,* Univ. Ariz. Press, Tucson, AZ, 1991.

Collier, M., R.H. Webb, and J.C. Schmidt, *Dams and rivers: A primer on the downstream effects of dams*, U.S. Geol. Surv. Circ. 1126, 1996.

Culp, J.M., G.J. Garry, and J. Scrimgeour, The effect of sample duration on the quantification of stream drift, *Freshwat. Biol.*, 31, 165-173, 1994.

Cummins, K.W., An evaluation of some techniques for the collection and analysis of benthic samples with special emphasis on lotic waters, *Am. Midl. Nat.*, 67, 477-504, 1962.

Duncan, S.W., and D.W. Blinn, Importance of physical variables on the seasonal dynamics of epilithic algae in a highly shaded canyon stream, *J. Phycol.*, 25, 455-461, 1989.

Fisher, S.G., L.J. Gray, N.B. Grimm, and D.E. Busch, Temporal succession in a desert stream ecosystem following flash flooding, *Ecol. Monogr.*, 52, 93-110, 1982.

Fisher, S.G., and W.L. Minckley, Chemical characteristics of a desert stream in flash flood, *J. Arid Environ.*, 1, 25-33, 1978.

Fry, B, R. Garritt, K. Thorpe, C. Neil, R. Michener, F. Mersh, and W. Brand, Cryoflow: cryofocusing nanomole amounts of CO_2, N_2, and SO_2 from an elemental analyzer for stable isotope analysis, *Rapid Comm. Mass. Spectrometry*, 10, 953-958, 1996.

Haden, A., *Benthic ecology of the Colorado River system through the Colorado Plateau region of the southwestern United States*, M.S. Thesis, N. Ariz. Univ., Flagstaff, AZ, 1997.

Leibfried, W.C., and D.W. Blinn, *The effects of steady versus fluctuating flows on aquatic macroinvertebrates in the Colorado River below Glen Canyon Dam, Arizona*, NTIS No. PB88206362/AS, 1986.

Minckley, W.L., Native fishes of the Grand Canyon region: an obituary?, in *Colorado River Ecology and Management*, ed. G.R. Marzolf, pp. 124-177, Natl. Acad. Press, Washington, D.C., 1991.

Minshall, G.W., Aquatic-insect-substratum relationships, in *The Ecology of Aquatic Insects*, ed. V.H. Resh and D.M. Rosenberg, pp. 358-400, Praeger Scientific, New York, 1984.

Morisawa, M., *Streams: Their dynamics and morphology*, McGraw Hill, NY, 1968.

National Research Council, *River resource management in the Grand Canyon*, Natl. Acad. Press, Washington, D.C., 1996.

Newcombe, C.P., and D.D. MacDonald, Effects of suspended sediments on aquatic ecosystems, *N. Amer. J. Fish. Manage.*, 11, 72-82, 1991.

Peterson, C.G, Response of benthic communities to natural physical disturbance in algal ecology, in *Freshwater Benthic Ecosystems*, ed. by R.J. Stevenson, M.L. Bothwell, and R.L. Lowe, pp. 375-402, Academic Press, NY, 1996.

Power, M.E., and A.J. Stewart, Disturbance and recovery of an algal assemblage following flooding in an Oklahoma stream, *Am. Midl. Nat.*, 117, 333-345, 1987.

Resh, V.H., A.V. Brown, A.P. Covich, M.E. Gurtz, H.W. Li, G.W. Minshall, S.R. Reice, A.L. Sheldon, J.B. Wallace, and R.C. Wissmar, The role of disturbance in stream ecology, *J. N. Am. Benthol. Soc.* 7, 433-455, 1988.

Robinson, C.T., and S.R. Rushforth, Effects of physical disturbance and canopy cover on attached diatom community structure in an Idaho stream, *Hydrobiologia*, 154, 49-59, 1987.

Schmidt, J.C., and D.M. Rubin, Regulated streamflow, fine-grained deposits, and effective discharge in canyons with abundant debris fans, in *Natural and anthropogenic influences in fluvial geomorphology*, ed. J.E. Costa, A.J. Miller, K.W. Potter, and P.R. Wilcock, pp. 177-195, Amer. Geophys. Union, 1995.

Shannon, J.P., D.W. Blinn, and L.E. Stevens, Trophic interactions and benthic animal community structure in the Colorado River, Arizona, USA, *Freshwat. Biol.*, 31, 213-220, 1994.

Shannon, J.P., D.W. Blinn, K.P. Wilson, P.L. Benenati, and G.E. Oberlin, I*nterim flow and beach building spike flow effects from Glen Canyon Dam on the aquatic food base in the Colorado River in Grand Canyon National Park, Arizona*, Final Rept. to Glen Canyon Environ. Studies, Flagstaff, AZ, 1996a.

Shannon, J.P., D.W. Blinn, P.L. Benenati, and K.P. Wilson, Organic drift in a regulated desert river, *Can. J. Fish. Aquat. Sci.*, 53, 1360-1369, 1996b.

Shaver, M.L., J.P. Shannon, K.P. Wilson, P.L. Benenati, and D.W. Blinn, Effects of suspended sediment and desiccation on the benthic tailwater community in the Colorado River, USA, *Hydrobiologia*, 357, 63-72.

Stanford, J.A., and J.V. Ward, The Colorado River system, in *The Ecology of River Systems*, ed. B. Davies and K.F. Walker, pp. 353-374, Dr. W. Junk Publishers, Dordrecht, The Netherlands, 1986.

Stanford, J.A., and J.V. Ward, Limnology of Lake Powell and the chemistry of the Colorado River, in *Colorado River and Dam Management*, ed. G.R. Marzolf, pp. 75-101, Natl. Acad. Press, Washington, D.C., 1991.

Statzner, B., and B. Higler, Stream hydraulics as a major determinant of benthic invertebrate zonation patterns, *Freshwat. Biol.*, 9, 251-262, 1986.

Steinman, A.D., and C.D. McIntire, Recovery of lotic periphyton communities after disturbance, *Environ. Manage.*, 14, 589-604, 1990.

Stevens, L.E., J.P Shannon, and D.W. Blinn, Colorado River benthic ecology in Grand Canyon, Arizona, USA: dam, tributary and geomorphic influences, *Regul. Rivers*, 13, 129-149, 1997.

Toner, M., and P. Keddy, River hydrology and riparian wetlands: a predictive model for ecological assembly, *Ecol. Appl.*, 7, 236-246, 1997.

Usher, H.D., and D.W. Blinn, Influence of various exposure periods on the biomass and chlorophyll *a* of *Cladophora glomerata* (Chlorophyta), *J. Phycol.*, 26, 244-249, 1990.

Wallace, J.B., Recovery of lotic macroinvertebrate communities from disturbance, *Environ. Manage.*,14, 605-620, 1990.

Wiele, S.M., and J.D. Smith, A reach-averaged model of diurnal discharge wave propagation down the Colorado River through the Grand Canyon, *Water Res. Res.*, 32, 1375-1386, 1996.

Wilkinson, L., SYSTAT: *The system for statistics*, Systat, Evanston, IL, 1989.

Yard, M., G.A. Haden, and W.S. Vernieu, *Photosynthetically available radiation (PAR) in the Colorado River: Glen Canyon and Grand Canyon*, Bur. Reclam., Glen Canyon Environ. Studies Rept., Flagstaff, AZ, 1995.

Yount, J.D., and G.J. Neimi, Recovery of lotic communities and ecosystems from disturbance - a narrative review of case studies, *Environ. Manage.*, 14, 547-570, 1990.

Dean W. Blinn, Joseph P. Shannon, Kevin P. Wilson, Chris O'Brien, and Peggy L. Benenati, Department of Biological Sciences, Northern Arizona University, Flagstaff, Arizona 86001; email: Dean.Blinn@nau.edu

Fish Abundance, Distribution, and Habitat Use

Timothy L. Hoffnagle

Arizona Game and Fish Department, Flagstaff, Arizona

Richard A. Valdez

SWCA, Inc., Logan, Utah

David W. Speas

Arizona Game and Fish Department, Phoenix, Arizona

The 1996 controlled flood in the Colorado River, Grand Canyon, was designed, in part, to improve conditions for juvenile native fishes by reshaping habitat and displacing non-native fishes. We examined changes in abundance and distributions of native and non-native fishes immediately before and after the controlled flood and recovery of affected species 2.5 and 6 months after. Catch-per-unit-effort (CPUE) of humpback chub and flannelmouth sucker did not differ in pre- versus post-flood periods. CPUE of plains killifish, bluehead sucker and fathead minnow decreased following the flood, and CPUE of speckled dace and rainbow trout increased. Juvenile humpback chub remained primarily along talus shorelines at all discharges, while at higher discharges, speckled dace shifted from mid-channel riffles to debris fans and talus and fathead minnows used primarily vegetated shorelines. There was evidence of some downstream displacement of plains killifish, fathead minnows and rainbow trout. Catch rates of all species showed seasonal variation following the flood, with summer recruitment of young-of-the-year, particularly fathead minnows and plains killifish. Although short-term reductions in catch rates of fathead minnows and plains killifish occurred, these populations returned to pre-flood densities by 6 months after the flood. Catch rates of all species before and after the flood were similar to those recorded in previous years. We determined that the controlled flood did not significantly alter native fish distributions or abundances through Grand Canyon.

1. INTRODUCTION

Floods were an integral part of the ecology of the Colorado River in Grand Canyon prior to the closure of Glen Canyon Dam [*Minckley*, 1991; *Valdez and Ryel*, 1995]. These late spring/early summer floods were important spawning cues for native fishes [*John*, 1963;

The Controlled Flood in Grand Canyon
Geophysical Monograph 110
Copyright 1999 by the American Geophysical Union

Harvey, 1987; *Valdez and Ryel*, 1995], and they restructured riverine habitats [*U. S. Department of the Interior*, 1996], particularly backwaters which are important rearing habitats for larval and juvenile fishes dispersing from spawning areas [*Valdez and Wick*, 1983; *Arizona Game and Fish Department*, 1996a]. Because of Glen Canyon Dam, only occasional large tributary floods, primarily from the Little Colorado River (LCR), now provide the Colorado River in Grand Canyon with the semblance of a natural hydrograph.

There is an abundance of literature on the effects of high flows on fish and fish communities in small streams [*Seegrist and Gard*, 1972; *Hoopes*, 1975; *Harrell*, 1978; *Ross and Baker*, 1983; *Matthews*, 1986]. Fishes native to the southwestern United States appear to be particularly resistant to flooding while non-native fishes are vulnerable [*Meffe and Minckley*, 1987; *Minckley and Meffe*, 1987]. Four of the original eight native species remain in the Colorado River and its tributaries in Grand Canyon National Park: bluehead sucker (*Catostomus discobolus*), flannelmouth sucker (*Catostomus latipinnis*), humpback chub (*Gila cypha*) and speckled dace (*Rhinichthys osculus*). An additional 25 exotic fish species have been reported from this reach of the Colorado River [*Valdez and Ryel*, 1995]; however, only seven are regularly found [*Valdez and Ryel*, 1995; *Arizona Game and Fish Department*, 1996a]: common carp (*Cyprinus carpio*), fathead minnow (*Pimephales promelas*), red shiner (*Cyprinella lutrensis*), plains killifish (*Fundulus zebrinus*), rainbow trout (*Oncorhynchus mykiss*), brown trout (*Salmo trutta*), and channel catfish (*Ictalurus punctatus*). The effect of a clear, cold water flood on the present fish assemblage of the Colorado River in Grand Canyon is unknown. It is also unknown how quickly affected fish will recover to pre-flood densities following such a flood.

Use of a controlled high release (i.e., flood) from Glen Canyon Dam to improve conditions for native fishes was first addressed by *Clarkson et al.* [1994], who discussed recommendations for operating Glen Canyon Dam to benefit native fishes. Beach/habitat-building flows, "...designed to rebuild high elevation sandbars, deposit nutrients, restore backwater channels and provide some of the dynamics of a natural system...", became an element of the preferred alternative of the Glen Canyon Dam Environmental Impact Statement [*U.S. Department of the Interior*, 1995]. This controlled flood was expected to reform backwater habitats for juvenile native fishes and reduce populations of some non-native fishes [*U.S. Department of the Interior*, 1996]. This 1274 m³/s flood occurred in late-March and early-April 1996, in part, to avoid detrimen-

tally affecting young-of-the-year (YOY) native fishes which descend into the main channel from natal tributaries later in spring. However, during the controlled flood, age 1 fish may be affected by changing habitat. While studies have documented effects of floods on fish communities in small streams, very little information is available concerning habitat use by small fishes during a flood. *Minckley* [1991] speculated that fish moved to the surface along channel margins to avoid high current velocities and the abrasive action of transported sediment.

It is rare that the timing of a flood is known sufficiently in advance to plan and conduct studies afield, as was done in this investigation. This chapter examines changes in abundance, distribution, and habitat use of native and non-native fishes before, during, and after the 1996 controlled flood and the recovery of fishes affected by the flood. The rising water levels of the flood were expected to inundate quiet-water habitats (e.g., backwaters and vegetated shorelines) used by small native and exotic fishes at lower flows, while creating new quiet-water habitats in flooded terrestrial vegetation and tributary mouths. It was unknown whether native and non-native fishes would be able to find areas of refuge or be dispersed downstream. We examined changes in distribution and abundance of fishes both near the confluence of the Colorado and Little Colorado rivers and throughout Grand Canyon. Specifically, we tested three hypotheses: 1) distribution and abundance of native and non-native fishes in the Colorado River, Grand Canyon, will not differ significantly in response to the controlled flood, 2) fishes affected by the flood will not recover to pre-flood densities by the end of the growing season, and 3) habitat selection by fathead minnow, speckled dace and juvenile humpback chub will not differ with river discharge.

2. METHODS

The study area was the Colorado River in Grand Canyon National Park, Arizona, from Lees Ferry (RM 0) to Diamond Creek (RM 225.6) [see map in *Webb et al.*, this volume]. This section of river was divided into 8 sampling reaches of varying length, based on known fish abundance and the availability of backwater habitat and spawning tributaries [*Arizona Game and Fish Department*, 1996a]. These reaches were used during pre-experimental flood, post-experimental flood, summer, and fall sampling trips which covered the entire study area. In addition, a reach extending from Awatubi Canyon (RM 58.1) to Lava Canyon Rapid (RM 65.5) was sampled during the controlled flood. This reach includes the confluence of the

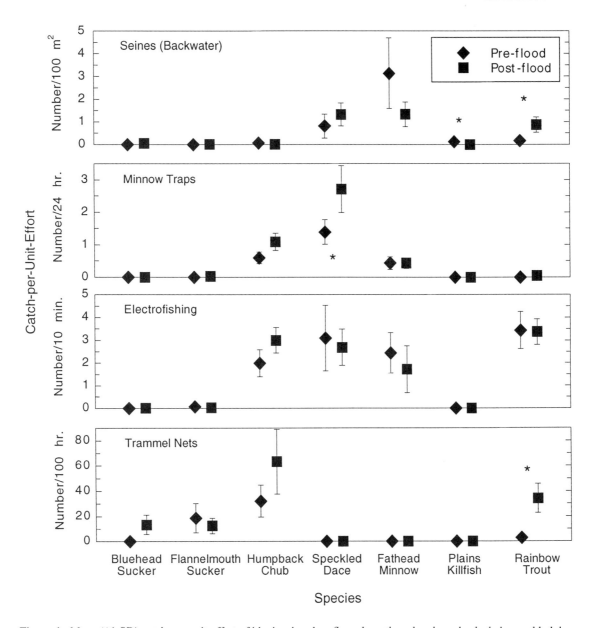

Figure 1. Mean (±1 SD) catch-per-unit-effort of bluehead sucker, flannelmouth sucker, humpback chub, speckled dace, fathead minnow, plains killifish, and rainbow trout in seines, minnow traps, electrofishing, and trammel nets during pre- and post-flood steady 226 m^3/s flows in 1996. * indicates significant difference between means at $\alpha = 0.05$.

Little Colorado and Colorado rivers (RM 61.5) which is an important rearing area, since all species of native fishes remaining in Grand Canyon spawn in the Little Colorado River [*Valdez and Ryel*, 1995; *Arizona Game and Fish Department*, 1996a].

Fish catches were compared for two time scales with each time scale using different, but overlapping, reaches of river [*Schmidt et al.*, Figure 4, this volume]. The first time scale was designed to evaluate effects of the controlled

flood experiment over a 6-month period. The experiment is defined as the controlled flood and the periods of low steady discharge immediately preceeding and following the flood. Pre-experiment (Trip 1; 28 February - 14 March 1996), post-experiment (Trip 3; 18 April - 3 May 1996), summer (Trip 4; 19 June - 4 July 1996) and fall (Trip 5; 8 - 23 September 1996) river trips were conducted from Lees Ferry to Diamond Creek under fluctuating flow conditions (i.e., normal dam operations) in conjunction with Arizona

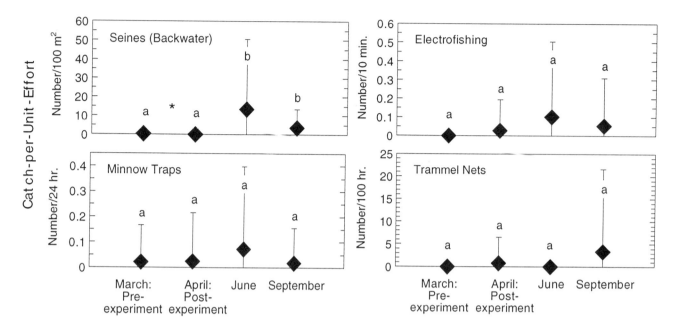

Figure 2. Mean (±1 SD) catch-per-unit-effort of bluehead sucker in seines, minnow traps, electrofishing, and trammel nets by AGFD during pre-flood (March), post-flood (April), June and September, 1996. Means with the same letter are not significantly different at α = 0.05. *indicates a significant difference (α = 0.05) within paired pre- versus post-flood comparison.

Game and Fish Department (AGFD) fish monitoring. Backwaters were seined throughout this entire stretch of the river, but electrofishing, minnow traps, and trammel nets were used only from Lees Ferry to National Canyon (RM 166.4), where the Hualapai Department of Natural Resources (HDNR) conducted sampling as part of a joint AGFD/HDNR monitoring program. The second time scale was designed to evaluate the immediate effects of the controlled high release (flood), itself. Sampling during Trip 2 was conducted during the steady low flows (226 m^3/s) immediately preceding (pre-flood; 23 - 27 March 1996) and following (post-flood; 4 - 7 April 1996) the flood flows and during the high flows (1274 m^3/s; flood; 27 March - 2 April 1996). All backwaters between RM 58.1 (Awatubi Canyon) and RM 65.5 (Lava Canyon Rapid) were sampled, but only during pre- and post-flood sampling periods since no backwaters were present during the flood. Electrofishing, minnow traps, and trammel nets were used during all three sampling periods. Sampling by electrofishing and minnow traps was conducted between RM 61.5 (mouth of the Little Colorado River) and RM 65.5. Trammel nets were used between RM 60.3 and 65.5.

Fish were collected using a variety of sampling gears. Bag seines (10 m x 2 m x 6.4 mm mesh; 2 m x 2 m x 2 m bag) were used in backwater habitats frequented by larval and juvenile fishes. Electrofishing and minnow traps (6.4

mm mesh) were used along talus, debris fan, and vegetated main channel shorelines inhabited mostly by small fishes. Trammel nets (22.9 m x 1.8 m x 3.8 cm inner mesh x 25.4 cm outer mesh) were deployed in deeper areas of the main channel, particularly in eddies, commonly inhabited by adult humpback chub, flannelmouth suckers, bluehead suckers and rainbow trout. All fish captured were identified to species, measured for total length (TL) in millimeters and weighed in grams. All fish were released alive at their site of capture. Data were recorded on field data sheets and transferred to electronic spreadsheets for analyses.

Minnow traps were deployed and electrofishing was conducted at standard transects on each trip. Trammel nets were deployed in suitable eddy complexes within standard sampling reaches. Electrofishing was conducted for 2-3 hr after dark and catch-per-unit-effort (CPUE) was calculated for each sample as the number of fish captured per 10 min of electrofishing time. Minnow traps were set overnight in groups of five traps and CPUE was calculated as the number of fish caught in 24 hours (day unit) for each group of five traps. Trammel nets were deployed at dusk and checked every 2 hr for a total set time of 4-6 hr. Trammel net CPUE was calculated for each sample as the number of fish caught per 23 m of net for 100 hr of netting. Nearly all available backwaters were sampled before and after the flood and the same sites were sampled, when possible,

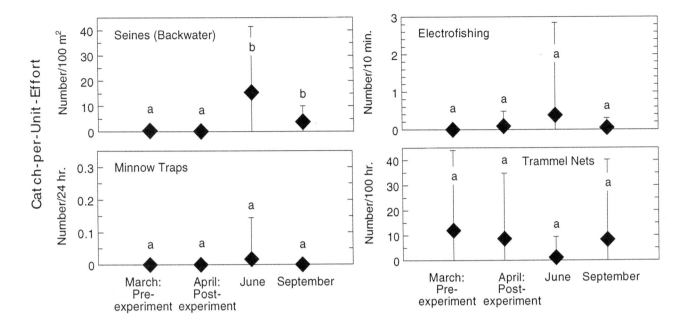

Figure 3. Mean (±1 SD) catch-per-unit-effort of flannelmouth sucker in seines, minnow traps, electrofishing, and trammel nets by AGFD during pre-flood (March), post-flood (April), June and September, 1996. Means with the same letter are not significantly different at α = 0.05.

before and after the experiment. Seining was conducted during daylight, and CPUE was calculated for each sample as the number of fish captured per 100 m² seined. Backwater number and location changed dramatically as a result of the flood and with changes in river discharge [*Hoffnagle et al.*, 1998]. New backwaters were created immediately after the flood, but variation in discharge and change in flow pattern following the experiment, destroyed many of these sample sites [*Brouder*, 1996; *Hoffnagle et al.*, 1998].

Catches-per-unit-effort were compared among habitats and sampling periods. CPUE data could not be normalized due to the high frequency of zero captures and were statistically analyzed using non-parametric tests [*Sokal and Rohlf*, 1981]. Comparisons of mean CPUE before and after the experiment were analyzed using the Mann-Whitney U test. Minnow trap and electrofishing CPUE's were used to compare habitat use for debris fans, talus and vegetated shorelines by small fishes before and after the flood. Comparisons of mean CPUE among shoreline types and river trips (for examination of recovery of fish populations) were analyzed with the Kruskal-Wallis test. Multiple Mann-Whitney U tests, with the Bonferroni correction, were used to discern differences among means for significant Kruskal-Wallis tests. Significance for all statistical tests was set at α=0.05.

3. RESULTS

Eleven species of fish were captured during the study, including all four remaining native species (i.e., humpback chub, flannelmouth sucker, bluehead sucker, speckled dace) and seven non-native species (i.e., common carp, fathead minnow, plains killifish, red shiner, redside shiner (*Richardsonius balteatus*), brown trout, and rainbow trout). The most commonly captured non-native species were fathead minnow, plains killifish and rainbow trout. Catch rates of all species for all gears were comparable to those reported from previous sampling in Grand Canyon [*Valdez and Ryel*, 1995; *Arizona Game and Fish Department*, 1996a; *Brouder et al.*, 1997].

3.1. Effects on Fish Distribution

Few differences in fish distribution were seen between pre- and post-experiment sampling periods, and those differences were only for non-native fathead minnows and rainbow trout. Mean minnow trap CPUE for fathead minnow significantly decreased (P≤0.0352) between pre- and post-experiment periods in the two reaches immediately below the LCR (i.e., Reaches 3 and 4). Mean CPUE in Reach 3 significantly decreased from 0.1 fish/day prior to the experiment to 0.01 fish/day afterwards. In Reach 4,

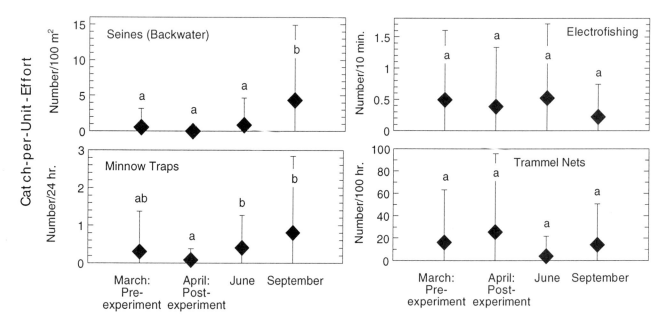

Figure 4. Mean (±1 SD) catch-per-unit-effort of humpback chub in seines, minnow traps, electrofishing, and trammel nets by AGFD during pre-flood (March), post-flood (April), June and September, 1996. Means with the same letter are not significantly different at α = 0.05.

fathead minnow mean CPUE decreased from 0.3 to 0.02 fish/day. Mean electrofishing CPUE for adult rainbow trout in Reach 4 decreased significantly (P=0.0104) from 4.1 fish/10 min, prior to the experiment, to 1.0 fish/10 min, afterwards.

3.2. Effects on Fish Abundance

There were few significant differences in mean CPUE between the periods of low steady flows before and after the flood (Figure 1). No differences were seen in mean CPUE for any species by electrofishing (P≥0.1656), but mean CPUE for plains killifish decreased in backwater seining (P=0.0347). Mean CPUE for speckled dace increased in minnow traps (P=0.0121), for adult rainbow trout in trammel nets (P=0.0097) and juvenile rainbow trout in backwater seining (P=0.0361).

Mean CPUE for speckled dace was significantly higher in minnow traps after the flood than before the flood, increasing from 1.4 to 2.7 fish/day. This increase in CPUE was attributed to an increase in numbers of speckled dace caught along talus shorelines (P=0.0018), where mean CPUE increased from 0.3 to 1.8 fish/day. Mean CPUE for plains killifish in backwaters decreased significantly from 0.12 to 0.00 fish/100 m² seined. Mean CPUE for adult rainbow trout in trammel nets set in main channel habitats increased significantly from 2.8 fish/100 hr before the flood

to 34.3 fish/100 hr after the flood. Mean CPUE for juvenile rainbow trout seined in backwaters also increased significantly from 0.2 fish/100 m² before the flood to 0.9 fish/100 m² after the flood.

There were few differences in catch rates between pre- and post-experiment river trips, with little evidence of native fishes being affected by the experimental flood (Figures 2 - 5). Non-native fishes, however, appear to have been affected by the experiment, as indicated by changes in CPUE in all three common species: fathead minnow, plains killifish and rainbow trout (Figures 6 - 8). Neither electrofishing (P≥0.0811) nor minnow trap (P≥0.1246) catch rates for any species differed by shoreline type throughout the reach sampled during pre- and post-experiment river trips.

Mean CPUE for native species with all gear types did not differ between pre- and post-experiment river trips (P≥0.0937), with one exception. Mean CPUE for bluehead sucker in backwaters decreased significantly (P=0.0443) from 0.5 fish/100 m² seined (22 fish) before the experiment to 0.05 fish/100 m² seined (4 fish) after the experiment. Bluehead sucker catch with all other gear types did not differ between sampling periods.

Mean CPUE for fathead minnow, plains killifish, and rainbow trout all differed between pre- and post-experiment samples, indicating that these species were affected by the experimental flood. Plains killifish mean CPUE in backwaters decreased significantly (P=0.0065) from 0.9

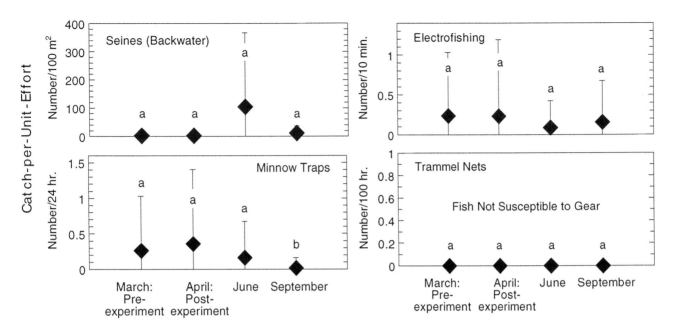

Figure 5. Mean (±1 SD) catch-per-unit-effort of speckled dace in seines, minnow traps, electrofishing, and trammel nets by AGFD during pre-flood (March), post-flood (April), June and September, 1996. Means with the same letter are not significantly different at α = 0.05.

fish/100 m² seined (43 fish) before the experiment to 0.0 fish/100 m² after the experiment. Conversely, mean CPUE for juvenile rainbow trout from backwater seining increased significantly from 0.04 fish/100 m² before to 0.3 fish/100 m² after the experiment, and mean CPUE for juvenile rainbow trout in minnow traps increased from 0.0 to 0.1 fish/day (P≤0.0146). Fathead minnows changed the most with significant decreases in mean CPUE, from pre- to post-experiment trips in two gear types. Mean CPUE for fathead minnow decreased significantly (P=0.0001) for minnow traps from 0.8 to 0.05 fish/day and for electrofishing (P=0.0185) from 1.6 to 0.3 fish/10 min, before and after the experiment, respectively.

3.3. Effects on Habitat Use

Catch rates were lowest during the flood for all species and gear types, but significantly lower only for half of the comparisons. Catch rates for nearly all species from both electrofishing and minnow traps varied with shoreline type; with all three species (i.e., humpback chub, speckled dace and fathead minnow) displaying some habitat segregation.

Electrofishing catch rate for juvenile humpback chub was lower during the flood (0.4 fish/10 min) than during either pre- (2.0 fish/10 min) or post-flood (3.0 fish/10 min) sampling periods (P=0.0001). For minnow traps, humpback

chub mean CPUE did not differ among pre-flood, flood and post-flood sampling periods (P=0.1963). Humpback chub CPUE by electrofishing was significantly higher (P=0.0300) along talus shorelines (2.9 fish/10 min) than vegetation (0.9 fish/10 min), but not debris fans (1.5 fish/10 min) which were similar to both talus and vegetation. Minnow trap CPUE for humpback chub was significantly higher (P=0.0004) along talus (1.3 fish/day) than in either vegetation (0.5 fish/day) or debris fans (0.3 fish/day). Electrofishing CPUE was significantly higher along talus shorelines than both debris fans and vegetation during both pre-flood (4.1 fish/10 min; P=0.0171) and post-flood (5.1 fish/10 min; P=0.0357) sampling periods (Figure 9). During the flood, electrofishing CPUE did not differ among shoreline types (P=0.4750). Minnow trap CPUE did not differ among shoreline types during the pre-flood (P=0.9690) or post-flood (P=0.2421) sampling periods (Figure 10). During the flood, however, humpback chub minnow trap CPUE was significantly higher along talus than debris fans or vegetated shorelines.

Electrofishing CPUE for speckled dace did not differ significantly among pre-flood, flood and post-flood sampling periods (P=0.2484). However, minnow trap CPUE for speckled dace was significantly higher (P=0.0074) after the flood (2.7 fish/day) than either before (1.4 fish/day) or during the flood (1.2 fish/day). Electro-

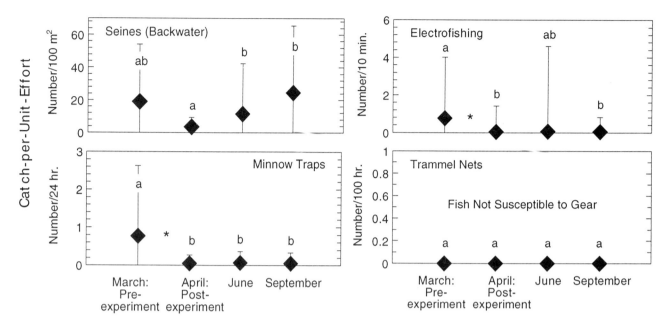

Figure 6. Mean (±1 SD) catch-per-unit-effort of fathead minnow in seines, minnow traps, electrofishing, and trammel nets by AGFD during pre-flood (March), post-flood (April), June and September, 1996. Means with the same letter are not significantly different at α = 0.05. *indicates a significant difference (α = 0.05) within paired pre- versus post-flood comparison.

fishing CPUE for speckled dace was significantly higher (P=0.0001) along debris fans (6.8 fish/10 min) than either talus or vegetated shorelines (1.1 fish/10 min, each). Speckled dace minnow trap CPUE also differed significantly (P=0.0001) among shoreline types: CPUE in debris fans (3.1 fish/day) was significantly higher than in talus (1.2 fish/day) or vegetation (0.6 fish/day) and CPUE in talus was higher than in vegetation. Within sampling periods, electrofishing CPUE for speckled dace did not differ (P=0.2908) among shorelines during the flood (Figure 9). Speckled dace CPUE was higher along debris fans than other shorelines during both the pre- and post-flood sampling periods (P≤0.0426). Minnow trap CPUE for speckled dace was always significantly higher (P≤0.0161) in debris fans than in vegetated shorelines and significantly higher than in talus during pre-flood and flood sampling periods (Figure 10).

Electrofishing CPUE for fathead minnow did not differ by sampling period (P=0.0599). However, minnow trap CPUE was significantly higher (P=0.0170) during both pre-flood and post-flood sampling periods (0.4 fish/day, each) than during the flood (0.1 fish/day). Electrofishing CPUE for fathead minnow did not differ by shoreline type (P=0.1128) but minnow trap CPUE was significantly higher (P=0.0391) among vegetation (0.5 fish/day) than along

either debris fans or talus shorelines (0.1 fish/day, each). Within sampling periods, fathead minnow CPUE varied little. There was no difference (P≥0.0605) in CPUE among shoreline types for electrofishing (Figure 9). In minnow traps, CPUE did not differ (P≥0.1732) among shorelines during the pre-flood or flood sampling periods (Figure 10). During the post-flood period, minnow trap CPUE was higher in vegetated shorelines than along talus; debris fan CPUE was similar to both talus and vegetation.

3.4. Recovery

Densities of only bluehead sucker, fathead minnow, plains killifish and rainbow trout were significantly affected by the flood and recovery of these affected fishes was rapid (Figures 2-8). Bluehead sucker mean CPUE in backwaters was higher in June and September than in March, before the flood (P=0.0002). Fathead minnow CPUE for electrofishing in June increased (P=0.0433) to equal that before the flood, while minnow trap CPUE remained low throughout the summer. However, backwater CPUE increased (P=0.0409) from post-flood levels. A few plains killifish were captured in backwaters in June, but catches in September exceeded all previous catches (P=0.0008). Catches of most other species were also higher in June and/or September,

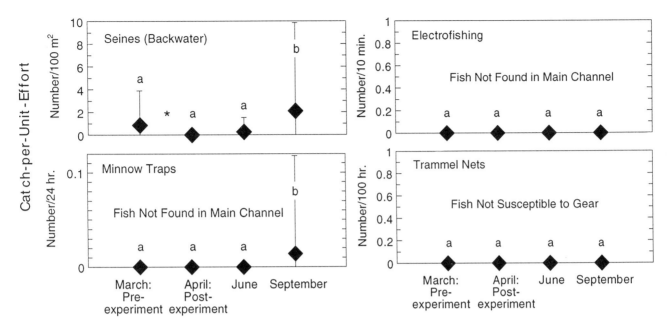

Figure 7. Mean (±1 SD) catch-per-unit-effort of plains killifish in seines, minnow traps, electrofishing, and trammel nets by AGFD during pre-flood (March), post-flood (April), June and September, 1996. Means with the same letter are not significantly different at α = 0.05. *indicates a significant difference (α = 0.05) within paired pre- versus post-flood comparison.

reflecting reproduction by these species, as well. Mean CPUE for juvenile rainbow trout returned to low levels in June, then increased in backwaters again during September (P=0.0008).

4. DISCUSSION

The 1996 controlled flood provided an opportunity to test the hypothesis that native southwestern riverine fishes respond to floods differently than non-native fishes. *Minckley and Meffe* [1987] hypothesized that morphological and/or behavioral attributes of fishes native to small and medium southwestern rivers enable them to maintain their position in streams during floods, as movement to low velocity refugia is frequently impossible in canyon bound reaches. Moreover, displacement to such refugia or downstream reaches is disadvantageous because such habitats are often ephemeral at best and mortality by stranding is probable. Non-native fishes, particularly taxa which evolved in systems with extensive flood plain habitat, are more likely to seek refuge from floods by moving onto flood plains or by drifting downstream to more quiet waters [*Hynes*, 1970; *Ross and Baker*, 1983]. The present study is the first time this hypothesis has been tested on a large southwestern river, such as the Colorado River through Grand Canyon.

The results of this study demonstrate that native Colorado River fishes were largely unaffected by the controlled flood, while non-native species were moderately affected. Although fathead minnows were not eliminated by the flood, significant decreases in their catch rates were observed, particularly in backwaters, which were inundated by the flood. Although a wide band of flooded and emergent riparian vegetation was present during the flood, fathead minnows did not appear to use this habitat extensively, suggesting that these fish were displaced downstream. Plains killifish, as expected, were greatly affected by the flood — no fish were collected immediately afterwards, probably as a result of inundation of backwaters and high current velocities that displaced these fish downstream. However, there was an increase in catch rates of adult and juvenile rainbow trout immediately following the flood, probably due to displacement from upstream areas or changes in habitat use from mid-channel to nearshore habitats. Catch rates of YOY bluehead sucker decreased in backwaters following the flood, as a result of inundation of habitats, although these fish were not found in alternative sites. Catch rates of speckled dace increased along debris fans as a result of shifts in habitat use from mid-channel riffles to proximate nearshore environments. All affected fishes recovered rapidly by reinvasion from tributary refugia and reproduction.

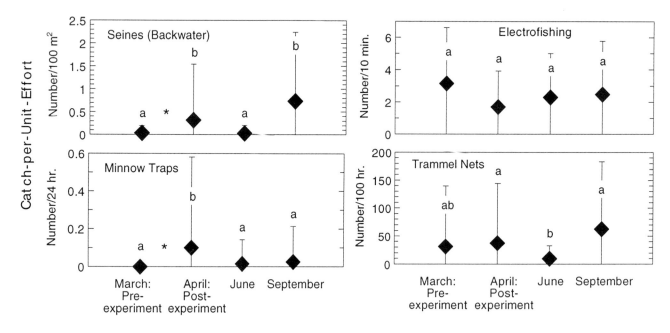

Figure 8. Mean (±1 SD) catch-per-unit-effort of rainbow trout in seines, minnow traps, electrofishing, and trammel nets by AGFD during pre-flood (March), post-flood (April), June and September, 1996. Means with the same letter are not significantly different at α = 0.05. *indicates a significant difference (α = 0.05) within paired pre- versus post-flood comparison.

Riverine habitats are frequently less heterogeneous during floods [*Harrel*, 1978] which may explain the increased habitat selectivity by the three species during the flood sampling period. Velocity, depth, sediment load, and turbidity increase with increasing river discharge [*Hynes*, 1970; *Bain et al.*, 1988] and substrate and cover features of rivers are frequently altered during floods. These habitat alterations are side-effects of increased discharge, which elicit changes in fish behavior that vary with age [*Schlosser*, 1985] and life history [*Harvey*, 1987; *Minckley and Meffe*, 1987] of individual taxa. *Valdez and Ryel* [1997] showed that nearshore and near surface activity by radio-tagged adult humpback chub in Grand Canyon was significantly increased at turbidity levels above 30 NTU. Similarly, *Arizona Game and Fish Department* [1996a] found that speckled dace, juvenile humpback chub and flannelmouth sucker were more commonly captured in shallow water when turbidity exceeded 30 NTU. Mean turbidity was significantly higher (P=0.0001) during the controlled flood (mean=58.2 NTU) than before (mean=8.64 NTU) or after (mean=10.0 NTU).

Despite the greater chances for activity by fishes during high flows, shoreline electrofishing catch rates were lower and more variable during the flood. We believe that increased turbidity and high current velocity interfered with electrofishing effectiveness during the flood. High turbidity

decreased visibility for netters and high current velocities interfered with boat handling. Efficiency and sampling variability of other gears, such as minnow traps and trammel nets, were also affected by flood conditions. Fish may use unbaited minnow traps as a source of cover. However, traps may not be used during periods of high turbidity because additional cover is not needed and traps may be less likely to be encountered. Additionally, minnow traps were frequently buried during the flood as a result of large shoreline sand deposits and trammel nets often clogged with debris dislodged by the high flows. Despite possible gear inefficiencies, catch rates of all species in all gears were comparable to those reported from previous sampling in Grand Canyon [*Valdez and Ryel*, 1995; *Arizona Game and Fish Department*, 1996a; *Brouder et al.*, 1997; *Hoffnagle et al.*, 1998].

The decrease in bluehead sucker CPUE is of interest since this is a native species well adapted to fast water [*Minckley*, 1991] and should be able to cope with a flood of this relatively low magnitude. Ontogenetic behavioral changes, observed in bluehead suckers in the Colorado River, Grand Canyon, from 1991 to 1994 [*Arizona Game and Fish Department*, 1996a], may explain this decrease in CPUE in backwaters. Bluehead suckers spend approximately the first year of life in schools, commonly with similar-sized flannelmouth suckers, in low-velocity

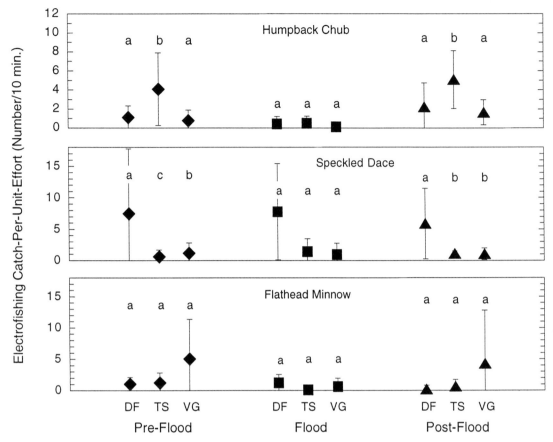

Figure 9. Mean (±1 SD) electrofishing catch-per-unit-effort for humpback chub, speckled dace, and fathead minnow along debris fan, talus, and vegetated shorelines during pre-flood, flood and post-flood sampling periods of the 1996 controlled flood in the Colorado River, Grand Canyon. Among habitat types, means with the same letter are not significantly different at α = 0.05. DF = debris fan shoreline; TS = talus shoreline; VG = vegetated shoreline.

habitats, such as backwaters, feeding primarily on benthic invertebrates [*Arizona Game and Fish Department*, 1996a]. As bluehead suckers reach approximately 50 mm TL (approximately one year of age), a cartilaginous scraper develops on the mandible and the fish leave the backwaters to feed by scraping algae and diatoms from rocks in riffles and other rocky areas [*Minckley*, 1991]. We believe that the dates of the 1996 controlled flood (March-April, 1996) approximately coincided with this shift in habitat by young-of-year (1995 year class) bluehead suckers and explains decreased densities in backwaters. Catch rates of bluehead sucker rebounded in June as larvae of the 1996 year class drifted out of spawning tributaries into the mainstem Colorado River and we conclude that it is unlikely that the flood affected the bluehead sucker population.

Catch rates of YOY flannelmouth sucker (1995 year class) did not differ as a result of the flood and it appears that this species was unaffected by backwater inundation

and increased velocities. Like the bluehead sucker, catches of young from the 1996 year class were noted as larvae drifted out of spawning tributaries shortly after the flood. Large numbers of bluehead and flannelmouth sucker larvae have been documented drifting within and from the Little Colorado River soon after hatching [*Robinson et al.*, 1996].

Humpback chub were also unaffected by the flood. Minnow trap catches of juvenile humpback chub along mainstem shorelines increased in June with early dispersal of young from the LCR. Later in the summer, shoreline and backwaters catches further increased as more YOY dispersed from the LCR during late summer monsoon flooding. Dispersal during monsoon floods is commonly seen in humpback chub in Grand Canyon [*Valdez and Ryel*, 1995, 1997; *Arizona Game and Fish Department*, 1996a; *Hoffnagle et al.*, 1998].

The increase in catch of speckled dace in minnow traps immediately after the flood was due to a sixfold increase in catches along talus shorelines. This increase was attributed

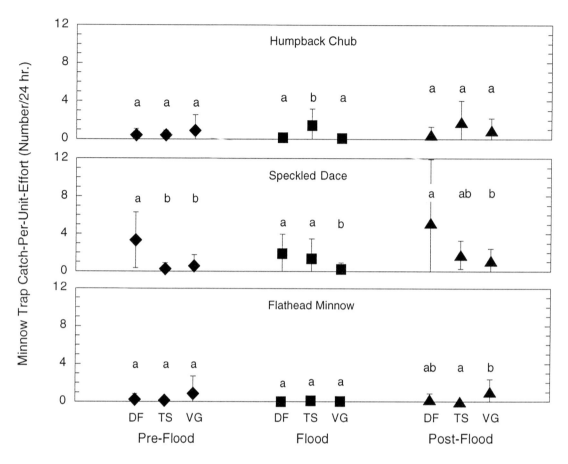

Figure 10. Mean (±1 SD) minnow trapping catch-per-unit-effort for humpback chub, speckled dace, and fathead minnow along debris fan, talus, and vegetated shorelines during pre-flood, flood and post-flood sampling periods, 1996. Among habitat types, means with the same letter are not significantly different at $\alpha = 0.05$. DF = debris fan shoreline; TS = talus shoreline; VG = vegetated shoreline.

to shifts in habitat use during the flood and not to an increased number of fish. Fish may have moved from mid-channel riffles to proximate shoreline habitats. The flood inundated mid-channel islands, eliminating associated shallow habitats and forcing fish to shoreline habitats along channel margins. Speckled dace commonly inhabit swift water in streams and rivers [*Minckley*, 1973], including the Paria River where the species survives floods of high discharge and turbidity [*Rinne and Minckley*, 1991] and Aravaipa Creek, where velocities reach 75 cm/s [*Rinne*, 1992]. Velocities along sandbars in the main channel Colorado River reached 52 cm/s, where speckled dace were the most common species captured species [*Arizona Game and Fish Department*, 1996a]. Speckled dace catch rates did not differ between pre- and post-experiment periods, indicating that the fish dispersed from the talus shorelines within days, presumably reoccupying other suitable habitats.

The fathead minnow is suspected to compete with and possibly prey on larval native fishes [*Valdez and Ryel*, 1995; *Arizona Game and Fish Department*, 1996a; *Tyus and Saunders*, 1996] and it was hypothesized that the flood would significantly reduce their population size. Although fathead minnow catch rates decreased in both minnow traps (tenfold) and by electrofishing (fivefold) immediately following the flood, the species recovered within 6 months, probably by reproduction and reinvasion from tributary streams. Backwater CPUE quickly increased following the experiment, particularly for small fish, indicating that reproduction had occurred in these habitats. Fathead minnow reproductive capacity is great; spawning occurs throughout summer, with up to 2622 eggs produced by a single female [*Carlander*, 1969, and references therein]. Also, since it is found as far north as Great Slave Lake, Canada [*Scott and Crossman*, 1979], it is unlikely that cold temperatures are debilitating to this species. However, while flooding of a low magnitude does not appear to eradicate this fish, it is apparent that flooding can temporarily reduce its numbers.

Although plains killifish were not commonly captured during pre-flood sampling, they are found primarily in calm, warm habitats of backwaters and tributaries and their populations in the mainstem Colorado River have been increasing in recent years [*Arizona Game and Fish Department*, 1996a]. It appears that plains killifish populations in the mainstem Colorado River were decimated by the controlled flood. During pre-flood sampling, 43 killifish were captured, but none were captured during post-flood sampling. *Cross* [1967, in *Minckley and Klaassen,* 1969] found plains killifish only in shallow water (<15 cm) and *Moore* [1968, in *Minckley,* 1973] found it in "saline, sandy-bottomed streams within its native range" in Texas and New Mexico. Although we found no information on swimming ability of this species, it certainly appears that they could not withstand current velocities found in the main channel of the Colorado River during this flood, particularly with the cold temperatures of the hypolimnetic releases.

Rainbow trout catch rates varied by gear and habitat, but suggest that this species was also affected by the controlled flood. Catches increased during the steady 226 m^3/s flows immediately after the flood in both eddies (adults caught in trammel nets) and in backwaters (juveniles and adults caught in seines). Rainbow trout catch also increased in backwater seining and in shoreline minnow traps during post-experiment sampling and they were commonly captured in areas of the river where they were rare prior to the experiment. Increases in rainbow trout catch are likely due to displacement of fish from upstream populations. *McKinney et al.* [1996] reported a decrease in the catch of rainbow trout <15 cm after the controlled flood in the Lees Ferry Reach, presumably as a result of downstream displacement. However, we also observed an increase in the catch of adults. *Seegrist and Gard* [1972] found that flooding greatly affected small rainbow trout as well as some large individuals. The origin of increased numbers of rainbow trout in downstream reaches is not known, but these fish could have originated from Lees Ferry or from local concentrations, such as Vasey's Paradise, Nankoweap Creek, Bright Angel Creek or Deer Creek.

Both plains killifish and fathead minnows are potential competitors of larval and juvenile native fishes, particularly humpback chub. There is considerable diet overlap between both of these non-native species and native fishes. Chironomids, ceratopogonids, cladocerans, simuliids and terrestrial insects were common in the gastrointestinal tract of each species throughout the Colorado River in Grand

Canyon [*Arizona Game and Fish Department,* 1996a]. While low-magnitude flooding did not appear to eradicate non-native fishes in Grand Canyon, it did temporarily reduced their numbers. Main channel flooding, in conjunction with flooding in tributaries may significantly reduce numbers of non-native fishes by reducing warm, sheltered, low-velocity refugia. Hence, properly-timed flooding may be a useful management tool to temporarily reduce competition among native and these non-native species. Reduction in numbers of these and other non-native fishes, including red shiners, which are also increasing in number, can benefit native species [*Hoffnagle et al.,* 1998]. However, floods of a significantly greater, and possibly infeasible, magnitude and/or duration will be required to significantly diminish non-native fish populations in Grand Canyon.

The effect of the 1996 controlled flood on Grand Canyon fishes cannot be determined from a single event. This controlled flood did not appear to impact native fish populations, but did temporarily diminish populations of non-native species. Mainstem flooding in conjunction with flooding in tributaries may be more effective in reducing non-native fish abundance. *Meffe and Minckley* [1987] and *Minckley and Meffe* [1987] have shown that native southwestern fishes are unaffected by flooding. Indeed, they may require predictable flooding as part of their life history. *John* [1963] found that photoperiod and temperature changes prepare speckled dace for spawning and that flooding cues the commencement of spawning. It is likely that a similar set of cues are used by the other native fishes. This flood probably affected native fishes only through its affect on habitat and food availability [see *Arizona Game and Fish Department,* 1996b; *Brouder et al.,* this volume], but these effects could not be discerned in the present short-term study. The success of this flood in providing backwater rearing habitats for native fishes in the late spring and early summer when larvae are drifting downstream is questionable [*Brouder,* 1996; *Brouder et al.,* this volume]. Low releases from Glen Canyon Dam immediately following the flood (i.e., 4 days at 226 m^3/s) and subsequent high releases following the experiment (i.e., 500-600 m^3/s; *Schmidt et al.* [Figure 4, this volume]) severely eroded the reattachment bars that created and deepened many of the backwaters. The eventual success of future controlled floods will be critically dependent on dam releases following the flood.

Acknowledgments. We acknowledge Dave Wegner, Mike Yard, Ted Melis and LeAnn Skrzynski of the Glen Canyon Environmental Studies for their work in coordinating the controlled flood and their assistance with logistical support. Mark Brouder and Tom Dresser (Arizona Game and Fish Department) and Brian Cowdell and Erika Prats (BIO/WEST, Inc.) and many volunteers helped collect data in the field. Mark Brouder, Ric Bradford, Tom Dresser, Bill Persons, Tony Robinson and Jody Walters (AGFD) and Tim Modde and Lynn Kaeding (USFWS) reviewed and improved this manuscript.

REFERENCES

Arizona Game and Fish Department, *Ecology of Grand Canyon backwaters*, 155 pp., Glen Canyon Environ. Studies, Final Rept., Bur. Reclam., Salt Lake City, UT, 1996a.

Arizona Game and Fish Department, *The effects of an experimental flood on the aquatic biota and their habitats in the Colorado River, Grand Canyon, Arizona*, 130 pp., Ariz. Game Fish Dept., Phoenix, AZ, 1996b.

Bain, M.B., J.T. Finn, and H.E. Brooke, Stream flow regulation and fish community structure, *Ecol.*, 69, 382-392, 1988.

Brouder, M.J., Number and area of backwaters, in *The effects of an experimental flood on the aquatic biota and their habitats in the Colorado River, Grand Canyon, Arizona*, ed. Arizona Game and Fish Department, pp. 4-1 to 4-7, Ariz. Game Fish Dept., Phoenix, AZ, 1996.

Brouder, M.J., T.L. Hoffnagle, M.A. Tuegel, and W.R. Persons, *Glen Canyon Environmental Studies Colorado River native fish study 1995 annual report*, 101 pp., Bur. Recl., Upper Colorado Region, Salt Lake City, UT, 1997.

Carlander, K.D., *Handbook of freshwater fishery biology*, 752 pp., Iowa State University Press, Ames, IA, 1969.

Clarkson, R.W., O.T. Gorman, D.M. Kubly, P.C. Marsh, and R.A. Valdez, Recommendations for operation of Glen Canyon Dam as a tool for management of native fishes (abstract), *Proc. Desert Fishes Coun.*, 25, 20, 1994.

Harrel, H.L. Response of the Devil's River (Texas) fish community to flooding, *Copeia*, 1978, 60-68, 1978.

Harvey, B.C., Susceptibility of young-of-the-year to downstream displacement by flooding, *Trans. Am. Fish. Soc.*, 116, 851-855, 1987.

Hoffnagle, T.L., M.J. Brouder, and D.W. Speas, *Glen Canyon Environmental Studies, 1996 Annual Report*, 120 pp., Arizona Game and Fish Department, Phoenix, AZ, 1998.

Hoopes, R.L., Flooding as a result of Hurricane Agnes and its effect on a native brook trout population in an infertile headwater stream in central Pennsylvania, *Trans. Am. Fish. Soc.*, 104, 96-99, 1975.

Hynes, H.B.N., *The Ecology of Running Waters*, Liverpool Univ. Press, Liverpool, UK, 1970.

John, K.R., The effect of torrential rains on the reproductive cycle of *Rhinichthys osculus* in the Chiricahua Mountains, Arizona, *Copeia*, 1963, 286-291, 1963.

Matthews, W.J., Fish faunal structure in an Ozark stream: stability persistence and a catastrophic flood, *Copeia*, 1986, 388-397, 1986.

Meffe, G.K., and W.L. Minckley, Persistence and stability of fish and invertebrate assemblages in a repeatedly disturbed Sonoran Desert stream, *Am. Midl. Nat.*, 117, 177-191, 1987.

McKinney, T.D., R.S. Rogers, A.D. Ayers, and W.R. Persons, Effects of experimental flooding on algae, macroinvertebrates, rainbow trout and flannelmouth suckers in the Glen Canyon Dam tailwater, in *The effects of an experimental flood on the aquatic biota and their habitats in the Colorado River, Grand Canyon, Arizona*, ed. Arizona Game and Fish Department, pp. 10-1 to 10-30, Arizona Game and Fish Department, Phoenix, AZ, 1996.

Minckley, C.O., and H.E. Klaassen, Life history of the plains killifish, *Fundulus kansae* (Garman), in the Smoky Hill River, Kansas, *Trans. Am. Fish. Soc.*, 98, 460-465, 1969.

Minckley, W.L., *Fishes of Arizona*, 293 pp., Ariz. Game Fish Dep., Phoenix, AZ, 1973.

Minckley, W.L., Native fishes of the Grand Canyon region: an obituary?, in *Colorado River Ecology and Dam Management*, ed. G.R. Marzolf, pp. 124-177, Natl. Acad. Press, Washington, DC, 1991.

Minckley, W.L., and G.K. Meffe, Differential selection by flooding in stream-fish communities of the arid American Southwest, in *Community and Evolutionary Ecology of North American Stream Fishes*, ed. W.J. Matthews and D.C. Heins, pp. 93-104, Univ. Okla. Press, Norman, OK, 1987.

Rinne, J.N., Physical habitat utilization of fish in a Sonoran Desert stream, Arizona, southwestern United States, *Ecol. Fresh. Fish*, 1, 35-41, 1992.

Rinne, J.N., and W.L. Minckley, *Native fishes of arid lands: a dwindling resource of the desert Southwest*, Gen. Tech. Rept. RM-206, U.S. Dept. of Ag., For. Serv., Rocky Mtn. Forest Range Exp. Sta., Fort Collins, CO, 1991.

Robinson, A.T., R.W. Clarkson, and R.E. Forrest, *Spatio-temporal distribution, habitat use, and drift of early life stage native fishes in the Little Colorado River, Grand Canyon, Arizona, 1991-1994*, Final Rept., Bur. Reclam., Glen Canyon Environ. Studies, Flagstaff, AZ, 1996.

Ross, S.T., and J.A. Baker, The response of fishes to periodic spring floods in a southeastern stream, *Am. Midl. Nat.*, 109, 1-14, 1983.

Schlosser, I.J., Flow regime, juvenile abundance, and the assemblage structure of stream fishes, *Ecol.*, 66, 1484-1490, 1985.

Scott, W.B., and E.J. Crossman, *Freshwater Fishes of Canada*, Fisheries Research Board of Canada, Ottawa, 1979.

Seegrist, D.W., and R. Gard, Effects of floods on trout in Sagehen Creek, California. *Trans. Am. Fish. Soc.*, 101, 478-482, 1972.

Sokal, R.R. and F.J. Rohlf, *Biometry*, 2nd ed., W.H. Freeman Co., New York, 1981.

Tyus, H.M., and J.F. Saunders, *Nonnative fishes in the upper Colorado River basin and a strategic plan for their control*. Rept. to U.S. Fish Wildl. Serv. (14-48-0006-95-923). Center Limno., Univ. Colorado, Boulder, CO, 1996.

U.S. Department of the Interior, *Operation of Glen Canyon Dam final environmental impact statement*, Bur. Reclam., Upper Colo. Reg., Salt Lake City, UT, 1995.

U.S. Department of the Interior, *Glen Canyon Dam beach/ habitat-building test flow, final environmental assessment and finding of no significant impact*, Bur. Reclam., Upper Colo. Reg., Salt Lake City, UT, 1996.

Valdez, R.A., and E.J. Wick, Natural vs. manmade backwaters as native fish habitat in *Aquatic resource management of the Colorado River ecosystem*, edited by V.D. Adams and V.A. Lamarra, pp. 519-536, Ann Arbor Science Publications, Ann Arbor, MI, 1983.

Valdez, R.A., and R.J. Ryel, *Life history and ecology of the humpback chub* (Gila cypha) *in the Colorado River, Grand Canyon, Arizona*, Final Rept., Bur. Reclam., Salt Lake City, UT, Contract No. 0-CS-40-09110, Bio/West Rept. No. TR-250-08, Logan, UT, 1995.

Valdez, R.A., and R.J. Ryel, Life history and ecology of the humpback chub in the Colorado River in Grand Canyon, Arizona, in *Proceedings of the Third Biennial Conference of Research on the Colorado Plateau*, ed. C. van Riper, III, and E.T. Deshler, pp. 3-31, Natl. Park Serv. Trans. Procs. Ser. NPS/ NRNAU/NRTP-97/12, 1997.

Timothy L. Hoffnagle, Research Branch, Arizona Game and Fish Department, 121 East Birch Street, Suite 307, Flagstaff, AZ 86001; email: thoffnag@flagmail.wr.usgs.gov

Richard A. Valdez, SWCA, Inc., 172 West 1275 South, Logan, UT 84321

David W. Speas, Research Branch, Arizona Game and Fish Department, 2221 West Greenway Road, Phoenix, AZ 85023

Flannelmouth Suckers: Movement in the Glen Canyon Reach and Spawning in the Paria River

Carole C. McIvor and Michele L. Thieme

U.S. Geological Survey, Tucson, Arizona

Flannelmouth sucker (*Catostomus latipinnis*) spawn March through April in the Paria River, a tributary of the Colorado River 25 km downstream of Glen Canyon Dam. We followed 50 ultrasonic-tagged, adult flannelmouth suckers in the Glen Canyon reach to determine if this native species was displaced downstream by the controlled flood. We also seined and made visual observations over known spawning areas in the Paria River during the flood, and surveyed for presence of young-of-the-year (YOY) during the summer rearing season to determine if the flood preempted or delayed spawning. Conservatively, forty-two (84%) tagged fish stayed within the Glen Canyon reach; the majority (33) used a low-velocity ponded habitat formed when high Colorado River flows ponded the lower 0.76 km of the Paria River. Spawning in the Paria River during 1996 was successful as evidenced by the presence of ripe fish over known spawning areas and by capture of a strong year-class of YOY in late spring and summer. Thus, we found no obvious deleterious effects of the controlled flood on movement of flannelmouth sucker in the Glen Canyon reach or on spawning in the Paria River.

1. INTRODUCTION

Floods can cause major alterations to stream fish habitat [e.g., *Matthews,* 1986] and to both absolute and relative fish abundances [*Matthews,* 1986; *Strange et al.,* 1992; *Jowett and Richardson,* 1994]. However, even following floods of unusually high magnitude that cause "catastrophic" damage to streams, recovery of fish assemblages often occurs within 8-12 months [*Moyle and Vondracek,* 1985; *Matthews,* 1986]. Patterns of change in fish abundances and assemblages due to floods can be predicted qualitatively. Adult fish are considerably less vulnerable to floods than young [*John,* 1964: a desert cyprinid; *Coon,* 1987: a riffle-dwelling darter (percid)] and larger fish species are

generally less adversely affected than small ones [*Collins et al.,* 1981]. Further, the timing of floods relative to species-specific times of spawning is crucial. If floods occur outside the period when young life-history stages are recruiting, negative effects on populations and assemblages will likely be minimal [*Matthews,* 1986; *Pearsons et al.,* 1992; *Strange et al.,* 1992]. Habitat complexity also ameliorates adverse effects of floods on fish: more structurally or hydraulically-complex stream reaches will likely lose fewer fishes to flooding, and have greater assemblage similarity before and after floods than will less complex stream reaches [*Fausch and Bramblett,* 1991; *Pearsons et al.,* 1992]. With regard to floods in the arid Southwest United States, *Minckley and Meffe* [1987] hypothesized that fishes native to the region can maintain their presence, and often their former abundance, in floods of two orders of magnitude greater than mean annual discharge. If *Minckley and Meffe's* [1987] flood hypothesis is correct, native fishes in Grand Canyon, would be expected to withstand a flood of considerably greater magnitude (50,500 m^3/s) than that of the controlled flood (1274 m^3/s).

Flannelmouth sucker is one of four remaining native fish species in the Colorado River in Grand Canyon. Based on surveys conducted over the last 5 decades, populations in Grand Canyon are considered stable. However, their range in Arizona has been reduced through extirpation in the lower Colorado River (below Lake Mead), the lower Salt River, the San Pedro River, and the Gila River [*Minckley,* 1973]. Additionally, there is concern about recruitment into the adult population in Grand Canyon. It appears that their lifetable is "inverted" with low numbers of young-of-year (YOY) and subadult fish (<360 mm TL) in comparison to adults [*Valdez and Ryel,* 1995; *Weiss et al.,* 1998]. This trend is especially pronounced in the upper reaches of Grand Canyon where summer (May through September) water temperatures are considerably colder (mean monthly range ca. 8-10°C) than historically (mean monthly range ca. 16-26°C) [*Valdez and Ryel,* 1995].

Therefore, it was important to determine the effects of the 1996 controlled flood on flannelmouth suckers for two reasons: (1) they are a species of concern [*Federal Register,* 1994]; and (2) the controlled flood occurred during their spawning season, prompting concerns over the potential loss of an entire year class. The objectives of this study were twofold: (1) to examine the effects of the controlled flood on movements of flannelmouth suckers within the Glen Canyon Reach immediately below Glen Canyon Dam; and (2) to determine if spawning occurred in the Paria River as it normally does each year in March and April. *A priori,* we identified three possible movement responses to the flood: fish would maintain position within the Glen Canyon reach in the mainstem Colorado River; fish would be displaced downriver; or fish would seek local shelter within the Glen Canyon reach. Likewise, with regard to spawning, a priori, we identified three possible outcomes: no effect on the occurrence or timing of spawning; delayed spawning; or no spawning in 1996.

2. METHODS

2.1. Study Area

We define the Glen Canyon reach of the Colorado River as the area from the base of Glen Canyon Dam at river mile -15.3 to the confluence of the Paria and Colorado Rivers at 0.9 Mile. This reach of river was chosen because of its importance as a staging area for flannelmouth suckers in the Paria River [*Weiss et al.,* 1998], and as a staging and spawning area at the -4.0 Mile gravel bar in the mainstem

[Ted McKinney, Arizona Game and Fish Department (AGDF), personal communication, 1996]. The Paria River is a small perennial river with an average base flow of only 0.77 m^3/s [*Topping,* 1997]. However, the river is prone to flash floods with a mean annual flood discharge of 88 m^3/s. In its lower 16 km, the Paria River is surrounded largely by an open floodplain with alluvial deposits. At base flow, the Paria River is uniformly shallow (<0.3 m deep) and occupies a channel not more than 20 m wide within the lower reaches. The Paria River does not constitute permanent habitat for adult or subadult flannelmouth suckers. Rather, it serves as spawning habitat for adults [*Weiss et al.,* 1998], and the lower kilometer is a nursery or rearing area for YOY from hatching through late summer [Mark Brouder, AGFD, personal communication, 1996; *Thieme,* 1997].

2.2. Available Wetted Area

Changes in wetted area of the Paria River mouth were documented using aerial videography during the pre-steady flow (226 m^3/s) and during the controlled flood (1274 m^3/s). Video tapes were analyzed using the Map & Image Processing System 3.30 (MIPS) software package (Micro Images, Inc.) to determine changes in surface area (m^2). In addition, length (m) and depth (m) of the pool in the impounded Paria River mouth were measured during the controlled flood.

2.3. Movement

To track movement, 50 adult flannelmouth suckers were surgically implanted with ultrasonic transmitters during the second week of March, 1996. Forty-five fish were captured by seining with a 9.1 m straight net of 3.8 cm mesh at the confluence of the Paria and Colorado rivers; 5 fish were obtained by electrofishing in the Glen Canyon reach of the Colorado River. Sonic tags were chosen instead of radio tags based on the ability of sonic tags to function in highly conductive waters like the Colorado River [mean daily conductivity at Lees Ferry in water year 1992 was 874 to 981 µS/cm; *Valdez and Ryel,* 1995]. Additionally, sonic tags were selected because they are known to function at depth and the flannelmouth sucker is a benthic fish. We used PRG-94 tags of frequencies from 72 to 83 kHz (SONOTRONICS, Tucson, AZ) that were 65-67 mm long, 18 mm in diameter, weighed 8 g in water (22 g in air) with a life expectancy of 24 months. Tags were crystal-controlled and hence did not drift between frequencies [*Winter,* 1983]. With one exception, only fish greater than 1100 grams received sonic implants, based on a recommendation that tags should not exceed 2% of a fish's body weight [*Nielsen, 1992*]. Fish ranged in size from 1032 g (450 mm TL) to 2274 g (597 mm TL).

Surgical procedures were performed in the field and followed protocols of *Valdez and Ryel* [1995] to minimize post-operative infections (i.e., use of sterile gloves, drape, and instruments). Fish were anesthetized with MS-222 (tricaine methane sulfonate), the transmitter was implanted into the peritoneal cavity, and the incision was closed with Chromic-gut sterile absorbable sutures. Fish were held after surgery until they recovered from MS-222 and were released at site of capture.

Mobile tracking was performed from a boat in the Glen Canyon reach at half-mile intervals from the base of Glen Canyon Dam (-15.5 Mile) to Lees Ferry (0.0 Mile) during both pre- and post-steady low flows (226 m³/s). Mobile tracking was performed daily during the flood by walking the banks of the 760-m long pool formed at the mouth of the Paria River. An assumption of this approach to determining fish displacement was that if a fish was located in the Glen Canyon reach within 48 hrs of the flood, it was considered not to have moved downriver during the flood, but remained in place or sought local shelter.

2.4. Spawning

To determine if spawning was interrupted or preempted by the flood, we attempted to collect adult flannelmouth suckers in the Paria River before and during the flood over known spawning areas [timing and location determined by *Weiss et al.*, 1998]. Realizing the limitations of point-in-time sampling of spawning fish, we also investigated recruitment of YOY fish to determine if spawning had occurred and if it was successful. We seined for YOY flannelmouth suckers with a 4.6-m long, 1.5-m deep, 6.4-mm mesh straight net at least once monthly throughout the summer at 10 sites previously designated by Arizona Game and Fish Department (AGFD) in the Paria River. Sites were located from the confluence of the Paria and Colorado rivers to 4.8 km upstream in the Paria.

3. RESULTS

3.1. Available Wetted Area

Amount of area available to adult flannelmouth suckers in the mouth of the Paria River changed dramatically during the flood. Surface area of the lower Paria River mouth increased 8-fold, from 1225 m² pre-flood to 9800 m² during

the flood. Floodwaters impounded the Paria River and formed a slackwater pool 760 m long that was up to 2.8 m deep (Figure 1). These photographs before and during the flood demonstrate the difference of wetted area available in the Paria River mouth under flow conditions of 226 m³/s versus 1274 m³/s in the mainstem Colorado River. These photographs are a conservative demonstration in the change in area within the Paria River mouth because the photographs were taken approximately 700 m upstream of the mouth, near the upper end of the impounded pool.

3.2. Movement

Forty-two of the 50 ultrasonic-tagged flannelmouth suckers (84%) were located within Glen Canyon reach either during the flood in the Paria River pool, or in the mainstem within 48 hrs after the flood, indicating that the vast majority of adult flannelmouth suckers were not displaced downriver. Of the 8 flannelmouth suckers undetected through sonic tracking, 3 were not relocated after tagging: we suspect tag malfunction, but cannot discount the possibility of shed tags. Three of the tagged fish undetected through sonic tracking were located within the Glen Canyon reach within 3 months, and 2 more were subsequently relocated at the Little Colorado River, 98.2 km downriver. It is possible that these latter 2 fish were moved downriver by the flood, although such movements have been documented [*Weiss et al.*, 1998].

Most flannelmouth suckers sought local shelter in the slackwater pool formed within the mouth of the Paria River. Thirty-three ultrasonic-tagged flannelmouth suckers were located in the pool, although this estimate is conservative due to the limitations of sonic telemetry. When 2 or more fish have tags of the same frequency but different codes, and are located in close proximity, it is nearly impossible to distinguish their unique identity. Overlapping codes were heard 8 times during sampling in the Paria River pool.

Five flannelmouth suckers were located in the mainstem Colorado River during the pre-flood steady flow (226 m³/s). Four of these 5 fish were subsequently located in the pool formed in the backed up Paria River during the flood. The fifth was located in the confluence of the Paria and Colorado rivers one day after the flood ended on April 4, 1996.

During post-flood sampling, 6 fish were located in the Glen Canyon reach, one as far up-river as -15.0 Mile. Additionally, a large school of sonic-tagged flannelmouth suckers were located at the confluence of the Paria and Colorado rivers, and none were located within the Paria River mouth, where they had been located during the flood. Thus, as flood waters receded, sonic-tagged flannelmouth

Figure 1. Habitat expansion in the Paria River pool during the 1996 controlled flood. Photos are about 700 m upstream of the confluence, looking upstream. A. Paria River at base flow, Colorado River at 226 m^3/s. B. Paria River at base flow, Colorado River at 1274 m^3/s, the peak of the controlled flood.

suckers quickly returned to the mainstem Colorado River and occupied areas from the base of Glen Canyon Dam to the mouth of the Paria River.

3.3. Spawning

Adult flannelmouth suckers in reproductive condition (tuberculated fins, some expressing gametes) were captured in the Paria River before (N = 16) and during (N = 9) the flood over known spawning areas [*Weiss et al.,* 1998]. Additional evidence of spawning success was the capture of 576 YOY individuals in the Paria River between May and September, 1996, a large number compared to data from 5 previous years of YOY sampling (Table 1). The majority of YOY flannelmouth suckers caught in the Paria River were captured in a pool formed by elevated releases from Glen Canyon Dam due to higher than normal snowmelt in the upper basin. These late spring and early summer releases were unrelated to the controlled flood.

4. DISCUSSION

We found that the 1996 controlled flood had no obvious detrimental effects on flannelmouth sucker movements in the Glen Canyon reach. We accounted for the presence of 42 of 50 ultrasonic-tagged fish in this 24-km reach either during the flood or within 48 hrs of the return to pre-flood conditions. Conservatively, 33 sonic-tagged flannelmouths moved from the flooded mainstem, where velocities were considerably higher than in non-flood conditions, into the low-velocity, ponded Paria River mouth. These fish remained in this newly-formed, expanded habitat during the peak of the flood, and returned to the mainstem as flood waters receded. The 1996 controlled flood did not displace tagged flannelmouth suckers downstream because a low velocity refuge (the ponded Paria River mouth) was available in the reach, and adult suckers were able to locate and take advantage of this refuge. Thus, of the possible responses that adult flannelmouth suckers might have shown in the controlled flood, we can reject two: maintaining position in the mainstem, and being displaced downriver.

Similarly, we found no deleterious effects of the controlled flood on spawning success in the Paria River. Ripe fish were captured and observed over known spawning areas [*Weiss et al.,* 1998] in the Paria River both during and after the flood. Further, CPUE of YOY flannelmouth sucker captured in the lower reaches of the Paria River in spring and summer, 1996 exceeded that of four of the 5 previous years, all non-flood years in the mainstem (Table 1). Thus, the controlled flood did not adversely affect

TABLE 1. Catch per unit effort (CPUE) for young-of-the-year (YOY) flannelmouth sucker in the Paria River

Year	CPUE (number/100 m^2)	Data Source
1991	1.7	AGFD, unpublished data
1992	8 total YOY, area unknown	*Weiss et al.,* 1998
1993	0	*Weiss et al.,* 1998
1994	14.5	AGFD, unpublished data
1995	0.4	AGFD, unpublished data
1996	12.6	*Thieme,* 1997

flannelmouth spawning success for two likely reasons: (1) the spawning location was in an unflooded tributary; and (2) the flood occurred during pre-spawning staging of adults before vulnerable larvae or small juveniles entered the Colorado River mainstem. Thus, we are able to reject two of the possible alternatives for effects of the flood on spawning — delayed spawning, or no spawning in 1996.

Our findings of no detrimental flood effects on flannelmouth sucker are in agreement with findings on effects of this controlled flood on other native fishes, i.e., movements of humpback chub [*Valdez and Hoffnagle,* this volume] and numbers of native fishes in several locations [*Hoffnagle et al.,* this volume]. We should point out, however, that our conclusions on movements of tagged flannelmouth sucker are based on surveys in the Glen Canyon reach before and after the flood, and during the flood only in the ponded Paria River mouth. Because of logistic constraints, we were unable to gain access to the Glen Canyon reach to locate fish during the 1274 m^3/s flood. However, our tracking efficiency during flood conditions would have been considerably less than at 226 m^3/s because ultrasonic signals are attenuated by both highly turbid and turbulent waters [*Winter,* 1983; *Thieme,* 1997]. The fact that we were able to account for 42 of 50 tagged fish during and immediately after the flood, even without access to the flooded mainstem, strengthens our conclusion that the majority of adult flannelmouth suckers were not displaced from the Glen Canyon reach. Similarly, *Gerking* [1950] found that 75 percent of 540 tagged adults of 9 fish species resisted downstream displacement during a "torrential" over-bank flood in Richland Creek, Indiana, a lowland floodplain stream. Of the 25% that had moved, fish were equally likely to have moved upstream as down.

The magnitude of the 1996 controlled flood relative to mean-annual discharge undoubtedly influenced our findings. Based on pre-dam flow records from a USGS gauge at Lees Ferry (1885-1962), average daily flows in early June at peak of snowmelt were 1698 m³/s, and peak daily flows were frequently over 2830 m³/s. The highest flow on record was 6225 m³/s on June 18, 1921. Maximum discharge since 1868 was about 8489 m³/s on July 7, 1884. Climatological evidence from tree rings indicates a flow of about 14,148 m³/s occurred in the 1600s. The highest post-dam flow was 2753 m³/s on June 29, 1983 [*Valdez and Ryel,* 1995]. These historical data help to put the 1996 controlled flow of 1274 m³/s in perspective: it was very modest indeed, only 3.14 times mean annual post-dam discharge, and 2.52 times mean annual pre-dam discharge. Viewed in this light, the lack of deleterious effects on staging adult flannelmouth sucker is readily understood, as this controlled flood is easily within the range of flows that were a normal part of the Colorado River environment in which the species evolved [*Minckley,* 1973; *Minckley and Meffe,* 1987].

In terms of adaptive management, a flood of the magnitude of the 1996 controlled flood clearly did not negatively affect flannelmouth suckers, and would not be expected to do so in the event of future controlled floods at the same time of year. In fact, this controlled flood provided a slackwater refuge for adults staging for the spawning run up the Paria River. Use of such a slackwater refuge is likely to be less energetically expensive than staging in the mainstem Colorado River. Additionally, we suggest that there is the potential for high spring and summer flows in the Colorado River to create impounded tributary mouths that could act as warmwater rearing or nursery areas for young-of-the-year of native species.

Acknowledgments. We wish to acknowledge the invaluable assistance of Debbie McGuinn-Robins of AGFD for oversight of field surgery, Ted McKinney and Andrew Ayers of AGFD and Mike Yard of GCES for electrofishing assistance, Zeb Hogan of University of Arizona for logistics and surgical assistance, and Marlin Gregor of SONOTRONICS for technical advice. Additionally, we thank all our volunteers who assisted in the field: Susan Dotson and Alex Almario of Glen Canyon National Recreation Area; and Rena Borkhataria, Gabe Paz, and Mike Perkins of the Arizona Cooperative Fish & Wildlife Unit's Minority Training Program. We thank three anonymous reviewers for helping with the manuscript. This research was funded through Arizona Game & Fish Department's Heritage Fund, Project Number 1-95038.

REFERENCES

Collins, J. P., C. Young, J. Howell, and W. L. Minckley, Impact of flooding in a Sonoran desert stream, including elimination of an endangered fish population (*Poeciliopsis o. occidentalis,* Poeciliidae), *Southwest Natur.* 26 (1), 415-423, 1981.

Coon, T.G., Responses of benthic riffle fishes to variation in stream discharge and temperature, in *Community and Evolu-tionary Ecology of North American Stream Fishes,* ed. W.J. Matthews and D.C. Heins, pp. 77-85, Univ. Okla. Press, Norman, OK, 1987.

Fausch, K., and R.G. Bramblett, Disturbance and fish communities in intermittent tributaries of a western Great Plains river, *Copeia* 1991 (3), 659-674, 1991.

Federal Register, Proposed Rules: Endangered and Threatened Wildlife and Plants; Animal Candidate Review for Listing as Endangered or Threatened Species, 59(219), 58982-58996, 1994.

Gerking, S.D., Stability of a stream fish population, *J. Wildl. Manage.* 14 (2), 193-202, 1950.

John, K.R., Survival of fish in intermittent streams of the Chiricahua Mountains, AZ, *Ecol.* 45 (1), 112-119, 1964.

Jowett, I.G., and J. Richardson, Comparison of habitat use by fish in normal and flooded river conditions, *New Zeal. J. Marine Freshwater Res.* 28:409-416, 1994.

Matthews, W.J., Fish faunal structure in an Ozark stream: stability, persistence and a catastrophic flood, *Copeia* 1986 (2), 388-397, 1986.

Minckley, W.L., *Fishes of Arizona,* Ariz. Game Fish Dept., Phoenix, AZ, 293 pp., 1973.

Minckley, W.L., and G.K. Meffe, Differential selection by flooding in stream-fish communities of the arid American Southwest, in *Community and Evolutionary Ecology of North American Stream Fishes,* ed. W.J. Matthews and D.C. Heins, pp. 93-104, Univ. Okla. Press, Norman, OK, 1987.

Moyle, P.B., and B. Vondracek, Persistence and structure of the fish assemblage in a small California stream, *Ecol.* 66 (1), 1-13, 1985.

Nielsen, L.A., *Methods of Marking Fish and Shellfish,* Amer. Fish. Soc., 208 pp., Bethesda, MD, 1992.

Pearsons, T.N., H.W. Li, and G.A. Lamberti, Influence of habitat complexity on resistance to flooding and resilience of stream fish assemblages, *Trans. Amer. Fish. Soc.* 121, 427-436, 1992.

Strange, E.M., P.B. Moyle, and T.C. Foin, Interactions between stochastic and deterministic processes in stream fish community assembly, *Environ. Biol. Fishes,* 36, 1-15, 1992.

Thieme, M.L., *Movement and recruitment of flannelmouth sucker,* (Catostomus latipinnis), *spawning in the Paria River, Arizona,* M.S. Thesis, Univ. Ariz., Tucson, AZ, 1997.

Topping, D.J., *Physics of flow, sediment transport, hydraulic geometry, and channel geomorphic adjustment during flash floods in an ephemeral river, the Paria River, Utah and Arizona,* Ph.D Thesis, Univ. Wash., Seattle, WA, 1997.

Valdez, R.A., and R.J. Ryel, *Life history and ecology of the humpback chub* (Gila cypha) *in the Colorado River, Grand Canyon, Arizona,* Rep. TR-250-08 to Bur. Reclam., Salt Lake City, UT, 1995.

Weiss, S.J., E.O. Otis, and O.E. Maughan, Spawning ecology of flannelmouth suckers *Catostomus latipinnis* (Catostomidae) in two small tributaries of the lower Colorado River, *Environ. Biol. Fishes* 52, 419-433, 1998.

Winter, J.D., Underwater biotelemetry, in *Fisheries Techniques,* ed. L.A. Nielsen, D.L. Johnson, and S.A. Lampton, pp. 371-395, Amer. Fish. Soc., Bethesda, MD, 1983.

Carole C. McIvor, Biological Resources Division, U.S. Geological Survey, Arizona Cooperative Fish and Wildlife Research Unit, 104 Biosciences East, University of Arizona, Tucson, AZ 85721; email: cmcivor@srnr.arizona.edu.

Michele L. Thieme, School of Renewable Natural Resources, Arizona Cooperative Fish and Wildlife Research Unit, 104 Biosciences East, University of Arizona, Tucson, AZ 85721. Present address: Maryland Department of Natural Resources, Matapeake Terminal, 301 Marine Academy Drive, Stevensville, MD 21666.

Movement, Habitat Use, and Diet of Adult Humpback Chub

Richard A. Valdez

SWCA, Inc., Logan, Utah

Timothy L. Hoffnagle

Arizona Game and Fish Department, Flagstaff, Arizona

The humpback chub (*Gila cypha*) is a big-river cyprinid fish endemic to the Colorado River, where river regulation has contributed to its endangerment. Flooding is essential to reshaping its habitat, redistributing nutrients, flushing terrestrial insects for food, and, in the post-dam river, controlling non-native competitors and predators. Effects of the 1996 controlled flood on movement and habitat use of adults were monitored with radiotelemetry, and diet was evaluated with a non-lethal stomach pump. Movement of 9 radio-tagged adults during the flood (mean, 0.40 km; range, 0-1.24 km) was not significantly different (P≤0.05) from movement in the month preceding the flood (mean, 1.26 km; range, 0.1-2.95 km), indicating no unusual movement or displacement of fish by the flood. Habitat used during the flood, as a percentage of radio-contacts (i.e., 73% eddies, 19% runs, 8% tributary inflows), was similar to that used under normal operations by 69 fish tracked during 1990-1992 (i.e., 74% eddies, 12% runs, 7% backwaters, 6% tributary inflows, 1% pools, <1% riffles). Diet of 43 adults showed dramatic shifts to items scoured by the flood. Simuliidae (68% ash-free dry weight) and Chironomidae (15%) dominated pre-flood diets; Amphipoda (31%), Simuliidae (25%), and terrestrial insects (i.e., beetles, ants, grasshoppers, 20%) were ingested during the flood; and Simuliidae (62%) and Amphipoda (18%) were eaten post-flood. While composition of the diet changed, biomass consumed was not significantly affected by the flood (P=0.9157). The controlled flood had no detrimental effects on movement, habitat use, or diet of adult humpback chub. Effects of habitat reshaping and nutrient redistribution can only be evaluated through long-term monitoring. Floods of higher magnitude or at a different time of year may have different effects on this endangered species and should be investigated before implementing controlled floods as an element of dam operations.

1. INTRODUCTION

The Colorado River in Grand Canyon supports one of only six known populations of humpback chub (*Gila cypha*) [*Valdez and Clemmer*, 1982], a big-river cyprinid fish that is endemic to the Colorado River Basin [*Miller*,

The Controlled Flood in Grand Canyon
Geophysical Monograph 110
Copyright 1999 by the American Geophysical Union

1946]. The humpback chub was listed as endangered by the U.S. Fish and Wildlife Service on 11 March 1967 (32 FR 4001) and is protected under provisions of the Endangered Species Act of 1973, as amended (Public Law 97-304). A recovery plan was approved on 22 August 1979 and revised on 5 May 1984 [*U.S. Fish and Wildlife Service*, 1990]. Critical habitat was designated on 21 March 1994 (59 FR 13374) and included 280 km of the Colorado River in Grand Canyon from Nautiloid Canyon downstream to Granite Park and the lower 12.8 km of the Little Colorado River (LCR)[see map in *Webb et al.*, this volume].

Humpback chub in Grand Canyon are considered one population consisting of 9 mainstem aggregations and one resident aggregation in the lower LCR. Estimated numbers of adults are 3750 in the mainstem [*Valdez and Ryel*, 1997] and 4346 in the LCR [*Douglas and Marsh*, 1996]. Reproduction occurs in spring only in the seasonally-warmed LCR and includes LCR residents, and mainstem adults staging and ascending from the proximate aggregation. Fish are precluded from spawning in the mainstem by cold water temperatures, although isolated spawning is suspected in warm springs and mouths of seasonally-warmed tributaries [*Valdez and Masslich*, in press].

The Colorado River in Grand Canyon has been regulated since completion of Glen Canyon Dam in 1963, transforming a turbid river with highly variable flow and temperature to regulated releases of cold, clear water [*U.S. Department of Interior*, 1995]. A major effect of the dam has been elimination of annual spring floods, which are important for restructuring the river channel, enhancing fish habitat, redistributing sediments and nutrients, and maintaining ecological diversity [*Stanford*, 1994]. Continued survival of the humpback chub and other native fishes in Grand Canyon is largely dependent on riverine conditions resulting from regulation by Glen Canyon Dam [*Minckley*, 1991]. Reoperating the dam to benefit this endangered species, in concert with other canyon resources, continues to be a priority of dam management [*National Research Council*, 1996].

The purpose of this study was to evaluate the effects of the 1996 controlled flood on movement, habitat use, and diet of adult humpback chub to determine if managed floods are detrimental or beneficial to the species. We tested the hypotheses that the controlled flood would not significantly affect these life history aspects, while other studies determined if the flood enhanced nursery backwater habitat [*Brouder et al.*, this volume] and affected abundance of native and non-native fishes [*Hoffnagle et al.*, this volume].

2. METHODS

2.1. Sample Design

The 1996 controlled flood consisted of pre- and post-flood steady releases of 226 m³/s for 4 days each and a 7-day steady flood release of 1274 m³/s [*Schmidt et al.*, this volume]. Sampling was confined to the vicinity of the LCR (river miles 57.0-65.4), the area with the largest aggregation of humpback chub in Grand Canyon. Data from previous studies of humpback chub in Grand Canyon were compared with one or all three releases and data were compared among the three releases. Field sampling methods were similar to those described by *Valdez et al.* [1993].

2.2. Fish Collections

Fish were captured with trammel nets that were 22.9 m long and 1.8 m deep, with 3.8-cm inside mesh and 30.5-cm outside mesh. Nets were tied from shore and extended diagonally downstream with weights at proximate and distal ends to position the net along the contour of the river bottom. Pools, runs, tributary inflows, and eddies were sampled. Humpback chub were measured for total length (TL), weighed, and examined for pre-existing marks or tags. Unmarked chub ≥150 mm TL were tagged with a passive integrated transponder (PIT tag; Biomark, Boise, ID) injected into the abdominal cavity [*Burdick and Hamman*, 1993] and released at respective capture locations. Chubs <150 mm TL were measured, weighed, and released. All data on fish captures were transferred to the Arizona Game and Fish Department for storage in a master data base on fishes of Grand Canyon.

2.3. Radiotelemetry

Adult humpback chub were captured about 1 month before start of the controlled flood (29 February - 2 March) and equipped with radio transmitters. Humpback chub selected for radio-implant (N=10) were transferred to an onshore surgical station and held in live pens in the river for no more than 2 hr prior to surgery. Only chubs weighing >550 g were considered for radio-implant to keep transmitter weight <2% of fish body weight (i.e., 11g/550 g = 2%) [*Bidgood*, 1980; *Marty and Summerfelt*, 1990]. Advanced Telemetry Systems (ATS, Isanti, MN) model 2 BEI 10-35 radio transmitters were surgically implanted into the abdominal cavity, according to procedures previously described by *Kaeding et al.* [1990] and *Valdez and Ryel* [1995]. Each transmitter weighed 11 g (6.0-cm long, 1.3-cm diameter) and had a battery life of 75-120 days. Transmitters emitted a radio signal in the 40 MHZ frequency range, and were separated by combinations of 10-KHZ intervals (i.e., 40.600, 40.610, etc.) and three pulse rates (i.e., 40, 60, 80 pulses/min) to identify fish individually by their radio signal. Each fish was released at its original capture location and tracked with ATS model R2000 programmable receivers with Larsen-Kulrod omni-directional whip antennas for search mode and Smith-Root (Smith-Root, Inc., Vancouver, WA) loop antennas for directional location.

2.4. Movement

Radio-tagged fish were contacted daily for 3-5 days following implant and release. During the flood, fish were contacted at least twice daily, with several monitored over 24-hr periods (i.e., fish locations were plotted every 30 min). Nine of the 10 radio-tagged fish were recontacted and monitored during the flood; one could not be relocated. Fish position was determined by triangulating radio signals from nearby shorelines. Labeled erect poles were used along triangulation lines to mark fish positions during 24-hr monitoring periods. Positions of fish were recorded on acetate sheets placed atop 1:1200 (scale: 1"=200') aerial photographs. Local movement was determined from field measurements using a range finder, and long-distance movements were computed from aerial photographs.

Movement was expressed as gross movement and net movement. Gross movement was the sum of all movements between contact points, while net movement was the direct distance between first and last contacts for a given monitoring period. Average movements were compared with a t-test [*Sokal and Rohlf*, 1987].

2.5. Habitat Use

Habitat use was expressed as a percentage of radio-contacts and total observation time within each of 6 habitat categories, including eddies, runs, tributary inflows, riffles, pools, and backwaters (i.e., eddy return channels). Habitat use was compared with similar radiotelemetry data from Grand Canyon [*Valdez and Ryel*, 1997], as well as from other populations of humpback chub [*Valdez and Clemmer*, 1982; *Kaeding et al.*, 1990].

2.6. Diet

Humpback chub selected for diet analysis were treated immediately after capture with a non-lethal stomach pump [*Wasowicz and Valdez*, 1994] to minimize regurgitation and further digestion of food items. Of 45 fish sampled, only 2 had no food items. Gut contents were sampled from 43 adults (250-450 mm TL, 143-815 g) during pre-flood (N=9), flood (N=16), and post-flood releases (N=18). None of the 43 fish treated with the stomach pump was implanted with a radio transmitter and each was released at its respective capture location. Gut content of each fish was collected individually in a large pan, transferred to a small labeled whirl-pack, and preserved in 70% ethanol. Gut contents were sorted in the laboratory, identified to the lowest possible taxon with the aid of *Pennak* [1989]. Ash-free dry weight (AFDW) was determined for each taxonomic group by incinerating the sample for 2 hr at 500° C. Analysis of variance (ANOVA) and Ryan-Einot-Gabriel-Welsh multiple F tests [*Day and Quinn*, 1989] were used to compare mean AFDW of taxonomic groups among fish captured during pre-flood, flood, and post-flood releases. Significance for all statistical tests was set at P=0.05.

3. RESULTS

3.1. Movement

Net movement of 9 radio-tagged adult humpback chub during the controlled flood (mean, 0.40 km; range, 0-1.24 km; 2-14 days) was not significantly different (P≤0.05) from movement in the month preceding the flood (mean, 1.26 km; range, 0.1-2.95 km; 26-39 days; Table 1). Gross movement during the flood was 1.03 km (range, 0-4.35 km). Movement observed during the flood was generally confined to small areas, usually within an eddy complex or a tributary inflow. The greatest movement observed during the flood was by a fish (PIT Tag No. 7F7B2B5760) first contacted in a recirculating eddy downstream of the LCR, where it remained during the flood. During descending flows, at about 988 m^3/s, the fish moved upstream and into a secondary, high-water channel of the LCR. We observed the fish ascending this channel during a 2-hr period followed by a descent 3 days later to the lower end of the channel. The only other extensive movement was by a fish believed to be displaced by changes in physical habitat, as described in the following section.

TABLE 1. Tag numbers, lengths and weights, and movements of radio-tagged adult humpback chub in the month preceding the controlled flood (release to first contact) and during the flood

PIT Tag Number	Radio Transmitter		Fish Length	Fish Weight	Release Locale	Release Time	Recontact Locale	Recontact Time	Net Movement[a] Before Flood	Movement[a] During Flood	
	Frequency/ Pulse	Wt (g)	(mm)	(g)	(mile)	(date)	(mile)	(dates)	(miles)	Net	Gross
7F7F276D65	40.600/40	9	394	507	64.70	29 Feb	64.60	26 Mar-8 Apr	+0.1	0	1.40
7F7D18082F	40.610/40	9	353	444	60.80	1 Mar	61.30	Apr 4	-0.5	0	0
7F7B2B5760	40.630/40	9	380	428	61.20	1 Mar	62.20	27 Mar-8 Apr	-1.0	+0.70	2.70
7F7F396046	40.650/40	11	405	547	61.20	1 Mar	63.03	26 Mar-8 Apr	-1.83	-0.77	0.77
7F7D31706D	40.660/80	11	391	622	64.70	2 Mar	64.30	2 Apr	+0.4	-	-
7F7A377B35	40.670/40	11	425	759	61.20	1 Mar	61.95	4-8 Apr	-0.75	0	0
7F7D181173	40.680/40	9	399	532	61.20	1 Mar	61.30	4-8 Apr	-0.10	0.10	0.10b
7F7F18382D	40.690/40	9	345	474	60.80	1 Mar	63.03	26 Mar-8 Apr	-1.83	-0.77	0.77
7F7F2C1314	40.700/40	11	368	576	60.80	2 Mar	61.30	24-26 Mar	-0.50	0	0b
7F7B552F62	40.710/60	11	384	516	61.20	1 Mar	-	-	-	-	-
AVERAGE MOVEMENTS:									1.26 km	0.40 km	1.03 km

[a] + = upstream
_ = downstream

3.2. Habitat Use

During the flood, adult humpback chub were captured primarily in large recirculating eddies and impounded tributary inflows (i.e., LCR). Catch rates of adult humpback chub with trammel nets in mainstem eddies were 27.3, 99.6, and 64.6 fish 30.5 m^{-1} 100 hr^{-1} during pre-flood, flood, and post-flood sampling. Catch rate in the impounded LCR inflow during the flood was 84.0 fish 30.5 m^{-1} 100 hr^{-1}.

During the controlled flood, seven radio-tagged adult humpback chub were contacted regularly in the mainstem and 2 were contacted in the LCR inflow. These 9 fish occupied eddies (73% of radio-contacts), runs (19%), and tributary inflows (i.e., LCR; 8%). The fish spent 97% of total time observed during the flood in eddies, 2% in runs, and 1% in tributary inflows. The fish contacted regularly in the mainstem all used the same type of habitat during the flood release. Each occupied the upstream end of a large recirculating eddy in a small triangular patch of quiet water formed by the interface of the mainstem downstream flow and the recirculating water reflecting off the shoreline near the separation point (as defined by *Rubin et al.*, 1990).

These triangular-shaped patches of water were usually 20-30 m on one side and were characterized by low velocity and low to moderate sediment deposition (Figure 1).

The only movement from this habitat was by a fish (PIT Tag No. 7F7F276D65) first located in the Carbon Creek eddy (64.7 Mile) at the low, steady, pre-flood release. The fish remained in the triangular patch of habitat through the first 4 days of the high release and then moved downstream about 1.1 km, where it remained in small shoreline runs and eddies until the end of the flood. The fish returned to its original location during the steady post- flood release.

3.3. Diet

Seven aquatic and 7 terrestrial food groups were found in guts of 43 adult humpback chub captured near the LCR confluence (Table 2). Simuliidae (blackflies) and Chironomidae (midges) remained dominant food items during pre-flood, flood, and post-flood releases, occurring in 94-100% of fish sampled. Occurrence of amphipods (*Gammarus lacustris*) was lowest during the pre-flood release (44%), but increased to 94% and 89% during the flood and post-

Habitat Use Polygons

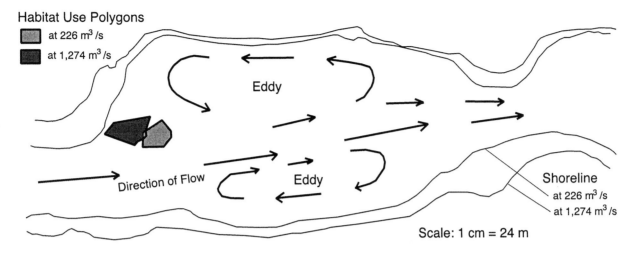

Figure 1. Habitat-use polygons for a radio-tagged adult humpback chub at RK 101.5 of the Colorado River. Polygons are for habitat use at 226 m³/s and at 1274 m³/s, 2 April (1522 hr) to 3 April (1610 hr), 1996. PIT tag no. 7F7F18382D.

flood releases, respectively. Incidence of terrestrial Coleoptera (beetles) also increased from 56% pre-flood, to 63% during the flood, and 100% post-flood. Terrestrial Diptera (true flies) increased from 33% pre-flood to 63% during the flood, but decreased to 17% post-flood. Human food remains from nearby upstream camps were not found pre-flood, but were consumed by 31% of the fish sampled during the flood and 6% after the flood. Formicidae (ants) were absent from guts during pre-flood and post-flood releases, but were found in 18.8% of guts during the flood. Lizards were found in 2 chubs, one in a fish during the flood and one in a fish post-flood.

No significant difference (P=0.6390) was observed in mean number of invertebrates in gut contents among the three sampling periods (Table 2). There were also no significant differences (P≥0.4226) in mean numbers of any invertebrate taxa found in gut contents among the three sampling periods.

No significant difference (P=0.9157) was observed in mean total biomass of gut contents of chub among sampling periods (Figure 2). Significantly greater AFDW biomass of Simuliidae (P=0.0399) and Chironomidae (P=0.0295) occurred in guts during pre-flood and post-flood periods. Conversely, significantly greater (P=0.0283) AFDW biomass of *G. lacustris* occurred during the flood than pre-flood. No significant difference in AFDW biomass was recorded for any other food category (P≥0.1893). When expressed as percent biomass AFDW (Figure 3), Simuliidae and Chironomidae were dominant pre- and post-flood, but *G. lacustris* and terrestrial insects composed significant percentages of the diet during the flood (P≤0.0082).

4. DISCUSSION AND CONCLUSIONS

The 1996 controlled flood did not significantly affect movement, habitat use, or diet of adult humpback chub in Grand Canyon, indicating that floods of comparable magnitude in March or April are not expected to detrimentally affect this endangered species. Using radiotelemetry, we found no significant difference (P≤0.05) in net movement of 9 adults during the flood (mean, 0.40 km; range, 0-1.24 km; 2-14 d) and of the same fish 1 month before the flood (mean, 1.26 km; range, 0.1-2.95 km; 26-39 d). This was comparable to net movement of 69 radio-tagged adults (mean, 1.49 km) in Grand Canyon in 1990-1992 [*Valdez and Ryel*, 1997], 8 radio-tagged adults (mean, 0.8 km) in Black Rocks, Colorado, in 1980-1981 [*Valdez and Clemmer*, 1982], and 10 radio-tagged adults (mean, 1.4 km) in Black Rocks in 1983-1984 [*Kaeding et al.*, 1990]. Gross movement by the 9 adults during the flood (mean, 1.03 km; range, 0-4.35 km) was less but comparable to gross movement of 69 adults monitored in 1990-1992 (mean, 5.13 km; range, 0.32-16.93 km; *Valdez and Ryel*, 1997). Movement during the controlled flood was consistent with other studies showing that the humpback chub is a sedentary species with high fidelity by adults for specific river locales. In contrast, average net movement by 43 radio-tagged adult Colorado squawfish (*Ptychocheilus lucius*) in the Green River in fall and spring was 31.8 km [*Archer et al.*, 1985] and average net movement of adult roundtail chub (*G. robusta*) in the Colorado River in spring and summer was 33.9 km [*Kaeding et al.*, 1990].

TABLE 2. Mean number of organisms per gut and percentage of adult humpback chub with food categories for pre-flood, flood, and post-flood periods and all fish during the controlled flood in the Colorado River, Grand Canyon, 1996. -- = mean number of organisms was not recorded. 0 = mean number or percentage was zero

Food Category	Pre-Flood (N=9)		Flood (N=16)		Post-Flood (N=18)		All Fish (N=43)	
	Mean No.	Percent	Mean No.	Percent	Mean No.	Percent	Mean No.	Percent
Aquatic Organisms								
Simuliidae (blackflies)	409.8	100	169.1	93.7	364.4	100	301.3	97.6
Chironomidae (midges)	37	100	17.1	93.7	24.2	94.4	24.2	93
Other Aquatic Insects	0.1	11.1	0.1	6.3	0.3	16.7	0.1	11.6
Ceratopogonidae (non-biting midges)	--	0	--	6.3	--	5.6	--	4.7
Culicidae (mosquitoes)	--	0	--	0	--	5.6	--	2.3
Hydracarina (water fleas)	--	11.1	--	0	--	5.6	--	4.7
Amphipoda								
Gammarus lacustris (amphipods)	7.2	44.4	23.1	93.7	6.2	88.9	12.7	79.1
Algae								
Cladophora glomerata (green algae)	--	11.1	--	0	--	11.1	--	9.3
Terrestrial Insects	10.1	66.7	11.3	93.8	3.5	100	7.8	90.7
Diptera (true flies - adults)	--	33.3	--	62.5	--	16.6	--	48.8
Culicidae (mosquitoes - adults)	--	22.2	--	0	--	5.6	--	6.9
Lepidoptera (butterflies, moths)	--	0	--	0	--	5.6	--	2.3
Orthoptera (grasshoppers)	--	44.4	--	0	--	0	--	2.3
Coleoptera (beetles)	--	55.6	--	62.5	--	100	--	72.1
Formicidae (ants)	--	0	--	18.8	--	0	--	11.6
Acarina (mites)	--	0	--	0	--	5.6	--	2.3
Miscellaneous								
Human food remains	--	0	--	31.3	--	5.6	--	13.9
Seeds or pods	--	11.1	--	0	--	0	--	2.3
Lizards	0	0	0.1	6.3	0.1	5.6	<0.1	4.7

Adult humpback chub congregated in large eddy complexes and in the LCR inflow during the flood. Trammel net catch rates of 27.3, 99.6, and 64.6 fish 30.5 m^{-1} 100 hr^{-1} during pre-flood, flood, and post-flood sampling of eddy complexes indicate that the fish were dispersed in the mainstem prior to the flood, but congregated during the flood and remained at relatively high concentrations post-flood. The pre-flood, mainstem catch rate of 27.3 fish is comparable to an average year-around catch rate of 21.5 fish during 1990-1992, while the catch rate of 84.0 fish in the LCR inflow is comparable to 100.4 fish during the peak of pre-spawning staging in 1991 [*Valdez and Ryel*, 1995]. This phenomenon of encouraging pre-spawning staging by adult humpback chub was discussed by *Clarkson et al.* [1994] as a beneficial effect of

spring floods that provides deep, warm, quiet water for staging fish and facilitates tributary access. We cannot surmise if the 1996 controlled flood benefitted spawning by the species, since the flood was of relatively short duration (i.e., 7 days) with no long-term monitoring. Congregations of humpback chub may have been stimulated by the flood or occurred coincidental as part of the natural life history of the species in Grand Canyon (as described by *Valdez and Ryel*, 1995). Attraction to tributary inflows may also be the result of ponding of warm tributary waters as refuge from the cold mainstem. Historically, average mainstem temperatures in late March and early April at Lees Ferry were 12-14° C, while post-dam temperatures are 8-9° C [*Valdez and Ryel*, 1995]. Below the LCR, mean mainstem temperatures were 9.5-10.2° C in March and April 1991-1994

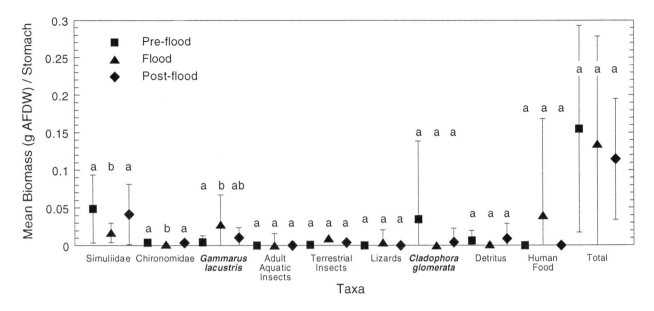

Figure 2. Mean biomass (± SD), as grams AFDW, of each food category present in guts of adult humpback chub examined during pre-flood, flood and post-flood periods of the 1996 controlled flood in the Colorado River, Grand Canyon.

[*Arizona Game and Fish Department*, 1996]. Hence, post-dam floods provide high magnitude flows of colder water temperatures than pre-dam floods, possibly affecting the physiology of warm water species, such as humpback chub. Floods of greater duration, followed by monitoring of spawning success, would be needed to more fully evaluate beneficial effects of tributary ponding.

Of the 10 adult humpback chub surgically equipped with radio transmitters 1 month prior to the flood, 9 were recontacted during the controlled flood. This recontact rate of 90% was similar to the recontact rate of 92% (69 of 75) reported in a previous study of humpback chub in the same area [*Valdez and Ryel*, 1997]. Few fish were contacted in the daytime during pre- and post-flood steady releases; this phenomenon has been attributed to use of deep areas and reduced fish movement caused by high water clarity and lack of turbidity as cover. The controlled flood increased water velocities and elevated turbidity from a mean of 8.6 NTU before the flood to 58.2 NTU during the flood, and back to 10.0 NTU following the flood [*Brouder et al.*, this volume]. This low water clarity probably made it difficult or impossible for sight-feeding fish, such as trout, to find food. However, adult humpback chub tend to be more active and more likely to be found nearshore during periods of low water clarity (i.e., ≥30 NTU) [*Arizona Game and Fish Department*, 1996; *Valdez and Ryel*, 1997].

Habitat use by adult humpback chub was also not dramatically affected by the controlled flood and was consistent with use prior to the experiment and in other populations. Use of large recirculating eddies by adults appears to be a strategy for coping with high water velocities and food requirements at virtually any flow level. As long as debris fans impinge upon the breadth of the river and cause flow recirculation, large eddies can persist with characteristic low-velocity vortices [*Griffiths et al.*, 1996; *Melis and Webb*, 1993], entraining drifting materials and providing a reliable source of food. Despite the relatively weak swimming ability of humpback chub [*Bulkley et al.*, 1982], the fusiform body, large falcate fins, and hydrodynamic nuchal hump provide this species with an ideal body for gliding in low velocity vortices while feeding in the mid-water column. This observation is supported by the close relationship between the occurrence of humpback chub in Grand Canyon and the longitudinal distribution of debris fans with associated recirculating eddies [*Converse*, 1995; *Valdez and Ryel*, 1995]. In other populations, humpback chub are found primarily in canyon-bound regions, associated with deep pools and eddy complexes, and rock and sand substrates [*Karp and Tyus*, 1990; *Valdez et al.*, 1990].

Movement by a fish from the Carbon Creek eddy at 64.7 Mile during the flood was unexpected and initially indicated abandonment of a recirculating eddy. Subsequent investigation of the geomorphologic dynamics of the eddy complex showed that the fish moved concurrent with deposition of large amounts of sediment during the flood, as

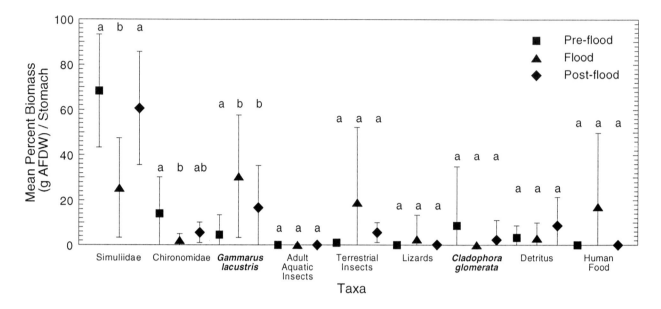

Figure 3. Mean percent biomass (± SD), as grams AFDW, of each food category present in guts of adult humpback chub examined during pre-flood, flood and post-flood periods of the 1996 controlled flood in the Colorado River, Grand Canyon.

recorded by multiple measures of channel bathymetry (Mark Gonzales, Glen Canyon Environmental Studies surveyor, pers. comm., 1996). We attribute this movement by the fish to a shower of heavy sediment in an otherwise low deposition zone [i.e., separation point, *Rubin et al.,* 1990], eaused by filling of the eddy complex and a shift in sediment distribution. This fish movement demonstrates the adaptability of humpback chub to arduous conditions during high flows.

We observed changes in food consumed by adult humpback chub during the three flow releases, which we believe reflect use of the most available foods delivered by the river. During the controlled flood, we found a variety of aquatic and terrestrial invertebrates, as well as algae, human food, seeds, and lizards in guts of adults. Simuliidae dominated diets during pre- and post-flood releases with 68% and 62% AFDW, respectively. However, *G. lacustris* (31%), Simuliidae (25%), and terrestrial insects (20%) dominated the diet during the flood, indicating that particularly *G. lacustris* and terrestrial insects increased in availability during the high release. Findings of other investigators during the controlled flood suggest that *G. lacustris* increased greatly in drift throughout the canyon. *McKinney et al.* [this volume] observed decreased densities in *G. lacustris* at Lees Ferry following the flood, indicating scour by the flood, while *Blinn et al.* [this volume] reported

an increase in *G. lacustris* in drift samples collected below the LCR. We also observed large numbers of *G. lacustris* stranded in isolated pools as flows subsided following the flood. It appears that large numbers of *G. lacustris* were dislodged by high current velocities during the high flows and these organisms were transported considerable distances downstream, making them available to feeding fish. The freshwater amphipod, *G. lacustris*, is an introduced invertebrate to Grand Canyon that is highly utilized by humpback chub, perhaps as a substitute for the reduction in other historic foods.

We attribute the increase in biomass of terrestrial invertebrates in the diet of humpback chub to inundation of the shoreline by the high flood release. Terrestrial organisms, such as beetles, true flies, ants, mites, grasshoppers, butterflies, moths, and small reptiles, are common in riparian areas along the river and are highly susceptible to becoming washed into the river with rising water levels. Diets of related roundtail chub and bonytail (*G. elegans*) from the unregulated Green River during the 1960s [*Vanicek,* 1967] contained a significant proportion of terrestrial insects. We believe that terrestrial insects were historically a common and vital component of the diet of humpback chub, being most available in spring runoff and during spates from late summer monsoonal rainstorms. Dams and flow regulation have eliminated or reduced most flooding and retained large amounts of woody material in reservoirs that historically accumulated as riverside debris piles, housing large numbers of terrestrial invertebrates.

Humpback chub in Grand Canyon appear to be opportunistic omnivores, consuming most small available items of apparent food value — a well adapted feeding strategy for the seasonally variable and highly turbid system in which the species evolved. The variety of food items consumed during the flood did not differ greatly from diets reported in other studies, but occurrence and numbers of these items appeared to vary with availability during the different releases. Other studies of humpback chub diets from Grand Canyon report "...mostly planktonic crustaceans and algae..." washed through the dam from Lake Powell [*Minckley*, 1973]; filamentous algae, mostly *Cladophora glomerata*, and larval Chironomidae [*Kubly*, 1990]; and Simuliidae (in 78% of fish sampled), Chironomidae (58%), *G. lacustris* (51%), terrestrial insects (24%) [*Valdez and Ryel*, 1997]. Simuliidae and Chironomidae were numerically dominant in diets of adult humpback chub from the LCR and *G. lacustris* was uncommon [*Kaeding and Zimmerman*, 1983]. *Tyus and Minckley* [1988] reported humpback chub in the Green and Yampa rivers consuming large numbers of Mormon crickets (*Anabrus simplex*) during annual emergences and migrations in late summer. Juvenile humpback chub consume a wider variety of food taxa than any other Grand Canyon fish of similar size, including zooplankton, benthic and pelagic aquatic invertebrates and terrestrial insects [*Arizona Game and Fish Department*, 1996].

We believe that the green alga, *C. glomerata*, did not occur in the diet of adult humpback chub during the flood because dislodged packets were quickly pulverized by the violent flows. *C. glomerata* is probably consumed by humpback chub incidentally, as small filamentous packets drifting in highly turbid water may be mistaken for an insect. Also, freshly dislodged packets may contain large numbers of Chironomidae and Simuliidae that the fish consume en mass. The amount of *C. glomerata* reported in diets of humpback chub can vary, depending on time of year (*in situ* algal production varies seasonally [*Angradi and Kubly*, 1993]), with dam operations (algal packets are dislodged more readily with increasing and high flows), and feeding mode of an individual fish (a given fish may feed more actively on items on substrates rather than in suspension). Hence, *Kubly* [1990] reported that *C. glomerata* dominated gut contents of humpback chub, while *Valdez and Ryel* [1995] reported that in some regions of Grand Canyon, *C. glomerata* composed about 26% of diet volume. During the controlled flood, *C. glomerata* occurred in only 9% of fish sampled and composed 0-9% of total biomass of gut contents. *C. glomerata* does not appear to be an important food item of humpback chub, but is certainly an important substrate and source of food and cover for many invertebrates.

Floods can be a valuable resource management tool in Grand Canyon by restructuring fish habitat, redistributing nutrients, and controlling non-native fishes. Our results reinforce the characterization that humpback chub are sedentary, opportunistic omnivores, adaptable to a variety of flow conditions and tolerant to managed floods. We found no significant differences in movement, habitat use, or diet of adults during pre-flood, flood, or post-flood sampling periods, nor did we find differences from other populations. Floods during March or April and of similar or greater magnitude to the 1996 controlled flood are not likely to be detrimental to adult humpback chub. It is unknown, however, if floods of sufficient magnitude to top debris fans are likely to modify hydraulics of eddy complexes and alter flow characteristics of habitats and availability of food items. Continued investigations into the effects of floods on humpback chub are imperative to insure that dam management options intended to benefit canyon resources (including humpback chub) do not adversely affect this endangered species.

Acknowledgments. The authors gratefully acknowledge the assistance of many individuals who helped with the field work and report assimilation for this study. We thank Dave Wegner and Mike Yard of the Glen Canyon Environmental Studies; Bill Persons and Tom Dresser of Arizona Game and Fish Department; Bryan Cowdell, Erika Cowdell, Helen Yard, and Laura Brunt of BIO/WEST; and professional river guides, Stewart Reeder and Greg Williams, for their support with the field work. Special thanks to Dr. Robert Muth, Tom Chart, and Yvette Converse for their review of the manuscript.

REFERENCES

Angradi, J.D., and D.M. Kubly, Effects of atmospheric exposure on *Chlorophyll a*, biomass and productivity of the epilithon of a tailwater river, *Regul. Rivers: Res. Manage.* 8: 345–358, 1993.

Archer, D.L., L.R. Kaeding, B.D. Burdick, and C.W. McAda, *A study of the endangered fishes of the upper Colorado River*, Final Rept. to N. Colo. Water Cons. Dist., U.S. Fish Wildl. Serv., Grand Junction, CO, 134 pp, 1985.

Arizona Game and Fish Department, *Ecology of Grand Canyon backwaters*, Final Rept. to Bur. Recl., Upper Colo. Reg., Glen Canyon Environ. Studies. Flagstaff, AZ, 155 pp, 1996.

Bidgood, B.G., *Field surgical procedure for implantation of radio tags in fish*, Alberta Fish Wildl. Div., Dept. Energy Natl. Res., Fish. Res. Rept. No. 20. 10 pp., 1980.

Bulkley, R.V., C.R. Berry, R. Pimentel, and T. Black, Tolerance and preferences of Colorado River endangered fishes to selected habitat parameters, in *Colorado River Fishery Project, Final Report, Part 3: Contracted studies*, U.S. Fish Wildl. Serv. and Bur. Recl., pp. 185–241, Salt Lake City, UT, 1982.

Burdick, B.D., and R.L. Hamman, *A study to evaluate several tagging and marking systems for Colorado squawfish, razorback sucker, and bonytail*, Final Rept. to U.S. Fish Wildl. Serv., Denver, CO, 56 pp, 1993.

Clarkson, R.W., O.T. Gorman, D.M. Kubly, P.C. Marsh, and R.A. Valdez, *Management of discharge, temperature, and sediment in Grand Canyon for native fishes*, Issue Paper, Glen Canyon Environ. Studies, Flagstaff, AZ., 1994.

Converse, Y., *A geomorphic assessment of subadult humpback chub habitat in the Colorado River of Grand Canyon*, M.S. Utah State Univ., Logan, UT., 1995.

Day, R.W., and G.P. Quinn, Comparisons of treatments after an analysis of variance in ecology, *Ecol. Mono.*, 59, 433-463, 1989.

Douglas, M.E., and P.C. Marsh, Population estimates/population movements of *Gila cypha*, an endangered cyprinid fish in the Grand Canyon region of Arizona, *Copeia*, 15-28, 1996.

Griffiths, P.G., R.H. Webb, and T.S. Melis, *Initiation and frequency of debris flows in Grand Canyon, Arizona*, U.S. Geol. Surv. Open-file Rept. 96-491, 35 pp., 1996.

Kaeding, L.R., B.D. Burdick, P.A. Schrader, and C.W. McAda, Temporal and spatial relations between the spawning of humpback chub and roundtail chub in the Upper Colorado River. *Trans. Amer. Fish. Soc.*, 119, 135-144, 1990.

Kaeding, L.R., and M.A. Zimmerman, Life history and ecology of the humpback chub in the Little Colorado and Colorado Rivers of the Grand Canyon, *Trans. Amer. Fish. Soc.*, 112, 577-594, 1983.

Karp, C.A., and H.M. Tyus, Humpback chub (*Gila cypha*) in the Yampa and Green Rivers with observations on other sympatric fishes, *Great Basin Natur.*, 119, 135-144, 1990.

Kubly, D.M., The endangered humpback chub (*Gila cypha*) in Arizona. A review of past and suggestions for future research. Arizona Game and Fish Department, Phoenix, 1990.

Marty, G.D., and R.C. Summerfelt, Wound healing in channel catfish by epithelialization and contraction of granulation tissue, *Trans. Amer. Fish. Soc.*, 119, 145-150, 1990.

Melis, T.S., and R.H. Webb, Debris flows in Grand Canyon National Park, Arizona–Magnitude, frequency and effects on the Colorado River, in *Hydraulic Engineering '93: New York, American Society of Civil Engineers, Proc. of the ASCE Conference, San Francisco, CA.*, ed. H.W. Shen, S.T. Su, and F. Wen, pp. 1290-1295, 1993.

Miller, R.R., *Gila cypha*, a remarkable new species of cyprinid fish from the Colorado River in Grand Canyon, Arizona, *J. Wash. Acad. Sci.*, 36, 409-415, 1946.

Minckley, W. L., *Fishes of Arizona*, Ariz. Game Fish Dept., Phoenix, AZ, 1973.

Minckley, W.L., Native fishes of the Grand Canyon region: an obituary? in *Colorado River Ecology and Dam Management.*

Proceedings of a Symposium, May 24-25, 1990, Santa Fe, New Mexico, Natl. Acad. Press, pp. 124-177, Washington, D.C., 1991.

National Research Council, *River Resource Management in the Grand Canyon*, Natl. Acad. Press, Washington, D.C., 1996.

Pennak, R.W., *Fresh-water invertebrates of the United States, Protozoa to Mollusca*, John Wiley & Sons, New York, 628 pp., 1989.

Rubin, D.M, J.C. Schmidt, and J.N. Moore, Origin, structure, and evolution of a reattachment bar, Colorado River, Grand Canyon, Arizona, *J. Sed. Pet.*, 60(6), 982-991, 1990.

Sokal, R.R., and F.J. Rohlf, *Introduction to biostatistics, second edition*, W.H. Freeman Company, New York, 363 pp., 1987.

Stanford, J. A., *Instream flows to assist the recovery of endangered fishes of the Upper Colorado River Basin*, Biol. Rept. 24, U.S. Dept. Interior, Natl. Biol. Surv., 47 pp, Washington, D.C., 1994.

Tyus, H.M. and W.L. Minckley, Migrating Mormon crickets, *Anabrus simplex* (Orthoptera:-Tettigoniidae), as food for stream fishes, *Great Basin Naturalist*, 48, 25-30, 1988.

U.S. Fish and Wildlife Service, *Humpback chub recovery plan*, U.S. Fish Wildl. Serv., Region 6, Denver, CO. 43 pp., 1990.

U.S. Department of the Interior, *Operation of Glen Canyon Dam - Final Environmental Impact Statement, Operation of Glen Canyon Dam*, Colorado River Storage Project, Coconino County, AZ, 337 pp + appendices, 1995.

Valdez, R.A., and G.C. Clemmer, Life history and prospects for recovery of the humpback and bonytail chub, in *Fishes of the upper Colorado River system: Present and future*, ed. W.H. Miller, H.M. Tyus, and C.A. Carlson, West. Div., Amer. Fish. Soc., pp. 109-119, Bethesda, MD., 1982.

Valdez, R.A., P.B. Holden, and T.B. Hardy, Habitat suitability index curves for humpback chub of the upper Colorado River basin, *Rivers* 1(1), 31-42, 1990.

Valdez, R.A., and W.J. Masslich, Evidence of reproduction by humpback chub in a warm spring in the Colorado River in Grand Canyon, Arizona, *Southwest. Natural.*, in press.

Valdez, R.A., W.J. Masslich, L. Crist, and W.C. Leibfried, Field methods for studying the Colorado River fishes in Grand Canyon National Parks, in *Proceedings of the First Biennial Conference on Research in Colorado Plateau National Parks*. National Park Service, Transactions and Proceedings Series NPS/NRNAU/NRTP-93/10, ed. C. van Riper, III, and E.T. Deshler, pp. 23-36, 1993.

Valdez, R.A. and R.J. Ryel, *Life history and ecology of the humpback chub* (Gila cypha) *in the Colorado River, Grand Canyon, Arizona*, Final Report to Bur. Recl., Salt Lake City, Utah, Contract No. 0-CS-40-09110, BIO/WEST Report No. TR-250-08, 286 pp., 1995.

Valdez, R.A., and R.J. Ryel, Life history and ecology of the humpback chub in the Colorado River in Grand Canyon, Arizona, in *Proceedings of the Third Biennial Conference of Research on the Colorado Plateau*, Natl. Park Serv. Trans. Procs. Ser. NPS/NRNAU/NRTP-97/12, ed. C. van Riper, III, and E.T. Deshler, pp. 3-31, 1997.

Vanicek, C.D., *Ecological studies of native Green River fishes below Flaming Gorge Dam, 1964-1966*, Ph.D. Dissertation, Utah State Univ., Logan, 124 pp., 1967.

Wasowicz, A., and R.A. Valdez, A nonlethal technique to recover gut contents of roundtail chub, *N. Amer. J. Fish. Manage.*, 14, 656-658, 1994.

Richard A. Valdez, SWCA, Inc., 172 W. 1275 S., Logan, UT 84321; email: valdezra@aol.com

Timothy L. Hoffnagle, Research Branch, Arizona Game and Fish Department, 121 E. Birch St., Suite 307, Flagstaff, AZ 86001

Riparian Vegetation Responses:
Snatching Defeat From The Jaws Of Victory And *Vice Versa*

Michael J.C. Kearsley and Tina J. Ayers

Northern Arizona University, Flagstaff, Arizona

The 1996 controlled flood failed to demonstrate five aspects of its primary vegetation-related management goal of removal of near-shore vegetation. First, when compared to pre-flood measurements, total vegetative cover was reduced only 20% and the areal extent of wetland and woodland/shrubland patches was not significantly different from the previous year when measured 6 months after the high flows. There was an immediate effect in terms of burial of some marshy areas under coarse sand, but most of these recovered within 6 months. Second, the controlled flood consistently affected only the lowest vegetation layer (grasses and herbs). Third, there was some effect on soil seed banks; sites lost roughly 45% of the seeds and 30% of the species richness of the pool of readily germinable seeds in the top 10 cm of the soil. Fourth, the loss of surface organic matter (duff) was significant in only 3 of the 9 sites, the other 6 showed no significant differences between years. There was no significant change across all sites. Finally, although there was no consistent effect on germination site quality in terms of mean soil grain sizes, there was a significant homogenization of substrates within and among sites due mostly to the loss or burial of fine sediments in return current channel settings. As documented in other chapters of this volume, the controlled flood was a success administratively and a successful demonstration of other management goals, especially in moving sediment from the channel bottom to high elevation deposits. Further, the flood was also a success in that it provided a relatively consequence-free opportunity to learn about flood hydrographs and vegetation in this system. Our data and those of others in this volume suggest that had the flows been successful in removing plants and reworking the underlying substrate in wetland patches, the recovery of vegetation would have been slowed considerably by the lack of fine, nutrient-rich sediments.

1. INTRODUCTION

Flooding is an important organizing force in terrestrial riparian habitats in arid areas of the southwestern U.S. [*Carothers and Aitchison*, 1976; *Franz and Bazzaz*, 1977; *Johnson et al.*, 1989; *Baker*, 1990]. Flood frequency, duration, and intensity often determine the identity and

The Controlled Flood in Grand Canyon
Geophysical Monograph 110

success of plant species growing near rivers in desert riparian areas [*Szaro and DeBano*, 1985; *Stromberg et al.*, 1993; *Auble et al.*, 1994]. By altering the flooding regime in riparian habitats, water diversions and dams have caused dramatic changes in plant communities [*Stromberg and Patten*, 1990; *Stromberg*, 1993; *Auble et al.*, 1994; *Kearsley and Ayers*, 1996a]. In fact, most southwestern riparian communities rely on periodic flooding to remove non-riparian and non-native species and create conditions conducive to the regeneration of the community [*Johnson et al.*, 1989; *Stromberg et al.*, 1993].

Since the completion of Glen Canyon Dam more than 30 yrs ago, major changes have taken place in riparian habitats in the Colorado River corridor between the dam and Lake Mead [*Webb et al.*, this volume]. With the stabilization of flows and control of most flooding, a diverse set of plant assemblages has become established in areas previously scoured bare by high annual flows. These areas, termed the "new high-water zone" [*Johnson and Carothers*, 1982], have undergone a series of changes associated with alterations to the dam release patterns and unplanned floods [*Carothers and Aitchison*, 1976; *Turner and Karpiscak*, 1980; *Stevens and Waring*, 1986; *Kearsley and Ayers*, 1996a; *Stevens et al.*, 1996]. Higher-elevation plants which formerly had access to groundwater during the annual spring flood (the "old high-water zone;" *Carothers and Aitchison* [1976], *Johnson and Carothers* [1982]), have been in decline due to the lack of flooding [*Carothers and Aitchison*, 1976; *Turner and Karpiscak*, 1980; *Webb*, 1996]

1.1. Vegetation-Related Goals of the Flood

The overall management goal of the 1996 controlled flood was to return flooding as a community organizing force in the river corridor of Grand Canyon [*Bureau of Reclamation*, 1995 p. 14]. Before the completion of Glen Canyon Dam, Grand Canyon saw flows in excess of 2265 m³/s on an annual basis. The magnitude, timing, and duration of this flood were determined mostly by the availability of water [*Schmidt et al.*, this volume]. Even though it was roughly half the magnitude of the pre-dam mean annual spring runoff floods, and a third of the 10-yr flood [*Turner and Karpiscak*, 1980], it was expected to be adequate to restore some of the natural dynamic forces which shaped the pre-dam physical and biotic systems.

With regard to vegetation, the primary goal of the experimental beach/habitat building flow was the removal of large amounts of new high-water zone plants [*Schmidt et al.*, this volume]. By removing vegetation and reworking the substrates in habitats close to the river, the Bureau of Recla-

mation and National Park Service hoped to reset riparian succession to an earlier stage and reopen some camping beaches which had become too vegetated to be useful [*Kearsley et al.*, this volume]. Second, the timing of the flood was expected to prevent the widespread germination of exotic plants. In many systems, the spread of exotic species is tied to altered disturbance regimes [*Diamond and Veitch*, 1981; *Décamps et al.*, 1995; *Tabacchi et al.*, 1996], although flood disturbance can have less of an effect in riparian systems [*Pyle*, 1995]. Most of the concern about exotics concerned the spread of tamarisk. Third, the flood was expected to reactivate high-flow eddies which would, in turn, rejuvenate return-current channel habitats by scouring existing vegetation and opening them to the main channel. These are among the most productive vegetation patches in Grand Canyon, primarily because they are close to the water table and accumulate fine, nutrient-rich sediments when tributary flash floods add silts and clays to the river [*Stevens et al.*, 1996]. Many of these habitats have been left perched and cut off from the main channel by coarse sediment accumulation at their mouths and the lowering of the upper limit of dam discharges associated with the Glen Canyon Dam Environmental Impact Statement and Record of Decision [*Bureau of Reclamation*, 1995].

The final stated management goal of the flood with regard to vegetation was to provide water to old high-water zone vegetation [*Bureau of Reclamation*, 1995, p. 14]. Undoubtedly surface water was closer to these plants and groundwater levels may have risen. However, determining the effects of the flood on these plants was beyond the scope of our project.

1.2. Flooding and riparian vegetation

Floods affect riparian habitats in several predictable ways. First, vegetation may be scoured out completely or damaged by some combination of swift-moving current, sediment, and entrained debris [*Minckley and Clark*, 1984; *Platts et al.*, 1985]. In the pre-dam Grand Canyon, high flows during the spring runoff would completely remove all vegetation on a nearly annual basis [*Turner and Karpiscak*, 1980; *Webb*, 1996]. Two important riparian habitat types, return-current channel marshes and riparian woodland/scrubland patches, were expected to be affected by the flood. Wetland patches tend to occur in the lowest elevation areas, closest to the river, and hence are subject to the strongest effects of flooding. They are also a rare but productive type of vegetation in the river corridor [*Stevens et al.*, 1996]. Riparian woodland and scrubland patches occur intermediate elevation channel margin settings, and are used as nesting areas by many species of birds [*Brown et al.*, 1983; *Brown and Johnson*, 1985; *Brown*, 1992].

Second, floods affect the distribution of seeds and alter the availability of readily germinable seeds in riparian habitats. In habitats where disturbance is large, frequent and predictable, the size and diversity of seed and propagule banks define the ability of sites to recover from major disturbances [*Thompson and Grime*, 1979; *Keddy and Reznicek*, 1982; *Thompson*, 1992; *Harmon and Franklin*, 1995]. Flooding and tributary inputs can distribute seeds of riparian species into areas where they have not been previously found [*Hansen*, 1918; *Staniforth and Cavers*, 1976; *Barrow*, 1992; *Pysek and Prach*, 1993]. Alternatively seeds can be scoured from the soil, prevented from germination by burial under sterile soil or organic matter [*Bertness and Ellison*, 1987; *Bertness*, 1992] or may be spread downstream with other organic debris [*Hansen*, 1918; *Schneider and Sharitz*, 1988]. Thus, floods can profoundly affect the distribution of seeds on or in the soil and thus affect both the ability of a site to recover from disturbance and the duration of that recovery.

Third, the impacts of flooding on substrates may have indirect effects on the recovery of vegetation after the disturbance. For example, ecologically significant characteristics of germination sites, including temperature regimes, soil texture, nutrients, and moisture holding capacity, and small-scale topography, may be severely altered by the scouring and deposition of sediment [*Horton et al.*, 1960; *Harper et al.*, 1965; *Brock*, 1986; *Stevens and Waring*, 1986]. In addition, the deposition or scour of soil during flooding may raise or lower the soil surface relative to the water table which would alter an important germination site characteristic. Finally, mats of plants and organic debris may be floated out of sites and may then be deposited in places where they affect future plant growth. In some systems, the presence of flood-borne debris has a beneficial effect on germination, serving as a kind of "mulch" [*Huenneke and Sharitz*, 1986] and plant growth is often most vigorous in areas where piles of organic debris are greatest [*Hansen*, 1918; *Smith and Kadlec*, 1985]. In other systems, burial by piles of debris results in the smothering of existing plants and the death of all seeds present [*Bertness and Ellison*, 1987; *Bertness*, 1992].

1.3. Research Questions

We approached the assessment of flood effects by addressing a series of questions related to the goals described earlier. First, what is the effect of the flood on extant vegetation? If the flood achieved its goal of removing much of the new high water-zone vegetation, there should be a large, measurable loss of vegetative cover, especially in wetland patches which are closest to the water and hence most susceptible to the effects of high flows. We also expected to find physical damage to woody vegetation and burial of herbaceous vegetation under new sediment. Second, what is the effect of the flood on the soil seed bank? If the flood removed surface vegetation, seeds and propagules left behind might allow for a rapid recolonization of sites after the flood. Third, are ecologically important aspects of germination sites affected by high flows such as these? Given the observed effects of the 1983 and 1986-1987 high flows [*Stevens and Waring*, 1986], we expected that soil texture, a correlate of moisture holding potential and soil nutrients, would shift towards coarser particles which are correlated with lower germination site quality, especially for woody species such as tamarisk and willow [*Stevens*, 1989]. And finally, how did these flows affect the topography of marshes and marshy areas? The 1996 controlled flood was expected to reactivate the return channels and scour the fine, nutrient-rich soil and wetland vegetation which have developed there over the previous 10 yrs. In addition, because the most recent high releases from Glen Canyon Dam in 1987 were similar in magnitude to these flows, we expected that the topography after these flows would also be similar.

2. METHODS

Data collection was based on the vegetation maps developed at the end of summer of 1995 from aerial photos and on-the-ground censuses [*Kearsley and Ayers*, 1996a] at nine sites located at river kilometers 69.3, 82.4, 89.3, 109.6, 114.9, 151.1, 197.6, 312.1, and 336.3. These maps divide the vegetation in these sites into polygons of relatively homogeneous vegetation which are distinct from patches surrounding them. The previously collected data for each polygon had information on plant species presence and percent plant species foliar cover, foliage vertical structure, and depth of the surface organic layer.

For each of the nine sites, we made 400% color enlargements of the 1:2400 aerial photos taken during the 1995 Memorial Day 225 m^3/s constant flows. To create the maps we laid Mylar on the enlargements and delineated vegetation polygons based on visual distinctness. We checked the boundaries of these polygons during an August 1995 field trip. Boundaries were moved, deleted, or added according to conditions encountered on the ground. For example, an area may have looked distinct from an adjacent area on the photograph but on the ground the difference may have been simply a change in soil texture or color.

The maps were digitized into the GCES GIS database using ARC-INFO software. The known coordinates of 8 or more reference points (large rocks or trees) per site were added to the maps and referenced by the software to rectify the polygon maps and bring them into a true plan view coordinate system. Thus, the spatial distortion of the maps caused by parallax errors in the original photographs and general distortion caused by the enlargement process [*Lillesand and Kiefer*, 1987, pp. 316-321, 328] was avoided. The error calculated from the rectification indicated an RMS errors of less than 1.5 m in the final maps for all sites. After rectification, ARC-INFO was used to calculate polygon areas for all of our sites.

2.1. Post-Flood Vegetation Maps

Because the Memorial Day 1996 aerial photos are black and white, which makes the interpretation of vegetation data difficult, our strategy for relocating polygon boundaries involved using the 1995 enlarged color photos and maps together with the black and whites as a guide for detecting polygon boundary changes. In the field, we revised polygon boundaries and made notes on copies of the 1995 Mylar maps. The fact that all the methods used to create the enlargements of both years' photos were identical (flight path, camera, altitude, print size, percent enlargement, etc.) allowed the overlaying of the 1995 map on the 1996 black and whites to compare gross scale changes in shorelines and approximate polygon boundaries.

When the Labor Day 1996 color aerial photos became available, the nine sites were remapped. The 1995 site perimeter was drawn on an sheet of Mylar laid over the 400% enlargements. Polygon boundaries were then drawn in based on the 1995 boundaries, changes to the Mylar maps in the field in 1996, and field notes on changes in species composition of polygons or the erosional loss of entire polygons.

2.2. Flood Effects on Vegetation

To measure flood effects on existing vegetation, data from the August-September 1996 monitoring trip was compared with the 1995 census data. Floristics and structure were sampled in the same way during both censuses. While walking through each polygon, a comprehensive species list was compiled, and estimates of percent foliar cover for all species were made in each of three to five randomly located, 3-m radius circular plots per polygon (radii of plots were reduced in unusually small polygons). Species which were present in the polygon but not encountered in any cover estimate samples were arbitrarily assigned a cover value of 0.001%. Percent cover estimates were repeatedly calibrated among observers by making all members of the field crew sample one or two polygons together before the rest of the day's sampling began. Because most between-observer variability occurs at the lower end of cover estimates [*Sykes et al.*, 1983; *Kennedy and Addison*, 1987], the calibration also involved using a 0.5 x 0.5 m square frame to provide a visual demonstration of approximately 1% cover in a 3 m radius plot (0.25 m^2/ 28.27 m^2 = 0.9%).

In order to measure gross effects of vegetation removal and/or burial by the flood, total vegetative cover in polygons in 1995 and 1996 was compared. For each polygon, all estimates of species cover in each of the samples were averaged to generate a total cover per sample. Statistical comparisons were made with the paired non-parametric Wilcoxon T. Because there were nine sites total, we used a Bonferroni-adjusted significance level of 0.0055 (= 0.05 / 9 sites; *Sokal and Rohlf*, 1995, p. 240). To test for overall effects, total cover in all sites was compared with a two-way ANOVA with site and flood as effects.

The areal extent of polygons dominated by two important vegetation types before and after the flood was compared: wetland patches and riparian woodland/ shrubland patches. Polygon vegetation data in each site from both years was classified simultaneously. Changes in polygon areas within vegetation type was used as a measure of overall vegetation change. It should be noted that the classification of vegetation types in this report was done solely for the analysis to be described here, and does not correspond to regional or national classification schemes.

The polygon data was classified using TWINSPAN, a polythetic divisive classification program [*Hill et al.*, 1975; *Hill*, 1979]. The method is especially suited to classification of vegetation data where sample attributes (e.g., species' abundances) have non-linear (e.g., Gaussian) responses to underlying environmental gradients [see review in [*Gaugh*, 1980, pp. 209-210] and has been used in this system in previous studies [*Kearsley and Ayers*, 1996a; *Stevens et al.*, 1996]. The TWINSPAN divisions were examined after 6 levels, those which did not make biological sense were ignored. For example, divisions based on he presence of 10% versus 20% cover of annual bromes were not used. The result was the classification of polygons in each site into five to eight distinct types.

The extent of polygons dominated by two types of vegetation were compared in 1996 and 1995. We focused our work on patches dominated by either obligate wetland species such as cattail (*Typha*), reed (*Phragmites*), rushes (*Juncus*), and bullrush (*Scirpus*) [*Reed*, 1988], and those dominated by riparian woodland/shrubland species such as willow (*Salix*), salt cedar (*Tamarix*), and/or species of *Baccharis*. To analyze changes in the extent of wetland patches, the area of each polygon dominated by wetland species in 1995 was compared to its area in 1996 within each site. If a polygon was classified as wetland in both years, we used area data from both years in the analysis. If a polygon changed to another type in 1996 or changed to a wetland type polygon in 1996 from something else in 1995, it was assigned a area of "0" in 1996 or 1995 respectively. To avoid problems with spatial independence, we combined data within sites before analyzing it. We compared the 2 yrs with a non-parametric paired comparison, the Wilcoxon T, with site totals as observations. Trends in riparian woodland/shrubland vegetation was analyzed in the same way.

To test for the effects of the high flows on different layers of the canopy, vegetation vertical structure of polygons was measured in 1996 and compared to 1995 measures. Eight point-contact, vertical structure measurements were taken in each polygon during floristic sampling in both years. The number of live vegetation contacts with a fiberglass survey rod was counted in each of the intervals 0 - 0.3 m, 0.3 - 1.0 m, 1.0 - 2.0 m, 2.0 - 4.0 m, and above 4.0 m. Because these were count data, they were square-root transformed to satisfy homoscedasticity assumptions [*Sokal and Rohlf*, 1995, pp. 415-417] before analyzing between-year differences with MANOVA for each site with flood and polygon as factors. After running the MANOVA, univariate ANOVA tests for each height increment were performed with flood and polygon as factors. We used a Bonferroni adjusted significance level of 0.01 (= 0.05 / 5 levels; *Sokal and Rohlf*, 1995, p. 280) to keep the overall error rate below 5%.

We did not attempt to measure tamarisk germination after the flood. Rather we looked for evidence of large tamarisk "lawns" in our post-flood floristic censuses and in the seed bank work described below. Neither are analyzed statistically, but are presented simply as qualitative results.

2.2. Seed Bank Effects

To test for the effects of the controlled flood on the pool of readily-germinable seeds, the direct germination method was used on samples collected immediately before the flood and one year later. This may have underestimated the immediate effects of the flood because seeds could be produced in the intervening year. However, collecting the second samples at the same time of year avoided the confounding effects of seed bank seasonality [*Thompson and Grime*, 1979]. Further, although this method does not

measure all seeds in the soil, it provides a reliable estimate of the an ecologically important measure of the regeneration capacity of a site [*Thompson and Grime*, 1979; *Gross*, 1990]. Three seed soil samples were collected from each polygon at randomly selected points in February 1996 and again in February 1997. Each sample contained approximately 200 g fresh weight of soil collected from the top 10 cm of the soil. Each sample was thoroughly mixed and approximately 150 cc was potted in 7 cm square pots which had a either a small square of paper or approximately 1 cm of sterile potting mix at the bottom to prevent soil loss during watering. Pots were then placed on a misting bench at the Northern Arizona University Research Greenhouse and kept moist. After 6 weeks, two of the daily supplemental waterings per week were replaced with half strength 10-10-10 soluble fertilizer to speed growth and development [*Young and Young*, 1986]. As seedlings became identifiable, we recorded their numbers and identity, and discarded them. Many common species could be identified based on vegetative characters before flowering took place (e.g. cattails and cudweed). Others which were less distinctive, such as species of fleabane (*Erigeron* sp.), centaury (*Centaurium* sp.), and many grasses could not be identified until flowers were present. Within each site, we compared the total number of seedlings and species diversity in pre- and post-flood samples using the paired, non-parametric Wilcoxon T. Because both measures were counts, the data were square-root transformed to satisfy homoscedasticity assumptions before comparing them across all sites using an ANOVA with site and year as factors [*Sokal and Rohlf*, 1995 pp. 415-417].

Finally, the seed bank data was inspected for species which showed unusual amounts of change between the two years. For example, some rare species in 1996 became abundant in 1997, and some more common species in 1996 disappeared. For these species, all samples in each site were pooled before comparing totals in 1996 and 1997 with a Wilcoxon T.

2.3. Physical habitat features

To measure changes in the depth of the duff layer in polygons, the duff depth data taken during the August 1996 monitoring trip was compared with the August 1995 data for each polygon. Duff depths were measured at 4 to 8 randomly located points per polygon both years. For polygons which had been entirely eroded by the high flows, duff depth was recorded as a "0" for depth in 1996. Mean duff depth was calculated for each polygon in each year and compared with a Wilcoxon T. To test for a system-wide effects of the high flows on surface organics, all sites' duff

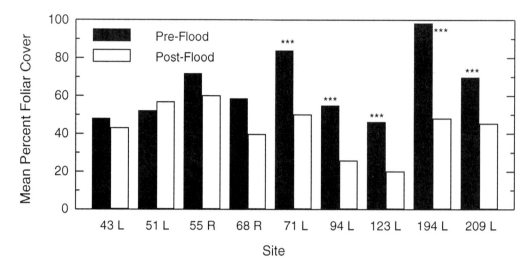

Figure 1. Effects of the high flows on mean total percent foliar cover in the nine mapping sites. Asterisks indicate significant differences with the Bonferroni adjustment to the Wilcoxon T test.

data was analyzed with an ANOVA in which flood, site, and polygon nested within site were factors.

In order to measure changes in substrate texture for a large number of polygons (ca. 350) before and after the flood without sieving each sample, we generated a subjective texture scale. The scale was based on appearance, feel, and response to wetting and rolling and ranged from 1 (fine) to 3 (coarse). Soils which were finer and rolled into tubes when wet, scored closer to one, those which were fine, but did not roll easily when wet scored in the middle of the scale and those which were coarse, and appeared to have particles roughly 1 mm in diameter scored higher on the scale. We measured scores to the nearest tenth of a point.

To make the scale useful, we quantified the actual particle-size distribution of scores by collecting a series of reference samples, up to 5 samples per tenth-point increment, in the interval between 1.8 and 3.0. We then sieved these into different size fractions, including clay/silt (< 0.062 mm), fine sand (0.062 - 0.125 mm), medium sand (0.125 - 0.250 mm), coarse sand (0.250 - 1.0 mm) and coarse sand and gravels (> 1.0 mm). Percentages in each fraction were averaged across samples in that increment to give a mean distribution for each score. This, in turn, was used to calculate mean particle size per tenth-point increment.

Before and after the flood, each polygon's substrate texture was assessed using the subjective scale. The texture of the seed soil samples were measured as they were potted to get three assessments of texture per polygon. To compare pre- and post-flood conditions, the scale measurements were converted to an average particle size using the reference sample data just described. To test for changes within sites, a two-way Welch ANOVA for equality of means when variances may not be equal [*Welch*, 1951] was used with polygon and flood as effects. All sites were compared simultaneously with the same test with site, flood, and polygon nested within site as factors. To test for differences in variability in pre- and post-flood soil texture a Levene's test for equality of variances [*Levene*, 1960] within sites was used. An all-site Levene's was performed to assess the effects of the 1996 controlled flood on substrate texture in all our sites simultaneously.

3. RESULTS

3.1. Flood Effects on Vegetation

Our results show that although the flood did remove or bury some plants, it failed to produce the large effects and habitat rejuvenation desired by planners. There were significant changes in total foliar cover, but little effect when examined more broadly as changes in the areas of patch types. First, the total foliar cover comparisons showed that polygons lost significant amounts of cover in seven of the nine sites (Figure 1). The two upstream sites, at 43 L and 51 L, showed non-significant changes in total cover. Downstream from that however, all sites lost significant amounts of cover. An overall test, across all sites, showed a significant loss of vegetation ($F_{(1,492)} = 42.2$, $p < 0.001$).

TABLE 1. Total number and areal extent of polygons dominated by obligate wetland species before (1995) and after (1996) the high flows

Site	1995 Number	Area (m²)	1996 Number	Area (m²)	% Area Change	Significance [1]
43 L	3	412	4	683	65.8	n.s.
51 L	12	5879	13	5190	-11.7	n.s.
55 R	12	3460	12	4854	40.3	n.s.
68 R	7	5342	7	5684	6.4	n.s.
71 L	7	3805	7	3499	-8.0	n.s.
94 L	3	579	5	3533	510.2	n.s.
123 L	4	1605	4	1407	-12.3	n.s.
194 L	7	2794	7	3121	11.7	n.s.
209 L	2	85	2	81	-4.7	n.s.

[1] Results of a Wilcoxon T test comparing areas before and after the flood.

TABLE 2. Total number and areal extent of polygons dominated by riparian woodland and shurbland vegetation before (1995) and after (1996) the high flows

Site	1995 Number	Area (m²)	1996 Number	Area (m²)	% Area Change	Significance [1]
43 L	16	12,173	18	11,928	- 2.1	n.s.
51 L	13	11,035	12	9278	- 15.9	n.s.
55 R	25	12,164	22	9259	- 23.9	n.s.
68 R	7	5342	8	5684	6.4	n.s.
71 L	11	10,771	11	13,279	23.3	n.s.
94 L	15	9389	15	9262	- 1.5	n.s.
123 L	12	8480	10	7881	- 7.1	n.s.
194 L	16	12,315	21	13,482	9.5	n.s.
209 L	15	4249	13	4062	- 4.4	n.s.

[1] Results of a Wilcoxon T test comparing areas before and after the flood.

On average, polygons lost approximately 20% of their vegetative cover (64.9% versus 43.4%).

When viewed on a broader perspective, in terms of changes in areas of different vegetation types, there was very little effect. There was no consistent difference in the areal extent of vegetation types within sites, even though the areas of individual patches varied considerably between years. Comparing the total area of patches dominated by wetland species in the 9 sites, cases were observed in which polygon areas increased, decreased, or remained the same (Table 1; Wilcoxon T = 7.5, d.f. = 8, p > 0.20). No sites showed an overall significant change in area covered by wetland vegetation. One of the sites (94 L) had a 5-fold increase in the total area of wetland plant patches, but 2 of the 5 wetland polygons at the site lost 10% and 70% of their original area. This pattern of variability in flood effects was common, and was most likely responsible for the finding of no significant impact.

Similarly, although there were considerable changes in the sizes of individual patches of riparian woodland and scrubland vegetation, no site had an overall significant change in the areal extent of these polygons (Table 2; Wilcoxon T = -4.5, d.f. = 8, p > 0.30). Some sites, such as Cardenas (71 L) gained as much as 23% in the total area of woody riparian polygons. Others, such as Kwagunt (55 R) lost 24% of their 1995 area of woody riparian vegetation. Within sites there were no significant changes either (Table 2).

An examination of the areal extent of the three wetland plant taxa also showed very little change from 1995 numbers (Figure 2). Across all sites, there was a slight, but non-significant increase in the total area of *Typha domingensis* (Wilcoxon T = 9.0, d.f. = 8, p > 0.1). Similarly, the increase in the total areal cover of *Phragmites australis* was not significant (Wilcoxon T = 8.0, d.f. = 8, p > 0.15). The combined taxa of all *Juncus* species covered slightly less area overall, but the change was not significant (Wilcoxon T = 3, d.f. = 8, p > 0.35).

The foliage vertical structure in our 9 sites changed very little. Only two of the sites, 55 R and 194 L, showed significant differences in the overall vertical structure from their 1995 measurements (Table 3). Although 4 of the other sites showed changes which differed at the 5% significance level, they were not significantly different from their 1995 levels after the conservative Bonferroni adjustment.

Univariate comparisons of individual levels within sites, demonstrated why the multivariate tests showed no change. The only canopy level which was consistently affected by the flood in sites was the herb/grass layer from 0.0 - 0.3 m (Table 3). This was obviously the result of the deposition of new sediment on top of grasses, low herbs, and seedlings of perennial seedlings. At one site, 209 L, there was a flush of growth of some exotic grasses, especially bermuda grass (*Cynodon dactylon*) which, coupled with very little deposition in that site, resulted in an increase in the number of contacts with vegetation in that site. There was a significant increase in the density of vegetation in the 0.3 - 1 m or the 1 - 2 m interval at the three downstream sites (123 L, 194 L, and 209 L; Table 3). We attribute some of this change to the positive effect the flood had on some species,

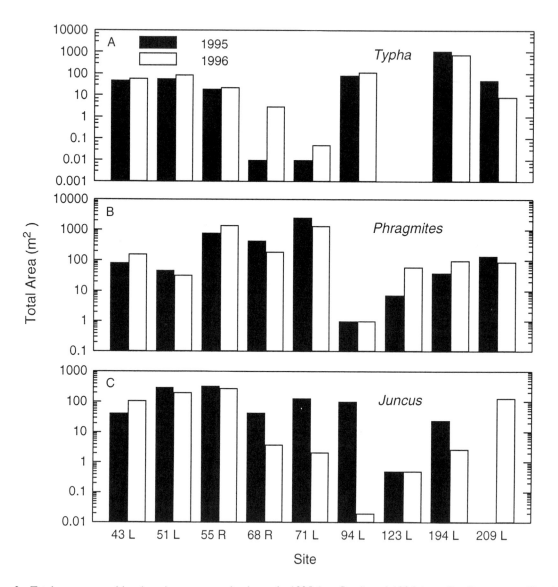

Figure 2. Total area covered by three important wetland taxa in 1995 (pre-flood) and 1996 (post-flood) censuses. The *Juncus* taxon includes *J. articulatus, J. balticus, J. ensifolius, J. longistylus,* and *J. torreyi.*

especially *T. domingensis*, and *P. australis*. For example, at 194 L, much of the *Typha* in the return-current channel had been gradually drying out and dying [*Kearsley and Ayers*, 1996]. After the flood, there were more *Typha* and *Phragmites* with fewer dead stalks than had been seen there in 3 yrs (personal observations). This phenomenon was observed at 43 L and 71 L, but only in a few patches.

3.2. Soil seed bank comparisons

There was a significant reduction in the number and species richness of the pool of readily germinable seeds in the top 10 cm of soil (Table 4). In all sites, the polygons seed banks lost, on average, 45% of the individuals present before the flood. Sites lost between 0 and 70% of their individuals on a per polygon basis. The paired Wilcoxon T tests are significant at $p < 0.05$ in four of the sites. There was a significant loss of seeds across all sites ($F_{(8,620)} = 47.88$, $p < 0.001$). Similarly, the loss of species diversity, on a per polygon basis, was significant. On average, sites lost 30% of their per sample species diversity (Table 4). Individual sites losses ranged from 0 to 60%, and in 4 of the sites, the losses were statistically significant at $p < 0.01$. The loss of species diversity across all sites was significant as well ($F_{(8,620)} = 32.19$, $p < 0.001$).

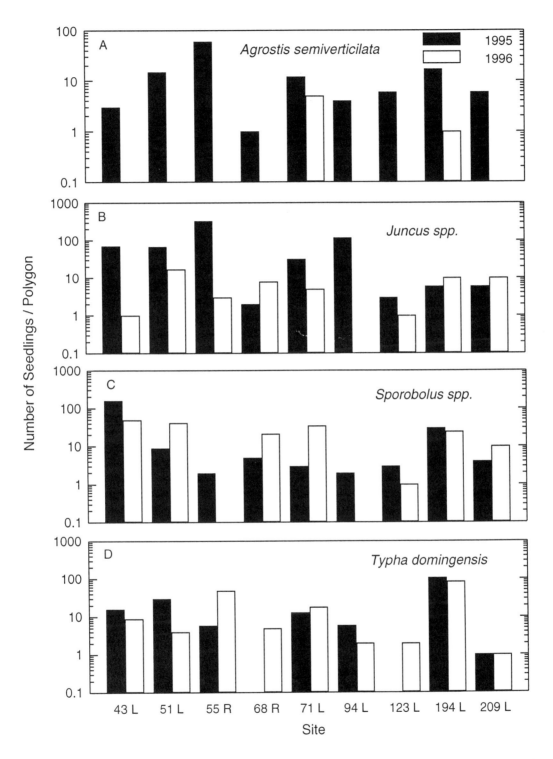

Figure 3. Species whose numbers changed unusually as a result of the 1996 controlled flood. a) *Agrostis semiverticillata* and b) *Juncus* spp. all but disappeared from the seed bank while c) *Sporobolus* spp. and d) *Typha domingensis* were mostly unaffected. Note the log scale of the y-axis.

TABLE 3. Change in vegetation vertical structure between 1995 and 1996. For each site, the MANOVA F-statistic, based on Wilks' Lambda, is listed, followed by site-wide means in each height increment in 1995 and 1996, the pooled standard error for that increment, and the probability of a difference that large, based on a paired t-test

Height (m)	1995 Mean	1996 Mean	Standard Error	Probability
43 L	MANOVA $F_{(5,29)} = 1.22$, p = 0.32			
0.0 - 0.3	3.05	1.23	0.53	0.01
0.3 - 1.0	1.54	1.22	0.17	0.20
1.0 - 2.0	1.27	1.14	0.16	0.58
2.0 - 4.0	0.75	0.87	.014	0.53
> 4.0	0.21	0.17	0.06	0.69
51 L	MANOVA $F_{(5,28)} = 2.10$, p = 0.10			
0.0 - 0.3	2.23	1.22	0.26	0.01
0.3 - 1.0	1.98	1.69	0.34	0.55
1.0 - 2.0	0.97	1.30	0.20	0.23
2.0 - 4.0	0.85	0.91	0.15	0.75
> 4.0	0.19	0.25	0.09	0.64
55 R	MANOVA $F_{(5,31)} = 7.43$, p < 0.0001			
0.0 - 0.3	6.60	1.70	0.66	0.001
0.3 - 1.0	3.74	2.26	0.38	0.0097
1.0 - 2.0	1.29	2.13	0.28	0.04
2.0 - 4.0	0.36	0.61	0.12	0.14
> 4.0	0.40	0.48	0.04	0.86
68 R	MANOVA $F_{(5,19)} = 1.09$, p = 0.40			
0.0 - 0.3	0.81	0.77	0.14	0.86
0.3 - 1.0	1.11	1.66	0.21	0.07
1.0 - 2.0	1.04	1.23	0.21	0.53
2.0 - 4.0	0.46	0.43	0.07	0.71
> 4.0	0.22	0.19	0.05	0.77
71 L	MANOVA $F_{(5,23)} = 3.15$, p = 0.05			
0.0 - 0.3	2.34	0.53	0.40	0.004
0.3 - 1.0	2.60	1.44	0.37	0.04
1.0 - 2.0	3.20	2.70	0.59	0.08
2.0 - 4.0	1.61	1.31	0.40	0.46
> 4.0	1.16	1.12	0.55	0.91
94 L	MANOVA $F_{(5,15)} = 4.16$, p = 0.014			
0.0 - 0.3	0.80	0.25	0.08	0.0004
0.3 - 1.0	0.86	0.72	0.14	0.51
1.0 - 2.0	1.08	1.32	0.23	0.47
2.0 - 4.0	0.94	1.38	0.20	0.13
> 4.0	0.14	0.18	0.07	0.65

TABLE 3. Change in vegetation vertical structure between 1995 and 1996—Continued

Height (m)	1995 Mean	1996 Mean	Standard Error	Probability
123 L	MANOVA F $_{(5,22)}$ = 3.86, p = 0.012			
0.0 - 0.3	1.31	0.47	0.16	0.0015
0.3 - 1.0	0.83	1.09	0.17	0.31
1.0 - 2.0	0.43	1.14	0.17	0.007
2.0 - 4.0	0.17	0.46	0.10	0.04
> 4.0	0.01	0.00	0.0006	0.33
194 L	MANOVA F $_{(5,27)}$ = 4.41, p = 0.0046			
0.0 - 0.3	2.80	1.32	0.26	0.0004
0.3 - 1.0	1.65	1.60	0.25	0.88
1.0 - 2.0	1.41	2.33	0.28	0.01
2.0 - 4.0	0.81	1.54	0.20	0.03
> 4.0	0.34	1.04	0.28	0.07
209 L	MANOVA F $_{(5,24)}$ = 3.59, p = 0.01			
0.0 - 0.3	1.57	2.25	0.17	0.01
0.3 - 1.0	1.26	3.34	0.23	0.002
1.0 - 2.0	0.69	1.33	0.20	0.03
2.0 - 4.0	0.07	0.62	0.24	0.11
> 4.0	0.03	0.28	0.11	0.15

Several species' seed banks showed far more or far less change than the overall mean loss of 45% of individuals. Two native species showed losses of more than 80% as a result of the controlled flood and two others showed almost no effects of the flood (Figure 3). First, waterbent grass, *Agrostis semiverticillata*, all but disappeared from seed banks as determined by our methods. In 1996, the seed banks of the 9 sites averaged between 1 and 50 *A. semiverticillata* germinations per polygon. In 1997, only two sites, Cardenas and Hualapai Acres (72 L and 194 L), had any individuals at all. The difference was statistically significant even with the low sample size of 9 sites (n = 9, T = -22.5, p < 0.005). Similarly but less dramatically, the number of individuals of *Juncus* spp. dropped between 1996 and 1997. In 1996, polygons in the 9 sites averaged between 2 and 333 individuals per polygon, averaging 71 germinations (median = 32). In 1997, the 9 sites averaged between 0 and 17 germinations, averaging approximately 6 per polygon (median = 5; n=9, T = -18.5, p< 0.01).

In contrast, the number of cattails (*Typha domingensis*) seeds did not change between the two censuses. Soil samples from the nine sites in 1996 averaged between 0 and 116 germinations per polygon, averaging 19.8 across all sites. In 1997, polygons averaged between 1 and 85 individuals, averaging 19.4 across all sites, a non-significant difference (n = 9, T = -2, n.s.). Similarly,

species of dropseed grasses (*Sporobolus* spp.), principally *S. cryptandrus* and *S. flexuosus*, were relatively unaffected by the 1996 controlled flood and intervening high flows. Sites averaged between 2 and 160 germinations per polygon, with an overall average of 24.2 individuals across all sites in 1996. In 1997, polygons contained between 0 and 50 individuals, with an overall average of 21.0 per polygon. The differences were not significant statistically (n = 9, T = 3, n.s.).

Because seed banks shrank by 45% overall between February 1996 and February 1997, the lack of change in the numbers of seeds of cattail and dropseed represented an increase in the proportion of these species in the site seed banks. In 1996, cattail seeds accounted for an average of 4% of site seed banks, while in 1997 they represented more than 13%, a significant increase (n = 9, t = 21.5, p < 0.005). Dropseed seeds comprised an average of approximately 5% of site seed banks in 1996 compared with an average of nearly 18% in 1997, a significant increase also (n = 9, T = 19.5, p < 0.001).

3.3. Physical habitat effects

The effects of the controlled flood on duff depth varied across sites. Three sites (43 L, 55 R, 209 L) lost significant

TABLE 4. Number of individuals and species germinating from soil samples before and after the high flows. Germination data were summed across samples within sites before being averaged within sites. Standard Error = pooled standard error

| | Number of Individuals | | | | Number of Species | | | |
Site	Pre-Flood	Post-Flood	Standard Error	Probability	Pre-Flood	Post-Flood	Standard Error	Probability
43 L	16.0	5.1	3.3	< 0.001	3.61	1.45	0.34	< 0.001
51 L	26.6	9.8	4.2	< 0.005	3.13	2.22	0.25	< 0.001
55 R	31.6	12.5	5.3	< 0.005	4.57	3.00	0.68	< 0.01
68 R	2.2	2.7	1.0	n.s.	0.88	1.12	0.33	n.s.
71 L	11.8	9.2	2.7	n.s.	3.00	2.86	0.50	n.s.
94 L	14.4	9.3	5.0	n.s.	2.00	1.27	0.35	n.s.
123 L	6.2	9.9	1.8	n.s.	2.20	2.20	0.38	n.s.
194 L	10.9	6.4	1.5	< 0.01	3.45	2.10	0.28	< 0.0001
209 L	4.3	2.5	0.9	n.s.	1.96	1.46	0.31	n.s.

amounts of leaf litter and debris during the flood (Figure 4). There both positive and negative changes in the leaf litter in other sites, none of which were significant. An overall test for changes in duff depth showed a slight, but non-significant effect of the flood (pre-flood = 0.532 cm, post-flood = 0.438 cm; $F_{(1,258)}$ = 1.13, p > 0.01).

Substrate texture changed significantly, but not directionally, as a result of the flood (Figure 5). Overall there was no significant change in the average particle size across all sites ($F_{(8,628)}$ = 0.87, n.s.). Individual sites at 55 R, 72 L, and 194 L did have significant differences between pre- and post-flood estimates, but they were not in a consistent direction. The sites at 55 R and 194 L were slightly, but significantly, coarser after the flood, while the third, 72 L, had slightly, but significantly, finer soils on average. The rest of the sites showed inconsistent, and non-significant changes in texture as a result of the flood (Figure 5).

The main flood effect on substrates was the homogenization across polygons within sites. With two exceptions (55 R and 68 R), all sites showed significant reductions in particle size variances, according to Levene's test (Table 5). Even at these two sites, there was some loss of variability. The all-site comparisons showed a significant loss of variation ($F_{(8,628)}$ = 18.9, p < 0.001). Much of this loss of variability came through the loss of the finest soils within sites (Figure 5). For example, at 94 L, the range of post-flood particle sizes begins at what had been the pre-flood 50th percentile of particle sizes, and the median post-flood particle size corresponds to roughly the 75th percentile of pre-flood values.

4. DISCUSSION

We have shown that the effects of the 1996 controlled flood on riparian vegetation small enough that the management goals for vegetation were not realized. The effects on extant vegetation in wetland and riparian woodland/shrubland patches were neither significant within sites nor consistent among sites. The deposition of sediment during the high flows [*Wiele et al.*, this volume; *Andrews et al.*, this volume; *Hazel et al.*, this volume] buried the lowest-growing species, grasses and small herbs, but this effect was either not great enough or long-lasting enough or both to reclassify most patches of vegetation as different types after the flood. The implications of our results for future planned flood releases and the use of floods as management tools are discussed below.

4.1. Flooding and Vegetation

One of the management aims of the high flows was to physically remove vegetation from the new high-water zone to reintroduce flood disturbance as a community organizing force in a place where it had historically played a dominant role [*Bureau of Reclamation*, 1995, p. 14]. Rather than removal of vegetation, much of it was buried [*Parnell et al.*, this volume; personal observation]. The fact that many species that defined our vegetation types were at least tolerant of burial, and some of which thrived after burial (especially *Typha*, *Phragmites*, and *Salix*), led to an almost undetectable effect on wetland vegetation in our study sites.

In other systems, flooding has had severe consequences for vegetation. Stromberg and her colleagues described moderate to severe damage to vegetation on low sandbars as a result of a 10-yr return flood [*Stromberg et al.*, 1993]. In other drainages in the southwestern United States, flooding has had a major impact on the composition of riparian assemblages [*Johnson et al.*, 1989]. The fact that such strong effects were not evident in the controlled flood is likely the result of the small size of the flood relative to historical floods. The annual return flood in Grand Canyon was in excess of 2265 m^3/s and, because it remains geomor-

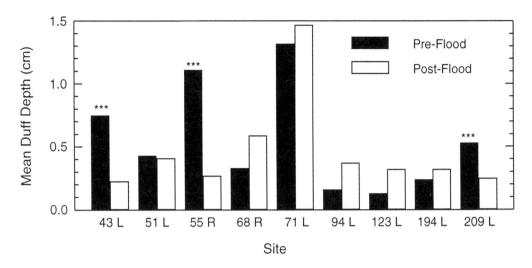

Figure 4. Effects of the high flows on mean duff depths in the 9 mapping sites. Although the statistical test was a paired comparison, numbers presented here are pre-flood means and post-flood means. *** = p < 0.001

phically nearly the same canyon it was before the construction of Glen Canyon Dam [*Webb et al.*, this volume], this is the range of the magnitude of the discharge expected to produce a significant effect on vegetation.

The lack of spread of tamarisk by the flood was a positive result of the flood. There was no massive tamarisk germination in our vegetation censuses; nowhere did we find extensive tamarisk lawns or even many tamarisk seedlings in the August-September 1996 censuses. Although there was anecdotal evidence from fishing guides in Glen Canyon that the flood had caused the spread of tamarisk, we saw only 2 tamarisk seedlings in all of our post-flood soil samples, and few seedlings in our census sites six months after the flood. Thus, seedling establishment was uncommon and most likely the result of the near-constant summer-long high flows rather than the controlled flood. Other anecdotal evidence exists for the spread of camel thorn, *Alhagi camelorum*, as a result of the flood [*M. Yeatts*, per. comm., 1996]. Again, because there were fewer *A. camelorum* seeds in post-flood than pre-flood samples, and there was no evidence of spread in our study sites, this appears to be a localized phenomenon.

One of the major effects documented in this study was the reduction of the number and diversity of readily germinable seeds in the top 10 cm of soil by burial under new sediment. *Hazel et al.* [this volume], *Schmidt et al.* [this volume], and *Kearsley and Ayers* [1996b] showed that the sites in this study were buried under 0.5 to 1.5 m of sediment with relatively little scouring, it is likely that the seeds we sampled before the flood are simply buried under

sediment mobilized from the channel and eddy deposits. Only 2 seedlings have been germinated from more than 45 samples taken from these areas [personal observation] so that the flood deposits were very likely sterile. Thus, although seeds were not physically removed from the system by scour, they will not contribute to the regeneration of these sites for two reasons. First, they may become inviable rather quickly; although some reports show that some species' seeds are very long-lived in the soil [*Oosting and Humphreys*, 1940; *Keddy and Reznicek*, 1982 and references in *Murdoch and Ellis*, 1992], the common species in these samples (e.g. *Bromus, Gnaphalium*) generally produce transient seed banks (Type I or II of [*Thompson and Grime*, 1979] which rapidly lose viability. And second, those that do not lose viability well remain buried under the new sediment until flows over powerplant capacity or aeolian processes bring them to the surface. Thus, although they are still present, they will not contribute to the regeneration of these sites.

The effects of the flood on the seed banks of individual species cannot be generalized based on taxonomic or physiognomic characters. The four species with the greatest amounts of change, both positive and negative, are all perennials and monocots. Two species which showed opposite effects (*Juncus* and *Typha*) are obligate wetland species [*Reed*, 1988], the other two are perennial bunch grasses, and all are natives. No obvious patterns, whether annuals *vs.* perennials, natives *vs.* exotics, wetland *vs.* upland are apparent in the data. Thus it would seem that differences between species are idiosyncratic. For example,

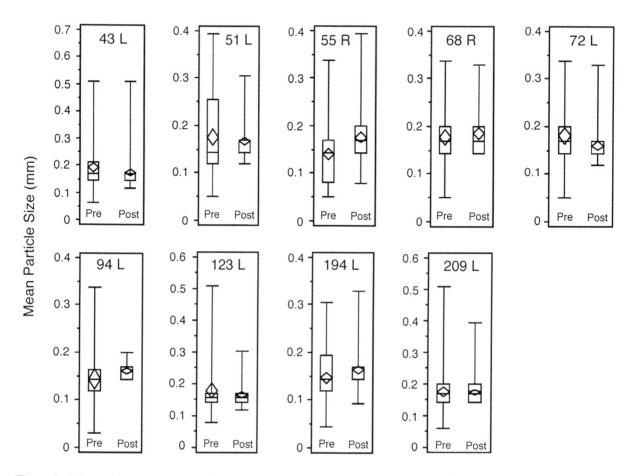

Figure 5. Effects of the high flows on substrate mean particle sizes in the 9 mapping sites. Site means and variances from before (left) and after (right) are based on polygon means. Diamonds indicate mean (center) and standard error (top and bottom), boxes represent 25th (lower line), 50th (middle line) and 75th (top line) percentiles of values. Lines outside boxes indicate the total range of values.

the *Juncus vs. Typha* difference resulting from the larger size of *Typha* allowing it to grow through the new sand and reproduce while *Juncus* remained buried.

Precipitation patterns in 1995 and 1996 may have contributed to the pattern of decreases in the numbers and diversity of seeds. Figure 6 shows monthly and total annual precipitation records from Phantom Ranch in 1995 and 1996, the years preceding the current study when seeds would have been added to the seed bank. The data (unpublished data provided by the National Climatic Data Center and K. Shinkle of Grand Canyon National Park) show that 1995 was a slightly wet year (roughly 1 standard deviation above average), and 1996 was an exceptionally dry year (more than 7 standard deviations below average). Thus, in 1995, which was the last year in which seeds were added to pre-flood seed banks, rainfall was far above average in the first four months. During the 1996 growing season, in which some 1997, post-flood sample seeds were produced,

early season rainfall was far below average. Thus, low numbers of early-season seeds would have been produced.

However, we believe that the effects of precipitation were of secondary importance to the effects of the flood flows for two reasons. First, any effects of rainfall in 1995 and 1996 would have been only a modification of the effects of burial by new sediment, and perhaps the dry conditions reinforced the effects of burial of both plants and seeds under sterile sand. Second, some of the species which seemed to have been only slightly impacted by the flood, including perennials like cattails, had root systems well below the soil surface and would be much less affected by increases in rainfall and more by high summer flows.

4.2. Flooding and Litter

As with the effects of the flood on vegetation, there was little effect on leaf litter. Although three sites lost signif-

TABLE 5. Effects of the 1996 controlled flood on soil texture. Figures represent site means and mean absolute deviation (for Levene's test) generated from averages of polygon samples. Asterisks indicate values which are significantly larger than their corresponding pre- or post-flood value based on Welch's ANOVA and Levene's test for equality of variances.

Site	MEAN PARTICLE SIZE (mm)		MEAN ABSOLUTE DEVIATION	
	Pre-Flood	Post-Flood	Pre-Flood	Post-Flood
43 L	0.195	.176	0.050*	0.031
51 L	0.175	.168	0.073*	0.026
55 R	0.146	.179*	0.426*	0.415
68 R	0.143	.178	0.050*	0.039
71 L	0.182*	.162	0.055*	0.025
94 L	0.173	.166	0.018	0.050
123 L	0.180	.169	0.050*	0.025
194 L	0.148	.164*	0.048*	0.029
209 L	0.178	.176	0.045*	0.027

icant amounts of litter, most sites showed non-significant changes, both positive and negative (Figure 4). This variability in outcome is consistent with other reports in which there is no consensus on the effects of flooding on leaf litter [Malanson, 1993, p. 170; Molles et al., 1995]. In these studies, the removal of litter seemed to depend on the severity of the flood relative to historical levels.

The lack of an effect on leaf litter is likely a result of both the small amount of litter present before the flood and the flood being insufficiently strong to cause litter transport out of our sites. Alternatively, riparian areas are very productive and create large amounts of leaf litter [Malanson, 1993, p. 166]. In other systems, an increase in productivity can be offset by increases in decomposition rates [Molles et al., 1995]. Thus the lack of an effect may be the result of an observed higher than normal productivity [personal observations] brought on by a pulse of high water and the high constant flows in the summer which followed the controlled flood. Here however, be believe the low levels of litter in many of the pre-flood polygons was the major cause of our finding no effect.

The fate of organic debris during this flood may have important implications for future flood designs. Casual observations that litter and debris were buried under sand along with vegetation [personal observations, Topping et al., this volume; D. Rubin, U.S. Geological Survey, per. comm., 1996] is of some interest. It indicates that even those sites where our measures showed a loss of litter may not have actually lost the nutrients contained in the litter to

the river. Rather, much of the carbon and nitrogen may have been effectively "banked" under the new deposits of sand [Parnell et al., this volume], slowing their entry into aquatic nutrient spirals [Heiler et al., 1995; Walker et al., 1995]. Thus, when future floods are planned, managers should decide if a rapid return of wetland and riparian vegetation is desired. If so, floods should be planned for short duration (3 or fewer days) to minimize the loss of organic matter which would speed recovery. Trends reported by Topping and his colleagues [Topping et al., this volume; Rubin et al., 1998] show that average grain size in deposits increases, and the amount of silts and clays decreases with time. If managers decide that unvegetated sandy beaches are the desired goal in the long term, floods should be run for longer periods of time.

4.3. Flooding and Substrates

The homogenization, rather than overall coarsening, of the substrates within sites was an unexpected result. A previous study on the effects of high flow events had concluded that the main effect on substrates was the coarsening of soils in their study site [Stevens and Waring, 1986]. In that study, no mention was made of the level of variability in soil texture before or after the flood.

The durations of the controlled flood and the earlier high flows is probably responsible for the difference between the results of that study and this one. Topping et al. [this volume] showed that the mean particle size in suspension in the water column was positively correlated with the time since high flows began. In other words, the longer it had been since the flows started, the coarser the average particle being carried and deposited. The high flows of 1983-1984 lasted much longer than the 10 days of the 1996 controlled flood reported on here. Thus a longer flood may have produced results more similar to those reported in earlier floods, in that only coarser sediments would have been present at the end of the flood, resulting in both coarsening and homogenization of soils.

The effects on seed germination of particle size and its correlates, nutrients and moisture holding capacity, thus have implications for future high flow designs. In general, riparian species have higher germination success in finer, moister, more nutrient-laden soils [Horton et al., 1960; Stevens and Waring, 1986]. Thus, we reiterate that the desired state of vegetation needs to be determined before the hydrograph of future floods is. If rapid recovery of riparian vegetation is a priority, floods of shorter duration that deposit more fines will create favorable germination sites for riparian species. Otherwise, the recovery of soil

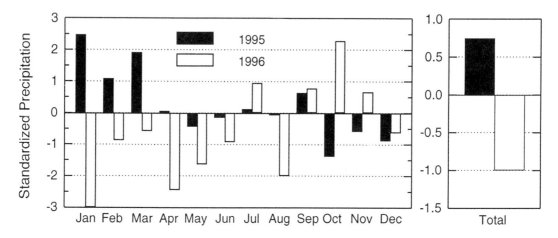

Figure 6. Standardized monthly (left) and annual (right) precipitation at Phantom Ranch in 1996 and 1996. Precipitation amounts are expressed as the number of standard deviations above or below monthly or annual means.

variability, especially in terms of the presence of fine soils, will depend on fine sediment inputs from tributaries coinciding with higher discharges from Glen Canyon Dam which would allow the deposition of fines in return current channels and other slack-water areas.

The large amount of deposition of coarse sediment in return-current channels was also an unexpected result. Although deposition of sand was one of the expectations of the flood flows, it was expected that return current channels would be reactivated and opened up to the main channel [*Bureau of Reclamation*, 1995, p. 14]. Instead, sand was pushed out of the main channel, across the upstream end of reattachment bars, and into return current channels where it remained [*Wiele et al.*, this volume; *Hazel et al.*, this volume]. This effectively buried much of the wetland vegetation in our study sites [*Parnell et al.*, this volume]. However, because many of these species were tolerant of burial, especially species of *Salix, Carex, Typha, Phragmites,* and *Juncus,* many of these patches had recovered enough within six months to be classified again as a wetland patch in 1996.

5. MANAGEMENT RECOMMENDATIONS

We believe that the use of flooding as a management tool is good practice. For many of the natural resources in Grand Canyon, including vegetation, native fish, and sandbar dynamics, flooding was a primary organizing force before the construction of Glen Canyon Dam [*Webb et al.*, this volume]. Annual spring floods on the order of 2265 m³/s and 10-yr return floods of 3400 m³/s removed vegetation, reworked and redistributed sediment, and shaped habitats for native fish, algae and invertebrates. The effect on

vegetation was to remove nearly everything below the 2265 m³/s stage elevation [*Turner and Karpiscak*, 1980; *Johnson and Carothers*, 1982; *Webb*, 1996]. Therefore, the reintroduction of this force has merit, especially in a place where one of the stated management goals is to, "restore altered ecosystems to their natural conditions" [*U.S. Department of the Interior*, 1995].

This flood failed to accomplish its goals in its effects on vegetation. Plants were buried rather than scoured and often flourished under the influences of the deposited sediment and post-flood near-constant high flows which kept the water table elevated [personal observations]. If the flood had been of sufficient force to remove vegetation and rework sediments, however, much or all the fine soils and their associated nutrients may have been removed from the system. By the end of the 1996 controlled flood, only the coarsest particles were being transported and deposited [*Topping et al.*, this volume]. The fine sediments associated with high nutrients in sandbars [*Parnell et al.*, this volume] might have been removed from the system and washed into Lake Mead by the end of the high flows.

Thus, this flood was a success in that it provided a relatively consequence-free opportunity to learn about how hydrographs of future floods should be planned to maximize the likelihood of achieving some pre-defined goal for vegetation resources. The extant vegetation, fine soils, and associated nutrients were not removed from the system so recovery has been rapid. However, it is now clear that higher flows should be part of future planed floods if removal of most vegetation is desired. Based on pre-dam flood conditions and the results of the 1983-1984 high flows, we expect that discharges of 2200 - 2500 m³/s will be sufficient to accomplish this. Second, it is also clear that the duration of the flood should be adjusted to reflect the desired end for vegetation. Planners can reduce the loss of fines and organic matter from the system and hence shorten the recovery period by shortening the flood. Otherwise, the recovery of vegetation in high productivity, low elevation areas such as return-current channels will be delayed because it depends on the deposition of fine, nutrient-rich soils during tributary inputs when dam discharges are sufficiently high to carry sediments into those areas. The process may take longer than the proposed 5 - 10 yr managed flood return cycle under the current dam operating parameters. Thus the development of wetland vegetation after a scour would be hastened by the having at least some fines and nutrients trapped in the eddy deposits in the initial post-flood settings.

We expect that the timing of future flows will be dependent mostly on considerations other than vegetation [*Schmidt et al.*, this volume, *D.L. Wegner*, per. comm., 1996]. There must be adequate levels of water above Glen Canyon Dam and sediment in the main channel of the Colorado River within the Park for future floods to be successful. Within these constraints, however, we would recommend that higher flows be run for shorter times so that vegetation will be scoured but recovery will be rapid.

Acknowledgments. The authors would like to express their sincere appreciation to all the people and agencies who made this flood possible. Specifically we wish to thank Dave Wegner for his tireless insistence on scientific support for management decisions over a 14-yr period as head of the Bureau of Reclamation's Glen Canyon Environmental Studies. We also wish to thank the GCES employees, especially Le Ann Skrzynski and Renee Davis, whose logistical miracles allowed us to do this work. May they someday recover their sanity. Finally, we would like to thank our field crew, D. "Bubba" Bechtel, W. Burger, A. Furgason, D. Huguly, S. Rhodes, M. Yeatts, and M. Weissman. May they never recover theirs.

REFERENCES

Auble, G.T., J.M. Friedman, and M.L. Scott, Relating riparian vegetation to present and future streamflows, *Ecol. Appl.,* 4, 544-554, 1994.

Baker, W.L., Climatic and hydrologic effects on the regeneration of *Populus angustifolia* James along the Animas river, Colorado, *J. Biogeography*, 17, 59-73, 1990.

Barrow, J.R., Use of floodwater to disperse grass and shrub seeds on native arid lands, in *Proceedings -Symposium on Ecology and Management of Riparian Shrub Communities. USDA Forest Service. General Technical Report INT-289; May 29, 1992; Sun Valley, ID*, ed. W.P. Clary, E.D. McAuthor, D. Behunah, and C.L. Wanbolt, pp., 167-169, Intermountain Research Station, Ogden, UT, 1992.

Bertness, M.D., The ecology of a New England saltmarsh, *Amer. Sci.*, 80, 260-268, 1992.

Bertness, M.D., and A.M. Ellison, Determinants of pattern in a New England saltmarsh plant community, *Ecol. Mono.*, 57, 129-147, 1987.

Brock, J.H., Velvet mesquite seedling development in three southwestern soils, *J. Range Manage.*, 39, 331-334, 1986.

Brown, B.T., Nesting chronology, density and habitat use of black-chinned hummingbirds along the Colorado River, Arizona, *J. Field Ornith.*, 63, 393-400, 1992.

Brown, B.T., and R.R. Johnson, Glen Canyon Dam, fluctuating water levels, and riparian breeding birds: the need for management compromise in the Colorado River in Grand Canyon, in *Riparian Ecosystems and Their Management: Reconciling Conflicting Uses,* ed. R.R. Johnson, C.D. Ziebell, D.R. Patton, P.F. Ffolliott, and R.H. Hamre, U.S. Dept. Agri., Forest Serv. General Tech. Rept. RM-120, pp., 76-80, Fort Collins, CO, 1985.

Brown, B.T., S.W. Carothers, and R.R. Johnson, Breeding range expansion of Bell's vireo in Grand Canyon, Arizona, *The Condor*, 85, 499-500, 1983.

Bureau of Reclamation, *Operation of Glen Canyon Dam, Final Environmental Impact Statement Summary*, Bur. Reclam., Salt Lake City, UT, 73 pp., 1996.

Carothers, S.W., and S.W. Aitchison, *An ecological survey of the riparian zones of the Colorado River between Lees Ferry and the Grand Cliffs: Final Report, Grand Canyon National Park. Colorado River Research Series, Contribution Number 38.*, 251 pp. Grand Canyon, Arizona, 1976.

Décamps, H., A.M. Planty-Tabacchi, and E. Tabacchi, Changes in the hydrological regime and invasions by plant species along riparian systems of the Adour River, France, *Regul. Rivers: Res. Manage.* 11, 23-33, 1995.

Diamond, J.M., and C.R. Veitch, Extinctions and introductions in the New Zealand avifauna: Cause and effect, *Sci.*, 211, 499-501, 1981.

Franz, E.H., and F.A. Bazzaz, Simulation of vegetation response to modified hydrologic regimes: A probabilistic model based on niche differentiation in a floodplain forest, *Ecol.*, 58, 176-183, 1977.

Gaugh, H., *Multivariate analysis in community ecology*, xiv + 425 pp., Cambridge Univ. Press, London, 1982.

Gross, K.L., A comparison of methods for estimating seed numbers in the soil, *J. Ecol.*, 78, 1079-1093, 1990.

Hansen, H.C., The invasion of a Missouri River alluvial flood plain, *Amer. Midl. Natur.*, 5, 196-201, 1918.

Harmon, J.M., and J.F. Franklin, *Seed rain and the seed bank of third- and fifth-order streams on the western slope of the Cascade Range,* U.S. Dept. Agri., Forest Serv., Pacific Northwest Res. Station Res. Paper PNW-RP-480, 27 pp., 1995.

Harper, J.L., J.T. Williams, and G.R. Sagar, The behavior of seeds in soil. I. The heterogeneity of soil surfaces and its role in determining the establishment of plants from seed, *J. Ecol.*, 53, 273-296, 1965.

Heiler, G., T. Hein, F. Schiemer, and G. Bornette, Hydrological connectivity and flood pulses as the central aspects for the integrity of a river flood-plain system, *Regul. Rivers: Res. Manage.*, 11, 351-361, 1995.

Hill, M.O., *TWINSPAN -- A FORTRAN program for arranging multivariate data in an ordered two-way table by classification of the individuals and attributes*, 60 pp., Microcomputer Power, Ithaca, NY, 1979.

Hill, M.O., R.G.H. Bunce, and M.W. Shaw, Indicator species analysis, a divisive polythetic method of classification, and its application to a survey of native pinewoods in Scotland, *J. Ecol.*, 63, 597-613, 1975.

Horton, J.S., F.C. Mounts, and J.M. Kraft, *Seed germination and seedling establishment of phreatophyte species*, Rocky Mtn. Forest Range Exp. Sta., U.S. Dept. Agri., Forest Serv., Fort Collins, CO, 26 pp., 1960.

Huenneke, L.F., and R.R. Sharitz, Microsite abundance and distribution of woody seedlings in a South Carolina cypress-tupelo swamp, *Amer. Midl. Natur.*, 115, 328-335, 1986.

Johnson, R.R., and S.W. Carothers, *Southwestern riparian habitats and recreation: Interrelationships and impacts in the Rocky Mountain Region*, Eisenhower Cons. Bull. 12, 118 pp., 1982.

Johnson, R.R., P.S. Bennett, and L.T. Haight, Southwestern woody riparian vegetation and succession: an evolutionary approach, in *California Riparian Systems Symposium*, U.S. Dept. Agri., Forest Ser. General Tech. Rept. PSW-110, pp., 135-139, Davis, CA, 1989.

Kearsley, M.J.C., and T.J. Ayers, *The effects of Interim Flows from Glen Canyon Dam on riparian vegetation in the Colorado River Corridor, Grand Canyon National Park, Arizona*, Final Rept., U.S. Dept. Interior, Natl. Park Serv. Coop. Agree. CA-8041-8-0002, 39 pp., 1996a.

Kearsley, M.J.C., and T.J. Ayers, *Effects of the 1996 Beach/Habitat Building Flows on riparian vegetation in Grand Canyon*, Final Rept., U.S. Dept. Interior, Bur. Reclam. Coop. Agree. CA 1425-96-FC-81-05006, 65 pp., 1996b.

Keddy, P.A., and A.A. Reznicek, The role of seed banks in the persistence of Ontario's coastal plain flora, *Amer. J. Bot.*, 69, 13-22, 1982.

Kennedy, K.A., and P.A. Addison, Some considerations for the use of visual estimates of plant cover in biomonitoring, *J. Ecol.*, 75, 151-157, 1987.

Levene, H., Robust tests for the equality of variances, in *Contributions to probability and statistics; essays in honor of Harold Hotelling*, ed. I. Olkin, S.G. Ghurye, W. Hoeffding, W.I.G. Madow, and H.B. Mann, pp., 278-292, Stanford Univ. Press, Stanford, CA, 1960.

Lillesand, T.M., and R.W. Kiefer, *Remote sensing and image interpretation*, 2nd ed., xiv + 721 pp., John Wiley and Sons, NY, 1987.

Malanson, G.P., *Riparian landscapes*, x + 296 pp. 1993.

Minckley, W.L., and T.O. Clark, Formation and destruction of a Gila River mesquite bosque community, *Desert Plants*, 6, 23-30, 1984.

Molles, M.C., Jr., C.S. Crawford, and L.M. Ellis, Effects of an experimental flood on litter dynamics in the middle Rio Grande riparian ecosystem, *Regul. Rivers: Res. Manage.*, 11, 275-281, 1995.

Murdoch, A.J., and R.H. Ellis, Longevity, viability, and dormancy, in *Seeds: the ecology of regeneration in plant communities*, ed. M. Fenner, pp. 193-229, CAB International, Wallingford, UK, 1992.

Oosting, H.J., and M.E. Humphreys, Buried viable seeds in a successional series of old fields and forest soils, *Bull. Torrey Bot. Club*, 67, 253-273, 1940.

Platts, W.S., K.A. Gebhardt, and W.L. Jackson, The effects of large storm events on Basin-Range riparian stream habitats, in *Riparian Ecosystems and Their Management: Reconciling Conflicting Uses*, ed. R.R. Johnson, C.D. Ziebell, D.R. Patton, P.F. Ffolliott, and R.H. Hamre, pp. 30-34, U.S. Dept. Agri., Forest Serv. General Tech. Rept. RM-120, Fort Collins, CO, 1985.

Pyle, L.L., Effects of Disturbance on Herbaceous Exotic Plant Species on the Floodplain of the Potomac River, *Amer. Midl. Natur.*, 134, 244-253, 1995.

Pysek, P., and K. Prach, Plant invasions and the role of riparian habitats: a comparison of four species alien to central Europe, *J. Biogeography*, 20, 413-420, 1993.

Reed, P.B., National list of plant species that occur in wetlands: southwest (region 7), U.S. Fish Wildl. Serv. Biol. Rept. 88 (26.7), 71 p., 1988.

Rubin, D.M., J.M. Nelson, and D.J. Topping, Relation of inversely graded deposits to suspended sediment grain-size evolution during the 1996 flood experiment in Grand Canyon, *Geol.*, 26, 99-102, 1998.

Schneider, R.L., and R.R. Sharitz, Hydrochory and regeneration in a bald cypress - water tupelo swamp forest, *Ecol.*, 69, 1055-1063, 1988.

Smith, L.M., and J.A. Kadlec, The effects of disturbance on marsh seed banks, *Can. J. Bot.*, 63, 2133-2137, 1985.

Sokal, R.B., and F.J. Rohlf, *Biometry*, 3rd ed., xix + 887 pp., W.H. Freeman and Company, San Francisco, CA, 1995.

Staniforth, R.J., and P.B. Cavers, An experimental study of water dispersal in *Polygonum* spp, *Can. J. Bot.*, 54, 2587-2596, 1976.

Stevens, L.E., *Mechanisms of riparian plant community organization and succession in the Grand Canyon, Arizona*, [Ph.D Dissertation], N. Ariz. Univ., Flagstaff, AZ, 1989.

Stevens, L.E., and G.L. Waring, *Effects of post-dam flooding on riparian substrates, vegetation, and invertebrate populations in the Colorado River corridor in Grand Canyon, Arizona*, U.S. Dept. Interior, Bur. Recl., Glen Canyon Environ. Studies Rept. GCES/19/87, NTIS PB88-183488, iv + 166 pp., 1986.

Stevens, L.E., J.C. Schmidt, T.J. Ayers, and B.T. Brown, Flow regulation, geomorphology, and Colorado River marsh development in the Grand Canyon, Arizona, *Ecol. Appl.*, 5, 1025-1039, 1996.

Stromberg, J.C., Fremont cottonwood-Goodding willow riparian forests: a review of their ecology, threats, and recovery potential, *J. Ariz.-Nev. Acad. Sci.*, 26, 97-110, 1993.

Stromberg, J.C., and D.T. Patten, Riparian vegetation in stream flow requirements: A case study from a diverted stream in the Eastern Sierra Nevada, California, USA, *Environ. Manage.,* 14, 184-194, 1990.

Stromberg, J.C., B.D. Richter, D.T. Patten, and L.G. Wolden, Response of a Sonoran riparian forest to a 10-year return flood, *Great Basin Natur.,* 53, 118-130, 1993.

Sykes, J.M., A.D. Horrill, and M.D. Mountford, Use of Visual Cover Assessments as Quantitative Estimators of Some British Woodland Taxa, *J. Ecol.,* 71, 437-450, 1983.

Szaro, R.C., and L.F. DeBano, The effects of streamflow modification on the development of a riparian ecosystem, in *Riparian ecosystems and their management: Reconciling conflicting uses,* ed. R.R. Johnson, C.D. Ziebell, D.R. Patton, P.F. Ffolliott, and R.H. Hamre, U.S. Dept. Agri., Forest Serv. General Tech. Rept. Rm-120, pp., 211-215, Fort Collins, CO, 1985.

Tabacchi, E., A.M. Planty-Tabacchi, M.J. Salinas, and H. Decamps, Landscape structure and diversity in riparian plant communities: A longitudinal comparative study, *Regul. Rivers: Res. Manage.,* 12, 367-390, 1996.

Thompson, K., The functional ecology of seed banks, in *Seeds: the ecology of regeneration in plant communities,* edited by M. Fenner, pp. 231-258, CAB International, Wallingford, UK, 1992.

Thompson, K., and J.P. Grime, Seasonal variation in the seed banks of herbaceous species in 10 contrasting habitats, *J. Ecol.,* 67, 893-921, 1979.

Turner, R.M., and M.M. Karpiscak, *Recent vegetation changes along the Colorado River between Glen Canyon Dam and Lake Mead, Arizona,* U.S. Geol. Surv. Prof. Paper 1132, 125 pp.,, 1980.

U.S. Department of the Interior, *Final general management plan and environmental impact statement, Grand Canyon National Park, Coconino and Mohave Counties, Arizona,* U.S. Dept. Interior, Natl. Park Serv., NPS D-298, 286 pp., 1995.

Walker, K.F., F. Sheldon, and J.T. Puckridge, A perspective on dryland river ecosystems, *Regul. Rivers: Res. Manage.,* 11, 85-104, 1995.

Webb, R.H., *Grand Canyon, a century of change,* 290 pp., Univ. Ariz. Press, Tucson, AZ, 1996

Welch, B.L., On the comparison of several mean values: an alternative approach, *Biometrika,* 38, 330-336, 1951.

Young, J.A., and C.G. Young, *Collecting, processing, and germinating seeds of wildland plants,* 236 pp., Timber Press, Portland, OR 1986.

Michael J.C. Kearsley and Tina J. Ayers, Biology Department Northern Arizona University, Box 5640, Flagstaff, Arizona 86001; email: mike.kearsley@nau.edu

Summary and Synthesis of Geomorphic Studies Conducted During the 1996 Controlled Flood in Grand Canyon

John C. Schmidt

Department of Geography and Earth Resources, Utah State University, Logan

The 1996 controlled flood demonstrated that a discharge of 1274 m^3/s was sufficient to form or rework alluvial deposits along the Colorado River in Grand Canyon. This flood was moderately large in relation to other floods that have occurred since Glen Canyon Dam was completed in 1963. The flood also provided the opportunity to make measurements of physical processes that occurred. The flood caused widespread deposition of sand to elevations 3-5 m above the stage of the administratively-determined minimum daytime dam release of 227 m^3/s and caused an increase in the area and volume of the high-elevation parts of sand bars, thereby increasing the number of new campsites. The low-elevation parts of the same eddies were eroded, as were adjacent channel pools. Recently-aggraded debris fans were reworked. Suspended-sediment transport measurements showed that suspended-sand concentrations decreased with time and that transport rates are highest when the channel bed is relatively fine and has a higher proportion of sand. The controlled flood was most effective during the first 4 days when mainstem transport rates and eddy deposition rates were high; debris-fan reworking entirely occurred during the first 4 hrs at one site. Modeling studies strongly indicate that resultant eddy-bar topography depends greatly on the concentration of mainstem suspended sediment and the initial topography of eddies. Large-scale erosion occurred on different days at different sites and had the potential to evacuate large proportions of newly-deposited sand from eddies. The final flood-formed topography of eddy bars was thus highly variable from site to site and from reach to reach.

1. INTRODUCTION

The 1996 controlled flood was an opportunity to demonstrate the role of floods in forming and reworking alluvial deposits along the Colorado River in Glen, Marble, and Grand canyons. The status of physical resources along the river was sufficiently different immediately after the flood that many river users and managers termed the flood a resource-management "success" [*Collier et al.,* 1997]. The flood also provided the opportunity to measure physical processes that cause sediment entrainment, transport, and deposition during flood in a river where channel constrictions and lateral zones of flow separation are common. These geomorphic measurements led research scientists to revise some fundamental ideas about geomorphic processes in Grand Canyon and other rivers where fan-eddy complexes, as described by *Schmidt and Rubin* [1995], are abundant. Thus, the flood was also a success in the sense that new scientific insights were gained.

Some research findings implied that basic management strategies utilized in the operations of Glen Canyon Dam should be revised. Thus, the flood was also an important step in learning to manage the river "adaptively." Managers were reminded that Colorado River management policies must be continually revised and refined in response to the evolving scientific understanding of physical and ecological processes. The purpose of this chapter is to summarize and synthesize research findings that concern geomorphic processes and to discuss implications of these findings for the management of the Colorado River in Grand Canyon. These research findings are also used to estimate a sand budget for the flood.

2. PHYSICAL RESOURCES OF THE COLORADO RIVER IN GRAND CANYON

Many of the physical resources of the Colorado River have societal value because of their use by the public, their intrinsic landscape attributes, or their ecological function [Webb et al., this volume]. Campsites are those parts of sand bars large and high enough to be used by boating parties without being inundated by the daily fluctuating dam releases caused by production of hydroelectric power (Figure 1). The challenge of whitewater boating in Grand Canyon is afforded by debris flows that deliver boulders to the river and by the Colorado River itself, which redistributes those boulders during flood. The size distribution of other fine-grained substrate determines the composition of riparian vegetation [Stevens, 1989] and the composition and abundance of benthic invertebrates [Brouder et al., this volume].

Areas of stagnant flow that typically exist in the lee of emergent reattachment bars that block recirculation into formerly active eddy return-current channels are potential nursery habitat for some native fishes. These habitats are called backwaters by ecologists working in Grand Canyon [Brouder et al., this volume]. The existence and size of these backwaters depends on two morphometric attributes of the sand bar within an eddy: the topography of the return-current channel and the topography of the adjacent reattachment bar. Deep return-current channels are inundated over a wider range of flows than are shallow channels, and higher reattachment bars block recirculation over a wider range of flows than do low-elevation bars. These "ecological" backwaters are different phenomena than are the backwaters described in hydraulics which are areas of ponded flow upstream from channel constrictions.

3. UNDERLYING SCIENTIFIC QUESTIONS ADDRESSED BY THE FLOOD

More than a decade of research in Grand Canyon, and elsewhere, provided the basis for the underlying geomorphic questions addressed in various controlled-flood research plans [Webb et al., this volume; Schmidt et al., this volume]. These questions were related to understanding the processes that form and maintain the physical resources described above. Thus, these questions concerned entrainment, transport, and deposition of coarse and fine sediment. Would large amounts of sand and finer sediment be scoured from the river bed and be deposited in eddies and elsewhere on the banks? Would the volume of newly-deposited sediment be greater at sites further downstream from the dam? Would net erosion of eddies and channel banks occur at sites near the dam, especially upstream from Lees Ferry? What would be the relative proportion of fine sediment deposited in eddies and along the banks in relation to the amount transported to Lake Mead reservoir? Would the rate of fine-sediment deposition decrease with time? Would the flood rework debris fan deposits, thereby widening rapids? Was the flood of sufficient magnitude and of appropriate duration to accomplish these tasks? How long would newly-created deposits of fine sediment last after the flood?

4. THE STATUS OF PHYSICAL RESOURCES IMMEDIATELY AFTER THE FLOOD

The 1996 controlled flood caused widespread deposition of sand at relatively high elevation on the river's banks at elevations 3 to 5 m above the stage of the administratively-determined minimum daytime dam release of 227 m^3/s. These new deposits were as much as 2 m thick, exceeding the average thickness of any previous flood deposits since 1983 [Hazel et al., this volume]. Much of this sand was deposited sufficiently high so that it was exposed above the stage of the typical maximum daytime dam release of 566 m^3/s. Hazel et al. [this volume] found that the average increase in volume of high-elevation sand was 164%, the average increase in area was 67%, and the average increase in thickness was 0.64 m. Schmidt et al. [this volume] found that most of this sand was deposited in eddies except that large amounts of sand were deposited as channel-margin levees in reaches where debris fans are infrequent; they determined that the total area of new deposits was between 10,500 and 21,000 m^2/km.

These new deposits increased the number and size of campsites throughout Grand Canyon. Kearsley et al. [this volume] found that the number of bars usable as campsites

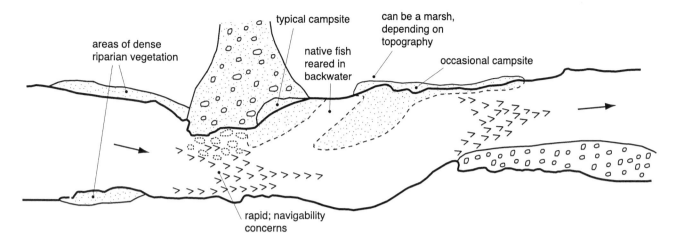

Figure 1. Diagram showing the distribution of physical resources in a typical fan-eddy complex.

increased from 218 to 299; 84 new campsites were created and 3 existing campsites were destroyed by the flood. *Thompson et al.* [1997] used photographic data collected by river guides to determine that 82% of the campsites increased in size in "critical" reaches where campsite carrying capacity is low. Most river guides found that camping was easier on these aggraded bars [*Thompson et al.*, 1997].

The new high-elevation deposits were composed of sand that had been scoured from the channel, from eddies further upstream, and from low-elevation parts of the same eddy in which high-elevation deposition occurred. *Hazel et al.* [this volume] determined that the average volume of sand in the low-elevation parts of eddies decreased by 5% and that the average volume of sediment in channel pools decreased by 11%. Fine sediment was winnowed from the river bed at the Grand Canyon gage [*Topping et al.*, this volume], and silt and clay was flushed from much of the bed [*Topping et al.*, this volume; *Smith*, this volume]. Silt and clay was not redeposited on the bed for at least 2 months at 3 widely-spaced sites in the canyon [*Blinn et al.*, this volume]. The concentration of suspended sand and of silt and clay decreased greatly during the first days of the flood [*Topping et al.*, this volume; *Smith*, this volume].

Recently-aggraded debris fans were reworked by the flood. The areas of 18 of these fans were reduced 2-42%, and their volumes were reduced 3-34% [*Webb et al.*, this volume]. Selective transport from the eroded edges of these fans caused the average grain size of the streamside edges of these fans to coarsen. The net result was that the width of the constricted channel in rapids increased slightly at most sites.

These changes had ecological implications. The declining concentration of suspended sediment caused the intensity of underwater light to increase after the first days of the flood [*Blinn et al.*, this volume]. Bed scour and winnowing of fine sediment from pools caused rooted macrophytes, such as *Elodea* and *Chara*, as well as tubificid and lumbriculid worms, to be preferentially transported downstream because these species prefer fine sediment substrate [*Blinn et al.*, this volume]. Newly-deposited sediment in backwaters was coarser than the pre-flood substrate, and benthic invertebrates were less numerous in the same backwaters than before the flood [*Brouder et al.*, this volume]. The number of backwaters potentially available to native fish at the low discharge of 227 m^3/s at which post-flood aerial photographs were taken increased immediately after the flood [*Brouder et al.*, this volume].

5. THE STATUS OF PHYSICAL RESOURCES 6 MONTHS AFTER THE FLOOD

There was widespread erosion of recently-formed deposits in the months after the flood, but the net "benefit" of the flood, in terms of new high-elevation sand bars, remained. *Kearsley et al.* [this volume] found that 37 of the 84 newly-formed campsites were no longer large enough to be used for camping 6 months after the flood, because vegetation had reestablished itself or the bars had eroded. *Hazel et al.* [this volume] found that the thickness of high-elevation sand had decreased 0.23 m in 6 months, but that the average change for entire eddies was only -0.02 m in the same period. Thus, most of the high-elevation sand that eroded was redeposited at low elevation in the same eddy,

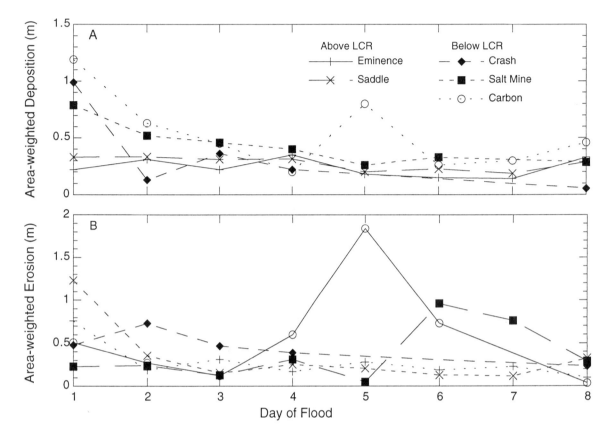

Figure 2. Graphs showing area-weighted (A) deposition and (B) erosion in 5 eddies at different times during the flood, calculated from the data of *Andrews et al.* [this volume]

because the change in thickness of the entire eddy was much less than the change in thickness at high elevation. Within 2 weeks, *Brouder et al.* [this volume] observed that backwaters created by the flood were no longer available as potential rearing habitat for native fishes.

6. SCIENTIFIC AND MANAGEMENT ISSUES THAT ARISE BECAUSE OF THE RESEARCH FINDINGS

6.1. Mainstem Sediment Transport, Computation of Sediment Budgets, and the Timing of Restoration Floods

Analysis of the suspended-sediment measurements demonstrates that some of the fundamental ideas underpinning management in Grand Canyon may need to be revised. One such idea is that the average annual rate of sediment delivery from the Paria River, Little Colorado River, and smaller tributaries downstream from Glen Canyon Dam exceeds the transport capacity of the Colorado River in those years when releases from the dam

are not unusually large and that accumulating sand can be stored on the bed for a number of years. To reach this idea, *Howard and Dolan* [1981] analyzed sediment-transport data for the Colorado River and its major tributaries for 1947-1970, assuming uniform sediment-delivery rates from ungaged tributaries and that measurement uncertainty was not significant. *Pemberton* [1987] and E. D. Andrews [cited by *Smillie et al.,* 1993] calculated transport relations from sand-transport data collected by the Geological Survey in 1983, 1985, and 1986 [*Garrett et al., 1993*]. These relations were used by *Randle et al.* [1993] to calculate that the bed and banks of the Colorado River had accumulated sediment between 1965 and 1982, because the mass of sediment entering Grand Canyon had exceeded the mass exported downstream. *Smillie et al.* [1993] calculated mass balances for different operational scenarios and suggested that sand accumulates in the river when tributary inflows are at least average and there are not wide daily fluctuations in discharge. Based on these findings, the Glen Canyon Dam EIS [*U.S. Department of the Interior*, 1995, fig. III-15] proposed that Beach/Habitat-Building Flows, such as the

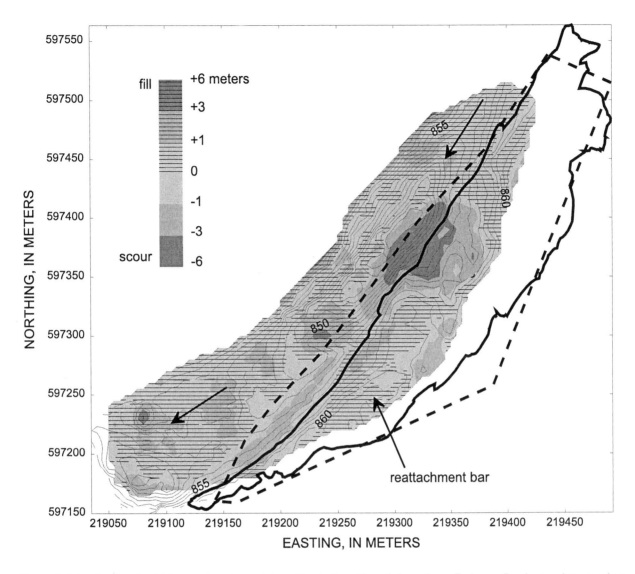

Figure 3. Map showing the thickness of erosion and deposition in the eddy and channel near Eminence Break camp between days 0 and 1 of the 1996 controlled flood. Thicknesses were computed from the original survey data of *Andrews et al.* [this volume]. Dashed outline is the eddy boundary used by *Andrews et al.* [this volume] in their calculations. Solid outline is the persistent eddy boundary calculated by *Schmidt et al.* [this volume]. Contours show topography for day 1 and are drawn at a 0.5 m contour interval. Note the large area of erosion at the upstream end of the reattachment bar. Note that there are no areas of thick deposition.

1996 controlled flood, be released but that sufficient time be allowed between these special flows to allow sediment to accumulate on the channel bed.

These calculations showing sediment accumulation in Grand Canyon were based on the assumption that the relation between sediment transport and discharge does not change significantly and that the scatter about the mean relation is not systematic. The assumption was also made that there have not been systematic variations in the amount of sand on the bed or in the size of that sand, because the

concentration and size of suspended sediment in transport is maintained by the distribution of those sizes on the bed.

Measurements during the 1996 controlled flood show that the relation between sediment transport and discharge is not stable. During the flood, the concentration of suspended sediment in the Colorado River at the mainstem gages upstream from the Little Colorado River [*Topping et al.,* this volume], near Grand Canyon [*Rubin et al.,* 1998; *Topping et al.,* this volume], and near National Canyon [*Smith,* this volume; *Topping et al.,* this volume] decreased with time. Thus, the daily transported load decreased during the flood even though discharge was steady. This finding is consistent with measurements of sediment transport in the Colorado River system during pre-dam floods [*Leopold and Maddock*, 1953].

Rubin et al. [1998] and *Topping et al.* [this volume] measured coarsening of the suspended load transported past the Grand Canyon gage between day 1 and either days 5 or 7 of the flood, depending on the sampler used. They showed that the size of bed sediment correspondingly coarsened with time and that eddy deposits between river miles 61 and 136 vertically coarsened. These data are impressive in their internal consistency. There is disagreement about temporal trends in the size of suspended sand further downstream at National Canyon and about the mechanism that determined those temporal trends [*Topping et al.*, this volume; *Smith*, this volume].

The high rate of suspended-sediment transport early in the flood implies that transport rates are highest when the bed has a large proportion of fines. Thus, it may not be possible to store large amounts of fine sediment on the bed for periods longer than a few years [*Topping et al.*, this volume]. Floods designed to build beaches may need to be scheduled soon after tributaries have delivered large loads of fine sediment to the mainstem, because the concentration of suspended sediment is likely to be highest at those times.

6.2. The Appropriate Duration of Restoration Floods

Many of the measurements made during the 1996 controlled flood show that only a few days are necessary to rework debris fans and to deposit eddy sand bars. Thus, the objectives of remobilizing coarse channel sediment and reconstructing fine-grained eddy bars and channel-margin deposits can be achieved by floods of shorter duration that the 1996 flood.

Field measurements and modeling studies demonstrate that deposition rates during the first few days of high discharge were high downstream from the Little Colorado River [*Wiele et al.*, this volume]. Rapid rates of channel-bed adjustment and high deposition rates in eddies had been documented in flume [*Schmidt et al.*, 1993] and modeling studies (J.M. Nelson and R.R. McDonald, U.S. Geological Survey, written commun., 1995) using simple channel geometries.

The area-weighted deposition in 3 eddies immediately downstream from the Little Colorado River declined greatly between days 0 and 4 (Figure 2a), and the area-weighted deposition in these eddies was more variable during the last

3 days of the flood. Area-weighted erosion and deposition were calculated as

$$D_a = V_{dep}/A_t, \text{ and} \qquad (1)$$

$$E_a = V_{ero}/A_t, \qquad (2)$$

where D_a = area-weighted deposition, E_a = area-weighted erosion, A_t = the total area surveyed in the eddy, V_{dep} = the volume of deposition, and V_{ero} = the volume of erosion. These calculations were made from the original survey data of *Andrews et al.* [this volume, referenced electronic data base] and the outlines of persistent eddies shown by *Schmidt et al.* [this volume], rather than the eddy outlines used by *Andrews et al.* [this volume], in order to facilitate comparison with other investigations. The primary difference between the boundaries of the eddies in the two investigations is that those of *Andrews et al.* [this volume] extend further into the channel and were estimated from field observations (Figure 3). The boundaries of *Schmidt et al.* [this volume] were computed from a photogrammetric method.

The field measurements are consistent with the predictions from models of the evolution of the channel and eddy bed in response to the high suspended-sediment transport of the winter 1993 flood [*Wiele et al.*, 1996] and the lower transport of the 1996 controlled flood. Initial deposition rates in main-channel pools and in eddies in 4 short reaches had been high in the Colorado River during the winter flood of 1993. The sediment-transport conditions during the 1993 flood were similar to high-concentration floods that had occurred prior to construction of Glen Canyon Dam [D.J. Topping, cited by *Wiele et al.*, 1996]. Deposition and erosion rates were sufficiently high such that channel bathymetry adjusted to maintain downstream transport of the large sediment loads within the first 3 days of that flood, and channel and eddy deposition rates were very low thereafter. The time necessary to achieve a new equilibrium condition was somewhat longer where the volume of the channel expansion was large. *Wiele et al.* [this volume] and *Wiele* [1998] showed that channel bed topography also adjusted quickly during the 1996 controlled flood. *Wiele et al.* [this volume] showed that eddy-bar deposition rates were highest during the first days of the flood, when the concentration of mainstem suspended sediment was high and the eddies were empty.

Wiele et al. [this volume] and *Wiele* [1998] thus showed that the resultant topography differs greatly depending on the concentration of sediment in transport. These results confirm the overriding sensitivity of post-flood bar form on the concentration of suspended sediment of the main flow (J.M. Nelson and R.R. McDonald, U.S. Geological Survey, written commun., 1995), especially on the side of the flow near the eddy. These findings are consistent also with measurements at 2 eddies in lower Marble Canyon upstream from the Little Colorado River, where the area-weighted deposition was low throughout the flood (Figure 2a).

Reworking of recently-aggraded debris fans also occurred rapidly and early during the 1996 controlled flood. *Pizzuto et al.* [this volume] showed that particle entrainment on debris fans occurs by two mechanisms: (1) slab failure by lateral erosion at the edge of the flow and (2) entrainment of individual particles from inundated parts of the fan. They showed that entrainment of individual particles occurred at 2 study sites, where dimensionless critical Shields stresses as large as 0.24 and 0.3 were measured. Lateral erosion is an important mechanism in debris-fan reworking, because slab failures impart initial movement to large particles that might not otherwise be entrained. *Webb et al.* [this volume] showed that reworked debris fans were coarser than they were before the flood, suggesting that bed coarsening results from winnowing of fine particles by selective entrainment. Additionally, large particles falling into the flow from bank failure eventually limit fan reworking. This process of fan armoring occurs rapidly and early and most reworking probably takes place within the first day of a flood. *Pizzuto et al.* [this volume] reported that reworking of the Prospect Canyon debris fan, which forms Lava Falls Rapid, ended approximately 4 hrs after the rise of the flood. On the basis of these observations, *Webb et al.* [this volume] stated, "Very short-duration and high-magnitude controlled floods would be highly effective in reworking aggraded debris fans."

6.3. Large-Scale Erosion during the Flood

The rapid erosion of sand bars was first documented by *Cluer et al.* [1994] and *Cluer* [1995] who photographed and observed large portions of reattachment bars erode into the main channel. This erosion occurred when the upper surface of the bar was exposed and discharge was less than powerplant capacity. *Cluer* [1995] attributed rapid erosion to destabilization of the base of reattachment bars by change in direction and magnitude of downstream flow as it exits a channel constriction. *Wiele et al.* [1996] showed that reorientation of the core of downstream flow within a channel expansion could occur when deposition in channel pools and in adjacent eddies was high. The processes observed by *Cluer et al.* [1994] had not occurred, however,

TABLE 1. Large erosion events in 5 eddies during the 1996 controlled flood in Grand Canyon

Eddy (day range)	Average thickness of erosion (m)	Proportion of eddy that eroded
Eminence (0-1)[1,4]	1.44	0.53
Saddle (0-1)[2]	1.82	0.68
Crash (1-2)[3,4]	1.12	0.66
Salt Mine (0-1)[1]	1.31	0.39
Salt Mine (3-4)[1,4]	1.53	0.39
Salt Mine (4-5)[3,4]	3.15	0.58
Salt Mine (5-6)[3]	1.23	0.60
Carbon (5-6)[3,4]	1.63	0.59
Carbon (6-7)[1,4]	1.49	0.51

[1] Upstream end of reattachment bar.
[2] Upstream half of reattachment bar.
[3] Reattachment-bar platform.
[4] Probable mass failure.

during high-transport conditions such as those during the first days of the 1996 controlled flood.

Rapid erosion of large proportions of reattachment bars occurred during the 1996 controlled flood. *Andrews et al.* [this volume] measured erosion of more than 50,000 m³ of sand from the upstream part of a reattachment bar at Salt Mine eddy between days 4 and 5 of the flood, when as much as 7 m of sediment were eroded from the bar. *Konieczki et al.* [1997] measured an average of 4 m of scour during a 2-hr erosional event that occurred during the last day of the flood at the same site. A similar event occurred at a reattachment bar immediately downstream from 122 Mile Creek (D.M. Rubin, U. S. Geological Survey, pers. commun., 1998).

These erosion events did not occur on any specific day; they occurred on different days at different sites (Figure 2b). Large areas of erosion occurred between days 0 and 1 in eddies at Eminence, Saddle, and Salt Mine, between days 1 and 2 at the eddy at Crash Canyon, between days 3 and 5 in the eddy at Salt Mine, between days 5 and 6 in the eddies at Salt Mine and Carbon Creek, and between days 6 and 7 in the eddy at Carbon Creek (Table 1).

Andrews et al. [this volume] suggested that mass failures of more than 10,000 m³ occurred at each of the 5 eddies that they measured and each measured reattachment bar probably had at least one erosion event during the flood. However, every erosion event was not necessarily a mass failure (Table 1). The onshore side of some areas of large erosion had a scalloped shape, characteristic of headwalls

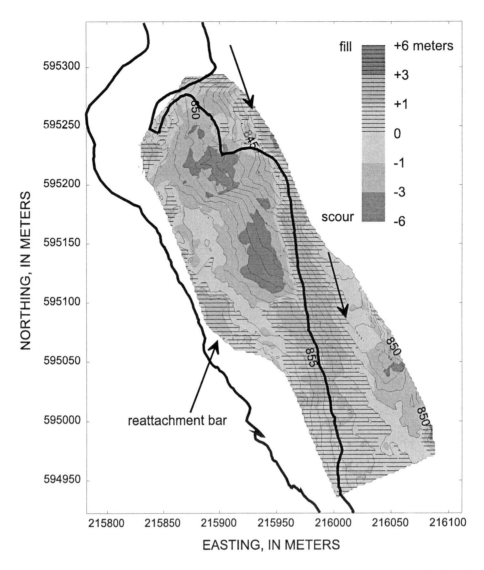

Figure 4. Map showing showing the thickness of erosion and deposition in the eddy and channel near Saddle Canyon between days 0 and 1 of the 1996 controlled flood. Thicknesses were computed from the original survey data of *Andrews et al.* [this volume]. Solid outline is the persistent eddy boundary calculated by *Schmidt et al.* [this volume]. Contours show topography for day 1 and are drawn at a 0.5 m contour interval. Note the large area of erosion at the upstream half of the reattachment bar.

of mass failures, similar to that illustrated by *Andrews et al.* [this volume, Figure 9]. These failures occurred typically at the upstream end of reattachment bars where the resultant topography created a deep depression between the separation and reattachment bars. Elsewhere, probable mass failures eroded most of the reattachment bar platform. At the time of their occurrence, mass failures at the 3 sites in Tapeats Gorge measured by *Andrews et al.* [this volume] affected between 39 and 66% of each persistent eddy.

Other areas of significant erosion may not have been due to mass failures, however, because the areas of erosion did not have definable headwall areas (Figure 4). Some of these areas of large erosion occurred between days 0 and 1, and it is therefore unlikely that they were preceded by significant deposition (Table 1). Those erosion events that occurred on the rising stage of the controlled flood may have been caused by the reorientation of flow directions as suggested by *Schmidt* [1990] and *Cluer* [1995].

A flood of shorter duration than that of 1996 may have been more effective in retaining sediment within eddies, because the probability of large-scale erosion events increased with time. At the 3 eddies in the Tapeats Gorge where there was relatively high initial suspended-sediment transport, area-weighted deposition greatly exceeded area-weighted erosion during the first 3 days of the flood (Figure 5a). The likelihood of area-weighted deposition exceeding area-weighted erosion decreased after day 4 (Figure 5b). Area-weighted deposition never greatly exceeded area-weighted erosion upstream from the Little Colorado River, and a longer flood may have been appropriate because deposition rates were low. Although the data collected by *Andrews et al.* [this volume] on which these conclusions are based are from only 5 sites, logistical constraints may make it prohibitively expensive for a larger number of sites to be measured on a daily basis in future studies. Thus, a different research approach may be needed to make further progress on this issue.

6.4. Variability in the Magnitude of Aggradation Among Sites

Variability in the magnitude of sand-bar change will force river managers to more precisely articulate their criteria for determining the "success" of any restoration flood. Large-scale measurements of sand-bar change made by *Kearsley et al.* [this volume], *Hazel et al.* [this volume], *Schmidt et al.* [this volume], and *Thompson et al.* [1997] demonstrate that every eddy in Grand Canyon did not aggrade in the same proportion and that a few eddies experienced net erosion. These measurements document that the processes measured by *Andrews et al.* [this volume] occurred throughout Grand Canyon, and that the different concentrations of suspended sediment that occurred in different reaches [*Topping et al.*, this volume] with different channel geometries [*Wiele et al.*, this volume] led to a wide range of topographic patterns in eddies throughout Grand Canyon.

Net erosion occurred in some eddies. *Kearsley et al.* [this volume] found that 3 campsites were destroyed by the flood, and *Thompson et al.* [this volume] identified another 3 sites that had been significantly eroded by the flood. *Hazel et al.* [this volume, Table 2] estimated large standard errors of mean bar change caused by the flood. The high-elevation part of 6 of the 35 sites that they surveyed decreased in area. *Schmidt et al.* [this volume] showed that the area of significant erosion exceeded the area of significant deposition at several sites.

Thus, the *average* response of bars to the 1996 controlled flood was high-elevation deposition in eddies, but there were specific sites where large amounts of erosion occurred. The distinction between average and site-specific response has led to revision of how scientists characterize bar response to changes in dam operations. *Kearsley el*

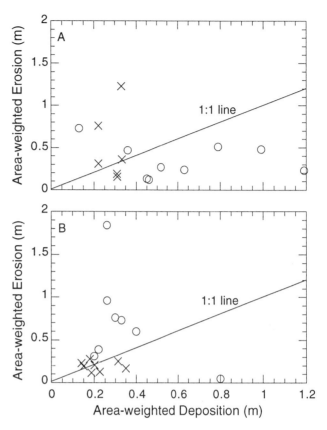

Figure 5. Graphs showing area-weighted deposition and area-weighted erosion for each eddy measured by *Andrews et al.* [this volume] for each day for (A) the first 3 days of the flood and (B) the last 4 days of the flood. Crosses are for sites upstream from the Little Colorado River and circles are for sites further downstream. Data above and below the 1:1 line indicate days and sites where erosion exceeded deposition or vice versa.

[this volume, Figure 10], *Hazel et al.* [this volume, Figure 2, 10], and *Schmidt et al.* [this volume, Figure 10, 11] reported changes in eddy bars by computing frequency distributions as well as mean bar response. Managers will have to decide if their desired end state is an average reach condition, an average reach response, or a desired change at specific sites.

6.5. Longitudinal Patterns of Change in Eddy Bars

There is disagreement about whether the flood caused similar responses everywhere in Grand Canyon, or if there was less deposition, and more evidence of sediment starvation, in the upstream parts of the river system. *Andrews et al.* [this volume] argued that the large volumes of sand that accumulated, and were evacuated, from the 5 eddies that they measured demonstrate that there was sufficient sediment available to build eddy bars everywhere. *Hazel et al.* [this volume] and *Thompson et al.* [1997] showed that there were large increases in the volume of newly-deposited sand in Marble Canyon and that the proportional increase in bars was greatest there.

However, area-weighted daily deposition was much lower in the 2 sites in lower Marble Canyon measured by *Andrews et al.* [this volume] than in the 3 sites further downstream (Figure 2). The sites in lower Marble Canyon were extensively eroded on the first day of the flood, and it took many days of slow aggradation to replace the sediment that had been initially eroded.

Other evidence also suggests that the extent of aggradation was less upstream from the Little Colorado River. More than 40% of eddies in the Tapeats Gorge and the Big Bend reaches downstream from the Little Colorado River filled to more than 50% of their capacity. Further upstream, less than 30% of eddies filled to this extent. Similarly, more than 25% of eddies downstream from the Little Colorado River had net normalized aggradation (NNA) values greater than 0.25; upstream, less than 15% of eddies had NNA values greater than 0.25 [*Schmidt et al.*, this volume]. The average net topographic change in eddies, averaging high-elevation deposition and low-elevation erosion, was negative for measured sites upstream from the Little Colorado River and positive for the sites between the Little Colorado River and the Grand Canyon gage [*Hazel et al.*, this volume].

These differences between the extent of deposition upstream and downstream from the Little Colorado River are probably due to the fact that suspended-sediment concentrations were much lower in Marble Canyon and that the magnitude of eddy deposition is dependent on the concentration of suspended sediment. *Topping et al.* [this volume] showed that the suspended-sediment concentrations at the gage upstream from the Little Colorado River were approximately half of the concentration downstream, and the total transport of sand from Marble Canyon was about half of that transported past the Grand Canyon gage. As argued by *Rubin et al.* [1994], low sediment concentrations in the main flow, limited sediment replenishment from only one large sediment-supplying tributary, and limited sediment storage on the bed have the potential to cause net scour in reaches nearest to Lees Ferry. These observations highlight the need to evaluate the sand balance of the 1996 controlled flood.

7. SAND BALANCE

One elusive goal of Grand Canyon researchers has been to provide managers an accurate budget of inflows,

outflows, and changes in storage of sand. Such a budget could provide essential insight into the mass of sand available for redistribution to high elevation. An accurate budget for the 1996 controlled flood could be used to analyze the efficiency with which high-elevation sand bars were deposited in relation to the mass of sand delivered to Lake Mead and to determine the source of sand deposited on the channel banks. Unfortunately, only imprecise sand budgets for the reaches between Lees Ferry and the Little Colorado River and between the Little Colorado River and the Grand Canyon gage can be developed. However, development of such budgets provides a conceptual framework within which the status of the current understanding of physical processes in Grand Canyon can be evaluated.

Based on the measurements of *Konieczki et al.* [1997], the total mass of sand transported to Lake Mead by the 1966 controlled flood was approximately $1.55 \cdot 10^6$ Mg, because the transport rates past the Grand Canyon and National Canyon gages were each approximately $1.55 \cdot 10^6$ Mg (D.J. Topping, U.S. Geological Survey, written commun., 1998). The mass transported past the Geological Survey gaging station upstream from the Little Colorado River was half this amount.

All of the transported sand must have come from the channel bed, eddies, or banks because no sand was released into the Colorado River from Lake Powell reservoir during the flood. A number of assumptions and generalizations must be made in order to estimate and partition these changes in storage, because the measurement of changes in sediment storage are not comprehensive.

Estimates of the net change in the mass of sand in eddies was based on multiplying the average change in measured eddies [*Andrews et al.*, this volume; *Hazel et al.*, this volume] by the estimated number of eddies and their average size (Table 2). It is necessary to distinguish the mass of sand deposited at high elevation from the net change in storage in eddies because the centers of large eddies eroded and thick deposition occurred at the margins of eddies and near their upstream and downstream ends [*Andrews et al.*, this volume; *Hazel et al.*, this volume; *Schmidt et al.*, this volume]. *Hazel et al.* [this volume] measured the average thickness of high-elevation deposits; the area of these thick deposits were estimated from *Schmidt et al.*'s [this volume] measurements of the area of significant deposition.

Although the sand budget for the 1996 controlled flood is imprecise (Table 3), the ratio of the mass of sand exported to the magnitude change in sand storage differs greatly for the two reaches. The sand exported from the reach upstream from the Little Colorado River was approximately equal to the mass deposited along the channel margins in eddies and

TABLE 2. Estimates of components involved in the sand budget of the 1996 controlled flood in Grand Canyon

Component	Lees Ferry to Little Colorado River	Little Colorado to Grand Canyon gage
Length (km)	102	43
Number of eddies/km[1,2]	4	4
Average eddy area (m^2)[1,2]	6600	8000
Total area of eddies (m^2)	$2.7 \cdot 10^6$	$1.4 \cdot 10^6$
Area of significant[1] deposition (m^2/km)	7000	8000
Change in thickness[3,4,7] of high-elevation sand (m)	0.62 (0.51-0.73)	0.55 (0.53-0.57)
Increase in mass of high[5,7] elevation sand (Mg)	$0.76 \cdot 10^6$ (0.63-$0.90 \cdot 10^6$)	$0.33 \cdot 10^6$ (0.31-$0.34 \cdot 10^6$)
Change in thickness of[3,4] sand in entire eddy (m)	-0.14 (-0.37 to +0.09)	0.15 (-0.23 to +0.53)
Change in mass of sand in[5] entire eddy (Mg)	$-0.65 \cdot 10^6$ (-1.7 to $0.42 \cdot 10^6$)	$0.36 \cdot 10^6$ -0.55 to $1.3 \cdot 10^6$)
Average new channel[1] margin deposit area (m^2/km)	3500	6000
Average thickness of[6] channel margin deposits (m)	0.2 (0.1-0.3)	0.2 (0.1-0.3)
Change in mass of channel[5] margin deposits (Mg)	$0.12 \cdot 10^6$ (0.06-$0.18 \cdot 10^6$)	$0.089 \cdot 10^6$ (0.04-$0.13 \cdot 10^6$)

[1] Data from *Schmidt et al.*, this volume.
[2] Data from *Bureau of Reclamation*, 1988, table A-3.
[3] Data from *Hazel et al.*, this volume.
[4] Standard error about the mean.
[5] Assumes density is 2.65 Mg/m^3 and porosity is 0.35.
[6] Assumed.
[7] Only where deposition was significant.

on the banks. In contrast, the mass of sand deposited in eddies and on the banks in the 43 km downstream from the Little Colorado River was approximately 25% of the total load transported from the reach.

The budgets also indicate that the source of the sand deposited along the channel banks differed in the two reaches. More of this sand probably was eroded from low-elevation parts of eddies than from the channel bed between Lees Ferry and the Little Colorado River. More of this sand probably was eroded from the bed than from low-elevation parts of eddies further downstream. These results are consistent with the conceptual model of *Rubin et al.* [1994], who proposed that eddies might be the primary sources of sand for redistribution to high elevation in upstream parts of Grand Canyon where there is limited sediment resupply.

There is insufficient data to accurately determine the proportion of the bed where significant scour occurred, but an order of magnitude estimate can be made from bed scour data of *Hazel et al.* [this volume]. The purpose of these estimates is to determine if bed scour was likely to have been a widespread occurrence. The proportion of the bed scoured to a given depth was calculated as:

$$r = m / (w \cdot d \cdot l \cdot \gamma_s \cdot n), \qquad (3)$$

where r = the proportion of the bed where scour equal to the value d occurred; m = the mass of sand estimated to have been scoured from the bed, calculated as shown in Tables 2 and 3; w = 100 m, an assumed average width of the river; d = the mean depth of scour for the each reach; l = the length of each reach; γ_s = the specific weight of sand (2.65 Mg/m^3); and n = the proportion of the total volume which is solids, assumed to be 0.65. *Hazel et al.* [this volume] showed that the channel bed scoured at 15 of the 17 sites that they measured upstream from the Grand Canyon gage. The mean depth of scour was 0.95 m upstream from the Little Colorado River and 0.80 m further downstream.

The proportion of the bed where significant scour occurred was probably small in both reaches. Only 5% of the channel between Lees Ferry and the Little Colorado River would have scoured 0.95 m to yield the 920,000 Mg estimated to have been derived from the bed in this reach (Table 3). Only 21% of the channel between the Little Colorado River and the Grand Canyon gage would have scoured 0.80 m to yield the 1,250,000 Mg estimated to have been derived from the bed in this reach.

These estimates of the proportion of the bed where significant bed scour occurred are reasonable, because they are less than estimates of the proportion of the bed covered by sand. According to one estimate (J.D. Smith and S.M. Wiele, U.S. Geological Survey, written commun., 1992), sand had to cover 13-28% of the bed, depending on reach geometry, in order to maintain equilibrium sand transport during research flows that were conducted in 1990 and 1991. The *Bureau of Reclamation* [1988, Table A-26] estimated that the proportion of the bed covered by gravel and finer sizes in 1984 was about 45% [*Wilson*, 1986].

8. IMPLICATIONS FOR FUTURE RESEARCH

The Glen Canyon Environmental Studies program provided a wealth of new knowledge that led to this controlled flood. The flood was performed, in part, as a river-management demonstration with substantial confidence that positive results were probable and that damage was improbable. In that respect, the flood was successful.

TABLE 3. Sand budget for the Colorado River in Grand Canyon during the 1996 controlled flood

| | LEES FERRY TO LITTLE COLORADO RIVER | | LITTLE COLORADO RIVER TO GRAND CANYON GAGING STATION | |
	Average (Mg)	Range (Mg)	Average (Mg)	Range (Mg)
Inflow	small	na	$0.8 \cdot 10^6$	na
High-elevation sand in eddies	$0.76 \cdot 10^6$	0.63 to $0.90 \cdot 10^6$	$0.33 \cdot 10^6$	0.31 to $0.34 \cdot 10^6$
High-elevation sand on banks	$0.12 \cdot 10^6$	0.06 to $0.18 \cdot 10^6$	$0.09 \cdot 10^6$	0.044 to $0.13 \cdot 10^6$
Net change in sand in eddies	$-0.65 \cdot 10^6$	-1.7 to $0.42 \cdot 10^6$	$0.36 \cdot 10^6$	-0.55 to $1.3 \cdot 10^6$
Sand at low-elevation in eddies	$-1.4 \cdot 10^6$	-2.8 to $-0.20 \cdot 10^6$	0	0 to $0.89 \cdot 10^6$
Sand in channel	$-0.92 \cdot 10^6$	-0.98 to $-0.86 \cdot 10^6$	$-1.25 \cdot 10^6$	-0.89 to $-2.1 \cdot 10^6$
Outflow	$0.8 \cdot 10^6$	na	$1.6 \cdot 10^6$	na

The flood was also conducted to test the validity of an evolving understanding of the river: that is, the controlled flood was experimental. If results were as predicted, then certainty of understanding was increased; if results differed from predictions, then revision of concepts was required and new understanding thereby emerged. Success, in these terms, was never in doubt.

That many of the predicted outcomes of the flood, such as widespread deposition of new sand bars, were realized is gratifying and is testimony to the scientific research that preceded the flood. That there were surprises was exciting, because new understanding was gained and the concept of how controlled flooding could be used in river management was thereby refined.

If managers determine that the changes in alluvial deposits that occurred in 1996 are desirable, then future controlled floods can be of shorter duration, because the early high rates of fine-sediment deposition rapidly constructed eddy sand bars and channel-margin deposits. The likelihood that large erosion events would remove newly-deposited sediment increased with time. However, other objectives may require longer duration floods, such as the removal of non-native riparian vegetation along the river banks. Flood-formed deposits will erode rapidly after they form.

Future floods will probably be timed to coincide with years when the Paria and Little Colorado rivers deliver large loads to the Colorado River because suspended sediment concentrations are likely to be highest soon after tributary floods. However, appropriate timing of floods may be constrained by the volume of water stored in Lake Powell and within the entire basin.

The longitudinal trends in suspended-sediment transport rate and longitudinal differences in sand storage in eddies and in the channel must be resolved by future research programs, because it is essential to know if Glen and Marble Canyons are adversely scoured by controlled floods. Thus, expanded monitoring and research will be necessary to provide more precise estimates of the mean change in sand storage in eddies upstream from the Little Colorado River, the relative proportion of stored sand that resides on the bed and in eddies, and the suspended-sand transport rate. The geometry of the Colorado River and its valley changes greatly near river mile 40 [Schmidt and Graf, 1990; Melis, 1997], and it may be appropriate to compute sand budgets for the upstream and downstream parts of Marble Canyon, respectively.

If managers desire to assess the status of river resources from a reach average perspective, then statistical characterizations are the appropriate tool. However, where site specific impacts are critical, detailed surveys and numerical modeling are appropriate strategies to predict channel change. The continued examination of the basic physical processes that control sediment transport and geomorphic adjustment on a reach-averaged and site-specific scale, which was forced by analysis of the research results from the 1996 controlled flood, indicates that continued research about the physical resources of the Colorado River is essential to the adaptive management of the resources of the Colorado River.

Acknowledgments. Earlier versions of this manuscript were reviewed by Ned Andrews, Paul Grams, Joe Hazel, Dick Marzolf, Jon Nelson, Dave Rubin, Jim Smith, David Topping, Bob Webb, Steve Wiele, and Peter Wilcock. The author thanks all of his Grand Canyon colleagues for years of intellectual stimulation, pleasure, and adventure.

REFERENCES

Bureau of Reclamation, *Glen Canyon environmental studies final report*, Salt Lake City, UT, 1988 (NTIS No. PB88-183348/AS).

Cluer, B.L., Cyclic fluvial processes and bias in environmental monitoring, Colorado River in Grand Canyon, *J. Geol.*, 103, 422-432, 1995.

Cluer, B.L., and L.R. Dexter, *Daily dynamics of Grand Canyon sandbars: monitoring with terrestrial photogrammetry*, final rept., 200 pp., U.S. Dept. Int., Natl. Park Serv., Flagstaff, AZ, 1994.

Collier, M.P., R.H. Webb, and E.D. Andrews, Experimental flooding in Grand Canyon, *Sci. Amer.*, 276, 82-89, 1997.

Garrett, W.B., E.K. Van DeVanter, and J.B. Graf, *Streamflow and sediment-transport data, Colorado River and three tributaries in Grand Canyon, Arizona, 1983 and 1985-86*, U.S. Geol. Survey Open-File Rept. 93-174, 624 pp. 1993.

Howard, A., and R. Dolan, Geomorphology of the Colorado River in the Grand Canyon, *J. Geol.*, 89 (3), 269-298, 1981.

Konieczki, A.D., J.B. Graf, and M.C. Carpenter, *Streamflow and sediment data collected to determine the effects of a controlled flood in March and April 1996 on the Colorado River between Lees Ferry and Diamond Creek, Arizona*, U.S. Geol. Survey Open-File Rept. 97-224, 55 pp., 1997.

Leopold, L.B., and T. Maddock, Jr., *The hydraulic geometry of stream channels and some physiographic implications*: U.S. Geol. Surv. Prof. Paper 252, 1953.

Melis, T.S., 1997, *Geomorphology of debris flows and alluvial fans in Grand Canyon National Park and their influences on the Colorado River below Glen Canyon Dam, Arizona*, [Ph.D dissertation], Tucson, Univ. Ariz., 1997.

Pemberton, E.L., *Sediment data collection and analysis for five stations on the Colorado River from Lees Ferry to Diamond Creek*, Bur. Recl. Glen Canyon Environ. Studies report, 159 pp., 1987 (NTIS No. PB88-183397/AS).

Randle, T.J., R.I. Strand, and A. Streifel, Engineering and environmental considerations of Grand Canyon sediment management, in *Engineering Solutions to Environmental Challenges: Thirteenth Annual USCOLD Lecture*, Chattanooga, U.S. Committee on Large Dams, Denver., CO, 1993.

Rubin, D.M., J.M. Nelson, and D.J. Topping, Relation of inversely graded deposits to suspended-sediment grain-size evolution during the 1996 flood experiment in Grand Canyon, *Geol.*, 26(2), 99-102, 1998.

Rubin, D.M., R.A. Anima, and R. Sanders, *Measurements of sand thicknesses in Grand Canyon, Arizona, and a conceptual model for characterizing changes in sand-bar volume through time and space*, U.S. Geol. Survey Open-File Rept. 94-597, 9 pp., 1994.

Schmidt, J.C., Recirculating flow and sedimentation in the Colorado River in the Grand Canyon, *J. Geol.*, 98, 709-724, 1990.

Schmidt, J.C., and J.B. Graf, *Aggradation and degradation of alluvial sand deposits, 1965-1986, Colorado River, Grand Canyon National Park*, U.S. Geol. Surv. Prof. Paper 1493, 1990.

Schmidt, J.C., and D.M. Rubin, Regulated streamflow, fine-grained deposits, and effective discharge in canyons with abundant debris fans, in *Natural and anthropogenic influences in fluvial geomorphology*, ed. J.E. Costa, A.J. Miller, K.W. Potter, and P.R. Wilcock, pp. 177-195, Amer. Geophys. Union, Geophys. Mono. 89, Washington, DC, 1995.

Schmidt, J.C., D.M. Rubin, and H. Ikeda, Flume simulation of recirculating flow and sedimentation, *Water Res. Res.*, 29 (8), 2925-2939, 1993.

Smillie, G.M., W.L. Jackson, and D. Tucker, *Colorado River sand budget: Lees Ferry to Little Colorado River*, U.S. Department of the Interior, Natl. Park Serv., Tech. Rept. NPS/NRWRD/NRTR-92/12, Washington, DC, 11 pp., 1993.

Stevens, L.E., *Mechanisms of riparian plant community organization and succession in the Grand Canyon, Arizona*, Ph. D. diss., N. Ariz. Univ., Flagstaff, AZ, 1989.

Thompson, K., K. Burke, and A. Potochnik, *Effects of the beach-habitat building flow and subsequent interim flows from Glen Canyon Dam on Grand Canyon camping beaches, 1996: a repeat photography study by Grand Canyon river guides (adopt-a-beach program)*, 11 pp., administrative report to Grand Canyon Monitoring and Research Center, Flagstaff, 1997.

U.S. Department of the Interior, *Final environmental impact statement, operation of Glen Canyon Dam, Colorado River storage project, Coconino County, Arizona*, 337 pp., Bur. Recl., Salt Lake City, UT, 1995.

Wiele, S.M., *Modeling of flood-deposited sand distributions in a reach of the Colorado River below the Little Colorado River, Grand Canyon, Arizona*, U.S. Geol. Surv. Water-Res. Invest. Rept. 97-4168, 15 pp., 1998.

Wiele, S.M., J.B. Graf, and J.D. Smith, Sand deposition in the Colorado River in the Grand Canyon from flooding of the Little Colorado River, *Water Res. Res.*, 32 (12), 3579-3596, 1996.

Wilson, R. P., Sonar patterns of Colorado riverbed, Grand Canyon, *Fourth Federal Interagency Sedimentation Conference*, 2, Las Vegas, NV, 1986.

John C. Schmidt, Department of Geography and Earth Resources, Utah State University, Logan, Utah, 84322-5240; email: jschmidt@cc.usu.edu

Biological Implications of the 1996 Controlled Flood

Richard A. Valdez

SWCA, Inc., Logan, Utah

Joseph P. Shannon and Dean W. Blinn

Department of Biological Sciences, Flagstaff, Arizona

The 1996 controlled flood provided evidence that elevated releases from Glen Canyon Dam can enhance short-term primary and secondary production of aquatic resources of the Colorado River in Grand Canyon National Park. The flood scoured substantial proportions of benthic algae and macroinvertebrates and removed fine sediments from the channel, which ultimately stimulated primary productivity and consumer biomass. Channel margin sand deposits buried riparian vegetation and leaf litter, entraining nutrients for later incorporation into the upper trophic levels. The flood restructured high-stage sand bars and associated eddy return channels (i.e., backwaters used as nurseries by native and non-native fish), but many were short-lived because reattachment bars were eroded shortly after the flood. The flood was of insufficient magnitude to permanently suppress non-native fish populations, even though there was significant population depletion at some collecting sites. Pre-spawning aggregations, spawning ascents of tributaries, and habitat use by native fishes were unaffected by the flood. Adult rainbow trout (*Oncorhynchus mykiss*) in the Lees Ferry tailwater fishery were also unaffected, but the proportion of juveniles <152 mm total length decreased by 10%; a strong year class following the flood indicated replacement through successful reproduction.

1. INTRODUCTION

Assemblages of aquatic plants and animals in the Colorado River of Grand Canyon have undergone dramatic changes since construction of Glen Canyon Dam in 1963 [*Stevens et al.*, 1995; 1997; *Blinn et al.*, 1998]. Clear, cold-water releases have established a new assemblage of diatoms and algal species [*Czarnecki and Blinn*, 1978; *Blinn et al.*, 1989; *Blinn and Cole*, 1991; *Hardwick et al.*, 1992], reduced the diversity of aquatic macroinvertebrates [*Shannon et al.* 1994; *Shannon et al.*, 1996], and precluded most mainstem reproduction by native fishes [*Valdez and Ryel*, 1997]. Daily flow fluctuations from power plants restrict primary producers to the wetted perimeter below the lowest controlled flow [*Blinn et al.*, 1995] and destabilize shoreline and backwater habitats for fish [*Converse et al.*, 1998]. Non-native plants, many present before the dam, have spread and thrived in the new environment, providing new habitats for native and non-native animal species, while also competing for resources with native forms. The existing "naturalized ecosystem" has become "...a blend of

TABLE 1. Responses of biological resources of the Colorado River in Grand Canyon to the 1996 controlled flood. 0 = no detectable response, + = desirable response, – = undesirable response. Responses were ranked 0 if post-flood conditions did not exceed pre-flood conditions.

Resource	Responses To Flood	
	Short-Term	Long-Term
Algae	– scouring decreased biomass	+ biomass exceeded pre-flood levels; however, may have also resulted from high steady flows with high water clarity
Macroinvertebrates	– scouring decreased densities & biomass	+ biomass exceeded pre-flood levels; however, may have also resulted from high steady flows with high water clarity
Fish Habitat	0 backwaters were restructured	– reattachment sand bars were eroded
Native Fish	0 no detectable effect	0 no detectable effect
Non-Native Fish	– significant decrease in some species	0 no detectable effect
Trout	– juveniles were displaced downstream	0 no detectable effect
Riparian Vegetation	– some vegetation was buried	0 no detectable effect
Kanab Ambersnail	– 10% of occupied habitat inundated	0 no detectable effect

the old and the new, a mixture of native and introduced organisms adapting to both natural and artificial processes" [*Carothers and Brown*, 1991].

One of the purposes of the 1996 controlled flood was to reinstate some of the natural dynamics to the aquatic ecosystem of the Colorado River in Grand Canyon. The reestablishment of emergent sand bars was designed to store valuable sand and sediments, while restructuring fish nursery habitats (i.e., backwaters) and redistributing nutrients stored as vegetation and debris along the river banks. The controlled flood provided the opportunity to test many hypotheses concerning biological processes in a regulated river located in an arid biome. Discharges of this magnitude have only been experienced one other time since impoundment in 1963 [*Kearsley et al.*, 1994]. For example, the controlled flood was used to test the hypothesis that high velocities and inundation of shallow, secluded habitats would suppress non-native fishes and promote habitat development and food availability for native species. At the same time, the controlled flood also revealed new, previously undefined biological relationships that have become important considerations for future use of floods in Grand Canyon.

In this paper, we provide an overview of the effects of the 1996 controlled flood on the biological resources of the Colorado River in Grand Canyon (Table 1). We also discuss the relative success or failure of the flood to alter these resources for management purposes, and we provide recommendations for future use of such floods.

2. BIOLOGICAL RESPONSES

2.1. Nutrient Composition

The 1996 controlled flood scoured the central river channel and redeposited sediments along channel margins, burying standing and allochthonous organic material under 0.2-1.95 m of fine sand [*Parnell et al.*, this volume]. Significant increases in dissolved carbon (28 mg/L), nitrogen (10 mg/L), and phosphorus (5 mg/L) in sand bar ground waters and in surface backwaters indicated that buried organic material was readily mineralized. This process may provide a constant source of nutrients to support primary and secondary production in backwaters and along the immediate shorelines of the river, both of which are critical nurseries for native fishes. It is not clear however, to what degree the elevated ground-water nutrients directly influenced production in the mainstem due to the confounding influences of the high dilution factor in the main channel and the relatively clear-water conditions that followed the flood event [*Blinn et al.*, this volume]. It was concluded that the 1996 controlled flood was of insufficient magnitude to radically scour standing riparian and terrestrial vegetation, but was large enough to bury shoreline materials. Stable isotope analyses, however, suggested that at least some riparian vegetation was entrained within the CPOM (coarse particulate organic matter) constituent of the mainstem [*Blinn et al.*, this volume].

2.2. Photosynthesis, Respiration, and Metabolism

Within the immediate tailwaters, the flood scoured the physiologically less-active plant tissue, moribund plant parts, and particulate organic matter associated with the phytobenthos, leaving behind a smaller but more efficient phytobenthic assemblage and a benthic community with reduced rates of microbial respiration [*Marzolf et al.*, this volume]. Open system community metabolism estimates, using diel pH and dissolved oxygen concentration, confirmed measurements of a 90% decrease in the standing crop of the benthic plant community in the dam tailwaters [*Marzolf et al.*, this volume]. Despite this loss, the photosynthetic signal remained strong, indicating high rates of primary productivity following the flood [*Brock et al.*, this volume]. The green alga *Cladophora* showed a higher rate of photosynthesis per unit area after the flood compared to before the flood [*Brock et al.*, this volume].

2.3. Algal and Macroinvertebrate Communities

The present aquatic food web of the Colorado River through Grand Canyon is highly modified from the more natural river conditions found elsewhere in the basin [*Hayden*, 1997]. The historic system of allochthonous energy has been replaced by locally-produced filamentous green algae and associated epiphytic diatoms [*Blinn and Cole*, 1991, *Blinn et al.*, this volume]. The diverse aquatic insect assemblage of mayflies, caddisflies, dobsonflies, and damselflies has been replaced by midges, blackflies, oligochaetes, gastropods, planaria, and an introduced crustacean, *Gammarus lacustris* [*Hayden*, 1997; *Shannon et al.*, 1994; *Stevens et al.*, 1997, *Oberlin, et al.*, 1998].

Direct measurement of the river substrate showed that over 90% of benthic plants and 50% or more of benthic invertebrates were scoured at the Lees Ferry reach; biota associated with unstable fine sediments were most vulnerable [*Blinn et al.*, this volume]. Most of the dissolved organic carbon (DOC) and particulate organic carbon (POC) passing through the river corridor were entrained in the hydrostatic wave of the flood, as indicated by significantly lower values of DOC and POC throughout the remainder of the flood [*Blinn et al.*, this volume].

Although the benthic plant community was the dominant drift constituent during normal dam operations, stable isotope analyses showed that riparian and upland vegetation made up most of the drift during the controlled flood [*Blinn et al.*, this volume; *Blinn et al.*, 1998]. These measurements confirmed observations of large quantities of suspended plant material and debris during the early stages of the flood. Also, windrows of stranded *Gammarus*

lacustris carcasses were observed along some shorelines following the flood.

Average biomass of benthic algae was similar to pre-dam levels only 1 month after the flood, showing rapid recovery of phytobenthos [*Blinn et al.*, this volume]. Benthic macroinvertebrates also recovered rapidly, reaching pre-flood standing crop levels within 2 months after the controlled flood. Subsequent biomass estimates for algae and macroinvertebrates have exceeded previously reported levels [*Usher and Blinn*, 1990; *Blinn et al.*, 1992; *Shannon et al.*, 1996; *Stevens et al.*, 1997].

The recovery rate of 2 months for primary producers and consumers in Grand Canyon was somewhat faster than the 3-month period reported following major disturbances in other aquatic ecosystems [*Yount and Niemi*, 1990]. This rapid recovery rate and subsequent high standing crops of aquatic benthos were largely attributed to high water clarity as a result of relatively high constant discharges from Glen Canyon Dam following the flood [*Blinn et al.*, this volume]. The greatest effect from the controlled flood was the flushing of fine sediments from the system, which along with consistently high flows, sustained high water clarity for photosynthesis.

The 1996 controlled flood was predicted to inundate 11-16% of available habitat of the endangered Kanab ambersnail (*Oxyloma haydeni kanabensis*) at Vasey's Paradise (RM 31.5; pers. comm. Larry Stevens, Grand Canyon Monitoring and Research Center, Flagstaff, AZ, 1996). As part of a mitigation program, 1242 snails below the projected high water line of the flood were relocated to a higher elevation. Approximately 11% of primary habitat of the Kanab ambersnail was scoured, eliminated, or severely damaged. This habitat, composed primarily of monkey flower (*Mimulus cardinalis*) and watercress (*Nasturtium officinale*), has been slow to recolonize in areas where litter and soil were scoured, leaving bare rock as a substrate.

2.4. Riparian Vegetation

The effects of the 1996 controlled flood on riparian vegetation were minimal, and effects on extant vegetation in wetland and riparian woodland/shrubland patches were neither significant within sites nor consistent among sites [*Kearsley and Ayers*, this volume]. Sedimentation along channel margins and in eddy deposition zones during high flows buried the lowest growing plant species, primarily grasses and small herbs. However, this effect was of insufficient magnitude, duration, or both to significantly restructure most patches of vegetation.

The management goal of physically removing vegetation from the new high-water zone of sandbars to reopen heavily vegetated campsites was not accomplished by the flood. Rather than removal of vegetation, much of it was buried [*Kearsley and Ayers*, this volume]. Since many of the plant species are at least tolerant of burial or thrive from burial (e.g., *Typha, Phragmites, Salix*), effects to riparian vegetation were immeasurably small.

A major effect of the flood on riparian communities was the burial of the seed bank by new sediment deposits [*Kearsley and Ayers*, this volume]. Although the seeds were not physically removed from riparian sites by scour, they were buried with plant stocks. It is suspected, that although the seeds are still present, their viability may be short-lived by burial, since the seed bank is primarily transient and associated with habitats where cycles of disturbance are predictable.

There was also little effect of the flood on leaf litter. Although the flood was of insufficient magnitude to scour and transport leaf litter, much of this material was buried under sediment deposits and stored under sand bars with vegetation. Although there was little loss of leaf litter from riparian areas, much of the carbon and nitrogen contained in this organic matter was stored for slow release into aquatic nutrient spirals.

Another effect of the flood was the homogenization, rather than coarsening, of the soils along riparian areas. It has been found that floods of longer duration result in increased coarsening of average particle size [*Topping et al.*, this volume]. Generally, riparian plant species have higher germination success in finer, moister, nutrient-laden soils [*Stevens and Waring*, 1986]. The 1996 controlled flood was of insufficient duration to significantly coarsen sediments and disrupt germination and future establishment of plants.

2.5. Fish Habitat

A primary objective of the controlled flood was restructuring of eddy return channels (i.e., backwaters) used as nursery habitat by native fishes, including flannelmouth sucker (*Catostomus latipinnis*), bluehead sucker (*C. discobolus*), speckled dace (*Rhinichthys osculus*), and the endangered humpback chub (*Gila cypha*). As expected, the flood created new backwaters and reworked or removed others [*Brouder et al.*, this volume]. The flood also scoured fine sediment and organic matter from backwaters which resulted in decreased densities of benthic macroinvertebrates.

Aerial videography and ground census yielded conflicting results on numbers of backwaters. Videography showed an increase of 31 to 39 backwaters between pre-flood and post-flood low steady flows (226 m^3/s)

[*Brouder et al.*, this volume]. A ground census showed a decrease from 68 backwaters before the flood to 42 afterwards, with 38 and 39 backwaters counted at 3 and 6 months, respectively, following the flood. This discrepancy is attributed to differences in discharge during counts, particularly for backwaters with low-elevation reattachment bars and the confounding effects of the steady low flows.

Many sand bars created by the flood were rapidly eroded during the low steady flows of 226 m^3/s and subsequent operations. Most newly created backwaters were temporary, and 2 weeks after the flood, many were no longer available as rearing habitat for young native fishes. Beach faces of many reattachment bars were undercut and eroded by the low post-flood steady flows.

Before the flood, several backwaters had filled with silt and detritus from the absence of high flows, eventually becoming marshes [*Stevens et al.*, 1995]. High velocities during the flood scoured silt from these backwaters, increasing their depth and volume [*Brouder et al.*, this volume].

It was unclear if decreases in densities and biomass of benthic macroinvertebrates in backwaters were attributed to inundation from the high flood flows or to desiccation from the low pre- and post-flood steady flows. Atmospheric exposure of benthos in the Colorado River in Grand Canyon was found to be a more severe disturbance than flooding because organisms were directly killed rather than displaced or buried [*Blinn et al.*, 1995]. Displacement allows for the chance of longitudinal recolonization. It appears that a substantial proportion of benthic macroinvertebrates in backwaters were eliminated by desiccation during the low pre-flood release [*Brouder et al.*, this volume].

2.6. Fish Populations

Responses by fish assemblages to the controlled flood varied longitudinally, depending on species, life stage, and magnitude of change in river stage in the inhabited area. There were no significant changes in numbers or condition factors of adult rainbow trout (*Oncorhynchus mykiss*) in the 27-km reach of the dam tailwaters [*McKinney et al.*, this volume], indicating no significant downstream displacement of trout and no significant depletion of food resources or feeding opportunities during the flood. Increased food intake during one week after the flood, despite a decline in macroinvertebrate numbers, likely reflected opportunistic feeding associated with greater drift rates of macroinvertebrates. The proportion of juvenile trout <152 mm total length [TL] declined by 10% in the week of the flood, suggesting downstream displacement of some smaller-sized trout. However, young-of-year trout were found 8 months after the flood, indicating that the flood did not prevent successful reproduction, and that natural reproduction replaced the juveniles lost from the tailwaters.

Native flannelmouth suckers in the dam tailwaters and in the vicinity of the Paria River (27 km from the dam) appeared unaffected by the flood [*McIvor and Thieme*, this volume]. Staging of pre-spawning aggregations near the mouth of the Paria River and subsequent ascent were observed, indicating no effect of the flood on tributary spawning. The flood impounded the Paria River inflow and provided good, deep, low-velocity habitat for staging adults and for larval flannelmouth suckers descending to the mainstem. The impounded area provided a thermal gradient for acclimation of young moving from the warm tributary (>16° C) to the cold mainstem (8° C), but the duration of the high flow was insufficient to span the spawning and hatching periodicity of the species in this tributary [*Weiss*, 1993]. This was the case as well at the LCR [*Gorman*, 1994], Shinumo Creek, and Bright Angel Creek [*Otis*, 1994].

In the vicinity of the LCR (115-130 km downstream of the dam), densities of juvenile rainbow trout caught along shorelines and in backwaters increased, confirming downstream displacement of these young fish by the flood [*Hoffnagle et al.*, this volume]. However, it was not determined if these fish originated from the dam tailwaters or from local populations associated with cold-water tributaries. There were no significant changes in densities of native speckled dace or juvenile humpback chub along shorelines as a result of the flood, although habitat shifting was observed for speckled dace, from mid-channel riffles to debris fans at high flow. Juvenile humpback chub remained primarily along talus slopes and debris fans, where habitat diversity persisted through the full range of flows before, during, and after the flood. Numbers of young and juvenile flannelmouth suckers and bluehead suckers were too low to detect significant changes in densities. Typical bluehead sucker habitat shift associated with life stage development occurred during the collection period.

Decreases in densities of non-native fathead minnows (*Pimephales promelas*) occurred along shorelines and in backwaters as a result of the flood, suggesting downstream displacement. However, densities returned to pre-flood levels in 8 months, as a result of reinvasion from tributaries and local reproduction. Non-native plains killifish (*Fundulus zebrinus*) were apparently eliminated by the flood from this region of the mainstem, but they reinvaded from tributaries following the flood.

The flood had little effect on adult fishes in the vicinity of the LCR. Large numbers of adult non-native carp (*Cyprinus carpio*) and rainbow trout, as well as native flannelmouth suckers and humpback chub, were found at the impounded inflow of the LCR and in flooded side canyons from the consistent high flows following the flood. These aggregations of fish were attributed to fish seeking refuge from high-velocity flows and to annual pre-spawning staging of native fishes [*Valdez and Ryel*, 1997], indicating that the flood did not interfere with spawning of humpback chub or flannelmouth suckers.

Large numbers of adult humpback chub, flannelmouth suckers, and rainbow trout were also found in large recirculating eddies of the main channel [*Hoffnagle et al.*, this volume]. Radio-tagged adult humpback chub moved to small triangular patches of low velocity near the upstream end of these eddies (separation point) during the high flows [*Valdez and Hoffnagle*, this volume]. Some fish vacated this habitat during dramatic shifts in sediment in eddy complexes and moved to alternative, more stable habitats. Densities of adult rainbow trout in this region increased, suggesting downstream displacement of these fish.

In the vicinity of Lava Falls Rapid (315-323 km downstream of the dam), densities of non-native fathead minnows increased, confirming displacement of this species from upstream areas. Densities of other non-native and native species remained the same, indicating no effects of the flood to local fish populations in this region.

Fish populations as far downstream as the Lake Mead inflow (400-428 km downstream of the dam) were detectably affected by the controlled flood. Increases in densities of fathead minnows were attributed to the possibility of downstream displacement, but could be explained by the variability of sampling methods and access by high densities of this species from local tributaries. Decreases in mainstem densities of non-native red shiners (*Cyprinella lutrensis*) in the Lake Mead inflow were probably the result of fish moving into local sheltered tributaries.

3. IMPLICATIONS FOR FUTURE FLOODS

The 1996 controlled flood was considered to be a moderate biological success as a management demonstration, because of limited desirable effects on some resources and the lack of significant undesirable effects on most resources (Table 1). Except for a small amount of damage to the habitat of the Kanab ambersnail, biological responses were either moderately positive or neutral. Long-term biotic effects were minimal. Knowledge gained from the results of the controlled flood is inarguably valuable for (a) evaluating the efficacy of floods in a regulated river, and (b) planning future floods to maximize managers' ability to alter or control aquatic resources. The results of the controlled flood clearly showed that flooding can benefit aquatic resources in Grand Canyon, given the proper timing, magnitude and duration, and shape of a flood hydrograph. Releases of higher magnitude, shorter duration, and timed for March-April are recommended for future controlled floods. Low steady releases before and after the high release are strongly discouraged because of their confounding effects on response evaluations.

3.1. Timing Of Controlled Floods

The 1996 controlled flood was held March 22 to April 7, 1996. This timing was considered to have minimal undesirable effects on many aquatic resources. During this time, most native fish were staged for pre-spawning at tributary mouths or already spawning in seasonally-warmed tributaries. Few larvae and young were in the mainstem. Significant decreases in benthic macroinvertebrates in backwaters had recovered by the time the young fish began to appear in the mainstem in June. Some of the newly restructured backwaters were also available to these young fish. Most rainbow trout had completed spawning activities, and eggs were incubating deep in gravels during the flood; the strong year class of trout following the flood indicated little effect from scouring of eggs and fry. The flood in late March and early April also preceded seeding of non-native tamarisk, minimizing the risk of further spreading this undesirable plant. This period also preceded nesting by the endangered willow flycatcher (*Empidonax traillii*), a bird that nests in dense cover of riparian willows.

Destruction of habitat used by the endangered Kanab ambersnail is likely to occur at any time of year, since these animals are relatively sedentary and elevated flows are likely to scour vegetation. However, periodic flooding could permanently scour vegetation and establish a new high-water line that would restrict distribution of the snails to elevations above most anticipated flood stages.

Based on the biological responses observed for the 1996 controlled flood, we conclude that the period of March or April is the best timing for future floods to maximize desirable effects and minimize undesirable effects to biological resources. March is typically the wettest period of the year in Grand Canyon National Park [*Kearsley and Ayers*, this volume]; however, in 1996 northern Arizona experienced a drought. Therefore the controlled flood was conducted under abnormal conditions. In order to better determine the biological effects of a flood with a similar

hydrograph and to realistically determine if this is the best time of the year, another controlled flood might be repeated in March/April under more normal climatic conditions. It is likely that antecedent and post-flood riverine conditions would be different with the more typical tributary input of suspended sediments than was experienced in 1996. Reduced light penetration due to a turbid flood would negatively affect the aquatic food base for a protracted period of time [*Shaver et al.*, 1997]. Further studies involving climatic patterns and Lake Powell reservoir storage elevation models are required to determine a useful frequency of controlled floods and the occurrence of dry conditions in the lower basin.

Although a flood in May or June would be more consistent with timing of historic flood events (historic peak flood is late May or early June), the potential effects could be detrimental to some components of the "naturalized ecosystem." Historic average annual temperatures varied from 17° C to 19° C during this runoff period, while post-dam temperatures are only 8° C to 9° C. A cold mass of water down the Colorado River could increase the risk of thermal shock to young native fishes in warmed backwaters, shorelines, and tributary inflows. Many larval and juvenile native fishes descend from natal tributaries to the mainstem during May-August [*Valdez and Ryel*, 1997].

Floods during the May-June period could also inundate nesting areas used by willow flycatchers and make food largely unavailable for adults to feed young fledglings [*Sogge et al.*, 1992]. In addition, these later floods could disperse large quantities of seeds of undesirable plants, such as tamarisk. Tamarisk invasion in the Green River of the upper Colorado River basin has reduced average river channel width by as much as 27% [*Graf*, 1978]; the deep, extensive root system stabilizes river banks and precludes redistribution of sand and sediment. Effects of high cold releases are not well known, and future releases need to be monitored under well-designed sampling programs.

One potential desirable effect of a May-June flood might be the impounding of tributary inflows to provide sheltered habitat with a thermal gradient for acclimation by young fish descending from natal tributaries [*Clarkson et al.*, 1992; *McIvor and Thieme*, this volume]. In order to accomplish this effect, high releases might be of lower magnitude but of greater duration (1-2 months) to provide a stable, persistent environment for young fish during the extended period of hatching. It is also at this time (May-June) that young of the year exotic salmonids, particularly brown trout, would be most vulnerable to flushing which might reduce competition with and predation on native fish in lower Marble and Grand Canyons [*Valdez and Ryel*, 1997].

3.2. Magnitude and Duration

The 1996 controlled flood reached a maximum of 1274 m^3/s for about 7 days. The lack of strong desirable biological responses and occurrence of some undesirable responses indicate that a release of higher magnitude might be preferred for future events. This flood was of insufficient magnitude to scour standing riparian and terrestrial vegetation, and failed to reestablish a new high water zone. Instead, the flood buried vegetation and much of it recovered shortly afterward, leaving a lush riparian zone with root stocks that bind sand bars, preclude redistribution of sediments, and impede recreational camping. While the flood appeared to be of sufficient magnitude to restructure backwaters, reattachment bars were quickly eroded, reducing the long-term value of the flood in providing nursery habitats for native fishes.

The most compelling evidence for a flood of higher magnitude was the failure of the 1274 m^3/s release to permanently suppress non-native fish populations in the mainstem. Instead of consistent high velocities across the channel, the flood impounded tributary mouths and shoreline vegetation, providing quiet sheltered habitats for escape by fishes. Higher-magnitude floods would be expected also to impound tributary mouths, but higher velocities would be expected to scour shoreline vegetation, reducing refugia for non-native fish species. *Minckley and Meffe* [1987] reported that native fishes dominated assemblages of southwestern rivers following floods approaching or exceeding two order of magnitude greater than mean discharge.

Historic mean discharge of the Colorado River in Grand Canyon is 505 m^3/s, and by this relationship, flows would have to approach or exceed 50,500 m^3/s (1.78 million ft^3/s) to suppress non-native fishes. With a maximum recorded discharge in Grand Canyon of 6228 m^3/s (220,000 ft^3/s) in June 1921, and flow regulation from Glen Canyon Dam, flows approaching this magnitude are unlikely. However, the 1996 controlled flood demonstrated that even a low magnitude flood can temporarily reduce numbers of fathead minnows and plains killifish (fish recovered by reinvasion from tributaries and local reproduction). Hence, a properly timed flood of similar or greater magnitude could temporarily reduce numbers of non-native fishes and reduce competition or predation on certain critical life stages of native fishes. Mechanical removal of non-native fishes in downstream reaches of tributaries prior to and during a controlled flood could also decrease the likelihood of reinvasion of the mainstem by these species.

3.3. Shape of Flood Hydrograph

The 1996 controlled flood of 1274 m^3/s for 7 days was preceded and followed by 4 days of 226 m^3/s. The purpose of the low steady flows was to expose and document the sand bars before and after the flood as a measure of the success of the flood in resuspending sediments and rebuilding sand bars. These low steady flows had undesirable effects on biological resources and confounded interpretation of responses to the flood. We strongly recommend against these low steady flows in association with future floods and suggest instead constant normal flows before and after controlled floods.

The low flows prior to the flood exposed and desiccated the phytobenthos, increasing the fragility of algal filaments to high flood velocities [*Blinn et al.*, 1995; *Blinn et al.*, this volume]. These low flows also drained and desiccated many backwaters, reducing numbers of benthic macroinvertebrates prior to the flood, and confounding interpretation of effects. The low flows after the flood undermined and eroded reattachment bars, making newly restructured backwaters short-lived and of little long-term value. Exceptionally high releases from the dam in the months following the flood also confounded many effects that may or may not have been caused directly by the controlled flood. Persistently high, relatively steady flows provided clear water that enhanced photosynthesis and net primary productivity, a phenomenon that allowed algal biomass and macroinvertebrates to recover quickly and might not be as dramatic under normal dam operations.

One desirable effect of the low steady pre- and post-flood releases was strengthening of sample designs that required collecting immediately before and after the flood, such as monitoring of fish populations. These steady releases controlled the all important variable of flow for evaluating flood effects. However, these low flows had undesirable biological effects. Therefore, constant pre- and post-flood flows of a moderate rate for several months would aid in defining flood effects. Our understanding of flood effects may eventually be sufficient to design ramping rates, flow magnitude, and duration of flood flows to maximize aquatic habitats to insure native fish recruitment.

REFERENCES

Blinn, D.W., and G.A. Cole, Algae and invertebrate biota in the Colorado River: comparison of pre- and post-dam conditions, in *Colorado River Ecology and Management*, ed. G.R.

Marzolf, pp. 102-123, Natl. Acad. Press, Washington, D.C., 1991.

Blinn, D.W., J.P. Shannon, P.L. Benenati, and K.P. Wilson, Algal ecology in tailwater stream communities: Colorado River below Glen Canyon Dam, Arizona, *J. Phycol.*, in press, 1998.

Blinn, D.W., L.E. Stevens, and J.P. Shannon, *The effects of Glen Canyon Dam on the aquatic foodbase in the Colorado River corridor in Grand Canyon, Arizona*, Glen Canyon Environ. Studies Tech. Rept., Flagstaff, AZ, 1992.

Blinn, D.W., J.P. Shannon, L.E. Stevens, and J.P. Carder, Consequences of fluctuating discharge for lotic communities, *J. N. Amer. Benthol. Soc.*, 14, 233-248, 1995.

Blinn, D.W., R. Truitt, and A. Pickart, Response of epiphytic diatom communities from the tailwaters of Glen Canyon Dam, Arizona, to elevated water temperature, *Regul. Rivers*, 4, 91-96, 1989.

Carothers, S.W., and B.T. Brown, *The Colorado River through Grand Canyon: Natural history and human change*, Univ. Ariz. Press, Tucson, AZ, 1991.

Clarkson, R.W., O.T. Gorman, D.M. Kubly, P.C. Marsh, and R.A. Valdez, *Management of discharge, temperature, and sediment in Grand Canyon for native fishes*, Issue Paper, Glen Canyon Environ. Studies, Flagstaff, AZ, 1992.

Converse, Y.K., C.P. Hawkins, and R.A. Valdez, Habitat relationships of subadult humpback chub in the Colorado River through Grand Canyon: Spatial variability and implications of flow regulation, *Regul. Rivers: Res. Manage.*, 14, 267-284, 1998.

Czarnecki, D., and D.W. Blinn, Diatoms of the Colorado River in Grand Canyon National Park and vicinity, in *Diatoms of the Southwestern USA*, Biblio. Phyco., 38, pp. 1-182, 1978.

Gorman, O.T., *Habitat use by the humpback chub, Gila cypha, in the Little Colorado River and other tributaries of the Colorado River*, Final Rept., Glen Canyon Environ. Studies, Phase II, U.S. Fish Wildl. Serv., Flagstaff, AZ, 1994.

Graf, W.L., Fluvial adjustments to the spread of tamarisk in the Colorado Plateau Region, *Geol. Soc. Am. Bull.*, 89, 1491-1501, 1978.

Hardwick, G.G., D.W. Blinn, and H.D. Usher, epiphytic diatoms on *Cladophora glomerata* in the Colorado River, Arizona: longitudinal and vertical distribution in a regulated river, *Southwest. Nat.*, 37, 148-156, 1992.

Hayden, G.A., *Benthic ecology of the Colorado River system through the Colorado Plateau region.*, M.S. Thesis, N. Ariz. Univ., Flagstaff, AZ, 1997.

Kearsley, L.H., J.C. Schmidt, and K.D. Warren, Effects of Glen Canyon Dam on Colorado River sand deposits used as campsites in Grand Canyon National Park, USA, *Regul. Rivers*, 9, 137-149, 1994.

Minckley, W.L., and G.K. Meffe, Differential selection by flooding in stream-fish communities of the arid American Southwest, in *Community and evolutionary ecology of North American stream fishes*, ed. W.J. Matthews and D.C. Heines, pp. 93-104, Univ. Oklahoma Press, Norman, OK 1987.

National Academy of Sciences, *Colorado River Ecology and Dam Management*, Proceedings of a Symposium, May 24-25, 1990, Santa Fe, New Mexico, Natl. Acad. Press, Washington, D.C., 1991.

Oberlin, G.E., J.P. Shannon, and D.W. Blinn, Watershed influence on the macroinvertebrate fauna of ten major tributaries of the Colorado River through Grand Canyon, Arizona. *Southwest. Nat.*, in press.

Otis, E.O., *Distribution, abundance, and composition of fishes in Bright Angel and Kanab Creeks, Grand Canyon National Park, Arizona*, M.S. Thesis, Univ. Ariz., Tucson, AZ, 1994.

Shaver, M.L., J.P. Shannon, K.P. Wilson, P.L. Benenati and D.W. Blinn, Effects of suspended sediment and desiccation on the benthic tailwater community in the Colorado River, USA, *Hydrobiologia*, 357, 63-72, 1997.

Shannon, J.P., D.W. Blinn, and L.E. Stevens, Trophic interactions and benthic animal community structure in the Colorado River, Arizona, USA, *Freshwat. Biol.*, 31, 213-220, 1994.

Shannon, J.P., D.W. Blinn, P.L. Benenati, and K.P. Wilson, Organic drift in a regulated desert stream, *Can. J. Fish. Aquat. Sci.*, 53, 1360-1369, 1996.

Sogge, M.K., T.J. Tibbitts, and S.J. Sferra, *Status of the southwestern willow flycatcher* (Empidonax traillii extimus) *along the Colorado River between Glen Canyon Dam and Lake Mead – 1993*, National Park Service summary report, 1993.

Stevens, L.E., J.C. Schmidt, T.J. Ayers, and B.T. Brown, Flow regulation, geomorphology, and Colorado River marsh development in the Grand Canyon, Arizona, *Ecol. Appl.*, 6, 1025-1039, 1995.

Stevens, L.E., and G.L. Waring, Effects of post-dam flooding on riparian substrate, vegetation, and invertebrate populations in the Colorado River corridor in Grand Canyon, in *Glen Canyon Environmental Studies Executive Summaries of Technical Reports*, pp. 257-270, Bur. Recl., Salt Lake City, UT, 1986.

Stevens, L.E., J.P. Shannon, and D.W. Blinn, Colorado River benthic ecology in Grand Canyon, Arizona, USA: Dam, tributary, and geomorphic influences, *Regul. Rivers*, 13, 129-149, 1997.

Usher, H.D., and D.W. Blinn, Influence of various exposure periods on the biomass and chlorophyll *a* of *Cladophora glomerata* (Chlorophyta), *J. Phycol.*, 26, 244-249, 1990.

Valdez, R.A., and R.J. Ryel, Life history and ecology of the humpback chub in the Colorado River in Grand Canyon, Arizona, in *Proceedings of the Third Biennial Conference of Research on the Colorado Plateau*, ed. C. van Riper III and E.T. Deshler, pp. 3-31, Trans. Proc. Series NPS/NRNAU/NRTP-97/12, National Park Service, 1997.

Weiss, S.J., *Spawning, movement and population structure of flannelmouth sucker in the Paria River*, M.S. Thesis, Univ. Ariz., Tucson, AZ, 1993.

Yount, J.D., and G.J. Niemi, Recovery of lotic communities and ecosystems from disturbances — a narrative review of case studies, *Environ. Manage.*, 14, 547-569, 1990.

Richard A. Valdez, SWCA, Inc., 172 W. 1275 S., Logan, UT 84321; email: valdezra@aol.com

Joseph P. Shannon, Department of Biological Sciences, P.O. Box 5640, Northern Arizona University, Flagstaff, AZ 86011

Dean W. Blinn, Department of Biological Sciences, P.O. Box 5640, Northern Arizona University, Flagstaff, AZ 86011

The Economic Cost of the 1996 Controlled Flood

David A. Harpman

U.S. Bureau of Reclamation, Lakewood, Colorado

The 7-day controlled flood released from Glen Canyon Dam in late March and early April of 1996 altered the water-release pattern across the entire water year and changed the timing and amount of hydropower produced both before and after the flood. Approximately 267 million cubic meters of water bypassed the powerplant during the flood, and the opportunity to generate 109,000 megawatt hours of electric energy was foregone. The resulting economic cost was approximately $2.5 million, a 3.3 percent decline in the economic value of hydropower generated during the water year. In addition, $1.5 million was spent on the research associated with the experiment. The estimated hydropower cost reflects hydrologic conditions during water year 1996, the timing and design of the experiment, and conditions in the electric-power market. The economic costs of future controlled floods may be less than or greater than those incurred during the 1996 controlled flood.

1. INTRODUCTION

A 7-day controlled flood was conducted in late March and early April of 1996 for research purposes [*Andrews et al.*, this volume]. This short-duration high release was designed to rebuild high elevation sandbars, deposit nutrients, restore backwater channels, and partially restore the dynamic nature of the riverine ecosystem. This paper describes the resultant economic cost of the controlled flood on the hydropower system. There were two sources of economic cost associated with the controlled flood—changes in the timing and amount of hydropower produced and the costs of the research. The focus of this chapter is on the former topic. The research projects carried out during the controlled flood and their estimated costs are described in *Bureau of Reclamation* [1996].

2. BACKGROUND AND SETTING

Glen Canyon Dam was completed by the U.S. Bureau of Reclamation (Reclamation) in 1963. It is located on the Colorado River upstream from Grand Canyon National Park. This 216 m high concrete arch dam controls a drainage basin of approximately 280,588 km². There are eight hydroelectric generators at the dam which can produce up to 1288 MW of electric power.

Glen Canyon Dam and Powerplant are part of the Colorado River Storage Project (CRSP), one of the Federal projects from which Western Area Power Administration (Western) markets power. The total annual amount of energy produced by the dam depends on actual water conditions. Western's Salt Lake City Area Integrated Projects (SLCA/IP) Office annually markets more than 4 billion kilowatt-hours (kWhr) from Glen Canyon Powerplant. The power produced at Glen Canyon Dam is ultimately sold to end-use consumers across a six-state area which includes Arizona, Colorado, New Mexico, Nevada, Utah, and Wyoming. The Glen Canyon powerplant represents approximately 3% of the summer capacity in this region.

The Controlled Flood in Grand Canyon
Geophysical Monograph 110
This paper not subject to U.S. copyright
Published in 1999 by the American Geophysical Union

The construction of Glen Canyon Dam spurred a nationwide protest [*Martin*, 1989] which continues to this day [*Brower*, 1997]. Public concerns over environmental impacts from proposed improvements in generator efficiencies in 1981 led to a series of scientific studies and ultimately to, "The Operation of Glen Canyon Dam Environmental Impact Statement" (GCDEIS) [*Bureau of Reclamation*, 1995a]. The purpose of the 1996 controlled flood was to test hypotheses made in the GCDEIS.

3. THE HYDROPOWER SYSTEM

Electric energy cannot be efficiently stored on a large scale using currently available technology: it must be produced as needed. Consequently, when a change in demand occurs, such as when an irrigation pump is turned on, somewhere in the interconnected power system the production of electric energy must be increased to satisfy this demand. In the language of the utility industry, the demand for electric energy is known as "load." Load varies on a monthly, weekly, daily, and hourly basis. During the year, the aggregate demand for electricity is highest in the winter and summer when heating and cooling needs, respectively, are greatest. Load is less in the spring and fall which are known as "shoulder months." During a given week, the demand for electricity is typically higher on weekdays, with less demand on weekends, particularly holiday weekends. During a given day, the aggregate demand for electricity is relatively low from midnight through the early morning hours, rises sharply during working hours, and falls off during the late evening.

Energy is most valuable when it's most in demand. This usually occurs during the day when people are awake and when industry and businesses are operating. The period when demand is highest is called the "onpeak period." In the West, the onpeak period is defined as the hours from 7:00 a.m. to 11:00 p.m., Monday through Saturday. All other hours are considered to be offpeak.

The maximum amount of electric energy that can be produced by a powerplant is called its capacity, typically measured in megawatts (MW). The capacity of thermal powerplants is determined by their design and is essentially fixed. In the case of hydroelectric powerplants, capacity varies with time because it is a function of reservoir elevation, the amount of water available for release, and design of the facility. The rate at which a powerplant can change from one generation level to another is called a "ramp rate." For hydropower plants, this is typically measured by the change in flow, measured in cubic meters per second (m^3/s), over a one-hour period. Ramp rates vary

widely depending on the type of powerplant, its design, and possible operational constraints.

Hydropower plants are relatively expensive to construct but their variable cost of operation is extremely low in comparison to thermal plants. Peaking hydropower plants, such as the one at Glen Canyon Dam, are designed to rapidly change generation levels in order to satisfy changes in the demand for electricity. Peaking hydropower plants are particularly valuable because they can be used to generate power during onpeak periods avoiding the cost of operating more expensive thermal plants such as gas turbine units. Hydropower plants are also more reliable than thermal plants, and do not generate emissions.

4. PREVIOUS STUDIES

The approach used in this *ex post* or "after the fact" economic analysis is essentially the same as that employed in the *ex ante* or "before the fact" analysis found in the Environmental Assessment of the controlled flood [*Bureau of Reclamation*, 1996]. The *ex ante* estimate of hydropower cost, which was based on projected water year 1996 hydrology and electricity prices, was $1.8 million. This *ex post* analysis uses actual water year 1996 hydrology and electricity prices.

This analysis is similar to previous analyses undertaken by Western Area Power Administration for controlled floods which were proposed, but not carried out. The estimated costs of these proposals were $1.4 to $2.1 million during water year 1994 [*Western Area Power Administration*, 1993c] and $3.0 to $3.7 million in water year 1995 [*Western Area Power Administration*, 1994]. The assumptions made about load data and prices in this analysis differ substantially from those used in the Western analyses. In addition, there are differences in the design of the controlled flood events analyzed and assumed hydrologic conditions.

5. DAM OPERATIONS DURING WATER YEAR 1996

In water year 1996, Glen Canyon Dam was operated under what are known as the Interim Operating Criteria. These criteria were established by the Secretary of the Interior in November 1991 and are described in further detail in *Bureau of Reclamation* [1991]. As summarized in Table 1, Interim Operating Criteria were designed to reduce daily flow fluctuations well below historic levels with the goal of protecting or enhancing downstream resources while allowing limited flexibility for power operations. Minimum flows, maximum flows, ramp rates, and

TABLE 1. Summary of Interim Operating Criteria

Minimum Releases (m^3/s)	Maximum Release (m^3/s)	Allowable Daily Fluctuations (m^{3}s^{-1} day^{-1})	Ramp Rates (m^3 s^{-1} hr^{-1})
227 [1] 142 [2]	566 [3]	142 [4] 170 [5] 227 [6]	71 up 42 down

[1] 7 a.m. to 7 p.m.
[2] Nighttime
[3] The maximum flow constraint could be exceeded when dictated by high monthly release volumes and special management flows such as controlled floods.
[4] for monthly release volumes of less than 740 million m^3
[5] for monthly release volumes between 740 million m^3 and 987 million m^3
[6] for monthly release volumes greater than 987 million m^3

allowable daily fluctuations were established to protect downstream resources until the final GCDEIS was completed [Bureau of Reclamation, 1995b] and a Record of Decision had been signed. As might be expected, the Interim Operating Criteria markedly constrain hydropower operations at Glen Canyon Dam.

Glen Canyon Dam and other CRSP reservoirs are operated in accordance with pertinent state and federal regulations, compacts, treaties, and operational objectives which are summarized in Nathanson [1980]. Collectively, these are known as "The Law of the River." Annual and monthly releases during water year 1996 were consistent with these objectives which include an 10,147.6 · 10^{6}m^3 minimum-objective annual release and equalized storage between Lake Powell and Lake Mead.

With the exception of the controlled-flood period, Glen Canyon Dam was operated in accordance with the Interim Operating Criteria during water year 1996 (October 1995 through September 1996). In order to accommodate the controlled flood, water volumes had to be redistributed from January and February to March and April. The actual water-release volumes in water year 1996 are shown in Table 2 (R.V. Peterson, Bureau of Reclamation, written communication 1996). The corresponding end-of-month reservoir elevations are shown in Harpman [1997, Appendix 3].

6. ECONOMIC ANALYSIS

The economic value of operating an existing hydropower plant is measured by the avoided cost of doing so. In this

TABLE 2. Monthly Release Volumes at Glen Canyon Dam With and Without the Controlled Flood

Month	Forecast Without Controlled Flood Release Volume (million m^3)	Actual With Controlled Flood Release Volume (million m^3)	Release Volume Difference (million m^3)
October	1108.5	1108.5	0
November	1061.6	1061.6	0
December	1128.2	1128.2	0
January	1356.3	1198.5	-157.8
February	1171.4	995.0	-176.3
March	1048.1	1384.7	+336.6
April	1017.2	1346.4	+329.2
May	1202.2	1295.9	+93.7
June	1233.0	1273.7	+40.7
July	1356.3	1213.3	-143.0
August	1356.3	1122.0	-243.3
September	1110.9	1022.2	-88.8
ANNUAL TOTAL	14,149.9	14,149.9	0

context, avoided cost is the difference between the cost of satisfying the demand for electric energy with and without operating the hydropower plant. Conceptually, avoided cost is the savings realized by supplying electric energy from a low-cost hydropower source rather than a higher-cost thermal source. These savings arise because the variable cost of operating a hydropower plant is relatively low in comparison to thermal units.

The economic value of operating an existing hydropower plant varies considerably with time of day. The variable cost of meeting demand varies on an hourly basis depending on the demand for electricity, the mix of plants being operated to meet demand and their output levels. During off-peak periods, demand is typically satisfied with lower cost coal, run-of-river hydropower, and nuclear units. During on-peak periods, additional load is met with more expensive sources such as gas turbine units. Consequently, the economic value of hydropower is greatest during the hours when the demand for electricity, and the variable cost of meeting this demand, is the highest.

6.1. Analysis Approach

Using hourly load data, monthly hydrology data, and the appropriate constraint set for the case being examined, a multi-period constrained optimization model is used to determine the optimal hourly pattern of monthly release and generation for each month in the water year. Next, using the spot market price data, the avoided cost or economic value of the simulated pattern of generation is evaluated for each hour in the water year. This procedure is carried out for both the with and without controlled flood cases. Finally, the hour-by-hour difference in economic value between the two cases is computed.

As detailed in *Wood and Wollenberg* [1996], given knowledge about existing generation resources, expected load, the amount of water available for release, regulatory constraints, and engineering limitations, the problem faced by the profit-maximizing hydropower producer is to generate as much power as possible during the onpeak hours, when it is most valuable. Hourly releases from the dam, q_t, are the variable under management control.

In total, the Interim Operating Criteria constraint set, shown in Table 1, are unique and outside the capability of most existing models. The peakshaving algorithm [*Staschus et al.,* 1990], one of two widely used approaches for simulating hydropower generation, allows for the efficient formulation and solution of this specialized problem. The model employed in this application uses the constrained peakshaving algorithm [*Environmental Defense Fund,* 1995] to reduce peaks in the hourly load curve, subject to operational and environmental constraints, by optimally releasing water for power generation. This model accounts for varying reservoir elevations and represents, in detail, the physical and engineering features of the Glen Canyon Dam and powerplant.

Three functions are used to formulate the model. Equation (1) calculates the electric energy produced in hour t by a release q at a given reservoir elevation, ele_t. In equation (1), the methods described in *Bureau of Reclamation* [1988, sections 3.38.2-3.38.5, and 1987, sections 9.1-9.2] are used to calculate effective head, H. The second function, ef [ ], is used to calculate the release, q_t, required to produce a given amount of electrical energy at a given reservoir elevation. This relationship is obtained by solving equation (1) for q_t.

$$fe\,[q_t, ele_t]\ =\ \frac{\Gamma \times eff \times q_t \times H\,(ele_t)}{1000}\,, \qquad (1)$$

where t = hour; Γ = 9.804, the specific weight of water at 10°C (kN/m^3); eff = 0.889, an efficiency factor (dimensionless); H = effective head (m), a function of reservoir elevation; q_t = discharge (m^3/s); and ele_t = reservoir elevation (m).

The third function, fv[·], converts q_t, to an equivalent water volume for a 1 hr period. The optimal series of hourly releases, $q_t(x)$, $\forall_t \in$ {1,2,3,...T}, is characterized by equation (2). Note that $q_t(x)$ is discontinuous and decreasing in x. In equation (2), expected aggregate load in hour t is L_t, the maximum generation release (capacity) is c, and x is an arbitrary level of release.

$$q_t(x)\ =\ \begin{cases} minf_t, & if\ ef\,[L_t] \leq x \\ ef\,[L_t] - x, & if\ x \leq ef\,[L_t] \leq x+c \\ c, & if\ ef\,[L_t] \geq x+c \end{cases} \qquad (2)$$

The peakshaving algorithm uses an iterative binary search routine to find an x which uniquely satisfies equation (3), subject to the set of Interim Operating Criteria constraints (4 through 9).

Equation (3) is the mass-balance equation that ensures that aggregate hourly releases equal the total amount of water available for release during the month:

$$\sum_{t=1}^{T} fv\,[q_t(x)]\ =\ mvol \qquad (3)$$

subject to equations (4) and (5), which are the up-ramp and down-ramp constraints, respectively:

$$q_t(x) - q_{t+1}(x) \leq uprate \qquad (4)$$

$$q_{t+1}(x) - q_t(x) \leq downrate \qquad (5)$$

Equation (6) is the maximum daily-change constraint. For the Interim Operating Criteria, this constraint varies with the amount of water released during the month.

$$\begin{aligned} max\,(q_t(x)...q_{t+k}(x)) \\ -min\,(q_t(x)...q_{t+k}(x)) \leq mdc \end{aligned} \qquad (6)$$

Equations (7) and (8) jointly define the maximum-flow constraint, which for the Interim Operating Criteria is the lesser of 566 m^3/s or the greatest amount of water that can be released given the elevation of the lake.

$$q_t(x) \leq c \qquad (7)$$

$$c\ =\ min\,(maxfc, pflow) \qquad (8)$$

Under the Interim Operating Criteria, the minimum-flow constraint varies by time of day and is described by equation (9):

$$q_t(x) \geq minf_t \qquad (9)$$

where T = the number of hours in the month; t = the hour during the month; q_t = water released through the turbines (m³/s) at hour t; L_t = expected aggregate load (mw) at hour t; maxfc = maximum flow constraint for the alternative (m³/s); $minf_t$ = minimum flow constraint in hour t for the alternative (m³/s); uprate = upramp rate (m³ s⁻¹ hr⁻¹); downrate = downramp rate (m³ s⁻¹ hr⁻¹); mdc = maximum daily change constraint for the alternative (m³ s⁻¹ day⁻¹); mvol = volume of water available for release during the month (m³); pflow = the maximum flow which can physically be passed through the generators at a given lake elevation (m³/s); and k = min(24, (T-t)).

In addition to constraint equations (4 through 9), there are a number of other physical and engineering constraints that are not shown.

6.2. Input Data and Sources

6.2.1. Hydrologic Data. The underlying hydrology, shown in Table 2, forms the basis for this analysis. As might be expected, both the with and without controlled flood hydrology critically influence estimates of economic cost. The second column in Table 2 reflects the actual monthly releases from Lake Powell during water year 1996. The annual release volumes for both the with and the without controlled flood case are the same by design to reflect compliance with "The Law of the River."

In the absence of the controlled flood, monthly releases from Glen Canyon Dam would have corresponded with the pattern of releases shown in column 1 of Table 2 (R.V. Peterson, Bureau of Reclamation, written communication 1996). As shown in Table 2, the October through December releases are the same since the final decision to proceed with the controlled flood was not made until after December. With the controlled flood, less water was released in January and February than otherwise would have been the case. This water was stored in Lake Powell and then released during March and April to create the controlled flood.

As originally planned, the only differences between the with and without controlled-flood monthly release patterns would have been in January, February, March, and April. However, following the controlled flood, the inflow forecast increased markedly thus increasing the probability of an uncontrolled spill. Under most conditions, this risk would have been judged to be within an acceptable range.

However, should such a spill have occurred immediately following the controlled flood, it would have greatly reduced the scientific value of the experiment, endangering a $1.5 million investment. After considering the options, a management decision was made to increase releases in the months of April, May, and June to reduce the risk of a spill and decrease the possibility that important experimental data would be lost. Inflows did not reach the magnitude of those forecast and releases were subsequently curtailed during the months of July and August to maintain compliance with the "Law of the River."

6.2.2. Aggregate Load Data. In this analysis, aggregate hourly load data were assumed to represent system demand during water year 1996. These aggregate-load data were constructed from 1994 hourly load data reported by Salt River Project, Platte River Power Authority, Colorado Springs Utilities, and Deseret Generation and Transmission. These publicly available data were obtained from information provided to the Federal Energy Regulatory Commission on form 714. The 1994 load data were escalated by 2% per annum to account for load growth and adjusted for the number of days and the pattern of weekdays and weekends in 1996.

6.2.3. Spot Market Price Data. Mean daily on-peak and off-peak spot market (non-firm) prices were used to value the simulated generation for this analysis. These data are specific to the Palo Verde and Westwing, Arizona, interchange. This location is a transaction accounting point for electric energy which is ultimately used elsewhere in the southwest. The price data for October 1995 through December 1995 were obtained from Economic Insight, Inc. The data for January 1996 through September 1996 were furnished for this analysis by the Dow Jones and Company, Inc., Energy Service. These data represent actual observations of electricity prices at a level of accuracy, spatial location, and disaggregation which was heretofore unavailable. Descriptive statistics for these data are found in *Harpman* [1997, Appendix 6].

7. RESULTS

There were four principle sources of hydropower impact during the controlled flood. First, during the 4 days of steady flows preceding the high release, on average, less power was generated than would ordinarily be the case (see Figure 1). Second, during the high release, the outlet works were used to release flows in excess of 850 m³/s, bypassing the powerplant. Water released through the outlet works is considered "spilled" and is unavailable to produce electric energy. In Figure 1, all releases above the 850 m³/s line were spilled. Third, during the high release, more power

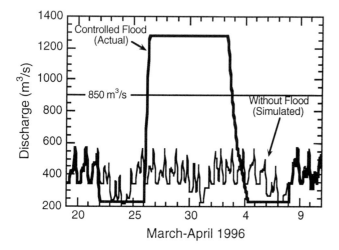

Figure 1. Actual and simulated releases from Glen Canyon Dam.

was generated than would have been the case without the experiment. Finally, during the 4 days of steady flows following the high release, on average, less power was generated than otherwise would have been the case.

Measured over the period 0200 hours on 22 March through 0200 hours on 8 April, the controlled flood required $504.1 \cdot 10^6 \, m^3$ of additional water. Of this, $267.4 \cdot 10^6 \, m^3$ of water were spilled. Approximately 109,000 MWhr less energy (2% less) was generated in water year 1996 as a result of the controlled flood. This difference is primarily due to the amount of water that was spilled.

The power system was also affected during other months in water year 1996. This resulted because water volumes were shifted from the months of January and February to March and April for the controlled flood. As previously described, there were also differences in monthly release volumes following the controlled flood. From a power perspective, the resulting pattern of monthly release volumes was less desirable.

There was no economic cost from the controlled flood during the months of October, November, and December. Compared to the without flood case, additional economic costs were incurred during the months of January, February, July, August, and September. Economic benefits were realized during the months of March, April, May, and June due to additional spot market sales. Across the water year, the *ex post* economic cost of the controlled flood is estimated to be $2.52 million. This represents a 3.3 percent reduction in the economic value of the power produced at Glen Canyon during water year 1996. While surely significant, the magnitude of this estimated cost is far below the dire predictions advanced by some flood opponents.

Some inferences about the cost of future flood events can be drawn from the 1996 flood event. It should be noted that unplanned April, May, and June releases contributed $1.22 million (48 percent) to the hydropower cost of this experiment [*Harpman, 1997*]. The research investment in future controlled floods is likely to be considerably less than for the 1996 event. Consequently, under circumstances similar to those of water year 1996, it is unlikely that similar management decisions would be made. Under these assumptions, a future controlled flood with the same design and power market conditions might be expected to cost approximately $1.3 million [*Harpman, 1997*].

8. LIMITATIONS OF THIS ANALYSIS

The short-run estimates of economic value presented here are sensitive to the quantity and pattern of water release across the year, the reservoir elevations used, and conditions in the electric power market that are reflected by spot market prices. Additionally, these estimates are based on an underlying optimization model. Unlike an optimization model, human operators do not have perfect foresight and are unable to perfectly anticipate weather and market conditions. In contrast with human operators, the simulation model is unaware of and incapable of simulating opportunities for energy interchange (trades) with other utilities. These two factors make it unlikely that simulated operations will exactly duplicate actual operations.

The modeling framework used here simulates the operation of Glen Canyon Dam in isolation from the other CRSP units. Admittedly, the opportunity to manage other CRSP units in a discretionary manner is limited. However, to the extent that operational flexibility exists, these units could be used to partially offset the power system impacts of changes in operations at Glen Canyon Dam. Finally, this analysis is restricted to direct power system costs. Although releases from Glen Canyon Dam have been shown to affect economic use value [*Bishop et al., 1987*], total economic value [*Welsh et al., 1995, Harpman, et al., 1995*] and air quality in the region [*Power Resources Committee, 1995*], these topics are not addressed here.

9. CONCLUSION

The 1996 controlled flood was conducted to test hypotheses about the dynamic nature of geomorphic processes and the aquatic and terrestrial habitats which are dependent on them. This experiment provided an unparalleled opportunity to measure large river sediment erosion, transport, and deposition processes, to observe the effects on the aquatic and terrestrial ecosystems, and to measure the economic effects of a controlled flood event on the power system. This experiment reduced the economic value of the power produced at Glen Canyon Dam during water year 1996 by $2.5 million (3.3 percent). Although not discussed in this report, an additional $1.5 million was expended on physical and biological research. Depending on the design of future beach/habitat-building flows, and the hydrologic and power market conditions at the time, the hydropower cost of future events may be less than or greater than those of the water year 1996 experiment. As with any such decision, the trade-off between downstream ecosystem effects and the economic cost should be carefully weighed.

Acknowledgments: This work was funded by the U.S. Bureau of Reclamation's General Investigation Program, Project Number 100-6520-0008-083-01-01 (9), and the Glen Canyon Environmental Studies Program. I am especially grateful to David L. Wegner, former Glen Canyon Environmental Studies Manager, for supporting this and other endeavors I have undertaken.

Disclaimer: The opinions expressed here do not necessarily reflect the policy or views of the U.S. Bureau of Reclamation.

REFERENCES

Bishop, R.C., K.J. Boyle, M.P. Welsh, R.M. Baumgartner, and P.C. Rathbun, *Glen Canyon Dam Releases and Downstream Recreation: An Analysis of User Preferences and Economic Values. Glen Canyon Environmental Studies Report #27/87*, 188 pp., Hagler Bailly Consult. Inc., Madison, WI, 1987. (NTIS No. PB88-183546/AS)

Brower, D.R., Let the river run through it, *Sierra*, Mar./Apr., pp. 42-43, 1997.

Bureau of Reclamation, *Colorado River Simulation System—System Overview*, 140 pp., Bur. Recl., Denver, CO, 1987.

Bureau of Reclamation, *Colorado River Simulation System—User's Manual*, 182 pp., Bur. Recl., Denver, CO, 1988.

Bureau of Reclamation, *Glen Canyon Dam Interim Operating Criteria: Finding of No Significant Impact and Environmental Assessment*, 71 pp., Bur. Recl., Salt Lake City, UT, 1991.

Bureau of Reclamation, *Operation of Glen Canyon Dam: Final Environmental Impact Statement*, 337 pp., Bur. Recl., Denver, CO, 1995.

Bureau of Reclamation, *Glen Canyon Dam Beach/habitat-building Test Flow: Final Environmental Assessment and Finding of No Significant Impact*, 68 pp., Bur. Recl., Salt Lake City, UT, 1996.

Environmental Defense Fund, *ELFIN User's Manual*, pp. VII-1-VII-4, Environ. Defense Fund, Oakland, CA, 1995.

Harpman, D.A., *Glen Canyon Dam Beach/Habitat-Building Test Flow-An Ex Post Analysis of Hydropower Cost*, 53 pp., Bur. Recl. Rept. #EC-97-01, Denver, CO, 1997 (NTIS No. PB97-159321).

Harpman, D.A., M.P. Welsh, and R.C. Bishop, Nonuse economic value: Emerging policy analysis tool, *Rivers* (4)4, 280-291, 1995.

Martin, R., *A Story That Stands like a Dam, Glen Canyon and the Struggle for the Soul of the West*, 354 pp., Henry Holt Co., New York, 1989.

Nathanson, M.N., *Updating the Hoover Dam documents*, 600 pp., U.S. Govt. Print. Off., Denver, CO, 1980.

Power Resources Committee, *Power System Impacts of Potential Changes in Glen Canyon Power Plant Operations, Phase III Final Report*, 305 pp., Stone Webster Manage. Consult., Inc., Denver, CO, 1995 (NTIS No. PB96-114004).

Staschus, K., A.M. Bell, and E. Cashman, Usable hydro capacity, electric utility production simulations, and reliability calculations, *IEEE Trans. Power Sys.*, 5(2), 531-538, 1990.

Welsh, M.P., R.C. Bishop, M.L. Phillips, and R.M. Baumgartner. *Glen Canyon Dam, Colorado River Storage Project, Arizona—Nonuse Value Study Final Report*, 400 pp., Hagler Bailly Consult., Madison, WI, 1995. (NTIS No. PB98-105406).

Western Area Power Administration, Adjusted final post-1989 allocation of power: Salt Lake City area integrated power projects, *Federal Register* 54(163), pp. 35235-35239, 1989.

Western Area Power Administration, *Financial Assessment High Discharge Experimental Flow Glen Canyon Dam (Qmax = 33,200) WY 1994*, 20 pp., West. Area Power Admin., Salt Lake City, UT, 1993a.

Western Area Power Administration, *Financial Assessment High Discharge Experimental Flow Glen Canyon Dam (Qmax = 45,000) WY 1994*, 20 pp., West. Area Power Admin., Salt Lake City, UT, 1993b.

Western Area Power Administration, *Financial Assessment High Discharge Experimental Flow Glen Canyon Dam (Qmax = 45,000) WY 1994* (Revised), 20 pp., West. Area Power Admin., Salt Lake City, UT, 1993c.

Western Area Power Administration, *Western Financial Assessment: High Discharge Experimental Flow (Qmax = 52,000 and Qmax = 45,000 cfs) WY 1995*, 25 pp., West. Area Power Admin., Salt Lake City, UT, 1994.

Wood, A.J. and B.F. Wollenberg, *Power Generation, Operation and Control*, 2nd ed., pp. 209-263, John Wiley, New York, 1996.

David A. Harpman, Bureau of Reclamation, P.O. Box 25007 (D-8270), Denver, Colorado 80225, (303) 445-2733; email: dharpman@do.usbr.gov

Flood Releases From Dams as Management Tools: Interactions Between Science and Management

G. Richard Marzolf

U.S. Geological Survey, Reston, Virginia

William L. Jackson

National Park Service, Fort Collins, Colorado

Timothy J. Randle

Bureau of Reclamation, Lakewood, Colorado

The linkage between management and science is sometimes strained. Management must be broadly attentive to the perceived desires of various resource users, while scientists are usually focused on natural phenomena, how they are controlled, and how perturbation effects change. These different perspectives commonly result in mismatching objectives and misunderstandings. The 1996 controlled flood represents a productive convergence of river science and dam management stimulated by the process of writing an environmental impact statement. The EIS task focused on effects the of dam operations on the river. The understanding and prediction of the effects required by the EIS was underpinned by scientific inquiry coordinated by the Bureau of Reclamation's Glen Canyon Environmental Studies (GCES) program. The controlled flood itself was a demonstration of a management tool while serving also as a manipulative experiment to test theoretical ideas about how the river works. It became clear that management's role involves the definition of the problems and goals and science's role is development of objective knowledge about natural phenomena.

1. INTRODUCTION

Floods and fires are damaging events when human development and resources are in their paths. The idea that events thought of as hazards could be part of a management strategy seems counter-intuitive because great effort is expended and high expense incurred to protect life and property from them. Nevertheless both floods and fires occur naturally and are important determinants of many features of natural ecosystem form and function.

For example, protection of tallgrass prairie in Kansas from fire causes loss of prairie because the vegetation changes from tall grasses to a cedar savannah. The prairie flora is adapted to fire [*Bragg and Hulbert*, 1976] and new equilibria of vegetation that includes trees are sought in the absence of fire. Spring burning, therefore, has become an important management tool for those who would maximize grass production for livestock grazing.

The Controlled Flood in Grand Canyon
Geophysical Monograph 110
This paper not subject to U.S. copyright
Published in 1999 by the American Geophysical Union

In a similar way we are learning that protection of a river reach from flooding causes the channel, its floodplain, and its biota, likewise, to seek new equilibria [*Stanford et al.*, 1996]. The absence of flooding since the construction of Glen Canyon Dam is now the normal condition in the Colorado River in Grand Canyon, a disturbance of the natural order by omission. This is conceptually parallel with protection from fire and suggests controlled flooding as a tool for river management.

The objective of this chapter is to review the interacting roles of science and management in the decision to implement the 1996 controlled flood. We will discuss also the idea of controlled flooding from Glen Canyon Dam as a way to meet management objectives in Grand Canyon National Park. As the 1996 controlled flood was being planned, it was perceived differently by different groups. Trout anglers and fishing guides were concerned that the fishery in Glen Canyon tailwater would suffer, endangered species interests were concerned about losses of snails and flycatcher habitat, hydropower interests complained of lost power revenues, and some native American tribes were concerned about various archeological and religious sites. It is not our intent to capture all of these diverse views but to develop the common ideas from various management and science perspectives as they emerged during the years leading up to the 1996 controlled flood. Our personal perspectives about the interacting roles of science and management changed, as did those of many of our colleagues involved in these events. We started with views from three bureaus with separate mandates in the Department of Interior and our views converged. We will attempt also to sharpen some of the most pressing policy issues from the perspectives of National Park Service mandates and Bureau of Reclamation responsibilities. We will not get everything that might be said into a short essay written so soon after the event, but we will present an account of some events that will prompt future discussion.

Natural resource managers are hopeful that the apparent success of this demonstration in the Grand Canyon will lead to greater opportunity for the use of controlled flooding to meet their management objectives. During a reporting session to review early results, one natural resource manager in the National Park Service said, "This was not just science for science's sake, it was science for the park." Another added, "We have management opportunities that we didn't know we had." There is great hope, and some evidence, that an important step has been taken in forging a link between objective science and river management.

1.1. What Science Can, and Cannot, Do

Science, as a method of learning about nature, has accumulated a large body of knowledge about ecosystem function, organizational patterns, and response to disturbance. Science's method uncovers new knowledge objectively, though not always certainly. In the case of effects of dam operations on downstream features of the river, the application of science is directed at resource management related to the protection and enhancement of natural values in the national parks. Here, in general terms, are some of the ways that scientific methods can be applied to river management [see also *Marzolf*, 1991].

1.1.1. Describe. Objective, quantitative measurement can document the status of central features of the river environment. These include the rate and volume of flow, the sediment load, the chemical and physical qualities of the water and sediments, the distribution and abundance of sediment deposits, and the flora and fauna of the river that are dependent on these features. These are a few of the features that might be measured; there are, of course, others. In this case a sensible approach has been to limit the measurement to variables that are related, according to conceptual understanding, to the control variable, the flow of the river from the reservoir through the dam. Such measurements also provide the ability to measure changes in these features through time, which is the basis of environmental monitoring.

1.1.2. Understand. The accumulation of objective facts through measurement provides the basis for the formation of general conceptual understanding through the assembly of facts into coherent patterns of how the system works. This approach is empirical and depends on inductive reasoning as the patterns emerge. Scientists proceed further by applying first principles of physics and mathematics, thus adding a deductive or theoretical element to understanding the relevant processes. Understanding achieves increasing certainty as the outcome of events in time are accurately predicted.

1.1.3. Predict. Prediction is the basis for using experiments in science; these are tests of the accuracy of understanding. In the present case, scientists felt that there was enough understanding of the relationship between flow and sediment-transport processes that if the flow was increased by a set amount for a set period, predicted results would be observed. In its simplest form, an experiment is a test of an idea about how the river works by challenging that idea with the facts. In this volume, many investigations relied on conceptual or numerical models (and the predictions that were made from them) to improve the level of scientific understanding.

1.1.4. Develop management options. If understanding is accurate enough that the outcome of events in nature can be predicted accurately, scientists can suggest manipulations to achieve the conditions desired by managers. Management suggestions become a form of prediction that leads to manipulation, and management actions become manipulative experiments. In this sense, the management objectives developed by the authors of the Environmental Impact Statement (EIS) on dam operations [*Bureau of Reclamation*, 1995] represented predictions that emerged from an impact analysis.

1.1.5. Evaluate success. Scientists can help determine if a particular management action has the desired effects. Thus, the manipulative experiment carries with it the opportunity to make post-manipulation measurements and to quantify the certainty with which changes can be predicted. The 1996 controlled flood was not only a demonstration of how to use controlled dam release to achieve objectives related to sediment distribution, in accord with what management agencies wanted, but it was also a manipulative experiment to test the accuracy of some rather detailed predictions, in accord with the scientific goals.

1.1.6. Evaluate feasibility. With adequate understanding about the effects of management actions, or manipulative experiments, scientists can help raise cautions about goals that might not be reached or about multiple goals that may be incompatible with one another. For example, it may not be possible to preserve features that are vulnerable to floods when attempting to manipulate other features that are enhanced by floods [*Schmidt et al.*, 1998].

1.1.7. Monitor long-term change. Variability among several measurements at one time must be understood if they are to be compared with measurements made at other times; that is, accurate comparison of resource states between one year and another requires sufficient precision to discriminate the two measures. The same might be said of comparing measurements before and after a flood manipulation. Accurate estimation often will involve repeated measurement followed by statistical analysis and comparison. Long-term trends might be detected with less precise measurements because trends may emerge despite the variance, but this approach takes longer or requires more measurements. In the case of the Colorado River in Grand Canyon, there has been great interest in long-term trends of change due to the presence and operation of Glen Canyon Dam. Now there will be great interest in long-term trends of change in response to changes in the operational rules for operating the dam. Objective science can provide the measurements with which to develop such information, but it must be done carefully and with great deliberation or it will yield questionable information and represent wasted money.

1.1.8. Interaction with managers. All these capabilities of science can provide assistance to river management. Specifically, science assists managers in four important ways. First, science provides the basis for defining the range of possible "conditions" given the range of available management options. Information on potential results is needed by management agencies to develop management objectives in terms of quantifiable resource conditions. Second, descriptive, conceptual, and process-based models assist management agencies by helping to predict responses to alternative management actions. Predictive information is needed as input to public policy and future resource management decisions. Third, monitoring assists managers by tracking the status of conditions through time to help answer the question, "Are management actions having their desired effects?" Finally, science should help managers explain to the public why resource objectives are, or are not, being achieved.

We believe that cost-effective science can be realized. Science can help management achieve more results at lower cost, saving more than enough to cover the cost of the science. This, however, will not happen by simply unleashing the power of science and, by letting scientists follow their own curiosities to a "logical" conclusion of meeting management goals. This begins by realizing that management goals cannot be set by scientific means; they require another way of knowing, one that applies human value to defining the desired conditions [*Schmidt et al.*, 1998]. Therefore, a basic requirement to assure that this role of science is fulfilled involves close and frequent iteration among the scientists and managers. Scientists are responsible for measuring, analyzing, interpreting, making predictions, and developing information. Managers are responsible for making decisions about what the management goals should be. In order for science to serve management and policy, it must be relevant and of high quality (accurate and precise) and objective (not influenced by value judgement).

There have been substantial conceptual developments during GCES (Phase II) in the relationships among hydrology, fluvial geomorphology, aquatic and riparian ecology (this volume). There is wide recognition that the challenge to couple predictive capability related to flow and sediment to system-wide responses is within reach. The flood regime, not an individual flood, is the more appropriate time scale for scientific investigation. The environmental impact statement recommendation of long-term monitoring and investigation was correct, and the prognosis for useful application of the new knowledge is high.

1.2. Processes and Time Scales: Toward a new View of Science and River Management in Grand Canyon

Many of the agreements that were required to implement the 1996 controlled flood were reached because the Secretary of Interior called for an Environmental Impact Statement (EIS) on the operation of the dam. It was a rare convergence of effort among energetic and divergent interests that brought this about. Here, we record a few of the perspectives that contributed to it.

The 1996 controlled flood marked a milestone in the integration of ecosystem science with operational decisions at Glen Canyon Dam. By examining our experience with this integration, we exposed some of the impediments to effective "applied" science. Recognizing them may suggest improvements for future science and river management initiatives.

We will: (1) describe some of the difficulties of integrating science and resource management; (2) explain how the Glen Canyon Dam Environmental Impact Statement focused science and management; (3) place limits on the role of science, (4) discuss the management implications of the controlled flood; and (5) discuss the scientific implications of the controlled flood, including suggesting new goals for science in support of dam management at Glen Canyon Dam.

1.2.1. Poor management focus contributes to poor science focus.
Before the Secretary's decision in 1989 to initiate an environmental impact statement (EIS) on the operation of Glen Canyon Dam, resource management agencies and their constituents had focussed on specific adverse resource impacts that were thought to be caused by operation of the dam. The popular hypothesis at the time was that the widely-fluctuating daily flows which resulted from load-following power production at the dam resulted in a series of undesirable resource responses including beach erosion, trout stranding, algal desiccation, impairment of native fish habitats, erosion of cultural resources, and impacts to emergent post-dam wetland and riparian resources. This hypothesis resulted from casual observation by river guides, sportsmen, and a few agency scientists. Widely fluctuating flows caused concern among

State, Federal, and Tribal resource management agencies; river users who fish in Glen Canyon, or raft the white-water in Grand Canyon; and Native American and environmental groups concerned about adverse consequences on cultural resources and downstream plants, animals, and their habitats.

Resource management agencies with an interest in and responsibility for Grand Canyon lacked a clear vision of the desired future conditions. They failed to define "target" or "objective" conditions for the downstream resources. Of course, without knowing that proactive management through dam operations was possible there was little motivation for such definition. Thus, there was little or no emphasis on understanding if or how river flows could be manipulated to achieve management objectives. Instead, the focus of management concern was on the simpler and more easily hypothesized adverse response of downstream resources to the dam and its operation. In most cases, the foci of early management and research were the individual resource attributes that had legal standing, such as threatened/endangered species or those attributes that had strong public constituencies, such as beaches for river runners or trout for sport fishing.

Although documenting the negative effects of fluctuating flows on individual resource attributes may have been the first step in the integration of science and resource management, documentation did not lead to the development of conceptual models of how the Colorado River ecosystem functioned nor did it predict change or responses to a changed flow regime. Furthermore, it did not lead to definitions of what downstream resource conditions were reasonably (or potentially) achievable, and it did not identify management alternatives for achieving those objectives.

1.2.2. Poor management integration contributes to poor science integration.
A second impediment to the development of an effective science-management partnership was the failure to effectively integrate resource management interests with the responsibilities of the management agencies. No single agency had Congressional authority to manage all resources. It is our view that some agencies tended to feel ownership of resources over which they had jurisdiction. As a result, they distrusted other agencies with interests that threatened theirs. This led to a situation where every agency or entity with a claim, jurisdictional or regulatory, to a river resource in Glen and Grand Canyons wanted a research role, a portion of the research budget, and a decision making role at the policy table.

As long as agencies and interests felt that they had a decision making influence and a fair share of the financial investment, a political peace was maintained. Consensus on management objectives never developed, nor is it likely in the future because of mutually exclusive objectives [*Schmidt et al.*, 1998]. Thus, the first phase of the GCES program, though integrated on paper, was really nothing more than the sum of the individual research interests of the participating agencies and entities [*National Academy of Sciences*, 1987]. Probably the greatest cost of this failure to integrate at the agency and scientist levels was the lack of cause-and-effect ecosystem models. Not only were biological studies poorly integrated, but there was a notable lack of integration of the hydrologic and sediment-transport studies with the biological studies. Since the 1996 controlled flood was an experiment in influencing ecological conditions by manipulating hydrology, there needed to be much stronger emphasis on studying the hydrology-biology linkages, rather than stand-alone hydrology or biology "state" variables. This, of course, is easier said than done, but we can't help but observe that agency fragmentation led to incoherent science even though GCES was supposed to be an interdisciplinary program.

By 1990, this fragmented, negative-effects approach to science and management resulted in the Secretary of Interior implementing constraints on daily release fluctuations from Glen Canyon Dam. Ironically, after 7 years of this negative-effects approach, the first management action taken was to further constrain flow variability in a river whose annual flow-variability had already been severely constrained by construction of the dam. In hind sight, the first management step may have been a backwards one. Indeed, some of the 1990 flow constraints were later relaxed.

1.3. The Glen Canyon Dam Environmental-Impact-Statement Process: Focus for Science and Management

The 1989 decision of the Secretary of Interior to conduct a full EIS on the operation of Glen Canyon Dam had monumental influence on the integration of science and resource management. Rather than simply studying whether dam operations had negative impacts on downstream resources, the EIS process dictated that there would be a formal effort, with public participation, to identify and evaluate better ways of operating the dam. Furthermore, it was decided that for the duration of the EIS process, the GCES Program should direct its efforts entirely to meeting the information needs of the EIS. Finally, to implement the EIS process and to produce a final Environmental Impact Statement, the Bureau of Recla-

TABLE 1. The cooperating agencies for the Glen Canyon Dam Environmental Impact Statement

U.S. Department of the Interior
    Bureau of Reclamation (lead agency)
    Bureau of Indian Affairs
    Fish and Wildlife Service
    National Park Service
U.S. Department of Energy
    Western Area Power Administration
Arizona Game and Fish Department
Hopi Tribe
Hualapai Tribe
Navajo Tribe
San Juan Southern Paiute Tribe
Southern Paiute Consortium
Zuni Pueblo

mation and its cooperating agencies established a technical, interdisciplinary, and interagency EIS Team.

Before this, in 1988, the Department of the Interior had determined that the GCES program should be continued to gather additional data on specific operational elements. This phase of studies initially was to be conducted over a four to five year period. However, the timetable and research approach had to be drastically altered to meet the Department's 24-month schedule for completing the EIS.

The research schedule was accelerated using special "research flows" within power plant capacity to provide more timely data for the EIS [*Chapter 1*, this volume]. These research flows consisted of a variety of water release patterns from the dam and data collection programs were conducted from June 1990 to July 1991. The Bureau of Reclamation, the lead agency responsible for producing the EIS, appointed a leader[1] for the EIS team. The many "cooperating agencies" in the GCES process became cooperating agencies in the EIS process and contributed subject matter experts who were responsible for preparing the document's technical chapters.

The EIS process, largely as implemented through the EIS Team, created several conditions that significantly influ-

[1] Thomas Slater, June 1990 - October 1991
Timothy J. Randle, November 1991 - October 1993
Gordon S. Lind, November 1993 - October 1994

enced the outcome. First, the EIS Team provided a communication forum for technical staffs of the agencies. Second, the EIS process (with considerable public input) forced development of a range of operational alternatives for the dam. Third, the EIS process encouraged and attempted to press the science community to bring scientific information and "professional opinion" to bear on the implications of alternative operations on downstream resources and related processes. Fourth, the EIS Team, collectively, had to digest and understand the technical input presented to them from scientists. Fifth, the EIS Team worked interactively to develop a range of reasonable alternatives for dam operations and to determine the consequences of alternative operations on the downstream river corridor. In short, the EIS process provided focus to both management and science, a formal mechanism to integrate science in support of management, a forum for communication between diverse agencies and interests, and a format for consensus-building.

Ironically, after almost 10 years of study, the GCES program was able to contribute very little to the evaluation of specific EIS alternatives. There was surprisingly little in the way of analytical- or conceptual-model development available in 1992 for use by the EIS team so that the comparative impacts of alternative flows on downstream resources could be understood or quantified. The focus and the pace of the GCES program was incompatible with the pace of the requirements of the National Environmental Policy Act (NEPA). This, then, was one of the pitfalls of attempting to conduct a science program on the schedule demanded by management. This forced the EIS team to develop its own analysis procedures and seek peer review from scientists after the fact. This mismatch will not be cured easily; the science cannot be forced and the management decisions cannot be delayed until all the science is complete. This issue must be approached with early and frequent iteration among scientists and managers so that the issues remain in focus and the conduct of irrelevant science, that wastes time and money, is minimized. Managers may need to rely on technical specialists, in an advisory role but not associated with the science program, to help ensure that the science program stays relevant and focussed.

The EIS process was being conducted during the constrained daily fluctuations of interim flows and observations during that period suggested that sandbars continued to erode and backwater habitats associated with sandbars continued to fill with sediment. It became clear that periodic ecosystem disturbance, or "beach/habitat-building" needed to be incorporated into operational management options being evaluated in the EIS. The EIS Preferred Alternative

rapidly and surely evolved toward endorsement of low-fluctuating daily flows to permit maximum accumulation of sediment in the river channel, punctuated by periods of high flows ("beach/habitat-building" flows and "habitat maintenance" flows) to redistribute accumulated sediments to the river margins where they would rebuild sandbars. The low fluctuating flows of the preferred alternative were similar to the interim operating criteria with slightly more flexibility for power operations. Thus, the EIS process formalized the concept of periodic high flows, and helped to develop a consensus among the agencies about the benefit of a controlled flood and particularly an experiment such as the 1996 flood.

Possibly as its final contribution, the EIS process highlighted some of the shortcomings of past monitoring and research and shed light on where science needed to evolve if it was to become more effectively integrated with management needs in the future [*Bureau of Reclamation,* 1995]. Specifically, it became clear that science needed to be more oriented towards the development of process models, so that the interactions between causes (flows) and responses (resource conditions) could be better understood, evaluated, and predicted. It also became clear that future monitoring and research needed better integration within the context of conceptual models of the Colorado River ecosystem in Grand Canyon. Finally, there seems to be a need to diminish some of the sense of agency ownership over resources and research and to develop mechanisms to attract the best talent and perspectives to the development of science in Grand Canyon. These needs, in part, resulted in the concepts for a program of Adaptive Management that isolate the science from political interference. Iteration with management keeps the science focussed. This program of adaptive management is described in the in the EIS that also includes a general organizational structure for its implementation.

1.4. Science and the Flood Experiment

The design of the 1996 flood was driven by recent advances and predictions from the flow and sediment transport models in this high-gradient, incised, gravel-bedded river [*Wiele et al.,* this volume] and by ideas about the kinetics of sand deposition in recirculation zones associated with debris fans at the mouths of tributaries [*Andrews et al.,* this volume; *Schmidt et al.,* this volume, *Topping et al.,* this volume]. There was, of course, great interest and substantial effort devoted to assessing the effects of the flood on biological components of the river and its riparian system. The scheduling of the flood in late March took into account the known timing of some biological events, such as spawning of native fish and germination requirements of non-native species, but, except for concern about the damage that the flood might cause to them, knowledge of none of these components suggested design features of the hydrograph.

There were two basic types of studies implemented in conjunction with the controlled flood: negative-impact studies, and process/response studies. Although there were some exceptions, most investigation of biological attributes such as aquatic food base, native and non-native fish populations, Kanab Ambersnails, and emergent marshes, process/response hypotheses were not well developed and the accompanying monitoring studies were designed primarily to document resource losses or changes that might occur.

For other resource attributes, there was expectation that there would be a flood response and that response would have significance in understanding ecosystem processes from the physical to the biological. These attributes and processes include such things as flood routing, sediment transport and storage, reduction in the size of aggraded debris fans, sediment redistribution and sandbar enlargement, organic matter transport, riparian vegetation scour, and water-quality changes. Conceptual or mathematical models had been proposed to formalize the understanding of water and sediment components of the river ecosystem. The experiment's objective was to make the measurements to validate and/or refine models, document processes and quantify resource responses.

2. A NEW VIEW OF SCIENCE AND RIVER MANAGEMENT IN GRAND CANYON: THE MANAGEMENT IMPLICATIONS OF THE FLOOD EXPERIMENT

From a management perspective, the 1996 controlled flood demonstrated that a high discharge lasting for a few days would redistribute sediment and rebuild sandbars without damage to non-sediment resources. Nevertheless, it is understood now that a flow regime that includes floods may generate long-term responses – both beneficial and damaging – even though short-term, or single-event, effects were not measurable. Research associated with the flood suggests further that (1) there may be other ways to design future high releases to accomplish more efficiently the sediment-management objectives, (2) there is a need to be able to predict the effects of alternative flood designs (for example, a controlled flood with a substantially larger peak discharge or shorter duration) on non-sediment resources, and (3) there remains a need to define the dynamic range of long-term sediment/sandbar responses that can be achieved

under various flood scenarios because sediment supplies are limited and delivered episodically to the Marble Canyon reach from the Paria River.

Finally, whereas the original concept of periodic "beach/habitat-building" floods was to restore and rejuvenate aquatic habitats lost to sedimentation during periods of normal or low-flow operations, it is now recognized that there may also be role that floods can play in preventing resource deterioration. The interval between floods was never an issue until after this first one. Specifically, floods can move sediments to high elevations along the channel margins prior to prolonged periods of high flow (such as would be associated with high runoff years) to minimize the export of stored sediments from the river channel and maximize sand storage in upstream reaches.

Without question, the most significant management implication of the 1996 controlled flood is that it initiated an era when high flows will be viewed as the primary dam-operation variable available for managing downstream resources. The science to support adaptive management will focus on controlled floods (management actions) as a long-term program and the long-term response of the canyon river to this "restored" flow regime. Challenges remain for defining appropriate political terms for the frequency of flood experiments and the resource circumstances that would trigger decisions to implement floods. It is conceivable, however, that a high degree of discretion in using floods to achieve management objectives (possibly with as great as annual frequency) could result in an opportunity to relax constraints on hydropower operations, thus permitting mutual attainment of the formerly-conflicting environmental and hydropower objectives. An important question to be addressed by the Adaptive Management Program is, "Can management objectives be achieved with greater reliance on the use of beach/habitat-building floods and less reliance on restricting daily hydropower operations?" The cost of the controlled flood was relatively low, less than 3 percent of the annual revenue from the sale of hydroelectric energy and power from Glen Canyon Dam [*Harpman*, this volume]. The cost of implementing future floods would be much lower than the steady flow restrictions on daily powerplant operations that some agencies (*e.g.*, the U.S. Fish and Wildlife Service) would impose and offers the potential of far greater benefits.

2.1. Scientific Implications of the Flood Experiment

As a scientific experiment the controlled flood was an unqualified success because things were learned things that were not previously known. Now, the ability exists to prescribe management ideas, to evaluate management options, and to predict the effects of future controlled floods. An additional and important scientific result of the controlled flood, however, was that it reinforced the need to evaluate resource responses in a broader system context and on a variety of time scales; for example, many features of sediment deposition might be controlled with flood events on hourly and daily time scales. The significance of habitats created during floods and their eventual evolution and fate may be controlled by processes that occur over a range of time scales from monthly to annual or more. In the case of individuals in native fish populations or, populations whose individuals have life spans of two to three decades, it should come as no surprise that short-term hydrologic events have little effect. Most native fish species evolved in a river where the hydrologic regime was characterized by annual flooding; thus, the flood regime, rather than the individual flood, is the more appropriate time scale for investigation.

The 1996 flood reinforced the need for a greater emphasis to be placed on developing descriptive-conceptual models of habitat succession following perturbations by floods. These models need to couple the interactions between physical and biological components. The flood also reinforced the need to better understand the role of floods on native ecosystems. Specifically, we need to develop a series of research hypotheses related to understanding how flow regimes that include floods influence the long-term dynamics of populations of native and non-native biological components. Scientists are calling for greater emphasis on system responses to regimes rather than on individual responses to events. Managers and scientists need to understand better why floods are important ecosystem processes, and worry less about the short-term impacts of floods.

2.2. Defining the New "Task at Hand"

We have concluded that to continue to foster strong linkages between applied science and management there needs to be a clearly articulated task-at-hand that defines both the management question and direction for science. During preparation of the EIS, the task at hand was to identify dam operation strategies and evaluate the comparative impacts on downstream resources. We propose that the new task at hand is to identify and evaluate alternative long-term flow regimes that include floods, to determine if resource management objectives can be achieved with less constraint on daily dam operations.

Some of the more specific science/management questions include that need to be addressed with research include:

• What is the optimal flood regime for rebuilding riparian and aquatic habitats during periods of abundant main channel sediment storage?

• Can floods be used to minimize sediment losses during sustained high flows during high flow years (sustained flows greater than power plant capacity).

• Can floods be utilized to conserve sediment in the system, even during periods when optimal conditions for habitat building do not exist?

• To what extent can the benefits of floods on physical and biological components of the system be influenced (positively or adversely) by daily operations?

• Are there alternative ways of achieving resource condition objectives that involve trade-offs between flood regimes and daily operations?

2.3. The Implications of the Economics of Hydropower Production for Science-Based Management

The economic costs of the controlled flood experiment [Harpman, this volume] include the direct costs of the research itself and the hydroelectric power benefits foregone. Lost hydropower benefits resulted from the change in monthly water release volumes across the water year 1996, the change in hourly hydropower operations during the flood, and the release of 267 million m^3 of water through the outlet works (that did not generate electricity). The estimated hydropower loss also reflects additional releases of water in the months immediately following the flood experiment. These unplanned releases were requested by GCES to further reduce the risk of an uncontrolled high flow (spill) later in water year 1996. This post-flood management decision nearly doubled the hydropower cost of the experiment. This post-flood operation may not be planned in future controlled floods; the decision being variable depending on year-to-year hydrologic variability, the seasonality of prescribed flooding, the accuracy of forecast models for annual inflow to Lake Powell, and the way the annual release volume is distributed monthly by the Bureau of Reclamation.

Cost estimates for the 1996 flood, especially the costs in May, June, and July [Department of the Interior, 1996] were not available until late in the water year. The magnitude of these post-flood costs was surprising. The unexpected cost was unfortunate because it was a source of tension among scientists, managers, policy-makers, and the public. Not clearly identified and understood early on, it caused misunderstanding because it seemed to diminish the relative benefits to science and river resources. If resource managers decide to reduce risk due to spills after controlled floods in the future, then cost will be increased substantially. On the other hand, if this action is not taken the cost will be considerably less.

3. CONCLUSION

The 1996 controlled flood, the central subject of this volume, was conducted to demonstrate management utility. At the same time, the flood was a manipulative experiment to test specific ideas about what had been learned about the physics of flow, sediment transport, and sediment deposition. As a management demonstration, the flood might have resulted in failure: that is, the expected beneficial effects might not have been realized. As a manipulative experiment, the flood could not fail, because no matter what happened new knowledge would have been gained as long as appropriate observations were made. Ideas would have been either reinforced and understood more certainly because the result was as expected and the causes and effects more clearly documented, or concepts would be rejected, and knowledge would have changed because the results were not as expected. In fact, science proceeds most certainly when incorrect ideas are rejected.

This may seem a fine philosophical point but it is highly relevant to management; that is, without taking a science-based risk, further degradation to the Grand Canyon is expected. The risks, therefore, should be taken by management, guided by as much scientific attention as possible.

Resource managers must define the resource goals; that is, define the problems. The science program must be objective, relevant, and focused on the problems. To accomplish focus and lead to predictability, the science program must seek to understand the cause-and-effect relationships, the functional linkages among system features that managers have identified as important resources. A science program that is predictive and well-integrated among physical, biological, and cultural resources will be the most successful. Predictive models – quantitative or qualitative – must be developed to help focus the scientific and monitoring program so that its progress will be of benefit to resource managers. Science

programs that are strictly empirical risk being too expensive and progress will be too slow to serve management's needs. Science programs that focus solely on negative effects of management's actions will only be of limited use. Such programs may document existing impacts, but will fail to suggest solutions based on causes rather than simple treatment of the empirical symptoms.

Acknowledgments. The authors thank Jerry Mitchell, Rick Gold, John Klein, and Bob Webb for relevant criticism, helpful suggestions, and encouragement.

REFERENCES

Bragg, T.B., and L.C. Hulbert, Woody plant invasion of unburned Kansas bluestem prairie, *J. Range Manage.*, 29, 19-23, 1976.

Bureau of Reclamation, *Glen Canyon Dam Environmental Impact Statement*, Bur. Recl., Salt Lake City, UT, 1995.

Department of the Interior, *Glen Canyon Dam Beach/ Habitat-Building Test Flow: Final Environmental Assessment and Finding of No Significant Impact*, Bur. Recl., Upper Colo. Reg., Salt Lake City, UT, February 1996.

Marzolf, G.R, The role of science in natural resource management: the case for the Colorado River, pp. 29-39 in *Colorado River Ecology and Dam Management*, National Acad. Sci. Press, Washington, D.C., 1991.

National Academy of Science, *River and Dam Management: A review of the Bureau of Reclamation's Glen Canyon Environmental Studies*, National Academy Press, Washington, D.C., 1987.

Schmidt, J.C., R.H. Webb, R.A. Valdez, G.R. Marzolf, and L.E. Stevens, The roles of science and values in river restoration in Grand Canyon, *Biosci.*, 48, 9, in press, 1998.

Stanford, J.A., J.V. Ward, W.J. Liss, C.A. Frissell, R.N. Williams, J.A. Lichatowich, and C.C. Coutant, A general protocol for restoration of regulated rivers, *Regul. Rivers Res. Manage.*, 12, 391-413, 1996.

G. Richard Marzolf, U.S. Geological Survey, 432 National Center, 12201 Sunrise Valley Dr., Reston, VA 20191; email: rmarzolf@usgs.gov

William L. Jackson, National Park Service, 1201 Oakridge dr., Suite 250, Ft. Collins, CO 80525

Timothy J. Randle, Bureau of Reclamation, 25007-D-117, Denver Federal Center, Lakewood, CO 80225